Advances in improving the safety and quality of eggs

It is widely recognised that agriculture is a significant contributor to global warming and climate change. Agriculture needs to reduce its environmental impact and adapt to current climate change whilst still feeding a growing population, i.e. become more 'climate-smart'. Burleigh Dodds Science Publishing is playing its part in achieving this by bringing together key research on making the production of the world's most important crops and livestock products more sustainable.

Based on extensive research, our publications specifically target the challenge of climate-smart agriculture. In this way we are using 'smart publishing' to help achieve climate-smart agriculture.

Burleigh Dodds Science Publishing is an independent and innovative publisher delivering high quality customer-focused agricultural science content in both print and online formats for the academic and research communities. Our aim is to build a foundation of knowledge on which researchers can build to meet the challenge of climate-smart agriculture.

For more information about Burleigh Dodds Science Publishing simply call us on +44 (0) 1223 839365, email info@bdspublishing.com or alternatively please visit our website at www.bdspublishing.com.

Related titles:

Achieving sustainable production of eggs Volume 1
Print (ISBN 978-1-78676-076-0); Online (ISBN 978-1-78676-078-4, 978-1-78676-079-1)

Achieving sustainable production of eggs Volume 2
Print (ISBN 978-1-78676-080-7); Online (ISBN 978-1-78676-082-1, 978-1-78676-083-8)

Advances in poultry genetics and genomics
Print (ISBN 978-1-78676-324-2); Online (ISBN 978-1-78676-326-6, 978-1-78676-327-3)

Embryo development and hatchery practice in poultry production
Print (ISBN 978-1-80146-252-5); Online (ISBN 978-1-80146-253-2, 978-1-80146-254-9)

Optimising poultry flock health
Print (ISBN 978-1-78676-887-2); Online (ISBN 978-1-78676-889-6, 978-1-78676-890-2)

Chapters are available individually from our online bookshop:
https://bdspublishing.com

climate•SMART•publishing
IN AGRICULTURAL SCIENCE

BURLEIGH DODDS SERIES IN AGRICULTURAL SCIENCE
NUMBER 165

Advances in improving the safety and quality of eggs

Lead editor: Professor Cristina Alamprese, University of Milan, Italy

Associate editors: Dr Yves Nys, formerly INRAE, France
Professor Yoshinori Mine, University of Guelph, Canada;
Dr Vincent Guyonnet, University of Montreal, Canada

burleigh dodds
SCIENCE PUBLISHING

Published by Burleigh Dodds Science Publishing Limited
82 High Street, Sawston, Cambridge CB22 3HJ, UK
www.bdspublishing.com

Burleigh Dodds Science Publishing, 1518 Walnut Street, Suite 900, Philadelphia, PA 19102-3406, USA

First published 2025 by Burleigh Dodds Science Publishing Limited

Library of Congress Control Number: 2025931712

British Library Cataloguing in Publication Data
A catalogue record for this book is available from the British Library

ISBN 978-1-80146-875-6 (print)
ISBN 978-1-80146-876-3 (ePub)
ISBN 978-1-80146-877-0 (PDF)
ISSN 2059-6936 (print)
ISSN 2059-6944 (online)

DOI: 10.19103/AS.2024.0150

Typeset by Deanta Global Publishing Services, Dublin, Ireland

BURLEIGH DODDS SERIES IN AGRICULTURAL SCIENCE
NUMBER 165

Advances in improving the safety and quality of eggs

Lead editor: Professor Cristina Alamprese, University of Milan, Italy

Associate editors: Dr Yves Nys, formerly INRAE, France
Professor Yoshinori Mine, University of Guelph, Canada;
Dr Vincent Guyonnet, University of Montreal, Canada

burleigh dodds
SCIENCE PUBLISHING

Published by Burleigh Dodds Science Publishing Limited
82 High Street, Sawston, Cambridge CB22 3HJ, UK
www.bdspublishing.com

Burleigh Dodds Science Publishing, 1518 Walnut Street, Suite 900, Philadelphia, PA 19102-3406, USA

First published 2025 by Burleigh Dodds Science Publishing Limited

Library of Congress Control Number: 2025931712

British Library Cataloguing in Publication Data
A catalogue record for this book is available from the British Library

ISBN 978-1-80146-875-6 (print)
ISBN 978-1-80146-876-3 (ePub)
ISBN 978-1-80146-877-0 (PDF)
ISSN 2059-6936 (print)
ISSN 2059-6944 (online)

DOI: 10.19103/AS.2024.0150

Typeset by Deanta Global Publishing Services, Dublin, Ireland

Contents

Series list

Title	Series number
Achieving sustainable cultivation of maize - Vol 1 From improved varieties to local applications Edited by: Dr Dave Watson, CGIAR Maize Research Program Manager, CIMMYT, Mexico	001
Achieving sustainable cultivation of maize - Vol 2 Cultivation techniques, pest and disease control Edited by: Dr Dave Watson, CGIAR Maize Research Program Manager, CIMMYT, Mexico	002
Achieving sustainable cultivation of rice - Vol 1 Breeding for higher yield and quality Edited by: Prof. Takuji Sasaki, Tokyo University of Agriculture, Japan	003
Achieving sustainable cultivation of rice - Vol 2 Cultivation, pest and disease management Edited by: Prof. Takuji Sasaki, Tokyo University of Agriculture, Japan	004
Achieving sustainable cultivation of wheat - Vol 1 Breeding, quality traits, pests and diseases Edited by: Prof. Peter Langridge, The University of Adelaide, Australia	005
Achieving sustainable cultivation of wheat - Vol 2 Cultivation techniques Edited by: Prof. Peter Langridge, The University of Adelaide, Australia	006
Achieving sustainable cultivation of tomatoes Edited by: Dr Autar Mattoo, USDA-ARS, USA and Prof. Avtar Handa, Purdue University, USA	007
Achieving sustainable production of milk - Vol 1 Milk composition, genetics and breeding Edited by: Dr Nico van Belzen, International Dairy Federation (IDF), Belgium	008
Achieving sustainable production of milk - Vol 2 Safety, quality and sustainability Edited by: Dr Nico van Belzen, International Dairy Federation (IDF), Belgium	009
Achieving sustainable production of milk - Vol 3 Dairy herd management and welfare Edited by: Prof. John Webster, University of Bristol, UK	010
Ensuring safety and quality in the production of beef - Vol 1 Safety Edited by: Prof. Gary Acuff, Texas A&M University, USA and Prof. James Dickson, Iowa State University, USA	011
Ensuring safety and quality in the production of beef - Vol 2 Quality Edited by: Prof. Michael Dikeman, Kansas State University, USA	012
Achieving sustainable production of poultry meat - Vol 1 Safety, quality and sustainability Edited by: Prof. Steven C. Ricke, University of Arkansas, USA	013
Achieving sustainable production of poultry meat - Vol 2 Breeding and nutrition Edited by: Prof. Todd Applegate, University of Georgia, USA	014

Introduction

As one of the most nutritious and affordable animal-sourced foods, eggs are an important part of the diet of many consumers. However, the table egg industry faces numerous challenges. Maintaining appropriate levels of safety across the supply chain continues to be challenging. Consumer expectations of sensory and nutritional quality have never been higher and, at the same time, there are new concerns about the environmental impact of egg production systems. This volume reviews recent research on key aspects of safety, quality and sustainability in egg production systems. It builds on the successful two-volume collection which Burleigh Dodds Science published in 2017: *Achieving sustainable production of eggs* (ed. Julie Roberts; Vol 1: Safety and quality; Vol 2: Welfare and sustainability).

Part 1 Pre-harvest safety

The first group of chapters focuses on the advances in pre-harvest safety.

Chapter 1 sets the scene by providing an overview of what we know about both mechanisms of eggshell contamination as well as horizontal and vertical routes to contamination of the inner part of the egg. It reviews current research on risk factors determining pathogen contamination of eggs, including *Salmonella*, *Campylobacter* and *Listeria monocytogenes*. Finally, it summarises key pre-harvest strategies (including biosecurity management) to prevent pathogen contamination.

Building on this foundation, Chapter 2 highlights the complexity and diversity of egg antimicrobial defences, what we know about their properties and factors affecting their efficacy. The chapter includes sections describing each protective layer (the eggshell, egg white and vitelline membrane), the main antimicrobial proteins identified in the egg, as well as factors affecting both physical and antimicrobial properties involved in egg defenses against pathogens.

As Chapter 3 points out, government regulations and voluntary sector practices are designed to monitor and prevent the occurrence of biological, chemical and physical hazards potentially affecting the safety of eggs. Building on themes in Chapter 1, the chapter reviews key areas of safety management such as building robust on-farm monitoring programs for key hazards such as *Salmonella* Enteritidis and (reflecting themes in Chapter 2) ensuring that handling of eggs safeguards the integrity of the shell, the best natural protection for eggs. The chapter concludes by emphasising the increasing importance of voluntary safety management programmes.

The last chapter in Part 1 reviews the latest technological and analytical advances to monitor layer flock health and welfare. Animal welfare and health are pivotal in meeting ethical standards and consumers' expectations to assure the sustainability of commercial poultry production. The chapter explores the use of state-of-the-art sensors and advanced data analytics, along with issues and future perspectives relating to their field applications.

Part 2 Post-harvest safety

The next group of chapters focuses on the safety of eggs once they leave the farm.

As Chapter 5 highlights, eggs are perishable products which start to lose quality immediately after the moment of laying. As the chapter points out, edible coatings have positive effects in maintaining egg quality and safety, including shelf life. The chapter explores the use of mineral oils, lipids, proteins and polysaccharides to develop egg coatings. It also discusses recent developments such as bioactive coatings to improve egg safety, as well as more complex coatings with a more elaborate combination of substances with more targeted properties to optimize shelf life.

Chapter 6 focusses on the key role of pasteurization treatments in maintaining the safety of eggs. The chapter reviews processes such as hot water bath and hot air pasteurization of shell eggs, batch and continuous pasteurization of liquid egg products (including high temperature/short time (HTST) equipment and ohmic heating), and dry heating pasteurization of egg white powders. It also considers the effects of these processes on egg nutritional quality as well as the broader environmental issues they raise.

Complementing the previous chapter, Chapter 7 reviews advances in non-thermal postharvest egg sterilisation able to ensure the microbial safety of the eggs and limit the potential of cross-contamination, whilst also preserving the quality and functional properties of shell eggs and derived egg products, such as egg albumin. The chapter reviews developments in non-thermal technologies such as irradiation, ultraviolet and pulsed light as well as cold plasma. It includes a detailed case study of the use of radio frequency technologies to pasteurize in-shell eggs.

Part 3 Quality

The next group of chapters focuses on advances in understanding, measuring and improving particular aspects of shell egg quality.

Chapter 8 provides a comprehensive overview of shell egg quality standards. It explores the multifaceted nature of egg quality, examining both external (eggshell) and internal (egg yolk and albumen) characteristics. External quality is assessed through factors like size, weight and shell composition, while internal quality focuses on the nutritional value and freshness of the yolk and albumen. The chapter also discusses the influence of genetic, nutritional, and environmental factors on egg quality. It compares international quality standards across major markets like the United States, Europe, and Japan, highlighting their practical implications for table egg producers.

Building on the previous chapter, Chapter 9 reviews recent advances in use of non-destructive techniques to evaluate shell egg quality and freshness, based on continued improvements in instrumentation and data analysis. Advantages of non-destructive techniques include high accuracy, speed of analysis and results as well as low environmental impact. The chapter reviews developments in spectroscopic approaches, machine vision systems, acoustic and electronic nose systems. It also looks forward to future developments to create an innovative egg industry 4.0.

Reflecting themes in Part 1, Chapter 10 reviews the latest research on the structure and composition of the eggshell, including how eggshell minerals are supplied and regulated to provide the large quantities required for shell formation. The chapter also describes the composition of the organic matrix which also plays a fundamental role, interacting with the mineral phase to give the eggshell its ultrastructure and resulting mechanical properties. The role and importance of amorphous calcium carbonate is underlined and detailed. The chapter concludes by describing how all this knowledge can be mobilized to improve the integrity of the eggshell, especially as hens age.

Chapter 11 reviews the range of egg quality traits from a genetic perspective. After reviewing the impressive advances in areas such as pure line breeding programs, the chapter discusses current research on genetic factors (including degrees of heritability) affecting external quality traits such as egg weight, color and resistance to breaking as well as a range of types of visible defect. It also looks at the genetics of internal quality (IQ) characteristics such as albumen quality, yolk weight and dry matter contents as well as a range of internal defects.

Echoing previous discussion of the importance of eggshell quality, Chapter 12 reviews what we know about the regulation of calcium metabolism in laying hens, which is critical to eggshell formation and strength. The chapter describes the process of eggshell calcification and high calcium requirements in the laying hen, including the effects on bone turnover. It then explores the practical implications of this research in optimising diets to promote calcium homeostasis in laying hens.

Switching the focus back to internal quality, Chapter 13 provides an update on ways of enriching table eggs with various vitamins (such as Vitamin A and Vitamin D) and trace minerals through modification of layer rations. It discusses how biofortified eggs can tackle current nutritional deficiencies, highlighting a case study on commercially producing vitamin D biofortified eggs in the UK. The chapter also explores fortification with iron, zinc, selenium and iodine.

Chapter 14 tackles the increasing importance of welfare in consumer perceptions of quality. Both within and beyond Europe, housing systems for egg production are transitioning from indoor cage (often still based on high stocking densities and selected genotypes) to cage-free systems as well as alternative systems, such as organic and free-range using resilient genotypes and extensive conditions. Based on a One-Welfare approach, this chapter addresses the consequences of this transition, both in terms of ongoing welfare concerns for animals and its impacts on egg quality. It includes assessments of welfare and egg quality in different commercial housing systems.

Part 4 Improving the sustainability of egg production

The last group of chapters focuses on improving the sustainability of egg production.

As Chapter 15 points out, whilst egg production has a lower impact on the environment than other animal production systems, it is still responsible for emissions and needs to contribute to net zero targets. The chapter uses life cycle assessment methodology (including various impact categories) to assess the environmental footprint of shell eggs, including the differences between cage, cage-free and organic production systems. These systems are assessed in terms of impacts such as greenhouse gas (GHG) emissions, water and land use. The chapter also considers potential mitigation strategies such as improving feed conversion and manure management.

Building on the previous chapter, Chapter 16 looks at the potential role of extending the laying period in improving the sustainability of egg production. Genetic selection for prolonged laying cycles not only enables higher egg output and profitability, but can also contribute to animal welfare and a reduced environmental footprint. As the chapter shows, it can be particularly beneficial in cage-free systems, as demonstrated by life cycle assessment analyses.

Chapter 17, as the last chapter in the book, focuses on particular types of emission from layer production systems. These include greenhouse gases (GHGs), ammonia, particulate matter (PM), odours and volatile organic compounds. The chapter reviews both animal-related factors affecting emissions as well as environmental factors such as housing system and type of management. It then summarises what we know about the pros and cons of different practical solutions for emissions mitigation in such areas as diet and genetics, manure and litter management as well as ventilation, filtration and acid scrubbing.

Part 1

Pre-harvest safety

Chapter 1

Advances in understanding pathogen contamination of eggs

Koen De Reu and Geertrui Rasschaert, ILVO, Flanders Research Institute for Agriculture, Fisheries and Food, Belgium

1 Introduction

Eggs are considered healthy because they are a source of proteins, vitamins, and unsaturated fatty acids. However, they may also be subject to microbial contamination. Eggs possess numerous antimicrobial self-defence mechanisms, which include mechanical defences such as the cuticle, shell and shell membranes as well as the antimicrobial properties of egg albumen, including lysozyme, avidin and ovotransferrin. Additionally, the pH of the albumen and the viscosity play a role. These protective mechanisms are in place to support the reproduction of the chicks and, in this way, also to prevent microbial contamination, resulting in an almost bacteria-free egg content in most cases. However, aside from the unavoidable bacterial contamination on the eggshell, contamination of the egg content can occur at every stage of the production chain, including contamination by foodborne pathogens. The pathogenic bacteria most associated with eggs are still *Salmonella*, especially *Salmonella*

http://dx.doi.org/10.19103/AS.2024.0150.01

enterica subspecies *enterica* serovar Enteritidis, from now on referred to as *Salmonella* Enteritidis.

2 Advances in general egg microbiology

Most advances in understanding general egg microbiology have been achieved in recent years by focusing more on culture-independent methods, although these have not been extensively investigated. In general, these results also confirm the insights previously obtained through classical culture methods, whether or not they were designed to select for specific microbial groups, genera, species etc.

2.1 General eggshell contamination

Eggshell contamination may already occur when eggs are formed at the end of the oviduct. Further contamination may occur when the egg passes through the highly contaminated cloaca area at the moment of lay, which is often illustrated by visible faecal contamination on the shell. Following oviposition, the shell also acquires contamination from all surfaces with which it comes into contact, e.g. faeces, soil, caging and nesting material (De Reu 2006; De Reu et al. 2008a). During gathering, egg grading, sorting, candling, packaging, transport and storage of eggs (cooled or not), further fluctuation in the eggshell contamination with total bacterial counts is possible, as studied by De Reu et al. (2005, 2006a).

First studies on bacterial eggshell contamination concerned research on hatching eggs, because trans-shell microbial contamination of hatching eggs may reduce hatchability (Quarles et al. 1970). Board and Tranter (1995) reported that the microbial eggshell contamination of hatching eggs ranges from 2 up to 7 log colony forming units (CFU) per eggshell. With the introduction of alternative housing systems for laying hens in the EU from 2002 (Council Directive 1999/74) and later worldwide, more research has focussed on the bacterial contamination of table eggs especially comparing between different housing systems. A review article of De Reu et al. (2008a) indicated that the contamination of eggshells with aerobic bacteria was generally higher for nest eggs from non-cage systems compared to nest eggs from furnished cages or eggs from conventional cages. Generally, eggshell contamination with total aerobic flora for nest eggs ranges from 4.5 log CFU/eggshell to 6 log CFU/eggshell (De Reu 2006; De Reu et al. 2005, 2006a, 2009; Mallet et al. 2006; Huneau-Salaün et al. 2010); outside nest eggs especially when laid on the floor have much higher counts (up to 7 log CFU/eggshell).

It is generally accepted that the microflora of the eggshell is dominated by gram-positive bacteria. De Reu et al. (2008a, 2008b, 2008c) reported that the natural eggshell contamination of table eggs was dominated by gram-positive *Staphylococcus*. *Staphyloccoccus* sp was also the most abundant species in the air of the poultry houses (De Reu et al. 2008a, 2008b, 2008c).

In a recent Moroccan study, eggs were sampled and analysed in a later stage, refrigerated eggs were bought in supermarkets and also non-refrigerated eggs at small markets. Plating on the selective medium Mac-Conkey agar to isolate gram-negative and enteric (normally found in the intestinal tract) bacteria, the most frequently isolated genera from eggshells were members of the Enterobacteriaceae, *Enterobacter*, *Klebsiella*, *Escherichia*, *Serratia*, *Shigella Salmonella* and *Rahnella*. Other gram-negative bacteria were *Pseudomonas* and *Proteus* (El Ftouhy et al. 2022).

Advances in the study of eggshell microbiology were realized using sequencing technologies, giving a broader, non-cultivation-based insight into the microbiological composition. Firmicutes is the major phylum, representing approximately 50% of total phyla, with the most dominant families being Clostridiaceae, Ruminococcaceae, Lachnospiraceae and Lactobacillaceae. Other important phyla – depending on the study, the housing system and even the season – are the Actinobacteria, Bacteroidetes, Proteobacteria and Fusobacteria (Neira et al. 2017; Shi et al. 2020; Gong et al. 2023). Not surprisingly, these phyla are also the most abundant in the microbiota of laying hens (Videnska et al. 2014; Trudeau et al. 2020). At lower taxonomic levels, major fluctuations can be seen between different studies. For example, according to a Chinese study, *Staphylococcus* was the dominant genus found on eggshell surfaces, followed by *Acinetobacter*, *Aerococcus*, *Psychrobacter* and *Lactobacillus* (Gong et al. 2023). In contrast, in the study of Neira et al. (2017) *Staphylococcus* was only detected in very low concentrations (<2%).

As mentioned above, several factors may impact the general eggshell contamination, starting already with the housing system of the laying hens. Studies indicate no or only limited differences in eggshell contamination with aerobic bacteria between eggs laid in the nest boxes of furnished cages and eggs from conventional cages. Between different laying hen farms, more differences can be found. In an international study, De Reu et al. (2009) found differences of more than 1 log in eggshell contamination with total count of aerobic bacteria between different producers as well for nest eggs from furnished cages (range from 4.24 log CFU/eggshell to 5.22 log CFU/eggshell) as for nest eggs from non-cage systems (range from 4.35 log CFU/eggshell to 5.51 log CFU/eggshell). On average, significant lower levels of contamination with total count of aerobic bacteria were found on eggshells from furnished cages compared to nest eggs from non-cage systems (4.75 log CFU/eggshell vs. 4.98 log CFU/eggshell). Concerning Enterobacteriaceae, no significantly difference in average eggshell contamination between both systems could be demonstrated. The same observations were made by a French study

(Huneau-Salaün et al. 2010), where the mean bacterial count on eggshells was significantly higher on eggs from on-floor systems (4.82 ± 0.51 log CFU/eggshell) than eggs from cage systems (4.57 ± 0.58 log CFU/eggshell). Contamination levels increased with the age of the hens, the concentration of airborne dust, manual egg packing, and the use of plastic packaging instead of recycled pulp egg flats. Also a later study of Jones et al. (2015) confirmed these findings in commercial housing in the US. Aviary system pooled eggshell samples had the highest levels of aerobic contamination (4.9 and 4.1 log CFU/mL, respectively) compared to conventional wire and enriched cage system (2.8 log cfu/mL and 2.6–3.5 log cfu/mL, respectively) eggshell pools.

The influence of the housing system on eggshell contamination with more specific groups of bacteria is variable. Specifically for coliforms, when comparing conventional cage and free-range production systems, Jones et al. (2011) found lower coliform counts for conventional cage shells. Similarly, Singh et al. (2009) reported coliforms being the lowest on conventional cage eggs compared to both nest box eggs and floor eggs from barn systems. As mentioned before De Reu et al. (2009) did not find differences in Enterobacteriaceae between eggshells from enriched cages, aviary and barn (without outdoor access), but floor eggs were not included in this study.

Besides classical microbiological cultivation approach, the advanced next-generation sequencing methods, also demonstrated the difference between housing systems. Although the taxa identified in both housing systems may be the same, the relative abundance can be completely different (Neira et al. 2017).

2.2 General egg content contamination

Different studies agree gram-negative bacteria are best equipped to overcome the antimicrobial defences of the egg content. As major contaminants of egg content, gram-negative bacteria as *Escherichia coli*, *Salmonella* and *Alcaligenes* and gram-positive genera like *Staphylococcus* and *Bacillus* were found by classic cultivation methods (De Reu et al. 2008a, 2008b, 2008c). In a more recent study, El-Ftouhy et al. (2022) isolated from egg contents gram-negative genera *Enterobacter, Escherichia, Klebsiella, Salmonella, Serratia, Shigella, Acinetobacter* and *Proteus*. The only gram-positive isolated genus was *Staphylococcus*. With culture-independent methods, the culture-dependent results are confirmed as the most frequently encountered gram-negative genera in egg content were taxa belonging to *Escherichia, Shigella* and further *Pseudomonas, Acinetobacter* and gram-positive genera *Staphylococcus* and *Lactobacillus* (Gong et al. 2023). Of course the most often reported egg content contaminant remains *Salmonella*. The observed prevalence of eggs with *Salmonella* positive contents is variable (see further in this chapter). There are a number of factors that explain this variability such as: the sample size, timing of sampling, hen ages, site(s) within the egg that are tested, techniques used, studies on eggs laid by artificially or naturally infected hens, country and farm management.

2.3 Mechanism of egg content contamination: horizontal and vertical transmission

There are two possible routes for bacterial contamination of the egg content: either vertically or horizontally. In the transovarian route (vertical transmission), the yolk, the albumen and/or the membranes are directly contaminated due to a bacterial infection of the reproductive organs, i.e. ovaries or oviduct tissue, before the eggs are covered by the shell (Messens et al. 2005; De Reu 2006). Vertical contamination can originate from infection of the ovaries of a laying hen by systemic infection, or from an ascending infection from the contaminated cloaca to the vagina and lower regions of the oviduct (De Reu 2006; De Reu et al. 2008a; Messens et al. 2005).

In horizontal transmission, microorganisms penetrate the eggshell through the pores, which may or may not be covered by the cuticle. At oviposition, the eggs are wet, and upon entering an environment about 20°C cooler than the hen's body temperature, they cool rapidly. This causes the egg contents to contract, creating negative pressure inside the egg, which draws contaminants through the shell (De Reu 2006; De Reu et al. 2008a; Messens et al. 2005). Also during later stages, during handling and storage horizontal transmission is possible. De Reu et al. (2006b) studied the role of different eggshell factors such as cuticle, pores, shell thickness and shell strength influencing eggshell penetration and whole egg contamination by different bacteria, including *Salmonella* Enteritidis. Cuticle deposition proved to be important in preventing horizontal transmission, with *Salmonella Enteritidis* being the primary invader of whole eggs, compared to both non-*Salmonella* strains and *Salmonella* Typhimurium.

3 Pathogens associated with eggs

Like other foods, eggs can be contaminated with human pathogenic bacteria. For decades, the most significant human pathogen associated with eggs has been *Salmonella*, particularly *Salmonella* Enteritidis. However, poultry can also be infected or colonized by other potential human pathogens, which may contaminate eggs. These will be discussed in this section of the chapter.

3.1 *Salmonella and* Salmonella Enteritidis

In the EU in 2022, the first and second most reported zoonoses in humans were campylobacteriosis and salmonellosis, respectively, caused by the foodborne pathogens *Campylobacter* and *Salmonella*. The EU trend for confirmed human cases of these two diseases remains stable during the last years (2015–2022) (EFSA and ECDC 2023). Where *Campylobacter* is most associated with poultry meat; eggs and egg products are the most suspected foodstuff linked to *Salmonella* foodborne outbreaks in the EU in 2022 (EFSA and ECDC 2023). The top five *Salmonella* serovars involved in human infections in the EU are: *S.* Enteritidis (67.3 %), *S.* Typhimurium (13.1%), monophasic *S.* Typhimurium (1,4,[5],12:i:-) (4.3%), *S.* Infantis (2.3%) and *S.* Derby (0.89%). Already for decades, the serovar *Salmonella* Enteritidis is most associated with table eggs.

An important change in the epidemiology of *Salmonella* occurred in the mid-1980s, when *Salmonella* Enteritidis became a major contaminant of eggs and egg products. This development was in most cases attributed to clean, intact, grade A table eggs. Although egg contamination with *Salmonella* Enteritidis is a sporadic event and although, in an infected flock, only a limited number of eggs have become contaminated, eggs and egg products became and remain the major source of foodborne salmonellosis (Raspoet et al. 2011). Research showed that *Salmonella* Enteritidis is much more capable of persistently colonizing the laying hen reproductive tract and to survives in the hostile egg white compared to other *Salmonella* serotypes (Van Immerseel 2010). Despite extensive efforts by various research groups, it remains a challenge to clearly understand why *Salmonella* Enteritidis is the predominant serovar associated with laying hens, particularly their reproductive tract and eggs.

The natural habitat of *Salmonella* is the gastro-intestinal tract. Besides the more frequent occurrence of asymptomatic carriers, colonized hens can excrete the bacteria in their faeces either continuously or intermittently. As a result, *Salmonella* dissemination into the environment is likely to occur (Wray and Wray 2000). During long periods (several months and even years) *Salmonella* can survive under various conditions in the environment outside its host (Garcia et al. 2010; Haysom and Sharp 2003).

Salmonella Enteritidis is particularly effective in spreading from the gastrointestinal tract via systemic spread to various organs, notably the ovary and oviduct. Additionally, the serovar can also access the oviduct through ascending infection from the cloaca (Jan and Baron 2017; Gantois et al. 2009). Certain studies report that *Salmonella* Enteritidis has a higher potential to colonize the ovary compared to other serovars (Okamura et al. 2001a, 2001b; Gantois et al. 2008; Gast et al. 2020). However, an equal colonization ability for Enteritidis and Typhimurium has also been demonstrated (Gantois et al. 2008; Howard et al. 2005). Nevertheless, Gast and Beard (1990) demonstrated with experimentally infected laying hens that *Salmonella* Enteritidis could not be isolated from yolk content, but only whole eggs (yolk + albumen). Therefore, it can be assumed that contamination of the yolk itself rarely occurs during egg formation. However, contamination of the yolk might occur after the egg has been laid via penetration of *Salmonella* through the vitelline membrane (Raspoet 2014).

The oviduct can be differentiated into the infundibulum, which captures the ovulatory follicles; the magnum, which produces the egg white; the isthmus, which deposits the egg shell membranes; the uterus, which forms the egg shell and the vagina, which is involved in oviposition. Colonization of the magnum, isthmus or uterus leads to the contamination of the egg white, egg shell membranes or the shell, respectively. The same could be observed as for the ovary, the serovar Enteritidis seems to have a higher potential for colonizing the oviduct than other serotypes, although studies are not always conclusive (Okamura et al. 2001a, 2001b; Gantois et al. 2008; Gast et al. 2004, 2020). It was demonstrated that *Salmonella* Enteritidis preferably invades the tubular glandular cells of the isthmal segment (De Buck et al. 2004). Once colonized, type I fimbrieae enable the adhesion of *Salmonella* Enteritidis. As type I fimbriae bind to the secretions of the isthmus, it was hypothesized that the *S.* Enteritidis binds to the secretions of the isthmus. Once in the reproductive tract, the *Salmonella* pathogenicity islands mediate the invasion and survival of *Salmonella* Enteritidis in the epithelial cells. Van Immerseel (2010) also hypothesized that stress-induced survival mechanisms enable the serotype Enteritidis to persistently colonize the oviduct without causing damage and excessive inflammation, and to cope with the antimicrobial compounds present in egg white. This hypothesis is also confirmed by different research groups (Baron et al. 2016).

3.2 *Survival of* Salmonella Enteritidis *in eggs*

The molecular mechanisms of the survival of *Salmonella* Enteritidis in egg white are also still not fully clear (Raspoet et al. 2019). Research has shown that the lipopolysaccharides of the membrane of the gram-negative bacterium play a role (Guard-Petter 2001; Parker et al. 2002; Gantois et al. 2009), but also lysozyme inhibitors as found by Callewaert et al. (2008) and protein and DNA damage repair mechanisms (Clavijo et al. 2006). Recently, Raspoet et al. (2019) also showed that the *Salmonella* Enteritidis TolC outer membrane channel is essential for survival in the egg white. De Vylder et al. (2013) also showed that *Salmonella* Enteritidis is superior in egg white survival compared with other *Salmonella* serotypes. These data again added evidence to the hypothesis that egg white survival is one of the reasons why *Salmonella* Enteritidis is more predominantly isolated from contaminated eggs, and helps explain why most reported egg-borne *Salmonella* outbreaks in humans are caused by *Salmonella* Enteritidis. De Reu et al. (2006b) also found that particularly *Salmonella* Enteritidis was a primary invader of whole eggs compared to *Staphylococcus warneri*, *Acinetobacter baumannii*, *Alcaligenes* sp., *Serratia marcescens*, *Carnobacterium* sp., *Pseudomonas* sp. On the other hand, they did not find more egg content contamination with *Salmonella* Enteritidis compared to *Salmonella* Typhimurium using artificial eggshell contamination.

Also of interest is the paper by Baron et al. (2016), which critically reviews the available evidence on the antimicrobial components of egg white. In addition, mechanisms employed by *Salmonella* Enteritidis to resist egg white exposure are also considered, along with various genetic studies that have shed light upon egg white resistance systems.

3.3 Occurrence of Salmonella contamination of eggs

As mentioned earlier in this chapter, *Salmonella* is still one of the most important isolated foodborne pathogens from eggs and especially *Salmonella* Enteritidis. In fact, *Salmonella* Enteritidis and eggs go hand in hand. As mentioned before the observed prevalence of eggs with *Salmonella* positive contents is variable.

In 2012, Martelli and Davies (2012) published an overview of studies on *Salmonella* serovars isolated from table eggs, incorporating data from 27 published and 2 unpublished surveys. In most of the studies where eggshells and contents were analyzed separately, *Salmonella* was predominantly isolated from the eggshells. Many different *Salmonella* serovars were isolated from the shells, but *Salmonella* Enteritidis, *Salmonella* Typhimurium and some other *Salmonella* serotypes (e.g. *S.* Infantis, *S.* Senftenberg, *S.* Goldcoast, *S.* Poona, *S.* Mbandaka, and *S.* Derby) were also found in the egg contents. The review also makes clear that the prevalence of *Salmonella* can vary a lot.

3.3.1 Salmonella on eggshells

The prevalence of *Salmonella* Enteritidis on eggshells is variable. In Spain, Perales and Audicana (1989) examined 372 eggs from flocks implicated with human cases of salmonellosis; *Salmonella* Enteritidis PT4 was found on 1.1% of the shells. On 998 shells of eggs not associated with human salmonellosis, *Salmonella* Enteritidis PT4 was found in five samples (0.5%). Humphrey et al. (1991), using artificially infected specific pathogen-free (SPF) hens with *Salmonella* Enteritidis PT4, found that eggshells could be *Salmonella*-positive even in the absence of faecal carriage. Infected birds continued to lay eggs with contaminated shells for up to 6 weeks after intestinal carriage had ceased. Eggs with contaminated shells were also laid by five birds that were faeces-negative throughout the course of the study. These results suggested – as explained above – the possibility that the shell gland or another part of the oviduct may be a site of infection. In a laying house where *Salmonella* was isolated from 72% of the environmental samples, 7.8% of the eggshells were found to be contaminated (Jones et al. 1995). A 2003 study by the UK Food Standards Agency found no significant differences in *Salmonella* contamination on eggshells based on the production system (Food Standard Agency UK 2004). Of 4753 retail samples of boxes containing 6 eggs each, the eggshells of 9 samples was contaminated. Statistical analysis revealed an overall prevalence of *Salmonella* in 0.34% of the boxes, or 1 box in every 290. Of the nine isolates, seven were *Salmonella Enteritidis*, with three being phage type 4. This prevalence was significantly lower compared to a previous UK survey from 1995 to 1996, which found 1 in 100 boxes to be positive. An Australian longitudinal study followed seven flocks with a history of *Salmonella* contamination until they reached 50 weeks of age. The prevalence of *Salmonella* on eggshells was 0.014%. The highest prevalence in eggs occurred at the onset of lay. Higher *Salmonella* prevalence was associated with lower body weight, higher egg production, larger egg weight and mass compared to the breed standard for age, and poorer feed conversion efficiency (Crabb et al. 2019).

Almost no information is available on the number of salmonellae on eggshells (Humphrey, 1994). In one old study of Baker et al. (1985), dirty 'duck' eggs were found with 5 log *Salmonella* CFU/egg, compared to less than 2 log CFU/egg 'clean' eggs.

3.3.2 *Salmonella* in the egg content

As for the eggshells, also the observed prevalence of eggs with *Salmonella*-positive contents can be variable. A number of factors explaining this variability are already mentioned before (the sample size, timing of sampling, site(s) within the egg that are tested, techniques used, investigations of eggs laid by artificially or naturally infected hens and so on (Humphrey 1994)).

In artificially infected hens, the percentage of infected eggs can range from 0% to 27.5% (Keller et al. 1995; Okamura et al. 2001a, 2001b). Studies on naturally infected layer flocks show mostly a prevalence below 3% (Kinde et al. 1996; Schlosser et al. 1999). Perales and Audicana (1989) found 0.1% of the contents positive from eggs from Spanish flocks implicated in a *Salmonella* Enteritidis PT4 outbreak. In a larger study by Humphrey et al. (1991), over 5700 eggs from 15 naturally colonized flocks were examined, with 32 eggs (0.6%) found to be contaminated. Generally, contamination levels were low (<20 CFU/egg). The prevalence of contamination in the egg contents was similar between eggs from battery and free-range systems, at 0.73% and 0.64%, respectively. Storage at room temperature had no significant effect on the prevalence of *Salmonella*-positive eggs, but those stored for more than 21 days at ambient temperature were statistically more heavily contaminated (>100 CFU/egg). When it was possible to identify the site of contamination in eggs, the albumen (80%) was more frequently positive than the yolk (13%). Gast and Beard (1990), using experimentally infected hens, showed that storage of eggs before testing influenced the rate of detection. Only 3% of freshly laid eggs from experimentally infected hens were identified as contaminated, whereas 16% were detected after storage for 7 days at room temperature.

In The Netherlands, 14 eggs on a total of 46 200 eggs (0.03%) sampled in 1998–1999 were *Salmonella* positive in their content (de Boer and Wit 2000). In Belgium, 2–10 eggs (pooled samples) on a total of 1304 eggs (0.15–0.77%) were *Salmonella* positive (Herman, unpublished results). Serotyping confirmed that the isolates were *Salmonella* Enteritidis and *Salmonella* Stanleyville, respectively. Using available data on the occurrence of *Salmonella* Enteritidis in US layer flocks and eggs, Ebel and Schlosser (2000) estimated, through a theoretical model, that the fraction of *Salmonella* Enteritidis contaminated eggs is 1 in every 20 000 eggs (0.005%) produced annually. In the UK, 30 pooled samples of 6 raw shell eggs tested positive for *Salmonella* out of 407 pooled samples, representing a prevalence of 7.4%. Three different serotypes were found: *Salmonella* Enteritidis, *Salmonella* Ohio and *Salmonella* Infantis (Mitchell et al. 2002). In the earlier-mentioned study of the UK Food Standards Agency in 2003, none of the 4753 pooled egg contents of retail samples were *Salmonella* positive (Food Standard Agency UK 2004). In a Canadian study by Poppe et al. (1998), between 0.07% and 0.4% of table eggs (n = 1512), including both eggshell and egg content, tested positive for *Salmonella*. *Salmonella* Agona was the isolated serotype. De Reu et al. (2008b, 2008c) found 0.18% (1/554 eggs) of table eggs *Salmonella* positive. The isolate was identified as *Salmonella gallinarum*. In a more recent study of the same research group, none of the 490 table eggs from 49 different producers were *Salmonella* positive in their content (De Reu, unpublished results). The average detection rate of *Salmonella*-positive eggs (content and shell) in China was 5.6%, with 6.6% of the eggs collected from poultry farms and 5.1% from marketplaces. Thirteen serotypes were identified, among which *Salmonella* Typhimurium, Indiana, Thompson and Enteritidis were the most abundant serotypes (Li et al. 2020). In a study of Singh et al. (2010), *Salmonella* prevalences were found of 3.84%, 4.00% and 7.40% on eggshell surfaces and egg albumin/yolk in North India with mostly *Salmonella* Typhimurium isolated. Bouda et al. (2019) reported 12% positive eggs (eggshell and content) in a West African breeder facility with *Salmonella* Derby dominating. In the Australian study of Crabb et al. (2019), the prevalence of albumen and yolk was 0.007%, and the concentration of salmonellae in all fractions was <1 CFU/mL.

The *Salmonella* concentration in the contents of freshly laid eggs from either naturally (Humphrey 1989; Humphrey et al. 1989, 1991; Mawer et al. 1989) or artificially infected hens (Gast and Beard, 1990) is usually low. One exception to the above findings is the isolation of $>10^7$ *Salmonella* Enteritidis CFU/g during outbreak investigations from the contents of a clean, intact egg thought to be 5 days old (Salvat et al. 1991).

Jones et al. (2016) investigated the influence of commercial laying hen housing systems on the incidence of *Salmonella* and *Campylobacter* in 120 hens aged 77 weeks in the US. Across all housing systems, approximately 7% of hens were colonized with *Salmonella*, while over 90% were colonized with *Campylobacter*. In conventional cage housing, a higher proportion of hens were colonized compared to those in enriched cages and aviary systems (13.33% vs. 3.33% and 5% for *Salmonella*, and 95% vs. 85% and 90.83% for *Campylobacter*). The EFSA scientific opinion on *Salmonella* control in poultry flocks and its public health impact (EFSA 2019) concluded that, overall, evidence suggests a lower occurrence of *Salmonella* in non-cage systems compared to 'cage systems'.

3.4 Other pathogens associated with eggs

3.4.1 Campylobacter

Campylobacter and more specifically *C. jejuni* and *C. coli* are commonly associated with poultry, resulting in the possibility that eggshells and egg contents can become contaminated as well. *Campylobacter* colonizes very frequently the gut of laying hens and once colonized the birds are excreting campylobacters at high levels (Rasschaert et al. 2007; Sulonen et al. 2007; Novoa Rama et al. 2018; Gharbi et al. 2022). Some authors report that campylobacters can also be isolated from the reproductive tract of breeder hens (Hiett et al. 2002a; Jones et al. 2016). Most studies, however, claim that vertical transmission does not play a role in *Campylobacter* contamination and that horizontal transmission is more likely. According to Cox et al. (2012), faecal contamination of eggshells may allow *Campylobacter* to survive in the membranes and even to be transmitted to offspring. Molecular testing found *Campylobacter* DNA in hatchery fluff and egg shell samples and developing embryos in case of fertilized eggs, but as this was demonstrated with DNA-based methods, the difference between live and dead campylobacters could not be made (Chuma et al. 1997; Hiett et al. 2002b). Doyle (1984) infected laying hens and found two shell surfaces *Campylobacter* positive, but no egg contents of 226 eggs from hens excreting *C. jejuni*. Under commercial settings, no *Campylobacter* was detected in any of the egg or egg-product samples or from water samples collected from the overflow of the egg washer in two US egg-processing facilities (Izat et al. 1988). *Campylobacter* was isolated from 4.1% of the shells of eggs collected at retail level in Germany but not from any of the egg yolk samples (Messelhäusser et al. 2011). Egg penetration studies showed that the organism did not penetrate into the content but could be isolated occasionally from the inner shell membranes. The same was observed under commercial settings, where Shanker et al. (1986) examined eggs from *Campylobacter*-colonized birds for *Campylobacter* penetration. Only from 2 eggs out of the 187 tested eggs, *Campylobacter* could be isolated. Likewise, Sahin et al. (2003) did not find *Campylobacter* in the shell membranes and egg content of 1000 eggs from a commercial hatchery nor in 500 eggs from commercial breeder flocks that were actively shedding *Campylobacter* in faeces. This was confirmed by Sulonen et al. (2007), where only 1 out of 360 eggs was contaminated with *Campylobacter* on the shell and none in the egg content. Clark and Bueschkens (1986) reported that *Campylobacter* could grow in inoculated yolks to levels up to 8 log CFU/ml, but that the egg albumen is highly toxic to *Campylobacter.* When *C. jejuni* was directly inoculated into the egg yolk and stored at 18°C, the bacteria could survive for up to 14 days. However, its viability was significantly reduced when injected into the albumen or the air sac. In freshly laid eggs from *Campylobacter*-infected layers, *C. jejuni* contamination was detected in 3 of 65 pooled eggs (5–10 eggs in each pool). However, the organism was not detected from any of the 80 pooled eggs (10 eggs/pool) (80 pools), after storage at 18°C for 7 days. These results suggest that survival of *Campylobacter* is probably a rare event (Sahin et al. 2003). Also via source attribution studies, the role of eggs for human Campylobacteriosis could not be confirmed (Mughini-Gras et al. 2022). To summarize, studies generally conclude that although *Campylobacter* can be found on the eggshell, it does not survive well and eggs do not play a role in human diseases.

3.4.2 Listeria monocytogenes

Eggs and egg products are not considered food vehicles for causing listeriosis. Nevertheless, it has been demonstrated that poultry flocks may be colonized with *Listeria (L.) monocytogenes*. In France, a prevalence of 15.5% was determined in laying hens with no significant difference between cage and floor-reared hens (Chemaly et al. 2008). A prevalence of 31% was found in another French study where only caged flocks were sampled (Aury et al. 2011). In contrast, in other studies, a much lower prevalence was seen; <2% in Germany with no significant difference between production methods (Schwaiger et al. 2010) and only sporadically in the US (Jones et al. 2012). This is probably more due to the sampling method than a true difference in prevalence.

Although hens can carry *L. monocytogenes*, their contribution to foodborne illness seems very limited (Pohjola et al. 2016; Mughini-Gras et al. 2022). *L. monocytogenes* is only sporadically isolated from eggshell surfaces. Also Guzmán-Gómez et al. (2013) detected *L. monocytogenes* via a sensitive PCR method on eggshell surfaces in a prevalence of 4.6% (pooled samples). Jones and Musgrove (2007) isolated frequently *Listeria* from the egg shells of restricted eggs, but this considered other *Listeria* species, most often *L. innocua*. They also sampled the egg content, and also only isolated once *L. innocua*. Brackett and Beuchat (1992) studied the survival of the organism on eggshells over a 6-week period at 5°C and 20°C. Both, low (2 log CFU/egg) and high (4 log CFU/egg) populations of *L. monocytogenes* on the surface of eggshells decreased to <10 CFU/egg after 6 days of storage at 5°C and 20°C. After 6 weeks of storage, the pathogen was still detectable but unquantifiable at both temperatures.

Listeria is known as an ubiquitous bacterium that may persist for longer periods of time in food processing environment and may be responsible for contamination of food products after, e.g., pasteurization. As a consequence, *L. monocytogenes* has been detected in egg-processing facilities over the years with variable numbers. Moore and Madden (1993) sampled in-line filters in a pasteurization plant and found approximately 72% to be positive for *Listeria*. The genera isolated were *L. innocua* (62%) and *L. monocytogenes* (38%). In addition, the pasteurized products were *Listeria* free, demonstrating the efficacy of the pasteurization process with regard to *Listeria*. In French egg-breaking plants, it was demonstrated that 8.5% of the raw egg products were contaminated by *L. monocytogenes*, but no pasteurized egg products were contaminated by *L. monocytogenes*. It was also demonstrated that *Listeria* sp. and *L. monocytogenes* could be isolated in the environment of these facilities with 65% and 8% of contaminated samples, respectively (Rivoal et al. 2013).

3.4.3 Other pathogens

Also other pathogens such as *S. aureus*, *B. cereus*, *E. coli* and *Yersinia enterocolitica* have been detected on eggshell surfaces and in some cases from the egg content. Due to the limited or no association with foodborne diseases, limited research is available.

According De Reu et al. (2008a, 2008b) *Staphylococcus aureus* is a predominant species on the eggshell, which can be explained by its dominance in the air and the environment at the moment of lay although no direct association with food safety issues is available. De Reu et al. (2006b) also found that the non-motile, clustering and gram-positive characteristics of *S. aureus* limits its eggshell penetration capacities and its survival in the egg content. In contrast, Al-Natour et al. (2012) demonstrated that – although only at high incubation temperatures of 25°C and 32°C – *S. aureus* was able to penetrate artificially contaminated eggshells and reach the egg white and yolk.

Koné et al. (2013) found on average 13% of pooled egg shell samples from 50 French layer farms were contaminated with mesophilic *Bacillus cereus* group bacteria and on average 2% of the samples were contaminated with psychrotophic *Bacillus cereus* group bacteria. Jan et al. (2011) tested psychrotophic *Bacillus cereus* group bacteria isolated from egg surfaces and pasteurized liquid whole egg products received from farms and egg-breaking facilities on their ability to grow in liquid whole egg liquid. The strains were inoculated in sterile liquid whole egg and incubated at 10°C and 30°C for 18 h to 5 days. All isolates were able to grow in liquid whole egg content from 3 log to 8 log after 18h at 30°C and to 7 log after 5 days at 30°C and 10°C. The same was observed at 6°C in another study (Techer et al. 2014). The authors demonstrated a differentiation between two groups within their isolate collection, one half of the isolates representing a safety risk and the other half a spoilage risk (Techer et al. 2014).

Schoeni and Doyle (1994) challenged 1-day-old laying hen chicks orally with *Escherichia coli* O157:H7. *E. coli* O157:H7 colonization persisted at least 10–11 months when chicks were administered 8 log bacteria. Eggs from five hens that were faecal shedders of *E. coli* O157:H7 until the end of the study (10–11 months) were studied for the presence of *E. coli* O157:H7. The organism was isolated from the shell of 14 of 101 (13.9%) eggs but not from the albumen or yolks. In experimental penetration studies, Al-Natour et al. (2012) demonstrated that *E. coli* O157:H7 is capable of penetrating the egg shell which was dependent on the inoculation dose and incubation temperature. At 10°C, it took more than 30 days to reach the albumen and yolk at a high inoculation dose of 5 log, whereas at 25°C and 32°C, this took significant less time, also at lower inoculation doses of 3 log.

Favier et al. (2005) evaluated a total of 352 eggs for the presence of *Yersinia enterocolitica* strains on the eggshell. No isolates were obtained by direct culture; however, eight *Y. enterocolitica* strains were recovered after enrichment, which represents a prevalence of 2.27% eggshell samples. *Y. enterocolitica* was not detected in 45 content samples.

4 Advances in factors influencing contamination risks with *Salmonella*

The contamination of eggs with *Salmonella* is influenced by numerous variables; only a few are discussed in the section below. Additional variables are described and explored in the section on control measures

4.1 The European approach to reduce Salmonella contamination in eggs

Enhanced *Salmonella* programmes in poultry were implemented in all EU Member States and targets were set for reducing *Salmonella* in poultry (breeding, laying hens, broilers and turkeys). More specifically in 2006, annual reduction targets for *Salmonella* Enteritidis and *Salmonella* Typhimurium were stated by Regulation No. 1168/2006 (Commission Regulation, 2006a) for all commercial laying hen flocks (both pullets and egg-producing layers). In addition, Regulation No. 1177/2006 (Commission Regulation 2006b) described requirements for the use of specific control methods of *Salmonella* in poultry, e.g., vaccination and antibiotic treatment. The ultimate target was to reduce *Salmonella* Enteritidis and *Salmonella* Typhimurium levels in egg-producing layers in each member state to 2% or lower by 2011. Regulation No. 1168/2006 also describes a detailed *Salmonella* sampling program for laying hens flocks (rearing and production flocks). To monitor the *Salmonella* status, laying hen flocks are sampled at day 1 (upon arrival) and at 16 weeks during rearing. After the hens are transferred to production farms, flocks are sampled at 24 weeks and subsequently every 15 weeks until the end of production. One-day-old layer chicks are checked for *Salmonella* presence by sampling the paper inlays of the delivery crates. At all other sampling moments, two samples are taken each consisting of either one pair of overshoes (non-cage systems) or 150 g pooled fresh faeces and analysed for *Salmonella*. Moreover, Regulation No. 1237/2007 (Commission Regulation 2007) describes that from any layer production flock positive for *Salmonella* Enteritidis or *Salmonella* Typhimurium, eggs must be placed on the market as B-eggs for heat treatment and cannot be commercialized as grade A eggs or table eggs. If laying hen flocks are positive, the poultry house has to be cleaned and disinfected after slaughter of the positive flock and a profound *Salmonella* control of the house is performed before a new flock can be started. Also laying hens need to be destructed or safe disposal of hens (e.g. heat-treated poultry meat) using logistic slaughter is needed. To ensure *Salmonella*-free 1-day-old chicks, the national *Salmonella* control programs in EU member states must also guarantee that adult breeding hen flocks are free from *Salmonella*. Target serovars are *Salmonella* Enteritidis, Typhimurium, Infantis, Virchow and Hadar as mentioned in Commission Regulation (EC) No. 200/2010 (Commission Regulation 2010). When positive, drastic measures as destruction or safe disposal of (hatching) eggs and birds is needed.

4.2 Risk factors influencing Salmonella contamination

Salmonella contamination of laying hen flocks and shell eggs is influenced by various management and environmental factors. A systematic review study of Denagamage et al. (2015) identified several risk factors associated with *Salmonella* in laying hen farms such as: the presence of previous *Salmonella* infection, lack of cleaning and disinfection, rodent presence, pests accessing the feed, induced molting, larger flock sizes (>30 000 hens), multi-age management, cage housing systems, in-line egg processing, rearing pullets on the floor, visitors in the layer houses, trucks near farms and air inlets. Risk factors specifically related to shell egg contamination include high levels of manure contamination, the middle and late phases of production, significant contamination of egg-handling equipment, large flock size (>30 000 hens) and high egg production rates (>96%). Dewaele (2012) reviewed literature for risk factors associated with the presence of Salmonella in laying hen flocks. The same risks were described: persistence and previous contamination of *Salmonella* on the farm (Carrique-Mas et al. 2009; Huneau-Salaün et al. 2009; Van Hoorebeke et al. 2010), larger flock size (>30 000 laying hens) (Chemaly et al. 2009; Mollenhorst et al. 2005; Namata et al. 2009; Snow et al. 2010), the presence of hens of different ages (Huneau-Salaün et al. 2009; Mollenhorst et al. 2005) and delivery trucks passing near (air) entrances of the poultry house (Chemaly et al. 2009; Huneau-Salaün et al. 2009; Snow et al. 2010). Additional risk factors included an older infrastructure (Van Hoorebeke et al. 2010), increased flock age (Huneau-Salaün et al. 2010; Namata et al. 2009), stress in birds (Humphrey 2006) and farmers involved in other animal production (Chemaly et al. 2009).

In the EFSA scientific opinion on *Salmonella* control in poultry flocks and its public health impact (EFSA 2019) is concluded that conflicting evidence is found in the literature and in member states data on the occurrence of *Salmonella* in laying hens when raised in cage systems compared to non-cage systems. Overall, evidence points to a lower occurrence in non-cage systems compared to cage systems. Whether this is linked to the housing system as such or whether it is caused by the associated change of furniture, break in the historical infection cycle or the reduced stocking density, is unclear. A conclusion on the effect of outdoor access or the impact of the shift from conventional to enriched cages on the *Salmonella* occurrence in laying hens at the EU level could not be reached. On the other hand, the scientific opinion clearly concludes that there is evidence that an increased stocking density, larger farms and stress result in increased occurrence, persistence and spread of *Salmonella* in laying hen flocks.

Different intervention options are also possible to limit or decontaminate *Salmonella* both on the eggshell surface and in the egg content. One of the options is egg washing. Egg washing improves the microbiological quality of table eggs and, in parallel, the eggshell contamination with Salmonella, if the washing process is done properly and is combined with the cool chain (Messens et al. 2011). On the other hand, egg washing is not permitted in the EU because of different disadvantages. The most important disadvantage is the damage to the cuticle as a physical barrier. Damage to the cuticle may favour the vertical transmission route by eggshell penetration and may favor moisture loss according to EFSA (2005) and thereby increase the risk to the consumer.

5 Advances in strategies for controlling *Salmonella* and other pathogen contamination: pre-harvest interventions

Controlling *Salmonella* in laying hens is challenging due to the numerous variables that must be managed. Practical experiences and research show that multiple intervention strategies are needed throughout the egg production chain to control Salmonella (Gast et al. 2022). Vaccination, breeding, genetic selection and gastrointestinal colonization control are possible intervention strategies, together with effective environmental management of housing systems. The latter is essential to minimize opportunities for the introduction, transmission and persistence of *Salmonella* in laying flocks (Gast 2011; Gast et al. 2022).

5.1 Vaccination

Vaccination of laying hens has been successfully used in EU member states to reduce egg contamination by *Salmonella* Enteritidis, decreasing human salmonellosis cases worldwide (Kilroy et al. 2016). Several vaccines are available: both live and killed vaccines, for either only Enteritidis or for both Enteritidis and Typhimurium. Still the protective effect of vaccination is dependent of the *Salmonella* challenge. Under high challenge, vaccination will not fully protect but will only reduce the internal egg contamination rate (Gantois et al. 2006). Vaccination aims to stimulate the natural immune response against *Salmonella* of the hens. Stimulating the naturally immune response reduces the duration and the severity of *Salmonella* infections and helps to prevent eventual re-infections (Gast 2011).

5.2 Breeding and genetic selection

During many years, breeders worked on realizing differences in resistance to *Salmonella* infection of different genetic chicken lines. This evolutionary research has resulted in a long-term progress in controlling the prevalence of *Salmonella* (Gast 2011) and is still going on.

5.3 Gastrointestinal colonization control

For decades, significant attention, efforts, and research have been focused on controlling *Salmonella* colonization in the gastrointestinal tract. The ability of the normal gut flora to inhibit or limit the colonization of *Salmonella* and other enteric pathogens forms the basis of preventive colonization control via prebiotics or probiotics (Gast 2011). Prebiotics are typically complex oligosaccharides that are not digested by the host, allowing them to reach the lower gut, where they promote the growth of resident gut microbiota. Well-known examples are inulin, galacto-oligosaccharides, fructo-oligosaccharides, xylo-oligosaccharides, pectin, beta-glucans and resistant starch. Probiotics are live bacteria that, when administered in sufficient quantities, can improve host health. This approach involves introducing specific bacterial cultures, which are part of the normal gut flora, to newly hatched birds to establish these beneficial organisms in the gut before exposure to pathogens like *Salmonella*. These bacteria can be administered as a well-known characterized bacterial species such as *Bacillus*, *Lactobacillus*, *Enterococcus*, *Bifidobacterium* and *Streptococcus* or can be administered as defined or undefined competitive exclusion cultures.

6 Advances in strategies for controlling *Salmonella* and other pathogen contamination: biosecurity management of housing and production

As explained above, the farm environment presents numerous pathways for pathogen spread. Therefore, specific intervention methods, including enhanced biosecurity measures, are essential on a layer farm to (a) prevent horizontal transmission of *Salmonella* from the environment to the flock and (b) minimize the spread of *Salmonella* within the farm (Dewaele 2012). Biosecurity includes all measures preventing pathogens, including *Salmonella*, from entering a herd (i.e. external biosecurity) and reducing the spread of pathogens within a herd (i.e. internal biosecurity). Biosecurity in animal production is key for both farm management (e.g. disease prevention) and meeting consumer demands concerning food safety issues as *Salmonella*. Good hygiene practices on farms can reduce the risk of introducing and spreading animal diseases and zoonoses, infectious agents that are transmittable from animals to humans. (Luyckx 2016).

6.1 External biosecurity

External biosecurity can be divided into different categories, such as purchase of animals; removal of manure and dead animals; feed, water and equipment supplies; personnel and visitors; (biological) vector control; and location of the farm (Gelaude et al. 2014; Laanen et al. 2014).

6.1.1 Purchase of animals

Laying hens purchased for rearing and production need to be free from *Salmonella*. To control this, monitoring of the *Salmonella* status is needed as discussed in an earlier paragraph of this chapter.

6.1.2 Dead animals

Dead animals are often a source of pathogens, as they may have died due to an infection and should be removed as quickly as possible (Gelaude et al. 2014). Additionally, collecting cadavers via transport lorries poses a risk for introducing pathogens such as *Salmonella*, as these vehicles travel from farm to farm (Doyle and Erickson 2006; Fritzemeier 2000; Hege et al. 2002). It is therefore recommended that vehicles, or at least their wheels, are disinfected before entering the farm (Casal et al. 2007; Gelaude et al. 2014; Lister 2008), and that cadaver bins are placed away from the laying hen houses, close to the public road and cleaned and disinfected regularly.

6.1.3 Feed, water and equipment supplies

According to EFSA and ECDC (2023), the *Salmonella* prevalence in compound poultry feed was 0.46% in the EU in 2022. Several studies have shown that contaminated feed was linked to the occurrence of pathogens in animal houses (Fink-Gremmels 2012). Not only must the feed be *Salmonella* free upon delivery but also the feeding system can be the cause of *Salmonella* contamination (e.g. feeder lines, silo) (Dewaele et al. 2012a, 2012b). Similarly, drinking water can harbour *Salmonella* and be the source of infection (Herman et al. 2003; Heyndrickx et al. 2002). Depending on its origin, water can be contaminated by dust, faeces of wildlife or rodents, making it crucial to store drinking water in closed reservoirs (Lister, 2008). Additionally, biofilms in reservoirs and water pipes can harbour *Salmonella* (Maes et al. 2019, 2020). Therefore, regular cleaning and disinfection of feeding and drinking water systems are essential, especially if *Salmonella* contamination in the flock is detected.

6.1.4 Personnel and visitors

People can act as vectors for pathogens, picking up infectious agents and transmitting them through physical contact (EFSA 2016; Heyndrickx et al. 2002). Therefore, access for visitors, including veterinarians, should be restricted, and the number of caretakers per animal house should be limited (Herman et al. 2003; Refrégier-Petton et al. 2001). Both visitors and farm workers must adhere to biosecurity measures, including handwashing and using farm-specific and even laying hen house-specific clothing and boots, as these practices are crucial for an effective *Salmonella* control program (Amass 2000; Dewaele 2012). Using disinfectant footbaths is essential. However, footbaths are often improperly used; they can become highly contaminated with organic material and may pose a greater risk of pathogen spread rather than serving as an effective preventive measure (Amass 2000). Simply walking through a footbath does not significantly reduce bacterial counts or *Salmonella*. Effective use requires scrubbing boots to remove manure, followed by soaking them in a clean bath of correctly diluted disinfectant for a specified period (Luyckx 2016). Possible solutions for improving farm hygiene include providing designated boots for specific farm areas or the use of disposable footwear/overshoes (Luyckx 2016).

6.1.5 Biological vector control

Several studies showed that vermin can be an important vector of pathogens (Dewaele et al. 2012b; Meerburg et al. 2007). The significance of rodents has been shown by several studies as they can act as transmitters to production animals on farms (Backhans and Fellström 2012; Dewaele et al. 2012a, 2012b; van de Giessen et al. 2009). House mice (*Mus musculus*) and brown rats (*Rattus norvegicus*) are the most common rodent species on farms (Backhans and Fellström 2012). Henzler and Opitz (1992) demonstrated that the prevalence of *Salmonella*-positive mice on *Salmonella* positive layer farms was 24%. Further, the faeces of one mouse yielded 5 log CFU per faecal dropping. The mice remained *Salmonella* positive for 10 months. According to Rose et al. (2000), the risk for *Salmonella* persistence on poultry farms was increased if rodents were observed by the farmer after cleaning and disinfection. A Japanese study demonstrated that the role of roof rats (*Rattus rattus*) in the epidemiology of *Salmonella* in layer farms was expanded from being a reservoir and an amplifier host to a shifting vessel of the most predominant serotypes (Camba et al. 2020). In order to limit rodent nesting, feed should be stored in a vermin-free place and buildings should be rodent proof. Other rodent control methods include rodenticides and traps. (Luyckx 2016).

More difficult to control are insects, which are potential reservoirs and transmitters of pathogenic bacteria such as *Salmonella* (Dewaele et al. 2012b; Holt et al. 2007; Khamesipour et al. 2018; Olsen and Hammack, 2000) and *Campylobacter* (Hald et al. 2004; Szalanski et al. 2004). Chickens fed *Salmonella*-contaminated lesser mealworms or flies under experimental conditions became *Salmonella* colonized (Holt et al. 2007). In contrast, release of contaminated flies in a room with *Salmonella*-free hens failed to result in colonization of the hens (Holt et al. 2007). In a recent Italian study, Bertelloni et al. (2023) demonstrated that the prevalence of *Salmonella*-colonized flies was low on laying farms compared to pig farms.

6.1.6 Pets

Salmonella Enteritidis has been isolated from cats (Kinde et al. 1996) and cat faeces (Davies and Breslin 2003) present on layer farms. Cats may acquire *Salmonella* Enteritidis infection by consuming infected mice, poultry carcasses (Davies and Breslin 2003) or by contact with the *Salmonella* Enteritidis-contaminated layer farm environment. In Sweden, it has been reported that many cats are colonized with *Salmonella* Typhimurium as cats may become infected after catching and eating colonized passerine birds (EURL Salmonella). In addition, *Salmonella* has been isolated from dog faeces (Dewaele et al.

2012b; Leonard et al. 2011). Given these observations, it is important to prevent contact between pets and farm animals (Luyckx 2016).

6.1.7 Wild birds/animals

Little research has been performed on the correlation between wild birds and *Salmonella* Enteritidis on laying hen farms. One study performed by Davies and Breslin (2001) found *Salmonella* Enteritidis contaminated wild bird droppings in the outside environment of layer farms infected with *Salmonella* Enteritidis, indicating that wild birds can act as vectors of *Salmonella* Enteritidis. Especially free-range farms are a higher risk (Bailey and Cosby 2005). Two recent studies in Italy and Switzerland showed that approximately 1% of the wild birds were found *Salmonella* positive (Ebani et al. 2021; Vogler et al. 2021).

6.1.8 Farm location, air and dust

Airborne transmission of pathogens is possible through several distance-related factors such as other near farms and backyard animals (Lister, 2008; Van Steenwinkel et al. 2011), animal transport on public roads (Graham et al. 2008; Vieira et al. 2009) and litter spread on nearby arable lands (Lister, 2008). Dust samples collected from poultry premises have been found to be contaminated with *Salmonella* (Namata et al. 2009; Huneau-Salaün et al. 2010; Dewaele et al. 2012a).

6.2 Internal biosecurity

Internal biosecurity measures aim at the prevention or reduction of the spread of pathogens within the herd. These measures can be divided into different categories, such as farm management, compartmentalization, implementation of working lines, disease management and cleaning and disinfection (Backhans et al. 2015; Gelaude et al. 2014; Postma et al. 2015; Luyckx 2016). In addition, several categories of external biosecurity (e.g. removal of dead animals and measures for pet animals, personnel and visitors) are also part of internal biosecurity measures.

6.2.1 Farm management

On farms, an all-in/all-out system (for each phase/unit) is recommended. This system involves emptying, cleaning and disinfecting premises between production cycles to limit the contact between the incoming animals and any dust, manure or debris from the previous round. Ideally, farm premises are divided into clean and dirty areas with clearly identifiable clean-dirty barriers.

The allocation of 'clean' and 'dirty' areas is farm dependent, but, in general, a 'clean' area includes the area around and part of the production site with restricted access. The 'dirty' area typically comprises the cadaver storage facility and the farm entrances used by employees, visitors and external transport vehicles. The design of the loading bay should ensure that external vehicles use the dirty entrance of the farm. Another example is the changing room on farms: the dirty zone is where the employees and visitors enter the room, store their personal clothing and shoes and wash/disinfect their hands. The clean zone is where farm (or house)-specific overalls and boots are put on. On farms with high hygiene standards, showering is mandatory before entering the clean zone. These two zones should preferably be separated by a physical barrier. Finally, cross-over between the dirty road/traffic (e.g. feed deliveries) and clean road/traffic (employee cars) should be avoided and the number of visitors should be limited to minimise the introduction of pathogens (Luyckx 2016).

6.2.2 Hygiene, cleaning and disinfection

Cleaning and disinfection (C&D) is an important aspect of internal biosecurity. Not only interior of the premises should be cleaned and disinfected but also the hardened environment (e.g. concrete) around the premises, as studies showed the presence of pathogens in the vicinity of animal houses (Schulz et al. 2012, 2004). In addition, *Salmonella* can survive for long periods under various conditions in the environment (Kramer et al. 2006). Effective C&D is therefore a crucial step in reducing the infection pressure in animal houses and preventing both endemic animal diseases and food-borne zoonoses as *Salmonella*, *Campylobacter* and *L. monocytogenes* (van de Giessen et al. 1998). Finally, also clothing, boots and equipment should be washed and disinfected, as they could be contaminated with pathogens. Improper disinfection could put the herd at risk of pathogen spread (Amass 2000).

Although one would think that a sufficiently long depopulation period is important, this has not been found to have a protective influence on the prevalence of *Salmonella* (Van Hoorebeke et al. 2010). Also in the pig nursery, Luyckx et al. (2016) found a longer vacancy period after cleaning and disinfection had no effect on the bacterial load in pig nursery units.

6.2.3 The henhouse housing system

It is often assumed that free-range poultry is more often *Salmonella* colonized due to the outdoor exposure than conventionally raised poultry (Bailey and Cosby 2005); however, this assumption is not always scientifically based (Van Overbeke et al. 2006; EFSA 2019).

6.2.4 Egg belts, the egg-collecting area and egg trays

In laying hen production, specific areas such as the egg conveyor belts and the egg-collecting area are critical. Dewaele et al. (2012a, 2012b) demonstrated that these areas play a key role in maintaining *Salmonella* Enteritidis contamination on laying hen farms, particularly if an all-in/all-out system is not implemented. A problem with multi-house farms is that they often have connections between houses to facilitate conveyor belts (Carrique-Mas and Davies 2008), which can enable the transfer of infections from one flock to another. Additionally, if an all-in/all-out system is not implemented, common areas and equipment may not be cleaned and disinfected between cycles, increasing the risk of cross-contamination (Dewaele et al. 2012a, 2012b). Also reusable egg trays are a risk factor if not well cleaned and disinfected between uses. Utrarachkij et al. (2012) found egg trays to be frequently contaminated with *Salmonella*, i.e, over a quarter (26.7%) of egg trays from farms and more than one-third (36.7%) of trays from the market.

7 Conclusion and future trends

Although eggshells are unavoidably loaded with bacteria, egg content contamination remains a rare event. The numerous antimicrobial self-defence mechanisms of eggs result in an almost bacteria-free egg content in most cases. In recent years, most advances in general egg microbiology have been made by using culture-independent methods but confirm essentially the insights previously obtained through classical culture methods. The pathogenic bacteria most associated with eggs are still *Salmonella* Enteritidis. Thanks to the considerable efforts made by egg producers and the competent authorities, especially in developed countries, *Salmonella* contamination of eggs is significantly reduced, resulting in a measurable impact on public health. Biosecurity, vaccination and professional skills of the farmer are of utmost importance to control *Salmonella* in laying hen flocks and must be maintained and further refined. Further research to understand the association between *Salmonella* Enteriditis and laying hens and eggs should still be encouraged to further enhance the food safety of a healthy egg.

8 Acknowledgement

I acknowledge the University of Ghent for the use of material from the PhD thesis 'Bacterial contamination and infection of shell eggs in the production chain.' De Reu K. (2006), ISBN: 90-5989-124-4, in preparing this chapter.

9 References

Al-Natour, M. Q., et al. (2012). *Escherichia coli* O157:H7 facilitates the penetration of Staphylococcus aureus into table eggs. *J. Food Sci.* 77(1), M29–M34. https://doi.org/10.1111/j.1750-3841.2011.02483.x

Amass, S. F. (2000). Evaluating the efficacy of boot baths in biosecurity protocols. *J. Swine Heal. Prod.* 8, 169–173.

Aury, K., et al. (2011). Risk factors for *Listeria monocytogenes* contamination in French laying hens and broiler flocks. *Prev. Vet. Med.* 98(4), 271–278. https://doi.org/10.1016/j.prevetmed.2010.11.017

Backhans, A. and Fellström, C. (2012). Rodents on pig and chicken farms - a potential threat to human and animal health. *Infect. Ecol. Epidemiol.* 2, 17093. https://doi.org/10.3402/iee.v2i0.17093

Backhans, A., et al. (2015). Biosecurity level and health management practices in 60 Swedish farrow-to-finish herds. *Acta Vet. Scand.* 57, 14. https://doi.org/10.1186/s13028-015-0103-5

Bailey, J. S. and Cosby, D. E. (2005). *Salmonella* prevalence in free-range and certified organic chickens. *J. Food Prot.* 68(11), 2451–2453. https://doi.org/10.4315/0362-028x-68.11.2451

Baker, R. C., et al. (1985). The frequency of salmonellae on duck eggs. *Poult. Sci.* 64, 646–652.

Baron, F., et al. (2016) Egg white versus *Salmonella* Enteritidis! A harsh medium meets a resilient pathogen. *Food Microbiol.* 53(Pt B), 82–93. https://doi.org/10.1016/j.fm.2015.09.009

Bertelloni, F., et al. (2023). House dlies (*Musca domestica*) from swine and poultry farms carrying antimicrobial resistant *Enterobacteriaceae* and *Salmonella*. *Vet. Sci.* 10(2), 118. https://doi.org/10.3390/vetsci10020118

Board, R. G. and Tranter, H. S. (1995). The microbiology of eggs. In Stadelman, W. J. and Cotterill, O. J. (Eds.), *Egg science and technology*, Food Products Press - The Haworth Press, Inc., New York, pp. 81–104.

Bouda, S. C., et al. (2019) Prevalence and antimicrobial resistance of salmonella enterica isolated from chicken and guineafowl in Burkina Faso. *International Journal of Microbiology and Biotechnology*, 4, 64–71. https://doi.org/10.11648/j.ijmb.20190403.12

Brackett, R. E. and Beuchat, L. R. (1992). Survival of *Listeria monocytogenes* on the Surface of Egg Shells and During Frying of Whole and Scrambled Eggs. *Journal of Food Protection*, 55(11), 862–865. https://doi.org/10.4315/0362-028X-55.11.862

Callewaert, L., et al. (2008). A new family of lysozyme inhibitors contributing to Lysozyme Tolerance in gram-negative bacteria. *PLoS Pathog.* 4(3), e1000019. https://doi.org/10.1371/journal.ppat.1000019

Camba, S. I., et al. (2020). The expanded role of roof-rats (*Rattus rattus*) in *Salmonella* spp. contamination of a commercial layer farm in East Japan. *Avian Dis.* 64(1), 46–52. https://doi.org/10.1637/0005-2086-64.1.46

Carrique-Mas, J. J. and Davies, R. H. (2008). Sampling and bacteriological detection of *Salmonella* in poultry and poultry premises: a review. *Rev. Sci. Tech.* 27, 665–677.

Carrique-Mas, J. J., et al. (2009). Persistence and clearance of different *Salmonella* serovars in buildings housing laying hens. *Epidemiol. Infect.* 137, 837–846.

Casal, J., et al. (2007). Biosecurity measures on swine farms in Spain: perceptions by farmers and their relationship to current on-farm measures. *Prev. Vet. Med.* 82, 138–150. https://doi.org/10.1016/j.prevetmed.2007.05.012

Chemaly, M., et al. (2009) Isolation of *Salmonella enterica* in laying-hen flocks and assessment of eggshell contamination in France. *J. Food Prot.* 72, 2071–2077.

Chemaly, M., et al. (2008). Prevalence of *Listeria monocytogenes* in poultry production in France. *J. Food Prot.* 71(10), 1996–2000. https://doi.org/10.4315/0362-028x-71.10.1996

Chuma, T., et al. (1997). Direct detection of *Campylobacter jejuni* in chicken cecal contents by PCR. *J. Vet. Med. Sci.* 59(1), 85–87. https://doi.org/10.1292/jvms.59.85

Clark, A. G. and Bueschkens, D. H. (1986). Survival and growth of *Campylobacter jejuni* in Egg Yolk and Albumen. *J. Food Prot.* 49(2), 135–141. https://doi.org/10.4315/0362-028X-49.2.135

Clavijo, R. I., et al. (2006). Identification of genes associated with survival of *Salmonella enterica* serovar Enteritidis in chicken egg albumen. *Appl. Environ. Microbiol.* 72(2), 1055–1064. https://doi.org/10.1128/AEM.72.2.1055-1064.2006

Crabb, H. K., et al. (2019). Does only the age of the hen matter in *Salmonella enterica* contamination of eggs?. *Food Microbiol.* 77, 1–9. https://doi.org/10.1016/j.fm.2018.08.006

Council Directive (1999). 1999/74/EC of 19 July 1999 laying down minimum standards for the protection of laying hens. *Off. J. Eur. Comm.* 203, 53–57.

Commission Regulation (EC) (2006a). No 1168/2006 of 31 July 2006 implementing Regulation (EC) No 2160/2003 as regards a Community target for the reduction of the prevalence of certain salmonella serotypes in laying hens of Gallus gallus and amending Regulation (EC) No 1003/2005.

Commission Regulation (EC) (2006b). No 1177/2006 of 1 August 2006 implementing Regulation (EC) No 2160/2003 of the European Parliament and of the Council as regards requirements for the use of specific control methods in the framework of the national programmes for the control of salmonella in poultry.

Commission Regulation (EC) (2007). No 1237/2007 of 23 October 2007 amending Regulation (EC) No 2160/2003 of the European Parliament and of the Council and Decision 2006/696/EC as regards the placing on the market of eggs from Salmonella infected flocks of laying hens.

Commission Regulation (EC) (2010). No 200/2010 of 10 March 2010 implementing Regulation (EC) No 2160/2003 of the European Parliament and of the Council as regards a Union target for the reduction of the prevalence of Salmonella serotypes in adult breeding flocks of Gallus gallus.

Cox, N. A., et al. (2012). Evidence for horizontal and vertical transmission in *Campylobacter* passage from hen to her progeny. *J. Food Prot.* 75(10), 1896–1902.

Davies, R. and Breslin, M. (2001). Environmental contamination and detection of *Salmonella* enterica serovar enteritidis in laying flocks. *Vet. Rec.* 149, 699–704.

Davies, R. and Breslin, M. (2003). Investigation of *Salmonella* contamination and disinfection in farm egg-packing plants. *J. Appl. Microbiol.* 94, 191–196. https://doi.org/10.1046/j.1365-2672.2003.01817.

de Boer, E. and Wit, B. (2000). Salmonella in eieren [Salmonella in eggs]. *Tijdschrift voor diergeneeskunde* 125(4), 126–128.

De Reu, K. (2006) Bacterial contamination and infection of shell eggs in the production chain. PhD Thesis, ISBN: 90-5989-124-4.

De Reu, K., et al. (2005). The use of total aerobic and Gram-negative flora for quality assurance in the production chain of consumption eggs. *Food Control* 16, 147–155.

De Reu, K., et al. (2006a). Bacterial eggshell contamination in the egg collection chains of different housing systems for laying hens. *Brit. Poultry Sci.* 47, 163–172.

De Reu, K., et al. (2006b). Eggshell factors influencing eggshell penetration and whole egg contamination by different bacteria, including Salmonella enteritidis. *Int. J. Food Microbiol.* 112(3), 253–260. https://doi.org/10.1016/j.ijfoodmicro.2006.04.011.

De Reu, K., et al. (2008a). Bacterial contamination of table eggs and the influence of housing systems. *World's Poultry Sci. J.* 64, 5–19.

De Reu, K., et al. (2008b). Estimation of the vertical and horizontal bacterial infection of hen's table eggs. Proceeding; XXIII World's Poultry Congress, Congress Proceedings CD p. 1–4, Brisbane, Australia, 30 June–4 July.

De Reu, K., et al. (2008c). Estimation of the vertical and horizontal bacterial infection of hen's table eggs. *World's Poult. Sci. Journal* 64(supplement 2), 142.

De Reu, K., et al. (2009). Bacteriological contamination, dirt, and cracks of eggshells in furnished cages and noncage systems for laying hens: an international on-farm comparison. *Poult. Sci.* 88(11), 2442–2448. https://doi.org/10.3382/ps.2009-00097

De Vylder, J., et al. (2013). *Salmonella* Enteritidis is superior in egg white survival compared with other Salmonella serotypes. *Poult. Sci.* 92(3), 842–845. https://doi.org/10.3382/ps.2012-02668

De Buck, J., et al. (2004). Colonization of the chicken reproductive tract and egg contamination by *Salmonella*. *J. Appl. Microbiol.* 97(2), 233–245.

Denagamage, T., et al. (2015). Risk factors associated with salmonella in laying hen farms: systematic review of observational studies. *Avian Dis.* 59(2), 291–302. https://doi.org/10.1637/10997-120214-Reg

Dewaele, I. (2012). Scientific support for *Salmonella* Enteritidis control on layer farms, PhD thesis.

Dewaele, I., et al. (2012a). Persistent Salmonella Enteritidis environmental contamination on layer farms in the context of an implemented national control program with obligatory vaccination. *Poult. Sci.* 91(2), 282–291. https://doi.org/10.3382/ps.2011-01673.

Dewaele, I., et al. (2012b) Polyphasic characterization of Salmonella Enteritidis isolates on persistently contaminated layer farms during the implementation of a national control program with obligatory vaccination: a longitudinal study. *Poult. Sci.* 91(11), 2727–2735. https://doi.org/10.3382/ps.2012-02218.

Doyle, M. P. (1984). Association of *Campylobacter jejuni* with laying hens and eggs. *Appl. Environ. Microbiol.* 47(3), 533–536. https://doi.org/10.1128/aem.47.3.533-536.1984

Doyle, M. P. and Erickson, M. C. (2006). Reducing the carriage of foodborne pathogens in livestock and poultry. *Poult. Sci.* 85, 960–973.

Ebani, V. V., et al. (2021). Survey on the presence of bacterial and parasitic zoonotic agents in the feces of wild birds. *Vet. Sci.* 8(9), 171. https://doi.org/10.3390/vetsci8090171

Ebel, E. and Schlosser, W. (2000). Estimating the annual fraction of eggs contaminated with Salmonella enteritidis in the United States. *Int. J. Food Microbiol.* 61(1), 51–62. https://doi.org/10.1016/s0168-1605(00)00375-5

EFSA. (2005). Opinion of the scientific panel on biological hazards on the request from the Commission related to the microbiological risks on washing of table eggs. *EFSA J.* 269, 1–39.

EFSA (2016). Vector-borne diseases [WWW Document]. http://www.efsa.europa.eu/en/topics/topic/vectorbornezoonoticdiseases (accessed 4.6.16).

EFSA (2019). EFSA BIOHAZ Panel. Scientific Opinion on the Salmonella control in poultry flocks and its public health impact. *EFSA J.* 17(2), 5596. https://doi.org/10.2903/j.efsa.2019.5596

EFSA and ECDC (European Food Safety Authority and European Centre for Disease Prevention and Control) (2023). The European Union One Health 2022 Zoonoses Report. *EFSA J.* 21(12), e8442. https://doi.org/10.2903/j.efsa.2023.8442

El Ftouhy, F. Z., et al. (2022). Bacteriological quality of table eggs in moroccan formal and informal sector. *Int. J. Food Sci.* 24, 6223404. https://doi.org/10.1155/2022/6223404

Favier, G. I., et al. (2005). Genotypic and phenotypic characteristics of Yersinia enterocolitica isolated from the surface of chicken eggshells obtained in Argentina. *J. Food Prot.* 68(9), 1812–1815. https://doi.org/10.4315/0362-028x-68.9.1812

Fink-Gremmels, J. (Ed.) (2012). *Animal feed contamination: effects on livestock and food safety*. Elsevier Science.

Food Standard Agency UK (2004). *Report of the survey of Salmonella contamination of UK produced shell eggs on retail sale*. London Food Standard Agency, 124 p.

Fritzemeier, J. (2000). Epidemiology of classical swine fever in Germany in the 1990s. *Vet. Microbiol.* 77, 29–41. https://doi.org/10.1016/S0378-1135(00)00254-6

Gantois, I., et al. (2006) Oral immunisation with TAD Salmonella vacE, TAD Salmonella vac T and the combined vaccine aids in protection against egg contamination. In: Proceedings of the International Symposium Salmonella and Salmonellosis. pp. 601–602.

Gantois, I., et al. (2008). A comparative study on the pathogenesis of egg contamination by different serotypes of Salmonella. *Avian Pathol.* 37(4), 399–406. https://doi.org/10.1080/03079450802216611

Gantois, I., et al. (2009) Mechanisms of egg contamination by *Salmonella* Enteritidis. *FEMS Microbiol. Rev.* 33(4), 718–738. https://doi.org/10.1111/j.1574-6976.2008.00161.x

Garcia, R., et al. (2010) Influence of temperature and predation on survival of *Salmonella enterica* serovar Typhimurium and expression of invA in soil and manure-amended soil. *Appl. Environ. Microbiol.* 76, 5025–5031.

Gast, R. K. (2011). Pre-harvest measures to control *Salmonella* in laying hens, Improving the safety and quality of eggs and egg products – volume 2: egg safety and nutritional quality. Immerseel, F., et al. (Eds), Woodhead Publishing Limited, pp. 120–145, ISBN 9780857090720. https://doi.org/10.1533/9780857093929.2.120

Gast, R. K. and Beard, C. W. (1990). Production of Salmonella enteritidis-contaminated eggs by experimentally infected hens. *Avian Dis.*, 34(2), 438–446.

Gast, R. K., et al. (2004). Colonization of reproductive organs and internal contamination of eggs after experimental infection of laying hens with Salmonella heidelberg and Salmonella enteritidis. *Avian Dis.* 48(4), 863–869. https://doi.org/10.1637/7204-05050R

Gast, R. K., et al. (2020). Research note: horizontal transmission and internal organ colonization by Salmonella Enteritidis and Salmonella Kentucky in experimentally infected laying hens in indoor cage-free housing. *Poult Sci.* 99(11), 6071–6074. https://doi.org/10.1016/j.psj.2020.08.006

Gast, R. K., et al. (2022). *Salmonella* in eggs and egg-laying chickens: pathways to effective control. *Crit Rev Microbiol.* 50(1), 39–63. https://doi.org/10.1080/1040841X.2022.2156772

Gelaude, P., et al. (2014). Biocheck.UGent: a quantitative tool to measure biosecurity at broiler farms and the relationship with technical performances and antimicrobial use. *Poult. Sci.* 93, 2740–2751. https://doi.org/10.3382/ps.2014-04002

Gharbi, M., et al. (2022). *Campylobacter* spp. in eggs and laying hens in the North-East of Tunisia: high prevalence and multidrug-resistance phenotypes. *Vet. Sci.* 9(3), 108. https://doi.org/10.3390/vetsci9030108

Gong, H., et al. (2023). Microbiota profiles of hen eggs from the different seasons and different sectors of Shanghai, China. *Microorganisms* 11(10), 2519. https://doi.org/10.3390/microorganisms11102519

Graham, J. P., et al. (2008). The animal-human interface and infectious disease in industrial food animal production: rethinking biosecurity and biocontainment. *Public Health Rep.* 123, 282–299.

Guard-Petter, J. (2001). The chicken, the egg and *Salmonella* enteritidis. *Environ. Microbiol.* 3(7), 421–430. https://doi.org/10.1046/j.1462-2920.2001.00213.x

Guzmán-Gómez, G., et al. (2013). Frequency of *Salmonella* and *Listeria monocytogenes* in five commercial brands of chicken eggs using a combined method of enrichment and nested-PCR. *J. Food Prot.* 76(3), 429–434. https://doi.org/10.4315/0362-028X.JFP-12-213

Hald, B., et al. (2004). Flies and *Campylobacter* infection of broiler flocks. *Emerg. Infect. Dis.* 10, 1490–2. https://doi.org/10.3201/eid1008.040129

Haysom, I. W. and Sharp, K. (2003). The survival and recovery of bacteria in vacuum cleaner dust. *J. R. Soc. Promot. Health* 123, 39–45.

Hege, R., et al. (2002). Incidence of reinfections with *Mycoplasma hyopneumoniae* and *Actinobacillus pleuropneumoniae* in pig farms located in respiratory-disease-free regions of Switzerland - identification and quantification of risk factors. *Acta Vet. Scand.* 43, 145–156. https://doi.org/10.1186/1751-0147-43-145

Henzler, D. J. and Opitz, H. M. (1992). The role of mice in the epizootiology of Salmonella enteritidis infection on chicken layer farms. *Avian Dis.* 36(3), 625–631.

Herman, L., et al. (2003). Routes for *Campylobacter* contamination of poultry meat: epidemiological study from hatchery to slaughterhouse. *Epidemiol. Infect.* 131, 1169–1180.

Heyndrickx, M., et al. (2002). Routes for *Salmonella* contamination of poultry meat: epidemiological study from hatchery to slaughterhouse. *Epidemiol. Infect.* 69, 7–13.

Hiett, K. L., et al. (2002a). Genotype analyses of *Campylobacter* isolated from distinct segments of the reproductive tracts of broiler breeder hens. *Curr. Microbiol.* 45(6), 400–404. https://doi.org/10.1007/s00284-002-3771-0

Hiett, K. L., et al. (2002b). Direct polymerase chain reaction detection of Campylobacter spp. in poultry hatchery samples. *Avian Dis.* 46(1), 219–223. https://doi.org/10.1637/0005-2086(2002)046[0219:DPCRDO]2.0.CO;2

Holt, P. S., et al. (2007). Isolation of *Salmonella enterica* serovar Enteritidis from houseflies (*Musca domestica*) found in rooms containing Salmonella serovar Enteritidis-challenged hens. *Appl. Environ. Microbiol.* 73(19), 6030–6035. https://doi.org/10.1128/AEM.00803-07

Howard, Z. R., et al. (2005). Ovarian laying hen follicular maturation and in vitro Salmonella internalization. *Vet. Microbiol.* 108(1–2), 95–100. https://doi.org/10.1016/j.vetmic.2005.03.002

Humprhrey, T. J. (1989). *Salmonella* Enteritidis phage type 4 and hens' eggs. *Lancet* I, 281.

Humphrey, T. J. (1994). Contamination of eggs with potential pathogens. In Board, R. G. and Fuller, R. (Eds.), *Microbiology of the avian egg*. Chapman & Hall, pp. 93–116.

Humphrey, T. J. (2006). Are happy chickens safer chickens? Poultry welfare and disease susceptibility. *Br. Poult. Sci.* 47, 379–391.

Humphrey, T. J., et al. (1989). *Salmonella enteritidis* phage type 4 from the contents of intact eggs: a study involving naturally infected eggs. *Epidemiol. Infect.* 103, 415–423.

Humphrey T. J., et al. (1991). Numbers of *Salmonella* Enteritidis in the contents of naturally contaminated hen's eggs. *Epidemiol. Infect.* 106, 489–496.

Huneau-Salaün, A., et al. (2009). Risk factors for *Salmonella enterica* subsp. *enterica* contamination in 519 French laying hen flocks at the end of the laying period. *Vet Med.* 89, 51–58.

Huneau-Salaün, A., et al. (2010). Factors influencing bacterial eggshell contamination in conventional cages, furnished cages and free-range systems for laying hens under commercial conditions. *Brit. Poult Sci.* 51(2), 163–169. https://doi.org/10.1080/00071668.2010.482462

Izat, A. L. and Gardner F. A. (1988). Incidence of *Campylobacter jejuni* in processed egg products. *Poult. Sci.* 67(10), 1431–1435. https://doi.org/10.3382/ps.0671431

Jan, S. and Baron, F. (2017). Mechanisms for the transmission of pathogens into eggs. https://doi.org/10.19103/AS.2016.0012.05. Book chapter in Achieving sustainable production of eggs – volume 1: Safety and Quality, Eds Julie Roberts, Burleigh Dodds Science Publishing.

Jan S., et al. (2011). Biodiversity of psychrotrophic bacteria of the *Bacillus cereus* group collected on farm and in egg product industry. *Food Microbiol.* 28(2), 261–265. https://doi.org/10.1016/j.fm.2010.05.029

Jones, D. R. and Musgrove, M. T. (2007). Pathogen prevalence and microbial levels associated with restricted shell eggs. *J. Food Prot.* 70(9), 2004–2007. https://doi.org/10.4315/0362-028x-70.9.2004

Jones, D. R., et al. (2011). Comparison of environmental and egg microbiology associated with conventional and free-range laying hen management. *Poult. Sci.* 90(9), 2063–2068. https://doi.org/10.3382/ps.2010-01139

Jones, D. R., et al. (2012) Prevalence of coliforms, Salmonella, Listeria, and Campylobacter associated with eggs and the environment of conventional cage and free-range egg production. *Poult. Sci.* 91(5), 1195–1202. https://doi.org/10.3382/ps.2011-01795

Jones, D. R., et al. (2015). Microbiological impact of three commercial laying hen housing systems. *Poult. Sci.* 94(3), 544–551. https://doi.org/10.3382/ps/peu010

Jones, D. R., et al. (2016). Influence of commercial laying hen housing systems on the incidence and identification of *Salmonella* and *Campylobacter*. *Poult. Sci.* 95(5), 1116–1124. https://doi.org/10.3382/ps/pew036

Jones, F. T., et al. (1995). Salmonella contamination in commercial eggs and an egg production facility. *Poult. Sci.* 74(4), 753–757. https://doi.org/10.3382/ps.0740753

Keller, L. H., et al. (1995). *Salmonella* enteritidis colonization of the reproductive tract and forming and freshly laid eggs of chickens. *Infect. Immun.* 63(7), 2443–2449. https://doi.org/10.1128/iai.63.7.2443-2449.1995

Khamesipour, F., et al. (2018). A systematic review of human pathogens carried by the housefly (Musca domestica L.). *BMC Public Health* 18(1), 1049. https://doi.org/10.1186/s12889-018-5934-3

Kilroy, S., et al. (2016). Prevention of egg contamination by Salmonella Enteritidis after oral vaccination of laying hens with Salmonella Enteritidis ΔtolC and ΔacrABacrEFmdtABC mutants. *Vet Res.* 47(82). https://doi.org/10.1186/s13567-016-0369-2. PMID: 27519174; PMCID: PMC4982998.

Kinde, H., et al. (1996). *Salmonella* Enteritidis, phase type 4 infection in a commercial layer flock in southern California: bacteriologic and epidemiologic findings. *Avian Dis.* 40, 665–671.

Koné, A. Z., et al. (2013). Identifying risk factors for eggshell contamination by *Bacillus cereus* group bacteria in French laying farms. *British Poult. Sci.* 54(3), 298–305. https://doi.org/10.1080/00071668.2013.783900

Kramer, A., et al. (2006). How long do nosocomial pathogens persist on inanimate surfaces? A systematic review. *BMC Infect. Dis.* 6, 130. https://doi.org/10.1186/1471-2334-6-130

Laanen, M., et al. (2014). Pig, cattle and poultry farmers with a known interest in research have comparable perspectives on disease prevention and on-farm biosecurity. *Prev. Vet. Med.* 115, 1–9. https://doi.org/10.1016/j.prevetmed.2014.03.015

Leonard, E. K., et al. (2011). Evaluation of pet-related management factors and the risk of Salmonella spp. carriage in pet dogs from volunteer households in Ontario (2005–2006). *Zoonoses Public Health* 58, 140–149. https://doi.org/10.1111/j.1863-2378.2009.01320.x

Li, W., et al. (2020). Prevalence, serotype, antibiotic susceptibility, and genotype of *Salmonella* in eggs from poultry farms and marketplaces in Yangling, Shaanxi Province, China. *Front. Microbiol.* 11, 1482. https://doi.org/10.3389/fmicb.2020.01482

Lister, S. A. (2008). *Biosecurity in poultry management in Poultry diseases*, 6th ed. Saunder Elsevier, Beijing, China.

Luyckx, K. (2016) Evaluation and implication of cleaning and disinfection of broiler houses and pig nursery units. PhD thesis, Ghent University. Faculty of Veterinary Medicine, Merelbeke, Belgium.

Luyckx, K., et al. (2016). A 10-day vacancy period after cleaning and disinfection has no effect on the bacterial load in pig nursery units. *BMC Vet. Res.* 12, 236. https://doi.org/10.1186/s12917-016-0850-1

Mallet, S., et al. (2006). Comparison of eggshell hygiene in two housing systems: standard and furnished cages. *British Poult. Sci.* 47(1), 30–35. https://doi.org/10.1080/00071660500468132

Maes, S., et al. (2019). Occurrence and characterisation of biofilms in drinking water systems of broiler houses. *BMC Microbiol.* 19(1), 77. https://doi.org/10.1186/s12866-019-1451-5

Maes, S., et al. (2020). *Pseudomonas putida* as a potential biocontrol agent against *Salmonella* Java biofilm formation in the drinking water system of broiler houses. *BMC Microbiol.* 20(1), 373. https://doi.org/10.1186/s12866-020-02046-5

Martelli F. and Davies R. (2012). Salmonella serovars isolated from table eggs: an overview. *Int. Food Res.* 45(2), 745–754. https://doi.org/10.1016/j.foodres.2011.03.054

Mawer, S. L., et al. (1989). *Salmonella enteritidis* phage type 4 and hens' eggs. *Lancet* I, 280–281.

Meerburg, B. G., et al. (2007). Controlling risks of pathogen transmission by flies on organic pig farms: a review. *Outlook Agric.* 36, 193–197. https://doi.org/10.5367/000000007781891432

Messelhäusser, U., et al. (2011). Occurrence of thermotolerant *Campylobacter* spp. on eggshells: a missing link for food-borne infections? *Appl. Environ. Microbiol.* 77(11), 3896–3897. https://doi.org/10.1128/AEM.00145-11

Messens, W., et al. (2005). Eggshell penetration by Salmonella: a review. *World's Poult. Sci. J.* 61(1), 71–86. https://doi.org/10.1079/WPS200443

Messens, W., et al. (2011). Egg decontamination by washing – volume 2: egg safety and nutritional quality. Immerseel F., et al. (Eds), Woodhead Publishing Limited, pp. 120–145, 163–180, ISBN 9780857090720.

Mitchell, R., et al. (2002). Public health investigations of *Salmonella* Enteritidis in raw shell eggs in England and Wales. *Eurosurveillance Weekly* 50.

Mollenhorst, H., et al. (2005). Risk factors for *Salmonella* Enteritidis infections in laying hens. *Poult. Sci.* 84, 1308–1313.

Moore, J. and Madden, R. H. (1993). Detection and incidence of *Listeria* Species in blended raw egg. *J. Food Prot.* 56(8), 652–654. https://doi.org/10.4315/0362-028X-56.8.652

Mughini-Gras, L., et al. (2022). A statistical modelling approach for source attribution meta-analysis of sporadic infection with foodborne pathogens. *Zoonoses Public Health* 69(5), 475–486. https://doi.org/10.1111/zph.12937

Namata, H., et al. (2009). Identification of risk factors for the prevalence and persistence of Salmonella in Belgian broiler chicken flocks. *Prev. Vet. Med.* 90(3–4), 211–222. https://doi.org/10.1016/j.prevetmed.2009.03.006

Neira, C., et al. (2017). Microbial diversity on commercial eggs as affected by the production system: a first approach using PGM. *Int. J. Food Microbiol.* 262, 3–7. https://doi.org/10.1016/j.ijfoodmicro.2017.09.008

Novoa Rama, E., et al. (2018). Prevalence, persistence, and antimicrobial resistance of *Campylobacter* spp. from eggs and laying hens housed in five commercial housing systems. *Foodborne Pathog. Dis.* 15(8), 506–516. https://doi.org/10.1089/fpd.2017.2404

Okamura, M., et al. (2001a). Differences among six Salmonella serovars in abilities to colonize reproductive organs and to contaminate eggs in laying hens. *Avian Dis.* 45(1), 61–69.

Okamura, M., et al. (2001b). Differences in abilities to colonize reproductive organs and to contaminate eggs in intravaginally inoculated hens and in vitro adherences to vaginal explants between *Salmonella* enteritidis and other *Salmonella* serovars. *Avian Dis.* 45(4), 962–971.

Olsen, A. R. and Hammack, T. S. (2000). Isolation of *Salmonella* spp. from the housefly, Musca domestica L., and the dump fly, Hydrotaea aenescens (Wiedemann) (Diptera: Muscidae), at caged-layer houses. *J. Food Prot.* 63, 958–960.

Parker, C. T., et al. (2002). Mitigation of avian reproductive tract function by Salmonella enteritidis producing high-molecular-mass lipopolysaccharide. *Environ. Microbiol.* 4(9), 538–545. https://doi.org/10.1046/j.1462-2920.2002.00333.x

Perales, I. and Audicana, A. (1989). The role of hens' eggs in outbreaks of salmonellosis in north Spain. *Int. J. Food Microbiol.* 8(2), 175–180. https://doi.org/10.1016/0168-1605(89)90072-x

Pohjola, L., et al. (2016), Zoonotic public health hazards in backyard chickens. *Zoonoses Public Health* 63, 420–430. https://doi.org/10.1111/zph.12247

Poppe, C., et al. (1998). Salmonella contamination of hatching and table eggs: a comparison. *Can. J. Vet. Res.* 62(3), 191–198.

Postma, M., et al. (2015). The biosecurity status and its associations with production and management characteristics in farrow-to-finish pig herds. *Animal* 10(3), 2016, 478–489. https://doi.org/10.1017/S1751731115002487

Quarles, C. L., et al. (1970). Bacterial contamination in poultry houses and its relationship to egg hatchability. *Poultry Sci.* 49, 60–66.

Raspoet, R. (2014). Survival strategies of Salmonella Enteritidis to cope with antibacterial factors in the chicken oviduct and in egg white. PhD thesis. Ghent University. Faculty of Veterinary Medicine, Merelbeke, Belgium.

Raspoet, R., et al. (2011) Internal contamination of eggs by *Salmonella* Enteritidis. Improving the safety and quality of eggs and egg products – volume 2: egg safety and nutritional quality. Immerseel F., et al. (Eds), Woodhead Publishing Limited, pp. 46–61.

Raspoet, R., et al. (2019). The Salmonella Enteritidis TolC outer membrane channel is essential for egg white survival. *Poult. Sci.* 98(5), 2281–2289. https://doi.org/10.3382/ps/pey584

Rasschaert, G., et al. (2007). Investigation of the concurrent colonization with Campylobacter and Salmonella in poultry flocks and assessment of the sampling site for status determination at slaughter. *Vet. Microbiol.* 123(1–3), 104–109. https://doi.org/10.1016/j.vetmic.2007.03.011

Refrégier-Petton, J., et al. (2001). Risk factors for *Campylobacter* spp. contamination in French broiler-chicken flocks at the end of the rearing period. *Prev. Vet. Med.* 50, 89–100.

Rivoal, K., et al. (2013). Detection of *Listeria* spp. in liquid egg products and in the egg breaking plants environment and tracking of *Listeria monocytogenes* by PFGE. *Int. J. Food Microbiol.* 166(1), 109–116. https://doi.org/10.1016/j.ijfoodmicro.2013.06.014

Rose, N., et al. (2000). Risk factors for Salmonella persistence after cleansing and disinfection in French broiler-chicken houses. *Prev. Vet. Med.* 44(1–2), 9–20. https://doi.org/10.1016/s0167-5877(00)00100-8

Sahin, O., et al. (2003). Detection and survival of *Campylobacter* in chicken eggs. *J. Appl. Microbiol.* 95(5), 1070–1079. https://doi.org/10.1046/j.1365-2672.2003.02083.x

Salvat, G., et al. (1991). Excretionate of *Salmonella enteritidis* in laying hens following a production of *Salm. enteritidis* contaminated eggs responsible for foodborne disease. In Mulder, R. W. A. W. and de Vries, A. W. (Eds.), *Quality of Poultry Products III, Safety and Marketing Aspects*. Spelderholt Centre for Poultry Research and Information Services, pp. 35–42.

Schlosser, W. D., et al. (1999). The *Salmonella enterica* serovar Enteritidis pilot project. In Saeed, A. M, et al. (Eds.), *Salmonella Enterica Serovar Enteritidis in Humans and Animals*. Iowa State University Press, pp. 353–365.

Schoeni, J. L. and Doyle, M. P. (1994). Variable colonization of chickens perorally inoculated with Escherichia coli O157:H7 and subsequent contamination of eggs. *Appl. Environ. Microbiol.* 60(8), 2958–2962. https://doi.org/10.1128/aem.60.8.2958-2962.1994

Schulz, J., et al. (2004). Staphylococci as an indicator for bacterial emissions from a broiler house. *Int. Soc. Anim. Hyg.* 2004, 75–78.

Schulz, J., et al. (2012). Longitudinal study of the contamination of air and of soil surfaces in the vicinity of pig barns by livestock-associated methicillin-resistant *Staphylococcus aureus*. *Appl. Environ. Microbiol.* 78, 5666–5671. https://doi.org/10.1128/AEM.00550-12

Schwaiger, K., et al. (2010). Comparative analysis on antibiotic resistance characteristics of *Listeria* spp. and *Enterococcus* spp. isolated from laying hens and eggs in conventional and organic keeping systems in Bavaria, Germany. *Zoonoses and Public Health* 57(3), 171–180. https://doi.org/10.1111/j.1863-2378.2008.01229.x

Shanker, S., et al. (1986). *Campylobacter jejuni* in broilers: the role of vertical transmission. *J. Hyg.* 96(2), 153–159. https://doi.org/10.1017/s002217240006592x

Shi, Y., et al. (2020). Microbiota Analysis of Eggshells in Different Areas and During Different Storage Time by Non-cultural Methods. *Curr. Microbiol.* 77(12), 3842–3850. https://doi.org/10.1007/s00284-020-02212-y

Singh, R., et al. (2009) Production performance and egg quality of four strains of laying hens kept in conventional cages and floor pens. *Poult. Sci.* 88(2), 256–264. https://doi.org/10.3382/ps.2008-00237

Singh, S., et al. (2010) Prevalence of Salmonella in chicken eggs collected from poultry farms and marketing channels and their antimicrobial resistance. *Food Res. Int.* 43(8), 2027–2030. https://doi.org/10.1016/j.foodres.2010.06.001

Snow, L. C., et al. (2010) Investigation of risk factors for *Salmonella* on commercial egg-laying farms in Great Britain, 2004–2005. *Vet. Rec.* 166, 579–586.

Sulonen, J., et al. (2007). *Campylobacter* in Finnish organic laying hens in autumn 2003 and spring 2004. *Poult. Sci.* 86(6), 1223–1228. https://doi.org/10.1093/ps/86.6.1223

Szalanski, A. L., et al. (2004). Detection of *Campylobacter* and *Escherichia coli* O157:H7 from filth flies by polymerase chain reaction. *Med. Vet. Entomol.* 18, 241–246. https://doi.org/10.1111/j.0269-283X.2004.00502.x

Techer, C., et al. (2014). Global overview of the risk linked to the Bacillus cereus group in the egg product industry: identification of food safety and food spoilage markers. *J. Appl. Microbiol.* 116(5), 1344–1358. https://doi.org/10.1111/jam.12462

Trudeau, S., et al. (2020). Contribution of the Broiler Breeders' Fecal Microbiota to the Establishment of the Eggshell Microbiota. *Front. Microbiol.* 11, 666. https://doi.org/10.3389/fmicb.2020.00666

Utrarachkij, F., et al. (2012). Possible horizontal transmission of Salmonella via reusable egg trays in Thailand. *Int. J. Food Microbiol.* 154(1–2), 73–78. https://doi.org/10.1016/j.ijfoodmicro.2011.12.024

van de Giessen, A. W., et al. (1998). Reduction of Campylobacter infections in broiler flocks by application of hygiene measures. *Epidemiol. Infect.* 121, 57–66. https://doi.org/10.1017/S0950268898008899

van de Giessen, A. W., et al. (2009). Occurrence of methicillin-resistant *Staphylococcus aureus* in rats living on pig farms. *Prev. Vet. Med.* 91, 270–273. https://doi.org/10.1016/j.prevetmed.2009.05.016

Van Hoorebeke, S., et al. (2010). The age of production system and previous *Salmonella* infections on-farm are risk factors for low-level *Salmonella* infections in laying hen flocks. *Poult. Sci.* 89, 1315–1319.

Van Immerseel, F. (2010). Stress-induced survival strategies enable *Salmonella* Enteritidis to persistently colonize the chicken oviduct tissue and cope with antimicrobial factors in egg white: A hypothesis to explain a pandemic. *Gut Pathog.* 2, 23. https://doi.org/10.1186/1757-4749-2-23

Van Overbeke, I., et al. (2006). A comparison survey of organic and conventional broiler chickens for infectious agents affecting health and food safety. *Avian Dis.* 50(2), 196–200.

Van Steenwinkel, S., et al. (2011). Assessing biosecurity practices, movements and densities of poultry sites across Belgium, resulting in different farm risk-groups for infectious disease introduction and spread. *Prev. Vet. Med.* 98, 259–270. https://doi.org/10.1016/j.prevetmed.2010.12.004

Videnska, P., et al. (2014). Characterization of egg laying hen and broiler fecal microbiota in poultry farms in Croatia, Czech Republic, Hungary and Slovenia. *PLoS One* 9(10), e110076. https://doi.org/10.1371/journal.pone.0110076

Vieira, A. R., et al. (2009). Human contacts and potential pathways of disease introduction on Georgia poultry farms. *Avian Dis.* 53, 55–62. https://doi.org/10.1637/8364-051608-Reg.1

Vogler, B. R., et al. (2021). Low occurrence of Salmonella spp. in wild birds from a Swiss rehabilitation centre. *Vet. Rec. Open* 8(1), e17. https://doi.org/10.1002/vro2.17

Wray, C. and Wray, A. (2000). *Salmonella in domestic animals.* CABI Publishing, 463 p.

Chapter 2

Advances in understanding and optimizing egg natural defenses against pathogens

Nicolas Guyot and Sophie Réhault-Godbert, INRAE, Université de Tours, Nouzilly, France

1 Introduction

Eggs are an important part of the human diet worldwide, widely recognized for their nutritional properties. Previously considered as the reference protein by the World Health Organization, eggs are a valuable source of macronutrients (proteins, lipids) and micronutrients (vitamins, minerals, trace elements) (Rehault-Godbert et al., 2019). It also offers the advantage of being low in calories and inexpensive, while having exceptional culinary potential. Besides its nutritional interest, the hygienic quality of eggs is also of crucial importance for the table egg industry and the consumer (public health).

Because of its biological role in avian reproduction, the egg contains not only nutrients but also numerous bioactive molecules and a sophisticated antimicrobial defense system, which are both required for the development and protection of the bird embryo outside the maternal body. These antimicrobial defenses are also intended to protect the egg yolk (vitellus). Indeed, given the nutrients present, and despite the presence of maternal immunoglobulin Y (produced and deposited in the yolk during vitellogenesis by the hen), the egg yolk is highly susceptible to microbial contamination because of its extremely favorable environment to the survival and multiplication of bacteria. Egg protections rely on a set of physical (shell and vitelline membrane) and

http://dx.doi.org/10.19103/AS.2024.0150.02

physicochemical (pH and viscosity of the egg white) barriers, as well as on a "chemical" protection based on a myriad of antimicrobial proteins and peptides (AMPPs) distributed throughout all egg compartments. When table eggs (non-fertile, non-embryonated eggs) are produced and stored in appropriate conditions, these natural defenses help to maintain the hygienic quality of the table egg for several weeks. However, for many years, eggs have been the main cause of salmonellosis (food-borne illness) involving *Salmonella enterica* serovar Enteritidis in Europe. *Salmonella* contamination of eggs can occur via two possible routes involving horizontal and/or vertical transmissions (Gantois et al., 2009). In horizontal transmission, eggs are contaminated by bacteria penetrating through the shell during or after laying, for example, following the deposition of feces from contaminated animals on the shell surface. In vertical transmission, there is a direct contamination of the egg yolk, vitelline membrane, egg white or eggshell (including shell membranes) before laying, during the process of egg formation, due to prior infection of the hen and the reproductive organs (ovary and/or oviduct) by *Salmonella* Enteritidis. Nevertheless, the predominance of one or other of these routes is still open to debate. Whatever the route of contamination, maintaining the physical and antimicrobial properties of all the structures surrounding the egg yolk is essential to limit the penetration, migration and multiplication of bacteria in the egg.

The aim of this chapter is to highlight the complexity and diversity of the egg antimicrobial defenses, their physical/physicochemical and antimicrobial properties (structure and composition) and the factors responsible for the variability of these properties. Each protective layer, namely the eggshell, the egg white and the vitelline membrane, will be individually described in independent sections that will be followed by a review of the main AMPPs identified in the egg.

2 Eggshell

2.1 Structure, composition and antimicrobial function

The chicken eggshell (300–400 μm thick) is an essential physical barrier against mechanical stress, dehydration and microbial penetration. Taking into account the eggshell membranes, it is composed of calcium carbonate (95%, calcite polymorph) and proteins (3.5%), polysaccharides and proteoglycans. The mechanical properties of the eggshell and the absence of any cracks are decisive factors in limiting horizontal contaminations and keeping the inside of the egg free from pathogens. The organization, orientation and morphology of calcite crystals, which determine the texture of the eggshell and therefore its mechanical properties, are major factors driving shell strength. From the outer surface inward, the eggshell consists of the cuticle, the calcified layer (composed of the palisade and mammillary layers) and finally the eggshell membranes (Figure 1). Each of these eggshell parts plays a role in the egg antibacterial protection. The shell surface of the laid egg is naturally exposed to the external environment and contains microorganisms originating from the digestive tract, acquired during the transit of the egg through the cloaca at oviposition, and from the laying environment. Interestingly, it is thought that this eggshell microbiome may also participate in egg defense via three potential mechanisms: competition between commensal microbiota and environmental pathogens for adhesion on the eggshell surface, competition for available nutrients and production of antimicrobial substances or metabolites (Kulshreshtha et al., 2022).

Although it contains relatively low amounts of proteins compared to the mineral fraction, the eggshell is composed of highly diversified proteins. About 900 unique proteins have been identified with various abundances by combining data from published eggshell proteomes (Gautron et al., 2019). Among them, a number of antibacterial proteins with diverse structures and activities are present throughout the eggshell (Moreau et al., 2022).

2.1.1 Cuticle

The cuticle is a relatively thin layer with variable thickness (0.5–12 μm) providing the first barrier against the penetration of pathogenic microbes through the shell (see Kulshreshtha et al., 2022 for a review on the cuticle). The completeness

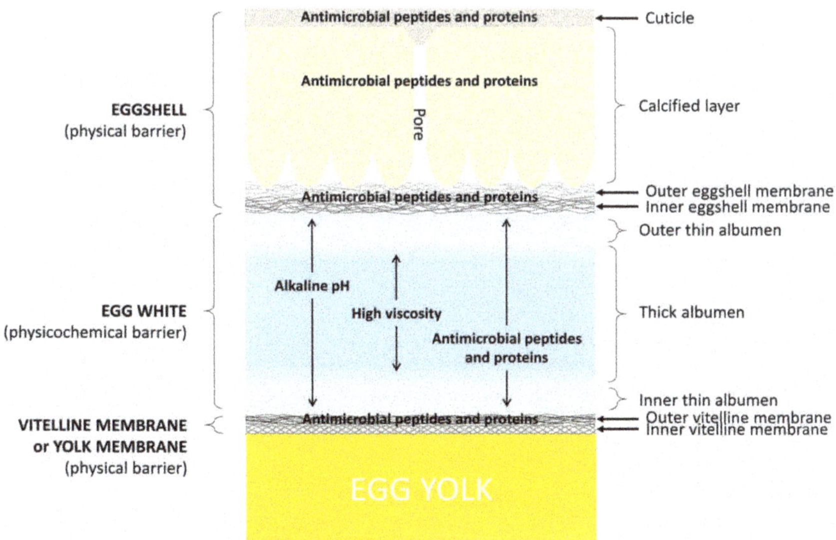

Figure 1 Schematic representation of the physical and chemical defenses of the egg (adapted from Réhault-Godbert et al., 2022)

of the cuticle itself, as the outermost layer of the egg, is essential to ensure that the egg remains pathogen-free. Indeed, it has been demonstrated that eggs with the best cuticle coverage are significantly protected from bacterial penetration (Bain et al., 2013). The cuticle forms a protective layer impervious to microbes and is subject to strong natural selection. Birds living in humid or microbiologically challenging environments (flamingos, pelicans) have thicker cuticles (Bain et al., 2013; Kusuda et al., 2011), suggesting some evolutionary pressure; however, in poultry, artificial incubation has impaired the selection pressure for this trait. This layer is made up of organic constituents that are deposited in the uterus during the last hour before the oviposition (Wilson et al., 2017). The outer part is richer in proteins while the inner part is richer in sulfated polysaccharides and phosphates (Rodriguez-Navarro et al., 2013). At oviposition, the cuticle is wet, but it dries in less than three minutes (Sparks and Board, 1985) and seals the shell pores, thereby limiting bacterial penetration through the shell. However, the cuticle becomes consolidated and mature around 6 h after oviposition, exhibiting optimum mechanical properties and better efficiency against bacterial penetration (Muñoz et al., 2015). The micro-fissures and micro-cracks that appear as a result of the drying process allow gaseous diffusion through the eggshell. The cuticle deposition on the eggshell is not homogeneous. Compared to the blunt end, the sharp end of the egg has a poorer cuticle quality and is associated with increased bacterial penetration (Chen et al., 2019). Therefore, the area where potential eggshell infection may occur can be of great importance. At least 47 different proteins were identified in the cuticle, including a number of proteins known to have antimicrobial activity (lysozyme C, ovotransferrin, ovocalyxin-32, cystatin, ovoinhibitor, ovocalyxin-36, ovocleidin-17, etc.) (Rose-Martel et al., 2012; Miksik et al., 2014). These antibacterial compounds (bacteriostatic and bactericidal) can provide antimicrobial protection, at least during the short period after oviposition until the cuticle is dry. Cuticle and outer eggshell protein extracts from domestic birds including chicken have antimicrobial activity against Gram-positive and Gram-negative bacteria (Wellman-Labadie et al., 2010; Wellman-Labadie et al., 2008b).

2.1.2 Calcified layers

Beneath the cuticle are several constitutive calcified layers, which are composed of calcium carbonate crystals and 2% to 3% of organic matrix. During eggshell formation, calcification is initiated in the red isthmus segment of the oviduct. The initial step is the deposition of organic nuclei on eggshell membranes (mammillary knobs), which are seeding sites for the initiation of shell calcification. During shell formation, outward-growing crystal units compete for available space, and only those with the fastest growth direction that is perpendicular to the eggshell membrane surface continue to grow. The crystalline texture of the shell results from a physico-chemical process mediated and controlled by the organic matrix proteins secreted temporally into the uterine fluid in a three-phase process: initiation, growth and termination (Gautron et al., 1997). These proteins play a key role in stabilizing the transient form of calcium carbonate (i.e. amorphous calcium carbonate), influencing the selection of the calcite polymorph and controlling the orientation of crystal growth (Marie et al., 2015; Rodriguez-Navarro et al., 2015). These interactions define the ultrastructure of the eggshell, which is responsible for its mechanical properties. The inner calcified layer (cone or mammillary layer) is composed of calcified columns firmly anchored in the outer shell membrane (initiation phase). The intermediate calcified layer is the palisade layer, which appears when adjacent cones fuse to form a compact layer (growth phase). This layer represents around 2/3 of the total eggshell thickness. Eggshell penetration by *Salmonella* can be facilitated with unfavorable ultrastructural variables of the mammillary layer including late fusion, type B bodies, type A bodies, poor cap quality, alignment, depression, erosion and cubics (Samiullah et al., 2013). The eggshell is crossed by pores, which form during the calcification process in the hen's uterus, and allow gas diffusion through the eggshell. On the surface of the palisade layer, a thin layer of small vertical crystals (vertical crystal layer) forms the last calcified layer (terminal phase), before the shell surface is coated with the cuticle layer.

The mean number of eggshell pores per egg is comprised between 5000 and 8000 (De Reu et al., 2006b). Eggshell pores have funnel-shaped openings (initially plugged with cuticle materials) facing outwards, and a narrow channel region joining the area above the eggshell membrane. The length of eggshell pores can vary from 200 to 360 µm and their average diameter is in the range of 10–25 µm (Kulshreshtha et al., 2021; Riley et al., 2014). These dimensions are compatible with the size of bacteria like *Salmonella enterica* (0.7–1.5 µm in diameter to 2–5 µm in length), which are able to penetrate and cross the pores, especially when the cuticle is removed or degraded. Of note, pore diameter and frequency vary between different regions of the eggshell. Wider pores are present at the equatorial region or blunt end of the egg, compared to the sharp end (Riley et al., 2014). However, although the eggshell pores constitute a route for bacteria to penetrate intact (uncracked) eggs, it seems that there is no correlation between the pore number and bacterial penetration (De Reu et al., 2006b; Messens et al., 2005a; Nascimento et al., 1992).

At least 520 proteins were identified in the calcified part of the eggshell (Mann et al., 2006). The shell matrix proteins are mainly involved in controlling the biomineralization process and/or in innate immune defense. Four proteins with known antibacterial activity (lysozyme, ovocalyxin-32, ovocleidin-17, cystatin) can be found in the top 10 most abundant proteins (Table 1). However,

Table 1 Chicken egg antimicrobial proteins and peptides (adapted from (Moreau et al., 2022))

				Proteomic identification[a]						Rank within compartment[b]			
							Eggshell						
Family/ protein type	Protein name (alternative name)	Gene-ID/ UniprotKB accession	Molecular weight/pI value	Egg yolk	Vitelline membrane	Egg white	Eggshell membrane proteome	Calcified layer proteome	Cuticle proteome	Egg yolk	Vitelline membrane	Egg white	Eggshell calcified layer
Muramidase	Lysozyme C	396218/P00698	14.3 kDa/11.0	*	*	*	*	*	*	18	2	3	2
	Lysozyme G	395708/P27042	20 kDa/10.4					*					Low
Chelator and chelator-binding molecules	Ovotransferrin (Conalbumin)	396241/P02789	78 kDa/7.3	*	*	*	*	*	*	10	5	2	15
	Extracellular fatty acid binding protein	396393/P21760	18 kDa/6.3	*	*	*	*	*			30	16	12
	Avidin	396260/P02701	17 kDa (monomer)/10.0	*	*	*	*	*			34	10	90
Antiproteases	Ovoinhibitor	416235/P10184	50 kDa/6.4	*	*	*	*	*	*	39	14	5	24
	Chicken cystatin (Ovocystatin)	396497/P01038	12 kDa/7.3	*	*	*	*	*	*	13	42	15	9
	Ovocalyxin-32 (OCX-32)	395209/Q90YI1	32 kDa/9.0		*	*	*	*	*		11	45	6
	Ovalbumin-related Protein-X (OVAX)	420898/R9TNA6	45 kDa/8,5	*	*	*	*	*	*	83	12	7	Low
	Ovostatin (Ovomacroglobulin)	396151/P20740	162 kDa/5.6	*	*	*	*	*		65	33	11	Low
BPI/LBP/ PLUNC	Ovocalyxin-36 (OCX-36)	419289/ Q53HW8	46 kDa/6.0		*		*	*	*		74		14
	TENP	395882/O42273	46 kDa/6.0	*	*	*	*	*	*	57	16	24	Low
Defensins	AvBD9 (gallinacin-9)	414343/Q6QLR1	4.4 kDa/8.9	*			*						
	AvBD10 (gallinacin-10)	414341/ Q6QLQ9	4.4 kDa/8.4			*	*	*					90

	AvBD11 (gallinacin-11, VMO2, VMO-II)	414876/Q6IV20	9.3 kDa/10.4		*	*	*	*			4	23	39
	AvBD12 (gallinacin-12)	414339/ Q6QLQ7	5.2 kDa/7.6			*							
	Ovodefensin A1 / gallin (OvoDA1, OvoDA1_2, OvoDA1_3)	422030/R4GI90 422031/D5GR60 422032/D5GR58	5 kDa/10.4		*	*	*				20	28	
	Ovodefensin B1	106780806/ A0A0A8WJ47	5.2 kDa/9.7		*								
Lectin-like	Ovocleidin-17 (OC-17)	121109246/ Q9PRS8	17 kDa/11.6	*	*	*	*	*	*		28		8
	Vitelline membrane outer layer protein 1 (VMO1)	418974/P41366	18 kDa/10.0	*	*	*	*	*		56	3	27	Low
Other	β-microseminoprotein-like 3 (MSMB3)	101750704/n.a	10 kDa (monomer)/10.4		*	*	*						
	Pleiotrophin	418125/P32760	15 kDa (monomer)/11.0		*	*		*				50	36

[a] AMPPs were identified from published proteomes of yolk (Mann and Mann, 2008), egg white (Mann and Mann, 2011; Pu et al., 2023), vitelline membrane (Mann, 2008; Bregeon et al., 2022), eggshell membrane (Ahmed et al., 2017; Makkar et al., 2015), calcified layer (Mann et al., 2006) and cuticle (Rose-Martel et al., 2012; Miksik et al., 2014).

[b] Rank within compartments was determined from the exponentially modified Protein Abundance Index (emPAI) values indicated in the proteomic studies of egg yolk (Mann and Mann, 2008), vitelline membrane (Mann, 2008), egg white (Mann and Mann, 2011) and egg calcified layer (Mann et al., 2006).

it is likely that the antimicrobials identified in the calcitic layer exert their protective roles only during the eggshell formation, not after the completion of the eggshell (Moreau et al., 2022). As they are immobilized/occluded in the mineralized structure of the eggshell, their antibacterial activity in the laid eggs is probably impaired.

2.1.3 Eggshell membranes

Eggshell membranes are the innermost layer of the eggshell and are synthesized in the white isthmus of the hen oviduct. They are composed of networks of highly cross-linked extracellular fibers that are parallel to the eggshell surface (Hincke et al., 2000). The outer shell membrane (50 μm thick) contains fibers that penetrate the mammillary cone tips of the calcified layer, while the inner membrane (20 μm thick) envelops the albumen and is not calcified. This inner shell membrane is lined by a very thin lysozyme-rich layer, called the "limiting membrane", in contact with the albumen (Hincke et al., 2000). Overall, these membranes are semi-permeable, allowing gas and water diffusion while acting as physical and chemical barriers to bacteria. Shell membrane fibers are composed of 10% collagens (mainly type X) and 70–75% other proteins and glycoproteins containing lysine-derived cross-links (Ahmed et al., 2017; Arias et al., 1991; Wong et al., 1984). CREMP (cysteine-rich eggshell membrane protein) is the main structural protein and is stabilized by disulfide cross-links (Ahmed et al., 2017; Kodali et al., 2011). About 300 distinct eggshell membrane proteins have been identified to date (Ahmed et al., 2017), including 18 antibacterial proteins (Moreau et al., 2022) (Table 1).

2.2 Factors affecting the physical and antimicrobial properties of eggshell

2.2.1 Factors associated with the hen and the rearing environment

Hen age during the course of a laying cycle is an important factor affecting eggshell properties. The eggshell strength deteriorates continuously during the egg-laying cycle (Hanlon et al., 2022; Samiullah et al., 2017; Roberts et al., 2013; Benavides-Reyes et al., 2021). This alteration can be explained in part by a progressive decrease in eggshell thickness, due to the increase in the egg mass during the laying period, while the eggshell mass remains relatively constant, but it is thought that ultrastructural and microstructural modifications of the eggshell have a greater impact on this process. Indeed, a decreased mammillary density and an increased size of crystal units have been observed between 33 and 67 weeks of age (Benavides-Reyes et al., 2021), which may explain the degradation of eggshell strength. In addition to shell weakening, a degradation of the cuticle can also be observed during the egg-laying cycle. The cuticle coverage of the egg was previously shown to be greater when hens are 44 weeks old than when they are 64 and 73 weeks old (Samiullah et al., 2017). With hen age, the cuticle becomes thinner and irregular and its chemical composition also changes with a decrease in amounts of polysaccharides and a depletion in lipids at the end of the laying cycle (Rodriguez-Navarro et al., 2013). All these modifications may affect the eggshell bacterial contamination and/or penetration. In this respect, the total number of bacteria on the eggshell surface and in eggshell pores increases as a function of hen age in free-range hens (Moyle et al., 2016). A study demonstrated that shell thickness, cuticle deposition and the pore number were found to decrease from 42 to 69 weeks of age; however, in spite of such changes, hen age did not significantly affect the eggshell penetration of *Salmonella* Enteritidis (Messens et al., 2005a). Consistently, another study also demonstrated that the pore number and cuticle deposition decreased between 39 and 67 weeks with no significant influence of the hen age on the bacterial penetration (De Reu et al., 2006a). The lack of a significant effect of hen age on the bacterial penetration in these studies might be due to antagonistic effects between low cuticle deposition (favoring bacteria penetration) and low pore number (possibly limiting bacteria penetration), even though the relationship between the pore number and the bacterial penetration is not clearly established. Interestingly, following experimental contamination of the egg surface with *Salmonella* Enteritidis or *Pseudomonas fluorescens*, bacterial contamination of egg content is higher in eggs from old hens than those from young hens (De Reu et al., 2006a; Jones et al., 2002). These data suggest that hen ageing impairs the protective systems of eggshell and/or internal egg parts (egg white or vitelline membrane). Given the lack of clear evidence for increased ability of bacteria to penetrate intact eggshell in the late laying cycle, the main issue is rather associated with weak eggshell and increased risk for eggshell breaking and crack formation. Indeed, eggshell cracks facilitate bacterial penetration and, in this respect, cracked eggs are obviously more prone to be contaminated with bacteria than intact ones (Ernst et al., 1998). It is noteworthy that eggshell strength can be improved after a molt induced at the end of laying cycle. This change in eggshell quality is associated with modifications in both the ultrastructure of the calcite crystals and the profile of organic matrix proteins (Ahmed et al., 2005). Depending on the type of bacteria, induced molting can also alter the ability of bacteria to contaminate egg contents. Following egg surface inoculation, the level of internal contamination by *Pseudomonas fluorescens* was transitory decreased after induced molting, while *Salmonella* Enteritidis contamination appeared to be unaffected (Jones et al., 2002). Interestingly, the resistance of eggshell membrane to bacterial penetration can be influenced by hen age.

A study performed on isolated pieces of eggshell membrane revealed that the time required for *Salmonella enterica* serovar Typhimurium to penetrate the eggshell membrane increased during the laying period of a commercial broiler breeder flock (Berrang et al., 1999b). However, although the number of open areas within the fiber network of the outer eggshell membrane was found to decrease with hen age, no significant relation was observed between the amount of open area and the penetration time.

It has been shown that eggshell thickness, strength and/or cuticle deposition can vary depending on the genotype of laying hens (De Ketelaere et al., 2002; Drabik et al., 2021; Ketta and Tumova, 2018; Lewko et al., 2021). However, in a study using eggs from ISA-Brown Warren and Bovans Goldline reared in furnished cages, no statistical differences were observed between the two strains with regards to eggshell thickness, cuticle score, pore number and eggshell strength (measured using dynamic stiffness and damping ratio), as well as eggshell penetration by *Salmonella* Enteritidis (Messens et al., 2007). In contrast, another study demonstrated that the penetration by *Salmonella* Enteritidis or *Pseudomonas fluorescens* is influenced by hen strain (Jones et al., 2002). However, in this study, egg characteristics for each strain was not indicated. Of note, a significant interaction between hen strain and hen age can be observed for physical/mechanical eggshell parameters (Lewko et al., 2021) and for bacterial penetration (Jones et al., 2002).

The housing system may affect both cuticle quality and bacterial adherence. Cuticle cover was previously shown to be higher in barn eggs, followed by free range and cage eggs (Samiullah et al., 2017). In a study comparing different housing systems, free-range eggs were characterized by longer eggshell pores, thicker pore plugs and thicker eggshell, as compared with conventional cage eggs (Kulshreshtha et al., 2021). In this work, adherence of *Salmonella* Typhimurium to free-range eggshells was significantly lower than that of conventional cage eggs.

Nutrition is an important factor for eggshell quality. Levels of dietary calcium, phosphorus, vitamin D3, microelements (e.g. Zn, Mn, Cu...) and/or feed additives (e.g. pre- and pro-biotics, organic acis, essential oils and plant extracts) can have an impact on eggshell strength and thickness (Swiatkiewicz et al., 2015).

The temperature of rearing environment is also known to influence eggshell quality. In particular, heat stress leads to a decrease in eggshell strength and thickness (Kim et al., 2024; Mignon-Grasteau et al., 2015), but the consequence of such an effect on bacterial penetration remains to be elucidated.

Additionally, eggshell thickness can be impaired by several infectious diseases, including mycoplasmosis, infectious bronchitis, egg drop syndrome, swollen head syndrome, and Newcastle disease (Roberts et al., 2011). In particular, the bacteria *Mycoplasma synoviae* can induce eggshell apex abnormalities, characterized by changes in the color and texture of the shell in the region of the air sac (blunt end of the egg). The presence of eggshell apex abnormalities is accompanied by reduced eggshell thickness and breaking strength (Feberwee et al., 2009; Jeon et al., 2014); however, the susceptibility of affected eggshells to bacterial penetration has not been evaluated in these studies.

2.2.2 Factors associated with the egg storage conditions

The cuticle layer can be removed or partly destroyed following a washing procedure aimed at cleaning eggshell surface (Samiullah et al., 2013; Wang and Slavik, 1998). Following such treatment, altered cuticle may facilitate microbial penetration. The use of sodium carbonate was shown to destroy the cuticle layer and increase eggshell penetration by *Salmonella* Enteritidis (Wang and Slavik, 1998). Hydroxide and hypochlorite-based solution can also remove large parts of the cuticle layer, but in spite of this, no significant differences in eggshell penetration by *Salmonella enterica* serovar Infantis were observed between washed and unwashed eggs (Samiullah et al., 2013).

The mechanical properties of the eggshell are not (or only slightly) affected during egg storage, regardless of the conditions of duration and temperature. However, it has been reported that the presence of moisture can be deleterious for eggshell solidity (Lott et al., 1976). The presence of moisture in the eggshell may also help bacteria to penetrate eggs, especially when there is a positive temperature differential between a warm egg and a cool environment (Berrang et al., 1999a). In fact, when warm eggs (i.e. newly-laid eggs) are immersed in a bacterial solution at a temperature below the temperature of eggs, a pressure applies from the exterior to the interior of eggs leading to the aspiration of external moisture and the drawing of the bacteria through the eggshell (Haines and Moran, 1940). In this regard, the transfer of "sweating" eggs with bacteria-containing moisture to a refrigerated environment might increase the risk of contamination. The sweating of egg surface (formation of condensation when eggs are removed from refrigerated to warmer environments) possibly affects bacterial penetration through the eggshell, although contradictory results are found in published studies. Indeed, a research work demonstrated that egg sweating is associated with increased bacterial penetration (De Reu et al., 2006a) while no effect was observed in other studies (Ernst et al., 1998; Gradl et al., 2017). This discrepancy probably results from the different experimental conditions used in these works regarding the sweating procedure, the inoculation method and the storage duration and temperature before the microbiological analyses and measurements.

Egg storage temperature and relative humidity may have an effect on bacterial penetration. Although certain studies have not revealed any effect

(Radkowski, 2002; Wang and Slavik, 1998), other studies have demonstrated that internal egg contamination after external exposure to bacteria is increased with increasing temperature and/or relative humidity (Messens et al., 2005b; Schoeni et al., 1995). Following *Salmonella* inoculation of the egg surface with contaminated faeces, contamination of egg contents was reduced when eggs were held at 4°C compared to 25°C (Schoeni et al., 1995).

The protective properties of the eggshell membrane against bacterial entry may also deteriorate during egg storage. Indeed, in comparison with freshly-laid eggs, the penetration of *Pseudomonas fluorescens* through the eggshell membrane was shown to be increased in stored eggs (7 days at 0°C) (Hartung and Stadelman, 1963).

3 Egg white

3.1 Composition, structure and antimicrobial function

Egg white (or albumen) is a biological fluid secreted in the oviduct by the magnum tissue during egg formation and essentially composed of water (~88%) and protein (~9.7–10.6%). At least 158 proteins have been identified in the egg white (Mann and Mann, 2011), including a number of antibacterial proteins. Twelve antimicrobial proteins are present in the top 30 most abundant proteins identified in the egg white proteome (Table 1). Egg white is generally resistant to bacterial contamination. Its antibacterial properties are multifactorial, relying not on only on the presence of the antimicrobial molecules but also on specific physicochemical properties like alkaline pH and viscosity (Baron et al., 2016). Egg white is in most cases, bactericidal towards Gram-positive bacteria, and bacteriostatic or bactericidal towards Gram-negative bacteria (Tranter and Board, 1984). The iron-free environment mediated by ovotransferrin (an iron-chelating protein) is an important antibacterial factor for Gram-negative bacteria. Consequently, the addition of exogenous iron at a concentration that is sufficient to saturate ovotransferrin allows Gram-negative bacteria to survive and grow in the egg white, which is not the case for Gram-positive bacteria (Tranter and Board, 1984). Actually, only a few bacteria, including *Salmonella* Enteritidis, are able to enter eggs naturally by vertical or horizontal contamination and survive in the egg white. Among the subspecies of *Salmonella enterica*, serotype Enteritidis has a greater ability to survive in egg white than other *Salmonella* serotypes (De Vylder et al., 2013). The antibacterial nature of egg white is highly dependent on its physicochemical and chemical properties (alkaline pH, viscosity, antimicrobial molecules) (Figure 1), but also on ambient temperature (Baron et al., 2016). While growth or bacteriostatic effects are usually observed at moderate temperatures, egg white exhibits bactericidal activity against *Salmonella* Enteritidis at 42°C (hen body temperature) or higher

(Alabdeh et al., 2011). Interestingly, the egg white protein fraction (>10 kDa) has a minor impact in this bactericidal effect at 45°C, which is mainly attributable to low molecular weight molecules, presumably peptides (Baron et al., 2020; Cochet et al., 2021). Egg white conditions induce numerous physiological changes in *Salmonella* Enteritidis. In particular, regulations of genes related to cellular damage repair, adaptation to osmotic stress and alkaline pH, and nutrient deprivation (especially iron and biotin restriction) can be observed in response to the egg white environment at 37°C or above (Baron et al., 2017; Huang et al., 2019). All these observations underline the harmful properties of egg white against bacteria and the adaptation strategies developed by *Salmonella* Enteritidis to survive in this milieu.

At oviposition, the pH of egg white is around 7.6, but it rises to 9 or more within a few days of storage, in a rate that depends on time and temperature. This rise in pH results from the loss of CO_2, present in the egg white as a dissolved gas or in ionic form, through the pores of the shell. In fact, the pH of egg white depends on the equilibrium between dissolved CO_2, HCO_3^-, CO_3^{2-} and proteins (Li-Chan et al., 1995). The concentrations of HCO_3^- and CO_3^{2-} are governed by the partial pressure of CO_2 in the egg environment. The alkaline environment present in egg white after storage is generally unfavorable to non-extremophilic bacteria, which usually grow at pH 5.5–9.0. Indeed, most bacteria must maintain an intracellular pH in the narrow range of 7.4–7.8, compatible with optimal functional and structural integrity of cytoplasmic proteins. Alkaline pH primarily impacts the respiratory status of bacterial cells, which must maintain the proton motive force and electrochemical potential across their membranes. Numerous strategies are used by bacteria to cope with alkaline environments, such as increased proton uptake and retention (e.g. via modulation of the ATP synthase and monovalent cation/proton antiporters), metabolic changes leading to increased acid production, and changes in cell surface layers, which contributes to cytoplasmic proton retention (Padan et al., 2005). In addition to accelerating proton import, high pH also represses energy-expensive systems such as flagella biosynthesis and chemotaxis (Maurer et al., 2005). Consequently, *Salmonella* Enteritidis lacks flagella at alkaline pH (Baron et al., 2016), which may limit its motility in the egg white and its access to yolk nutrients.

Egg white is a heterogeneous liquid, composed of three distinct superimposed layers, each characterized by distinct viscosities. The outer and inner thin layers (in contact with the eggshell and egg yolk, respectively) are separated by a thick layer of egg white (Figure 1). Egg white also contains a structure formed by the chalaziferous layer, which surrounds the vitelline membrane and forms twisted filament structures (chalazae) that connect the yolk to each egg pole. In fresh eggs, at 20°C and a shear rate of 34.2 s^{-1}, the apparent viscosity of thick egg white is around 40 times higher than that of thin egg white (Lang and Rha, 1982). It forms a capsule around the yolk and, together with the chalazes, helps to keep the yolk in a centered position in the freshly laid egg. Given that the motility of flagellated bacteria is affected by viscous environments (Schneider and Doetsch, 1974), it is likely that the gelatinous structure of the thick egg white limits the motility of bacteria and their migration towards the yolk. Indeed, the swimming speed and flagellar turnover rate of *Salmonella* Typhimurium are significantly altered in thick albumen, compared to thin albumen (Okuno et al., 2019). Both thin and thick egg white are repellent to *Salmonella* Typhimurium and cause reversals of flagellar rotation (Okuno et al., 2019). The gelatinous structure of thick egg white is due to the presence of ovomucin, a sulfated glycosylated protein of very high molecular weight, and its association with lysozyme. Ovomucin is composed of two subunits: a low-carbohydrate α-subunit (MUC5B) and a carbohydrate-rich β-subunit (MUC6). Compared to thin egg white, thick egg white contains higher amounts of α-ovomucin, β-ovomucin and lysozyme (Pu et al., 2023). The abundance of the two ovomucin subunits was found to be increased by 1.43-fold (MUC5B) and 9.7-fold (MUC6) in the thick egg white (Pu et al., 2023).

3.2 Factors affecting the physicochemical and antimicrobial properties of egg white

3.2.1 Factors associated with the hen and the rearing environment

The viscosity and/or quantity of thick egg white can be roughly estimated with the measurement of some egg white quality indicators, such as the height of the thick albumen that can be translated into Haugh units (the formula for calculating Haugh units takes into account the weight of the egg). These two quality parameters are known to decrease with hen age, during the laying cycle (Banaszewska et al., 2019; Samiullah et al., 2014; Silversides et al., 2006; Silversides and Scott, 2001; Singh et al., 2009; Roberts et al., 2013; Benavides-Reyes et al., 2021). The concentration of egg white antimicrobials can also change with hen age. Two studies have demonstrated that lysozyme content and activity increase with hen age, as demonstrated between 30 and 60 weeks (Banaszewska et al., 2019) and between 33 and 53 weeks (Lewko et al., 2021). A study showed that 26-week-old hens produced eggs with higher ovotransferrin concentration than 51 week-old hens (Vlckova et al., 2019), but Jabalera et al. showed that the concentration of ovotransferrin increases with hen age between 35 and 100 weeks of age (Jabalera et al., 2022). These latter authors demonstrated that the age-dependent increase in ovotransferrin concentration contributes to the higher antibacterial activities against both *Staphylococcus aureus* and *Salmonella* Typhimurium in the egg white of 100 vs. 35-week-old hens (Jabalera et al., 2022); however, of note, this change in antibacterial activities was observed in Hy-Line brown eggs, not in Hy-Line white eggs. Sellier et al found a weak but significant increase in the concentration of lysozyme and ovotransferrin with age, as well as increased antibacterial activities of egg white against *Salmonella* Enteritidis and *Staphylococcus aureus* in 51–54 versus 22–25 week-old hens (Sellier et al., 2007). From these studies, it seems that the decrease in egg white physical properties during the laying period can be compensated by an increase in the concentration of egg white antimicrobials, thereby enhancing its antibacterial potential. These results suggest that egg white from old hens may be less prone to contamination by *Salmonella*, provided that the eggshell remains intact.

A variability in egg white quality is often observed depending on hen strains. For instance, the albumen height of Babcock B-300 eggs is higher than that of ISA-Brown and Brown Leghorn eggs (Silversides et al., 2006). The albumen height of ISA-White eggs is higher than that of ISA-Brown eggs, regardless of storage times, but the egg white pH is not affected by the strain (Scott and Silversides, 2000; Silversides and Scott, 2001). The hen genotype can also influence the antimicrobial content of the egg white. A proteomic analysis performed on the egg white from five traditional breeds revealed differences in the abundance of some AMPPs, including ovalbumin-related protein X, ovoinhibitor, Vitelline Membrane Outer layer protein 1 (VMO1), ovodefensin B1 and ovocalyxin-32 (Bilkova et al., 2018). In a study comparing 23 traditional breeds, significant differences in antimicrobial content were observed according to the color of eggshell, categorized as green, white, tinted, brown and dark brown eggs (Javurkova et al., 2019). Indeed, lysozyme and ovotransferrin concentrations were found to be higher in tinted eggs than in white and brown eggs, and higher lysozyme concentration was also measured in dark brown eggs compared to white and brown eggs. Another study demonstrated that lysozyme content and activity were higher in Rhodes Island Red, Yellowleg Partridge and Greenleg Partridge eggs, than in Leghorn white-shelled eggs (Lewko et al., 2021), while Jabalera et al. did not find any significant difference in lysozyme and ovotransferrin concentration between Hy-Line White and Hy-Line Brown lines (Jabalera et al., 2022). However, in this work, higher antibacterial activities against both *Staphylococcus aureus* and *Salmonella* Typhimurium have been detected in egg white from Hy-Line white hen line compared to Hy-Line brown (Jabalera et al., 2022); however, this difference in the antibacterial effect may be age-dependent since it was observed only in eggs laid by hens of 35 weeks of age, not in eggs laid by hens of 100 weeks of age.

The albumen height and/or Haugh units can be influenced by the rearing system; however, certain interactions between the rearing system and hen age (Samiullah et al., 2014), between the rearing system and the genotype (Kucukyilmaz et al., 2012) and between the rearing system, the hen age and the genotype (Singh et al., 2009) make it difficult to interpret. A study demonstrated that albumen height and Haugh units of Hy-Line Brown eggs were higher for cage eggs than free-range ones (Samiullah et al., 2014). In another study, these two parameters were higher in organic versus conventional cage rearing systems for ATAK-S brown laying hens, while no difference was observed for these parameters between these two rearing systems for Lohmann LSL white laying hens (Kucukyilmaz et al., 2012). Despite interactions between the rearing system, the hen age and the genotype, it has been suggested that the albumen height is higher in cages than in floor pens (Singh et al., 2009), especially at the beginning of the laying period. Concerning the egg white antimicrobials, higher lysozyme content has been detected in free-range eggs compared to enriched cage eggs (Vlckova et al., 2019) or to litter eggs (Krawczyk and Gornowicz, 2010).

Interestingly, the immune stimulation of laying hens is potentially able to improve egg white antibacterial activities. In a study with hens treated with intravenous *Salmonella* Enteritidis LPS administration, it has been shown that such a treatment induced an increase in the anti-*Staphylococcus aureus* activity of egg white proteins, while the anti-*Escherichia coli* activity remained unchanged (Bedrani et al., 2013a). Another study compared the antibacterial activities of egg white proteins from axenic hens, specific pathogen-free (SPF) hens, and hens raised under standard rearing conditions. Results of this work demonstrated that the antibacterial activity of egg white proteins against certain types of pathogenic bacteria (*Staphylococcus aureus*, *Streptococcus uberis*) increased, depending on the immune status of the hens and the degree of contamination of the living environment (Bedrani et al., 2013b). In the light of these studies, it seems possible to reinforce egg antibacterial activities with immune stimulation, but this reinforcement is of low amplitude and only concerns certain types of activities.

Of note, albumen height and Haugh units can be reduced by a number of other factors including among others, infectious diseases, especially infectious bronchitis virus (Roberts et al., 2011), heat stress (Kim et al., 2024) and gossypol-containing cottonseed meal (He et al., 2017; Yuan et al., 2014). However, the effects of these factors on the egg white antibacterial activities remain unclear.

3.2.2 Factors associated with egg storage conditions

Egg storage conditions strongly affect egg white quality, especially egg white pH, thick albumen height and Haugh units. A pH increase and a decrease in the thick albumen height are classically observed in eggs stored at room temperature for several days (Scott and Silversides, 2000; Silversides and Scott, 2001). Egg white viscosity is highest at lay, and gradually decreases as pH increases. Egg white alkalinization during egg storage is time- and temperature-dependent. A study has shown that, after 4 weeks, the egg white pH is 9.19 at 4°C, and 9.35 at 22°C; then, a slight decline is observed, with pH values reaching 8.89 at 4°C, and 9.20 at 22°C, after 6 weeks of storage (Chen et al., 2005). The rise of egg white pH is due to the diffusion of CO_2 through the porous eggshell (see above) and results in the liquefaction of the thick egg white. This liquefaction, known as egg-white thinning, may facilitate bacterial motility and migration towards the nutrient-rich yolk, but on the other hand, bacterial growth of some strains can be challenged with alkaline pH. The role of CO_2 loss in the egg white thinning can be demonstrated with the use of a CO_2-enriched atmosphere during egg storage. Rocculi et al. have demonstrated that, when eggs are stored in 100% CO_2 environment, both egg white pH and Haugh units can be stabilized at values similar to fresh eggs during storage for up to 28 days at 25°C (Rocculi et al., 2009).

The temperature storage has a major effect on the fate of *Salmonella* Enteritidis in the egg white. Following experimental inoculation of the egg white, growth of *Salmonella* Enteritidis occurs in eggs stored at 22–25°C, but the bacteria are inhibited (or even killed) when eggs are stored at 4°C or 10°C (Chen et al., 2005; Schoeni et al., 1995). Therefore, storage at refrigerated temperatures is beneficial for both the safety and the internal quality of eggs. Interestingly, at 42°C (hen body temperature) or higher, egg white exhibits bactericidal activity against *Salmonella* Enteritidis (Alabdeh et al., 2011).

Storage time and temperature of eggs before the inoculation are also important. Indeed, some changes in the antimicrobial activity of egg white occur during storage. A study has revealed that the growth of *Salmonella* is enhanced in fresh compared to stored egg white (Messens et al., 2004). These results were confirmed by another study showing that egg white exhibits a higher anti-SE activity after a few days of storage at 20°C and 37°C compared to egg white of fresh eggs (Rehault-Godbert et al., 2010). This stimulation of egg white antibacterial properties can be partly explained by the pH increase upon storage.

4 Vitelline membrane

4.1 Composition, structure and antimicrobial function

The vitelline membrane (or yolk membrane or perivitelline layer) is a thin protein layer of about 10 μm in thickness. It surrounds the yolk and provides a physical separation between this bacterial-sensitive source of nutrients and the egg white. As such, the vitelline membrane participates in the antibacterial protection of eggs. This biological barrier also contains numerous antibacterial proteins and limits the diffusion of nutrients from the yolk to the egg white and the penetration of bacteria from the egg white to the yolk.

In the hen egg, the vitelline membrane is essentially composed of proteins. Hundreds of proteins have been identified by mass spectrometry (Mann, 2008; Zhou et al., 2020; Bregeon et al., 2022). It consists of two superimposed layers, the inner and outer vitelline membranes (Figure 1), in contact with the yolk contents and egg white, respectively. These layers are separated by a very thin granular "continuous membrane" of 50 to 100 nm in thickness (Bellairs et al., 1963). The inner vitelline membrane (1.0–3.5 μm thick) is initially produced in the ovarian follicle prior to ovulation and consists of an entangled network of fibers (200–600 nm in diameter), which are themselves composed of fibrils (15–120 nm in diameter) (Bellairs et al., 1963). These fibers, mainly composed of Zona Pellucida Proteins (ZP1 and ZP3 proteins) (Okumura et al., 2015), appear flat in scanning electron microscopy and intertwine to form a mesh. The outer vitelline membrane is synthesized in the proximal part of the oviduct (infundibulum) after ovulation and is deposited onto the inner vitelline membrane. This layer is 3 to 8.5 μm thick and is composed of a varying number of superimposed sublayers (Bellairs et al., 1963). Each sublayer consists of a latticework of fine fibrils (as little as 15 nm in diameter) running in straight lines for distances up to 1 μm or more and forming swellings (300 nm in diameter) at the contact points between fibrils (Bellairs et al., 1963). The more abundant proteins of the chicken outer vitelline membrane are ovomucin and the antibacterial proteins lysozyme, VMO1 and VMO2 (now known as avian beta-defensin-11, AvBD11) (Back et al., 1982; De Boeck and Stockx, 1986a; Kido et al., 1992). The innate immune properties of these compounds (see Section 5) strongly support the antimicrobial function of the outer vitelline membrane, as recently evidenced by the proteomic analysis of each of the outer and inner layers (Bregeon et al., 2022). Besides its antibacterial properties, it has been hypothesized that lysozyme also contributes to the structure of the vitelline membrane through electrostatic interactions with ovomucin (Back et al., 1982; De Boeck and Stockx, 1986b). As demonstrated in quail, VMO2/AvBD11 may also have a structural role in the outer vitelline membrane by mediating the binding of the outer layer to the inner layer, possibly via an interaction with the ZP1 and ZP3 proteins (Rahman et al., 2009). Maintaining the integrity of the vitelline membrane is crucial in preventing or limiting the penetration of bacteria, from egg white into the yolk.

4.2 Factors affecting the physical and antimicrobial properties of vitelline membrane

4.2.1 Factors associated with the hen and the rearing environment

The mechanical properties of the vitelline membrane are usually assessed by measuring its breaking strength and deformation, and can be roughly estimated using the yolk index (measurement of the ratio between the yolk height on the yolk diameter after breaking the egg on a flat surface).

Yolk index and vitelline membrane strength are influenced by hen age. These two parameters are higher in eggs from young hens (25–26 weeks) compared to those from older hens (69–70 weeks) (Marzec et al., 2019), suggesting that the mechanical properties of the vitelline membrane degrade as hens are getting older. However, in spite of such physical alterations, the penetration of the vitelline membrane by *Salmonella* Enteritidis may not be dramatically affected, as no significant differences were observed between eggs from hens at 33 and 51 weeks of age (Gast et al., 2010).

The vitelline membrane strength and yolk index can be influenced by the hen strain. These two parameters were higher in Bovans white compared to Hy-Line W 36 (Keener et al., 2006). The penetration of *Salmonella* Enteritidis through the vitelline membrane can also vary depending on the laying hen lines (Gast et al., 2010).

The mechanical properties of vitelline membrane can be influenced by the hen nutrition. A study has demonstrated that hens fed a wheat diet have a higher vitelline membrane strength compared to hens fed a corn diet (Woodward et al., 1987), although the explanation for such an effect is not clearly understood.

The vitelline membrane strength is degraded when hens are exposed to high temperatures, but supplementation of the hen diet with vitamin E can alleviate the degradation induced by the heat stress (Kirunda et al., 2001).

4.2.2 Factors associated with the egg storage conditions

The physical and mechanical properties of the vitelline membrane (yolk index and breaking strength) are strongly influenced by the duration and/or temperature of egg storage (Chen et al., 2005; Feeney et al., 1956; Fromm and Matrone, 1962; Jones et al., 2018). The more time and temperature increase, the more quickly these quality parameters deteriorate. This degradation is likely to result from the egg white alkalinization and water transfer from the albumen to the yolk. This modification of the physicochemical environment may have direct effects on the structure of the vitelline membrane. Indeed, a dissipation of the fibers on the surface of the vitelline membrane can be observed during egg storage (Fromm, 1967; Kirunda and McKee, 2000). Storing eggs at ambient or high temperatures also leads to a decrease in the membrane weight and protein content (Fromm, 1964; Fromm, 1967; Zhou et al., 2020), which suggests that a loss of material occurs when eggs age. At the molecular level, the protein composition of the vitelline membrane is modified during prolonged egg storage. The levels of VMO1 and AvBD11 proteins decrease during storage at ambient (Schäfer et al., 1998) and high (Zhou et al., 2020) temperatures, while it is not the case for lysozyme. All these structural and chemical changes might facilitate the transfer of nutrients from the yolk to the egg white, and the penetration into the yolk of any bacteria present in the vicinity of the vitelline membrane. In experimental conditions based on inoculations of the yolk surface, the rate of penetration of *Salmonella* Enteritidis across the vitelline membrane is shown to be temperature-dependent (Gast and Holt, 2001; Gast et al., 2006), probably because of the decreasing integrity of the vitelline membrane. It can be rapid at ambient or high temperatures, but by lowering storage temperature, the bacterial penetration can be significantly reduced or even prevented. These observations thus suggest that prompt refrigeration of eggs is beneficial to prevent degradation of the vitelline membrane and limit yolk contamination.

5 Antimicrobial proteins and peptides (AMPPs)

5.1 Lysozyme

C-type (chicken-type) lysozyme, or lysozyme C, is a cationic protein of 14.3 kDa, highly abundant in the egg white, vitelline membrane and calcified eggshell layer (Table 1). In hen egg white, it accounts for 3.4% of the total egg white protein (Kovacs-Nolan et al., 2005). In the vitelline membrane, lysozyme is a major component of the outer layer, where it represents 32% (Kido et al., 1992) to 60% (Back et al., 1982) depending on studies. As with the major egg white proteins, oviduct expression of the lysozyme-coding gene is controlled by steroid hormones (Palmiter, 1972). Lysozyme C is an enzyme (EC 3.2.1.17) catalyzing the hydrolysis of the β-(1,4)-glycosidic bond between the N-acetylmuramic acid (NAM) and N-acetylglucosamine (NAG) residues of peptidoglycan, a bacterial cell wall polymer composed of sugars and short peptide units. Its enzymatic activity (muramidase) is responsible for the well-known and potent bacteriolytic effect of lysozyme. Lysozyme C has antibacterial properties against Gram-positive and Gram-negative bacteria through different mechanisms, which may or may not depend on its enzymatic activity. The muramidase activity, responsible for the hydrolysis of peptidoglycan, is more effective against Gram-positive bacteria, which have a thick outer cell wall of peptidoglycan, than against Gram-negative bacteria, for which the peptidoglycan is generally thinner and protected by an outer lipopolysaccharide membrane. Some bacteria have developed resistance mechanisms to lysozyme either by modifying their peptidoglycan (N-deacetylation, O-acetylation) (Davis and Weiser, 2011; Vollmer, 2008) or by producing lysozyme inhibitors (Callewaert et al., 2012). An elegant demonstration of a muramidase-independent mechanism was described thanks to the production of a recombinant lysozyme mutant lacking muramidase activity (using site-directed mutagenesis). Although being enzymatically inactive, this lysozyme mutant retained substantial bactericidal activity against *Staphyloccocus aureus* and *Bacillus subtilis* (Ibrahim et al., 2001). This alternative mechanism may involve the induction of bacterial peptidoglycan hydrolases (autolysins) (Laible and Germaine, 1985) and/or membrane-disrupting effects (Ibrahim et al., 2001). Lysozyme C also exhibits antibacterial activity against Gram-negative strains (Pellegrini et al., 1992) and is able to interact with the bacterial cell wall to increase the permeability of both the outer and inner membranes, as demonstrated with *Escherichia coli* (Derde et al., 2013). The antibacterial effect against *Escherichia coli* and other Gram-negative bacteria is supposed to involve the cationic nature of lysozyme. Indeed, this feature is often considered to be a key factor in the binding of certain cationic antimicrobial peptides to the bacterial cell wall and plasma membrane, and in the destabilization of these cell structures.

Another lysozyme, namely G-type (goose-type) lysozyme, or lysozyme G, has been identified in the calcified layer of the chicken egg, but at much lower amounts than lysozyme C (Table 1). In spite of amino acid sequence differences, this 20 kDa protein shares some structural similarities with lysozyme C and possesses muramidase activity (Moreau et al., 2022).

5.2 Chelator and chelator-binding proteins

5.2.1 Ovotransferrin

Ovotransferrin, formerly known as conalbumin, is an iron-chelating glycoprotein of approximately 78–80 kDa, highly concentrated in the egg white where it is the second most abundant protein (Table 1). Its proportion in the egg white is 12% (Kovacs-Nolan et al., 2005). Ovotransferrin synthesis in the oviduct is regulated by estrogen and progesterone (Lee et al., 1978). This protein possesses metal ion binding properties. It contains two iron-binding sites located in two different lobes, namely the N- and C-terminal lobes. In each lobe, the Fe^{3+} ion binds reversibly with high affinity along with a carbonate anion (CO_3^{2-}). The apparent binding constants for ferric ions were estimated as 1.5×10^{14} M^{-1} for the N lobe, and 1.5×10^{18} M^{-1} for the C lobe, at 25°C and pH 7.5 (Lin et al., 1994), giving K_D (dissociation constant) values of about 10^{-14} and 10^{-18} M, respectively. Ovotransferrin exhibits antibacterial properties that occur either through an indirect mechanism (by its ability to chelate iron that is an important nutrient for the growth of some bacterial strains, including *Salmonella* Enteritidis) or directly through its ability to interact with the bacterial cell surface and induce membrane perturbation (Legros et al., 2021) or membrane permeation (Aguilera et al., 2003). Ovotransferrin has a bacteriostatic effect on the growth of *S. enterica* in rich medium (BHI), which is partly inhibited when the protein is in the holo (iron-saturated) form; however, bactericidal activity is observed against the same bacteria in phosphate-buffered saline without Ca^{2+}/Mg^{2+}, regardless of the iron status of ovotransferrin (apo or holo) (Superti et al., 2007). It has been suggested that ovotransferrin can chelate divalent ions present on the outer membrane of Gram-negative bacteria and thereby induce membrane destabilization, similar to other members of the transferrin family (Legros et al., 2021). Ovotransferrin can also permeate the outer membrane of *Escherichia coli* to reach the inner membrane and induce selective permeation of ions (Aguilera et al., 2003). Bactericidal activities of ovotransferrin have been reported against *Staphylococcus aureus* and *Bacillus cereus* (Baron et al., 2014; Ibrahim et al., 1998). Interestingly, the mechanism involved in the bactericidal properties of ovotransferrin may be pH-dependent. Regarding the antibacterial effect on *Bacillus cereus*, some experimental observations suggest indeed that iron deprivation would occur at neutral pH while a membrane perturbation mechanism would be rather involved at alkaline pH (Baron et al., 2014).

5.2.2 Extracellular fatty acid-binding protein (Ex-FABP)

Ex-FABP (extracellular fatty acid-binding protein), formerly known as Ch21 protein, is an 18 kDa protein belonging to the lipocalin family, the members of which bind fatty acids and various ligands including small hydrophobic molecules (steroids, bilins, retinoids, etc.) and proteins. This protein is present in the vitelline membrane, egg white and eggshell (Table 1). Ex-FABP is a siderophore-binding protein involved in the sequestration of various ferric siderophores. Siderophores are small, high-affinity iron-chelating molecules, produced particularly by bacteria during iron restriction to scavenge and internalize extracellular iron ions required for their growth. The interaction of siderophores with ferric iron is characterized by both an exceptional stability and affinity, with K_D values reaching 10^{-30} M or even lower in some cases (Neilands, 1995). Indeed, the affinity of siderophores for iron is so strong that they can dissociate the iron bound to ovotransferrin or transferrins (Legros et al., 2021). Siderophore-binding proteins can thus counteract this bacterial siderophore strategy. Ex-FABP has been shown to bind specifically and tightly to ferric complexes of bacterial siderophores: enterobactin (K_D = 0.22 nM), parabactin (K_D = 42 nM), bacillibactin (K_D = 14 nM), and monoglucosylated enterobactin

(K_D = 0.07 nM) (Correnti et al., 2011). Ex-FABP exhibits bacteriostatic activities via its ability to bind to ferric siderophores. Such activities have been reported *in vitro* against various bacteria, including *Escherichia coli* (Correnti et al., 2011; Garenaux et al., 2013), *Bacillus subtilis* (Correnti et al., 2011), and a salmochelin-deficient mutant of *Salmonella* Enteritidis (Julien et al., 2020). Growth inhibition of *Escherichia coli* is abolished when using a mutant of native Ex-FABP that has been modified within the binding site, or by the addition of stoichiometric amounts of iron (Correnti et al., 2011). These data demonstrate that the bacteriostatic activity of Ex-FABP is mediated by sequestration of the iron-siderophore complex. However, no inhibitory activity was observed against wild-type *Pseudomonas aeruginosa* (Correnti et al., 2011) or *Salmonella* Enteritidis (Julien et al., 2020), probably because these bacteria secrete siderophores that are not sequestered by Ex-FABP (such as salmochelin in *Salmonella*).

5.2.3 Avidin

Chicken avidin is a cationic homotetrameric glycoprotein (≈ 67 kDa) with a remarkably high affinity for biotin. The dissociation constant of the avidin-biotin complex is approximately 10^{-15} M (Green, 1963). Avidin is particularly enriched in egg white (Table 1), where it constitutes 0.05% of total proteins (Kovacs-Nolan et al., 2005). The gene expression of avidin in the oviduct can be induced by progesterone or bacterial infection (Kunnas et al., 1993, Tuohimaa et al., 1989). This protein is thought to inhibit the growth of biotin-requiring microorganisms in eggs (Board and Fuller, 1974). Under experimental conditions, avidin slightly inhibits the growth of an *Escherichia coli* mutant requiring exogenous biotin (Campbell et al., 1972). Avidin also binds to various Gram-negative and Gram-positive bacteria, including *Escherichia coli*, *Klebsiella pneumoniae*, *Serratia marcescens*, *Pseudomonas aeruginosa*, *Staphylococcus aureus* and *Staphylococcus epidermis*, independently of its biotin-binding properties (Korpela et al., 1984). The interaction of avidin with the *Escherichia coli* cell wall is mediated by the outer membrane porin (OmpF/OmpC) (Korpela et al., 1984).

5.3 Antiproteases

Many protease inhibitors (or antiproteases) are effective antimicrobials against pathogens and are therefore associated with host defense (Supuran et al., 2002). Their mechanism of action is rather indirect and depends in part on the ability of their reactive inhibitory site to trap microbial proteases and inhibit their deleterious activity. Indeed, many of these microbial proteases are major virulence factors that are essential to the survival of viruses, bacteria, yeasts, and parasites, and their dissemination in host organisms. Microbial proteases are often associated with increased pathogenicity, and there is increasing evidence that some pathogen-derived proteases and host protease inhibitors have co-evolved (Jongsma and Beekwilder, 2011). Like proteases, protease inhibitors are assigned to families and clans based on amino acid similarities and their three-dimensional structure, respectively (Rawlings et al., 2018). Protease inhibitors inhibit enzymes using a family-specific mechanism. Five protease inhibitors with reported antibacterial activity have been identified in the chicken egg (Table 1): ovoinhibitor, chicken cystatin, ovocalyxin-32 (OCX-32), ovalbumin-related protein X (OVAX), and ovostatin.

5.3.1 Ovoinhibitor

Ovoinhibitor (50 kDa) is highly abundant in the egg white, where it represents 1.5% of total proteins (Kovacs-Nolan et al., 2005). It belongs to a family of serine protease inhibitors that include inhibitors possessing one or more Kazal-type units. Each Kazal-type subunit belongs to the I1 family and contains six cysteine residues engaged in disulfide bonds according to a specific pattern. Like most small serine protease inhibitors, Kazal-type inhibitors interact with their cognate protease through the "standard mechanism" (Krowarsch et al., 2003), in which the inhibitory reactive site specifically interacts with the active site of the cognate protease. This high-affinity interaction leads to the formation of a tight complex and may be accompanied by slow but reversible hydrolysis of the inhibitory reactive site loop. Ovoinhibitor contains seven Kazal domains that can inhibit chymotrypsin and trypsin-like proteases. The inhibitor purified from egg white exhibits antimicrobial activities against *B. thuringiensis* (Bourin et al., 2011) ; however, the contribution (or not) of the inhibitory sites of ovoinhibitor in this effect remains undefined.

5.3.2 Cystatin

Chicken cystatin (12 kDa), or ovocystatin, is present in all egg compartments (Table 1). In the egg white, it represents 0.05% of total proteins (Kovacs-Nolan et al., 2005). Cystatin belongs to the I25 family of protease inhibitors. This protein contains two disulfide bonds and inhibits cysteine proteases such as papain and lysosomal cathepsins B, H, L. The mechanism of cystatin inhibition does not obey the "standard mechanism" of Kazal-type inhibitors described above (Bode et al., 1988). The chicken cystatin has a broad antimicrobial spectrum that is evidenced by its capacity to inhibit proteases from viruses, parasites and bacteria (Blankenvoorde et al., 1998; Korant et al., 1985). Antibacterial activities have been demonstrated against *Porphyromonoas gingivalis* (Blankenvoorde et al., 1996, Blankenvoorde et al., 1998), *Escherichia coli* (Szpak et al., 2014;

Wesierska et al., 2005) and various bacteria including *Acinetobacter lwoffii*, *Oligella* sp. and *Pseudomonas aeruginosa* (Wesierska et al., 2005).

5.3.3 Ovocalyxin-32

Ovocalyxin-32 (OCX-32) is a 32-kDa phosphoprotein mainly present in the eggshell and in the vitelline membrane (Table 1). In particular, it is one of the most abundant constituents of the cuticle (Rose-Martel et al., 2012). It belongs to the I47 (latexin) family of protease inhibitors. OCX-32 contains two cystatin-like domains (structural fold similar to ovocystatin) and is phosphorylated on two amino acid residues (Mann et al., 2007; Sun et al., 2020). Recombinant OCX-32 is capable of inhibiting bovine carboxypeptidase (Xing et al., 2007), but inhibitory activities against cysteine proteases has not yet been evaluated. Recombinant OCX-32 has been reported to impair the growth of *Bacillus subtilis* (Xing et al., 2007).

5.3.4 Ovalbumin-related protein X

Ovalbumin-related protein X (OVAX) is a 45 kDa mainly found in the egg white and the vitelline membrane (Table 1). OVAX gene expression is oviduct-specific and is induced by estrogen (Lim et al., 2011). OVAX is primarily expressed in the magnum (responsible for egg white formation) (Da Silva et al., 2015). OVAX belongs to the serpin family (SERine Protease Inhibitor) I4 and more specifically to the subgroup of clade B serpins, also called ov-serpins (Benarafa and Remold-O'Donnell, 2005; Rawlings et al., 2004). Serpins are a superfamily of proteins with similar structures. Many serpins are protease inhibitors that inhibit proteases via an unusual mechanism, in which they form an irreversible complex with their target protease, accompanied by a large conformational change in the inhibitor structure. However, like ovalbumin, OVAX has no inhibitory activity against serine proteases (Rehault-Godbert et al., 2013). Purified OVAX has heparin-binding properties and exhibits antibacterial activity against *Listeria monocytogenes* and *Salmonella* Enteritidis (Rehault-Godbert et al., 2013). Interestingly, this activity can be blocked in the presence of heparin, a negatively charged glycosaminoglycan. A cationic domain present in OVAX, but not in ovalbumin or OVAY structures (Da Silva et al., 2015), might be involved in the antibacterial activity, as it could explain the interaction of OVAX with the negatively charged bacterial cell wall.

5.3.5 Ovostatin

Ovostatin (or ovomacroglobulin) is a high molecular weight glycosylated protein (780 kDa) mainly present in the egg white where it represents 0.5% of total proteins (Kovacs-Nolan et al., 2005). It belongs to the alpha2-macroglobulin family I39, the members of which undergo a unique interaction with various types of proteases (cysteine, serine, aspartic proteases, and metalloproteases). Ovostatin is a homotetramer (four similar chains of 195 kDa) of which two pairs are linked by disulfide bonds (dimers) and assembled non-covalently (Nagase et al., 1983). The inhibitory mechanism of macroglobulins begins with the cleavage of the "bait" region of the macroglobulin by the protease, which induces a massive conformational change of the inhibitor, while trapping the targeted protease in a "venus flytrap" mechanism (Garcia-Ferrer et al., 2017; Marrero et al., 2012). Another specificity of the macroglobulin-protease complex is the formation of a covalent thiol-ester bond between the protease and the inhibitor, while the active site of the enzyme remains accessible. Therefore, small peptide substrates can still be hydrolyzed by the complexed protease, unlike protein substrates (or antibodies) that cannot reach the protease active site due to steric hindrance (Barrett, 1981). Ovostatin has been shown to inhibit the growth of protease-producing bacteria such as *Serratia marcescens* and *Pseudomonas aeruginosa* (Miyagawa et al., 1991). These data suggest that the antibacterial effect of ovostatin against bacteria involves the inhibitory site. In addition, ovostatin may participate in host immunity by binding to host antimicrobial compounds, thereby protecting them from degradation by microbial proteases (Craig-Barnes et al., 2010).

5.4 Defensins

Defensins are cysteine-rich antimicrobial cationic peptides found in a wide range of living organisms (vertebrates, invertebrates, plants, fungi) (Shafee et al., 2017). Avian defensins belong to the β-family of vertebrate defensins and are grouped into two different gene clusters, namely the avian β-defensins (abbreviated AvBD) and the ovodefensins (OvoD), both located on the chromosome 3 of the chicken genome. Four AvBDs (AvBD9, AvBD10, AvBD11, AvBD12) and two ovodefensins (OvoDA1/gallin, OvoDB1) have been identified in the chicken egg by proteomic analyses (Table 1). All these defensins are cationic (positively charged at pH 7). The cationic properties of defensins and, more generally, antimicrobial peptides, are thought to favor their interaction with negatively charged bacterial membranes.

AvBD11 and OvoDA1 (formerly named "gallin") are mainly present in the vitelline membrane and egg white (Table 1). In particular, AvBD11 (formerly named VMO2) is a major component of the outer vitelline membrane layer along with ovomucin, lysozyme and VMO1 (Kido et al., 1992). Oviduct expression of AvBD11 and OvoDA1 genes is regulated by sex hormones such as estrogen and progesterone (Lim et al., 2013; Whenham et al., 2015). In contrast, AvBD9 and AvBD10 have been identified in the eggshell, but not in egg white and vitelline membrane proteomes (Table 1).

AvBD11 (9.3 kDa) is unique among avian defensins in that it possesses two β-defensin domains (Guyotet al., 2020, Herve-Grepinet et al., 2010). The purified protein, isolated from chicken vitelline membrane, binds heparin and has antimicrobial activity against various microorganisms, including bacteria (Gram-positive and Gram-negative), *Eimeria tenella* sporozoites, and H1N1 influenza virus (Guyot et al., 2016; Guyot et al., 2020; Herve-Grepinet et al., 2010). Interestingly, the antibacterial activity of AvBD11 is mainly due to the N-terminal β-defensin domain. Experiments using synthetic domains demonstrated that the N-terminal domain exhibit activities against *Salmonella* Enteritidis and *Listeria monocytogenes* similar to the native protein, while no activity was detected with the C-terminal domain (Guyot et al., 2020).

OvoDA1 (5 kDa) has a single β-defensin domain like most beta-defensins, but some specific structural features suggest this egg defensin, and probably the other ovodefensins, form a new structural subfamily of β-defensins (Herve et al., 2014). Like AvBDs, OvoDs are also antibacterial (Whenham et al., 2015). In particular, chicken OvoDA1 is active against *Escherichia coli* and *Staphylococcus aureus* (Gram-negative and Gram-positive, respectively) (Gong et al., 2010; Herve et al., 2014; Whenham et al., 2015), but not against *Salmonella* Enteritidis or *Salmonella* Typhimurium (Whenham et al., 2015; Herve et al., 2014). Interestingly, the mature form of this defensin is encoded by three paralogous genes resulting from a relatively recent duplication (Gong et al., 2010). Chicken OvoDB1 (5.2 kDa) also exhibits some antibacterial activities, as previously observed against *Escherichia coli* DH5α, but no or very weak activity was detected against *Staphylococcus aureus*, avian pathogenic *Escherichia coli*, *Salmonella* Enteritidis or *Salmonella* Typhimurium (Whenham et al., 2015). In chicken, OvoDs are oviduct-specific and are thought to be involved in the antimicrobial protection of the egg (Whenham et al., 2015).

In contrast to AvBD11, both AvBD9 and AvBD10 are monodomain beta-defensins (4.4 kDa each) and possess antibacterial properties. Synthetic chicken AvBD9 was shown to exhibit antimicrobial activities against various bacteria and yeasts (*Candida albicans* and *Saccharomyces cerevisiae*) (van Dijk et al., 2007). Recombinant chicken AvBD9 and AvBD10 have bactericidal activities against *Escherichia coli* (Gram-negative) and *Enterococcus faecalis* (Gram-positive), and it seems that the tryptophan residue located at the C-terminus of the AvBD9 gene contributes to its bactericidal activity (Mowbray et al., 2018). Recombinant chicken AvBD12 possesses antibacterial activities against *Staphylococcus aureus*, *Escherichia coli* and *Salmonella* Typhimurium (Zhao et al., 2014; Yang et al., 2016).

5.5 BPI/LBP/PLUNC family proteins

5.5.1 Ovocalyxin-36

Ovocalyxin-36 (OCX-36) is a protein mainly present in the eggshell (Table 1). This protein is composed of two lipid-binding domains, BPI1 (N-terminal BPI/LBP/CETP domain) and BPI2 (C-terminal BPI/LBP/CETP domain). OCX-36 shares significant structural similarity with mammalian lipopolysaccharide-binding proteins (LBPs), bactericidal permeability-increasing proteins (BPIs) and palate, lung and nasal epithelium clone family proteins (PLUNCs), which are key components of the innate immune system and act as a first line of host defense (Gautron et al., 2007; Gautron et al., 2011). Purified OCX-36 protein is able to bind bacterial lipopolysaccharide (LPS) and lipoteichoic acid (LTA) and inhibits bacterial growth in *Staphylococcus aureus* (Cordeiro et al., 2013).

5.5.2 TENP

TENP (transiently expressed in neural precursors), also known as BPI fold-containing family B, member 2 (BPIFB2) is a 47 kDa protein mainly present in the albumen and vitelline membrane (Table 1). Initially referred as ovoglobulin G2, TENP represents approximately 0.1–0.5% (Guerin-Dubiard et al., 2006) or 1.0–4.0 % (Kinoshita et al., 2016) of egg white proteins. In the magnum (oviduct segment responsible for albumen secretion), TENP expression is confined to the tubular glands and is induced by estrogen or progesterone (Whenham et al., 2014). TENP purified from emu egg white shows antibacterial activity against Gram-positive bacteria, notably *Micrococcus luteus* and *Bacillus subtilis*, but not against *Escherichia coli* and *Salmonella* Typhimurium Gram-negative bacteria (Maehashi et al., 2014).

5.6 Lectin-like proteins

5.6.1 Ovocleidin-17

Ovocleidin-17 (OC-17) is a 17 kDa protein, highly abundant in the organic matrix of the eggshell (Table 1) from which it was first purified (Hincke et al., 1995). OC-17 contains three disulfide bonds, two phosphorylated serine residues and possesses a C-type lectin (CTL) domain. CTL proteins constitute a large protein family encompassing at least seven subgroups such as hyalectans, asialoglycoprotein receptors, collectins, selectins, natural killer group transmembrane receptors, macrophage mannose receptors and simple lectins (Zelensky and Gready, 2005). In addition to its role in the biomineralization process (Reyes-Grajeda et al., 2004), chicken OC-17 also exhibits antimicrobial activity. Purified ovocleidin-17 can bind bacterial polysaccharides and is bactericidal against *Bacillus subtilis*, *Staphylococcus aureus* and *Pseudomonas aeruginosa* (Wellman-Labadie et al., 2008a). Interestingly, its antimicrobial activity is enhanced in the presence of calcium.

5.6.2 Vitelline membrane outer layer protein 1

Vitelline membrane outer layer protein 1 (VMO1) is a cationic protein of around 18 kDa, highly abundant in the vitelline membrane (Table 1). It is one of the main components of the vitelline membrane outer layer along with ovomucin, lysozyme and AvBD11 (Kido et al., 1992). This protein is synthesized in the oviduct, in particular at the infundibulum-magnum junction (Uyeda et al., 1994) responsible for the formation of the outer vitelline membrane and/or albumen. Its synthesis can be induced by estrogen (Lim and Song, 2015), like most major egg white proteins. VMO1 has a beta-prism fold containing four disulfide bonds. The beta-prism fold is found in plant lectins related to jacalin, known to specifically bind mannose and galactose (Rabijns et al., 2005). Experimental evidences suggest that VMO1 has carbohydrate-binding properties (Kido et al., 1995). VMO1 also binds heparin (negatively charged glycosaminoglycan) and has antibacterial activity against *Listeria monocytogenes* but not against *Salmonella* Enteritidis (Guyot et al., 2016). Despite this antibacterial effect, it is believed that the main physiological function of VMO1 has not yet been discovered.

5.7 Others

5.7.1 Beta-microseminoprotein-like (MSMB3)

The "beta-microseminoprotein-like" protein (MSMB3) is a protein found in the vitelline membrane, egg white and eggshell (Table 1). Beta-microseminoproteins are small, nonglycosylated, disulfide-rich proteins that have been identified in many animal species. In chicken, "beta-microseminoprotein-like" protein is encoded by the MSMB3 gene on chromosome 6, together with two other beta-microseminoprotein-coding genes, namely MSMB1 and MSMB2 (Coste et al., 2021). Chicken MSMB3 can be found as a homodimer in solution (Coste et al., 2021); however, it is not yet known whether the biologically active species is the monomer, dimer, or both forms. The physiological function of the chicken beta-microseminoproteins is not known, but MSMB3 purified from egg white exhibits some antibacterial activities against *Listeria monocytogenes* and *Salmonella* Enteritidis (Guyot et al., 2016). Like OVAX, the antibacterial activity of MSMB3 is blocked by heparin, suggesting that the heparin-binding site of MSMB3 is important for its interaction with bacteria. Compared to MSMB1 and MSMB2, MSMB3 is highly cationic (Coste et al., 2021), which may partly explain its antibacterial activity.

5.7.2 Pleiotrophin

Pleiotrophin is a cationic 15 kDa protein with heparin-binding properties identified in the vitelline membrane, egg white and eggshell (Table 1). Its expression in the oviduct is regulated by estrogen (Lee et al., 2012). The amino acid sequence of pleiotrophin is highly conserved (>90% identity) between chicken and mammals (Hampton et al., 1992; Raulo et al., 1994), which thus suggests broad evolutionary conservation of its structure and biological functions between birds and mammals. Pleiotrophin is a growth factor/cytokine known for its implication in many biological functions including neural development, angiogenesis, inflammation/injury, bone development, adipocyte differentiation, and mammary epithelial cell differentiation (Wang, 2020). Moreover, antibacterial activities were reported for the human and chicken pleiotrophin (Guyot et al., 2016; Svensson et al., 2010) suggesting it may also play a role in host immunity. Purified pleiotrophin isolated from the chicken vitelline membrane exhibits antibacterial effects against *Listeria monocytogenes* and *Salmonella* Enteritidis (Guyot et al., 2016). It is thought that the heparin-binding site of pleiotrophin is involved in these effects.

6 Conclusion and future trends

In conclusion, the egg possesses effective and complementary antimicrobial protection systems that are the eggshell, the egg white, the vitelline membrane, and a myriad of antimicrobial molecules mainly concentrated in the egg white. The antimicrobial content of eggs has been shaped over the course of evolution, probably by microbial pressure in the nesting habitat of the wild bird (the red jungle fowl that is the ancestor of domestic chickens), to cope with microbial challenges and potential contamination. Although variations probably appeared after domestication and selection of productive traits (egg versus meat), egg antimicrobial defenses remain effective against many bacteria. In fact, few bacteria are able to penetrate and proliferate naturally inside the egg. The main microbiological threat related to the egg is *Salmonella*, particularly *Salmonella* Enteritidis, which is able to colonize the hen's reproductive tissues and survive the physicochemical conditions of egg white, which are generally considered unfavorable for the survival of most bacteria. The main factors in the variability of egg natural defenses identified in the literature to date are the hen age and/or genetics of the hens (mainly for eggshell and egg white), and the conditions of storage including time and temperature (mainly detrimental to egg white and vitelline membrane). In a context of transition of rearing systems of laying hens (from cages to alternative free-range systems) and lengthening of the hen laying period, it is necessary to continue the assessment of the effects of the rearing environment and hen age, and even the interactions between these two factors, on the status of the egg natural defenses. Moreover, it should also be noted that technological advances over the last 20 years (sequencing of the chicken genome, use of high-throughput techniques) allowed the identification of about a thousand proteins in eggs. Eggs are an immense source of bioactive molecules, but only a small number have been characterized for their activity to date. A huge work remains to be done to identify and characterize new egg antimicrobials, especially within the proteins of intermediate or low abundance or after partial proteolysis that may occur naturally during egg storage or when the egg is fertilized and incubated. It is hypothesized that many egg molecules are in a latent form and may express their biological activity only during the incubation of a fertilized egg. The characterization of these molecules throughout incubation and embryonic development may help to identify their physiological function and reveal new activities that could be of interest for human health. One of the next challenges will also be to study and understand the effects of the egg white physicochemical environment (which varies during storage and incubation) on the activity of AMPPs, and to characterize the synergies and complementarities between these activities.

7 References

Aguilera, O., et al. 2003. Transferrins selectively cause ion efflux through bacterial and artificial membranes. *FEBS Lett*, 548, 5–10.

Ahmed, A. M., et al. 2005. Changes in eggshell mechanical properties, crystallographic texture and in matrix proteins induced by moult in hens. *Br Poult Sci*, 46, 268–279.

Ahmed, T. A., et al. 2017. In-depth comparative analysis of the chicken eggshell membrane proteome. *J Proteomics*, 155, 49–62.

Alabdeh, M., et al. 2011. Role of incubation conditions and protein fraction on the antimicrobial activity of egg white against *Salmonella* Enteritidis and *Escherichia coli*. *J Food Prot*, 74, 24–31.

Arias, J. L., et al. 1991. Collagens of the chicken eggshell membranes. *Connect Tissue Res*, 26, 37–45.

Back, J. F., et al. 1982. Proteins of the outer layer of the vitelline membrane of hen's eggs. *Biochim Biophys Acta*, 705, 12–19.

Bain, M. M., et al. 2013. Enhancing the egg's natural defence against bacterial penetration by increasing cuticle deposition. *Anim Genet*, 44, 661–668.

Banaszewska, D., et al. 2019. The impact of breeder age on egg quality and lysozyme activity. *Turk J Vet Anim Sci*, 43, 583–589.

Baron, F., et al. 2014. Ovotransferrin plays a major role in the strong bactericidal effect of egg white against the Bacillus cereus group. *J Food Prot*, 77, 955–962.

Baron, F., et al. 2016. Egg white versus *Salmonella* Enteritidis! A harsh medium meets a resilient pathogen. *Food Microbiol*, 53, 82–93.

Baron, F., et al. 2017. Global gene-expression analysis of the response of *Salmonella* Enteritidis to egg white exposure reveals multiple egg white-imposed stress responses. *Front Microbiol*, 8, 829.

Baron, F., et al. 2020. Egg-white proteins have a minor impact on the bactericidal action of egg white toward *Salmonella* Enteritidis at 45°C. *Front Microbiol*, 11, 584986.
Barrett, A. J. 1981. Alpha 2-macroglobulin. *Methods Enzymol*, 80 Pt C, 737–754.
Bedrani, L., et al. 2013a. Systemic administration of lipopolysaccharide in laying hens stimulates antimicrobial properties of egg white against *Staphylococcus aureus*. *Vet Immunol Immunopathol*, 152, 225–236.
Bedrani, L., et al. 2013b. Passive maternal exposure to environmental microbes selectively modulates the innate defences of chicken egg white by increasing some of its antibacterial activities. *BMC Microbiol*, 13, 128.
Bellairs, R., et al. 1963. The vitelline membrane of the hen's egg: a chemical and electron microscopical study. *J Ultrastruct Res*, 8, 339–359.
Benarafa, C. & Remold-O'donnell, E. 2005. The ovalbumin serpins revisited: perspective from the chicken genome of clade B serpin evolution in vertebrates. *Proc Natl Acad Sci U S A*, 102, 11367–11372.
Benavides-Reyes, C., et al. 2021. Research Note: Changes in eggshell quality and microstructure related to hen age during a production cycle. *Poult Sci*, 100, 101287.
Berrang, M. E., et al. 1999a. Bacterial penetration of the eggshell and shell membranes of the chicken hatching egg: a review. *J Appl Poult Res*, 8, 499–504.
Berrang, M. E., et al. 1999b. Eggshell membrane structure and penetration by *Salmonella* Typhimurium. *J Food Prot*, 62, 73–76.
Bilkova, B., et al. 2018. Domestic fowl breed variation in egg white protein expression: application of proteomics and transcriptomics. *J Agric Food Chem*, 66, 11854–11863.
Blankenvoorde, M. F., et al. 1996. Inhibition of the growth and cysteine proteinase activity of *Porphyromonas gingivalis* by human salivary cystatin S and chicken cystatin. *Biol Chem*, 377, 847–850.
Blankenvoorde, M. F., et al. 1998. Cystatin and cystatin-derived peptides have antibacterial activity against the pathogen *Porphyromonas gingivalis*. *Biol Chem*, 379, 1371–1375.
Board, R. G. & Fuller, R. 1974. Non-specific antimicrobial defences of the avian egg, embryo and neonate. *Biol Rev Camb Philos Soc*, 49, 15–49.
Bode, W., et al. 1988. The 2.0 A X-ray crystal structure of chicken egg white cystatin and its possible mode of interaction with cysteine proteinases. *EMBO J*, 7, 2593–2599.
Bourin, M., et al. 2011. Antimicrobial potential of egg yolk ovoinhibitor, a multidomain Kazal-like inhibitor of chicken egg. *J Agric Food Chem*, 59, 12368–12374.
Bregeon, M., et al. 2022. Multifaceted roles of the egg perivitelline layer in avian reproduction: Functional insights from the proteomes of chicken egg inner and outer sublayers. *J Proteomics*, 258, 104489.
Callewaert, L., et al. 2012. Guards of the great wall: bacterial lysozyme inhibitors. *Trends Microbiol*, 20, 501–510.
Campbell, A., et al. 1972. A mutant of *Escherichia coli* that requires high concentrations of biotin. *Proc Natl Acad Sci U S A*, 69, 676–680.
Chen, J., et al. 2005. Outgrowth of salmonellae and the physical property of albumen and vitelline membrane as influenced by egg storage conditions. *J Food Prot*, 68, 2553–2558.
Chen, X., et al. 2019. Impact of cuticle quality and eggshell thickness on egg antibacterial efficiency. *Poult Sci*, 98, 940–948.
Cochet, M. F., et al. 2021. Identification of new antimicrobial peptides that contribute to the bactericidal activity of egg white against *Salmonella enterica* serovar Enteritidis at 45°C. *J Agric Food Chem*, 69, 2118–2128.

Cordeiro, C. M., et al. 2013. Ovocalyxin-36 is a pattern recognition protein in chicken eggshell membranes. *PLoS One*, 8, e84112.

Correnti, C., et al. 2011. Galline Ex-Fabp is an antibacterial siderocalin and a lysophosphatidic acid sensor functioning through dual ligand specificities. *Structure*, 19, 1796–1806.

Coste, F., et al. 2021. Three-dimensional structures of avian beta-microseminoproteins: insight from the chicken egg-specific beta-microseminoprotein 3 paralog. *FEBS Open Bio*, 11, 1739–1756.

Craig-Barnes, H. A., et al. 2010. Surfactant protein D interacts with alpha2-macroglobulin and increases its innate immune potential. *J Biol Chem*, 285, 13461–13470.

Da Silva, M., et al. 2015. The family secrets of avian egg-specific ovalbumin and its related proteins Y and X. *Biol Reprod*, 93, 71.

Davis, K. M. & Weiser, J. N. 2011. Modifications to the peptidoglycan backbone help bacteria to establish infection. *Infect Immun*, 79, 562–570.

De Boeck, S. & Stockx, J. 1986a. Egg white lysozyme is the major protein of the hen's egg vitelline membrane. *Int J Biochem*, 18, 617–622.

De Boeck, S. & Stockx, J. 1986b. Mode of interaction between lysozyme and the other proteins of the hen's egg vitelline membrane. *Int J Biochem*, 18, 623–628.

De Ketelaere, B., et al. 2002. Measuring the eggshell strength of 6 different genetic strains of laying hens: techniques and comparisons. *Br Poult Sci*, 43, 238–244.

De Reu, K., et al. 2006a. Influence of eggshell condensation on eggshell penetration and whole egg contamination with *Salmonella enterica* serovar Enteritidis. *J Food Prot*, 69, 1539–1545.

De Reu, K., et al. 2006b. Eggshell factors influencing eggshell penetration and whole egg contamination by different bacteria, including *Salmonella* Enteritidis. *Int J Food Microbiol*, 112, 253–260.

De Vylder, J., et al. 2013. *Salmonella* Enteritidis is superior in egg white survival compared with other *Salmonella* serotypes. *Poult Sci*, 92, 842–845.

Derde, M., et al. 2013. Hen egg white lysozyme permeabilizes *Escherichia coli* outer and inner membranes. *J Agric Food Chem*, 61, 9922–9929.

Drabik, K., et al. 2021. The variability of quality traits of table eggs and eggshell mineral composition depending on hens' breed and eggshell color. *Animals (Basel)*, 11, 1204.

Ernst, R. A., et al. 1998. Effect of sweating on shell penetration of *Salmonella* Enteritidis. *J Appl Poult Res*, 7, 81–84.

Feberwee, A., et al. 2009. Induction of eggshell apex abnormalities by *Mycoplasma synoviae*: field and experimental studies. *Avian Pathol*, 38, 77–85.

Feeney, R. E., et al. 1956. Studies of the kinetics and mechanisms of yolk deterioration in shell eggs. *Poult Sci*, 35, 1061–1066.

Fromm, D. 1964. Strength distribution weight and some histological aspects of vitelline membrane of hens egg yolk. *Poult Sci*, 43, 1240–1247.

Fromm, D. 1967. Some physical and chemical changes in vitelline membrane of hens egg during storage. *J Food Sci*, 32, 52-56.

Fromm, D. & Matrone, G. 1962. A rapid method for evaluating strength of vitelline membrane of hens egg yolk. *Poult Sci*, 41, 1516-1521.

Gantois, I., et al. 2009. Mechanisms of egg contamination by *Salmonella* Enteritidis. *FEMS Microbiol Rev*, 33, 718–738.

Garcia-Ferrer, I., et al. 2017. alpha(2)-macroglobulins: structure and function. *Subcell Biochem*, 83, 149–183.

Garenaux, A., et al. 2013. Avian lipocalin expression in chickens following *Escherichia coli* infection and inhibition of avian pathogenic *Escherichia coli* growth by Ex-FABP. *Vet Immunol Immunopathol*, 152, 156–167.

Gast, R. K. & Holt, P. S. 2001. Assessing the frequency and consequences of *Salmonella* Enteritidis deposition on the egg yolk membrane. *Poult Sci*, 80, 997–1002.

Gast, R. K., et al. 2006. Effect of refrigeration on *in vitro* penetration of *Salmonella* Enteritidis through the egg yolk membrane. *J Food Prot*, 69, 1426–1429.

Gast, R. K., et al. 2010. *In vitro* penetration of *Salmonella* Enteritidis through yolk membranes of eggs from 6 genetically distinct commercial lines of laying hens. *Poult Sci*, 89, 1732–1736.

Gautron, J., et al. 1997. Precursor matrix proteins in the uterine fluid change with stages of eggshell formation in hens. *Connect Tissue Res*, 36, 195–210.

Gautron, J., et al. 2007. Cloning of ovocalyxin-36, a novel chicken eggshell protein related to lipopolysaccharide-binding proteins, bactericidal permeability-increasing proteins, and plunc family proteins. *J Biol Chem*, 282, 5273–5286.

Gautron, J., et al. 2011. Ovocalyxin-36 and other Lbp/Bpi/Plunc-like proteins as molecular actors of the mechanisms of the avian egg natural defences. *Biochem Soc Trans*, 39, 971–976.

Gautron, J., et al. 2019. Bioactive minor egg components. In: Jianping Wu (ed.) *Eggs as Functional Foods and Nutraceuticals for Human Health*. The Royal Society of Chemistry. https://books.rsc.org/books/edited-volume/768/chapter-abstract/494002/Bioactive-Minor-Egg-Components?redirectedFrom=fulltext.

Gong, D., et al. 2010. Gallin; an antimicrobial peptide member of a new avian defensin family, the ovodefensins, has been subject to recent gene duplication. *BMC Immunol*, 11, 12.

Gradl, J. A., et al. 2017. Assessing the impact of egg sweating on *Salmonella* Enteritidis penetration into shell eggs. *Poult Sci*, 96, 2393–2399.

Green, N. M. 1963. Avidin. 1. The use of (14-C)biotin for kinetic studies and for assay. *Biochem J*, 89, 585–591.

Guerin-Dubiard, C., et al. 2006. Proteomic analysis of hen egg white. *J Agric Food Chem*, 54, 3901–3910.

Guyot, N., et al. 2016. Proteomic analysis of egg white heparin-binding proteins: towards the identification of natural antibacterial molecules. *Sci Rep*, 6, 27974.

Guyot, N., et al. 2020. Structure, function, and evolution of Gga-AVBD11, the archetype of the structural avian-double-beta-defensin family. *Proc Natl Acad Sci U S A*, 117, 337–345.

Haines, R. B. & Moran, T. 1940. Porosity of, and bacterial invasion through, the shell of the hen's egg. *J Hyg (Lond)*, 40, 453–461.

Hampton, B. S., et al. 1992. Structural and functional characterization of full-length heparin-binding growth associated molecule. *Mol Biol Cell*, 3, 85–93.

Hanlon, C., et al. 2022. Bone and eggshell quality throughout an extended laying cycle in three strains of layers spanning 50 years of selection. *Poult Sci*, 101, 101672.

Hartung, T. E. & Stadelman, W. J. 1963. *Pseudomonas fluorescens* penetration of egg shell membranes as influenced by shell porosity, age of egg and degree of bacterial challenge. *Poult Sci*, 42, 147-150.

He, T., et al. 2017. Proteomic comparison by itraq combined with mass spectrometry of egg white proteins in laying hens (*Gallus gallus*) fed with soybean meal and cottonseed meal. *PLoS One*, 12, e0182886.

Herve-Grepinet, V., et al. 2010. Purification and characterization of avian beta-defensin 11, an antimicrobial peptide of the hen egg. *Antimicrob Agents Chemother*, 54, 4401–4409.

Herve, V., et al. 2014. Three-dimensional NMR structure of Hen Egg Gallin (Chicken Ovodefensin) reveals a new variation of the beta-defensin fold. *J Biol Chem*, 289, 7211–7220.

Hincke, M. T., et al. 1995. Purification and immunochemistry of a soluble matrix protein of the chicken eggshell (ovocleidin 17). *Calcif Tissue Int*, 56, 578–583.

Hincke, M. T., et al. 2000. Identification and localization of lysozyme as a component of eggshell membranes and eggshell matrix. *Matrix Biol*, 19, 443–453.

Huang, X., et al. 2019. Transcriptional sequencing uncovers survival mechanisms of *Salmonella enterica* serovar Enteritidis in antibacterial egg white. *mSphere*, 4, e00700–18.

Ibrahim, H. R., et al. 1998. Identification of a distinct antibacterial domain within the N-lobe of ovotransferrin. *Biochim Biophys Acta*, 1401, 289–303.

Ibrahim, H. R., et al. 2001. Genetic evidence that antibacterial activity of lysozyme is independent of its catalytic function. *FEBS Lett*, 506, 27–32.

Jabalera, Y., et al. 2022. Antimicrobial defenses of table eggs: Importance of antibacterial proteins in egg white as a function of hen age in an extended production cycle. *Food Microbiol*, 107, 104068.

Javurkova, V. G., et al. 2019. Concentration of egg white antimicrobial and immunomodulatory proteins is related to eggshell pigmentation across traditional chicken breeds. *Poult Sci*, 98, 6931–6941.

Jeon, E. O., et al. 2014. Eggshell apex abnormalities associated with *Mycoplasma synoviae* infection in layers. *J Vet Sci*, 15, 579–582.

Jones, D. R., et al. 2002. Microbial contamination in inoculated shell eggs: I. Effects of layer strain and hen age. *Poult Sci*, 81, 715–720.

Jones, D. R., et al. 2018. Hen genetic strain and extended cold storage influence on physical egg quality from cage-free aviary housing system. *Poult Sci*, 97, 2347–2355.

Jongsma, M. A. & Beekwilder, J. 2011. Co-evolution of insect proteases and plant protease inhibitors. *Curr Protein Pept Sci*, 12, 437–447.

Julien, L. A., et al. 2020. The three lipocalins of egg-white: only Ex-FABP inhibits siderophore-dependent iron sequestration by *Salmonella* Enteritidis. *Front Microbiol*, 11, 913.

Keener, K. M., et al. 2006. Effect of testing temperature on internal egg quality measurements. *Poult Sci*, 85, 550–555.

Ketta, M. & Tumova, E. 2018. Eggshell characteristics and cuticle deposition in three laying hen genotypes housed in enriched cages and on litter. *Czech J Anim Sci*, 63, 11–16.

Kido, S., et al. 1992. Isolation of a novel protein from the outer layer of the vitelline membrane. *Biochem J*, 286 (Pt 1), 17–22.

Kido, S., et al. 1995. Characterization of vitelline membrane outer layer protein I, VMO-I: amino acid sequence and structural stability. *J Biochem*, 117, 1183–1191.

Kim, H. R., et al. 2024. Effects of heat stress on the laying performance, egg quality, and physiological response of laying hens. *Animals (Basel)*, 14.

Kinoshita, K., et al. 2016. Identification of TENP as the gene encoding chicken egg white ovoglobulin G2 and demonstration of its high genetic variability in chickens. *PLoS One*, 11, e0159571.

Kirunda, D. F. & Mckee, S. R. 2000. Relating quality characteristics of aged eggs and fresh eggs to vitelline membrane strength as determined by a texture analyzer. *Poult Sci*, 79, 1189–1193.

Kirunda, D. F., et al. 2001. The efficacy of vitamin E (DL-alpha-tocopheryl acetate) supplementation in hen diets to alleviate egg quality deterioration associated with high temperature exposure. *Poult Sci*, 80, 1378–1383.

Kodali, V. K., et al. 2011. A novel disulfide-rich protein motif from avian eggshell membranes. *PLoS One*, 6, e18187.

Korant, B. D., et al. 1985. Cystatin, a protein inhibitor of cysteine proteases alters viral protein cleavages in infected human cells. *Biochem Biophys Res Commun*, 127, 1072–1076.

Korpela, J., et al. 1984. Binding of avidin to bacteria and to the outer-membrane porin of *Escherichia coli*. *FEMS Microbiology Letters*, 22, 3–10.

Kovacs-Nolan, J., et al. 2005. Advances in the value of eggs and egg components for human health. *J Agric Food Chem*, 53, 8421–8431.

Krawczyk, J. & Gornowicz, E. 2010. Quality of eggs from hens kept in two different free-range systems in comparison with a barn system. *Archiv Fur Geflugelkunde*, 74, 151–157.

Krowarsch, D., et al. 2003. Canonical protein inhibitors of serine proteases. *Cell Mol Life Sci*, 60, 2427–2444.

Kucukyilmaz, K., et al. 2012. Effects of rearing systems on performance, egg characteristics and immune response in two layer hen genotype. *Asian-Australas J Anim Sci*, 25, 559–568.

Kulshreshtha, G., et al. 2021. Impact of different layer housing systems on eggshell cuticle quality and *Salmonella* adherence in table eggs. *Foods*, 10, 2559.

Kulshreshtha, G., et al. 2022. Properties, genetics and innate immune function of the cuticle in egg-laying species. *Front Immunol*, 13, 838525.

Kunnas, T. A., et al. 1993. Induction of chicken avidin and related mRNAs after bacterial infection. *Biochim Biophys Acta*, 1216, 441–445.

Kusuda, S., et al. 2011. Diversity of the cuticle layer of avian eggshells. *J Poult Sci*, 48, 119–124.

Laible, N. J. & Germaine, G. R. 1985. Bactericidal activity of human lysozyme, muramidase-inactive lysozyme, and cationic polypeptides against *Streptococcus sanguis* and *Streptococcus faecalis*: inhibition by chitin oligosaccharides. *Infect Immun*, 48, 720–728.

Lang, E. R. & Rha, C. 1982. Apparent shear viscosity of native egg white. *Int J Food Sci Technol*, 17, 595–606.

Lee, D. C., et al. 1978. The action of estrogen and progesterone on the expression of the transferrin gene: a comparison of the response in chick liver and oviduct. *J Biol Chem*, 253, 3494–3503.

Lee, J. Y., et al. 2012. Chicken pleiotrophin: regulation of tissue specific expression by estrogen in the oviduct and distinct expression pattern in the ovarian carcinomas. *PLoS One*, 7, e34215.

Legros, J., et al. 2021. The role of ovotransferrin in egg-white antimicrobial activity: a Review. *Foods*, 10, 823.

Lewko, L., et al. 2021. Effect of genotype and some shell quality traits on lysozyme content and activity in the albumen of eggs from hens under the biodiversity conservation program. *Poult Sci*, 100, 100863.

Li-Chan, E. C. Y., et al. 1995. The chemistry of eggs and egg products. In: Stadelman, W. J. & Cotterill, O. J. (eds.) *Egg Science and Technology*. 4th ed. Binghamton, NY: The Haworth Press, Inc.

Lim, W. & Song, G. 2015. Differential expression of vitelline membrane outer layer protein 1: hormonal regulation of expression in the oviduct and in ovarian carcinomas from laying hens. *Mol Cell Endocrinol*, 399, 250–258.

Lim, W., et al. 2011. Avian Serpinb11 gene: characteristics, tissue-specific expression, and regulation of expression by estrogen. *Biol Reprod*, 85, 1260–1268.

Lim, W., et al. 2013. Expression and regulation of beta-defensin 11 in the oviduct in response to estrogen and in ovarian tumors of chickens. *Mol Cell Endocrinol*, 366, 1–8.

Lin, L. N., et al. 1994. Calorimetric studies of serum transferrin and ovotransferrin: estimates of domain interactions, and study of the kinetic complexities of ferric ion binding. *Biochemistry*, 33, 1881–1888.

Lott, B. D., et al. 1976. Effect of relative humidity on egg shell breaking strength. *Poult Sci*, 55, 665–668.

Maehashi, K., et al. 2014. Biochemical and functional characterization of transiently expressed in neural precursor (TENP) protein in emu egg white. *J Agric Food Chem*, 62, 5156–5162.

Makkar, S., et al. 2015. Chicken egg shell membrane associated proteins and peptides. *J Agric Food Chem*, 63, 9888–9898.

Mann, K. 2008. Proteomic analysis of the chicken egg vitelline membrane. *Proteomics*, 8, 2322–2332.

Mann, K. & Mann, M. 2008. The chicken egg yolk plasma and granule proteomes. *Proteomics*, 8, 178–191.

Mann, K. & Mann, M. 2011. In-depth analysis of the chicken egg white proteome using an LTQ Orbitrap Velos. *Proteome Sci*, 9, 7.

Mann, K., et al. 2006. Proteomic analysis of the acid-soluble organic matrix of the chicken calcified eggshell layer. *Proteomics*, 6, 3801–3810.

Mann, K., et al. 2007. Phosphoproteins of the chicken eggshell calcified layer. *Proteomics*, 7, 106–115.

Marie, P., et al. 2015. Quantitative proteomics provides new insights into chicken eggshell matrix protein functions during the primary events of mineralisation and the active calcification phase. *J Proteomics*, 126, 140–154.

Marrero, A., et al. 2012. The crystal structure of human alpha2-macroglobulin reveals a unique molecular cage. *Angew Chem Int Ed Engl*, 51, 3340–3344.

Marzec, A., et al. 2019. Effect of hens age and storage time on functional and physiochemical properties of eggs. *J Appl Poult Res*, 28, 290–300.

Maurer, L. M., et al. 2005. pH regulates genes for flagellar motility, catabolism, and oxidative stress in *Escherichia coli* K-12. *J Bacteriol*, 187, 304–319.

Messens, W., et al. 2004. Growth of serovars in hens' egg albumen as affected by storage prior to inoculation. *Food Microbiol*, 21, 25–32.

Messens, W., et al. 2005a. Eggshell characteristics and penetration by *Salmonella enterica* serovar Enteritidis through the production period of a layer flock. *Br Poult Sci*, 46, 694–700.

Messens, W., et al. 2005b. Eggshell penetration by *Salmonella*: a review. *World's Poult Sci J*, 61, 71–85.

Messens, W., et al. 2007. Eggshell penetration of various types of hens' eggs by *Salmonella enterica* serovar Enteritidis. *J Food Prot*, 70, 623–628.

Mignon-Grasteau, S., et al. 2015. Robustness to chronic heat stress in laying hens: a meta-analysis. *Poult Sci*, 94, 586–600.

Miksik, I., et al. 2014. Proteomic analysis of chicken eggshell cuticle membrane layer. *Anal Bioanal Chem*, 406, 7633–7640.

Miyagawa, S., et al. 1991. Effects of protease inhibitors on growth of *Serratia marcescens* and *Pseudomonas aeruginosa*. *Microb Pathog*, 11, 137–141.

Moreau, T., et al. 2022. Antimicrobial proteins and peptides in avian eggshell: structural diversity and potential roles in biomineralization. *Front Immunol*, 13, 946428.

Mowbray, C. A., et al. 2018. Gene expression of AVBD6-10 in broiler chickens is independent of AVBD6, 9, and 10 peptide potency. *Vet Immunol Immunopathol*, 202, 31–40.

Moyle, T., et al. 2016. Bacterial contamination of eggs and behaviour of poultry flocks in the free range environment. *Comp Immunol Microbiol Infect Dis*, 49, 88–94.

Muñoz, A., et al. 2015. Importance of eggshell cuticle composition and maturity for avoiding trans-shell contamination in chicken eggs. *Food Control*, 55, 31–38.

Nagase, H., et al. 1983. Ovostatin: a novel proteinase inhibitor from chicken egg white. I. Purification, physicochemical properties, and tissue distribution of ovostatin. *J Biol Chem*, 258, 7481–7489.

Nascimento, V. P., et al. 1992. Relationship between shell structure and movement of *Salmonella* Enteritidis across the eggshell wall. *Br Poult Sci*, 33, 37–48.

Neilands, J. B. 1995. Siderophores: structure and function of microbial iron transport compounds. *J Biol Chem*, 270, 26723–26726.

Okumura, H., et al. 2015. Identification of distinctive interdomain interactions among ZP-N, ZP-C and other domains of zona pellucida glycoproteins underlying association of chicken egg-coat matrix. *FEBS Open Bio*, 5, 454–465.

Okuno, K., et al. 2019. *Salmonella* Typhimurium is attracted to egg yolk and repelled by albumen. *Curr Microbiol*, 76, 393–397.

Padan, E., et al. 2005. Alkaline pH homeostasis in bacteria: new insights. *Biochim Biophys Acta*, 1717, 67–88.

Palmiter, R. D. 1972. Regulation of protein synthesis in chick oviduct. I. Independent regulation of ovalbumin, conalbumin, ovomucoid, and lysozyme induction. *J Biol Chem*, 247, 6450–6461.

Pellegrini, A., et al. 1992. Bactericidal activities of lysozyme and aprotinin against gram-negative and gram-positive bacteria related to their basic character. *J Appl Bacteriol*, 72, 180–187.

Pu, J., et al. 2023. Quantitative proteomic analysis of chicken egg white and its components. *Food Res Int*, 170, 113019.

Rabijns, A., et al. 2005. Structural analysis of the jacalin-related lectin MornigaM from the black mulberry (*Morus nigra*) in complex with mannose. *FEBS J*, 272, 3725–3732.

Radkowski, M. 2002. Effect of moisture and temperature on survival of Enteritidis on shell eggs. *Archiv Fur Geflugelkunde*, 66, 119–123.

Rahman, M. A., et al. 2009. VMO-II mediates the binding of the chalaziferous layer with the vitelline membrane in quail eggs. *J Poult Sci*, 46, 240–248.

Raulo, E., et al. 1994. Isolation of a neuronal cell surface receptor of heparin binding growth-associated molecule (HB-GAM). Identification as N-syndecan (syndecan-3). *J Biol Chem*, 269, 12999–13004.

Rawlings, N. D., et al. 2004. Evolutionary families of peptidase inhibitors. *Biochem J*, 378, 705–716.

Rawlings, N. D., et al. 2018. The MEROPS database of proteolytic enzymes, their substrates and inhibitors in 2017 and a comparison with peptidases in the PANTHER database. *Nucleic Acids Res*, 46, D624-D632.

Rehault-Godbert, S., et al. 2010. Effect of temperature and time of storage on protein stability and anti-*Salmonella* activity of egg white. *J Food Prot*, 73, 1604–1612.

Rehault-Godbert, S., et al. 2013. Ovalbumin-related protein X is a heparin-binding ov-serpin exhibiting antimicrobial activities. *J Biol Chem*, 288, 17285–17295.

Rehault-Godbert, S., et al. 2019. The golden egg: nutritional value, bioactivities, and emerging benefits for human health. *Nutrients*, 11, 684.

Réhault-Godbert, S., et al. 2022. Chapter 13 - Innate defenses of the avian egg. In: Kaspers, B., et al. (eds.) *Avian Immunology*. 3rd ed. Boston, MA: Academic Press.

Reyes-Grajeda, J. P., et al. 2004. Crystal structure of ovocleidin-17, a major protein of the calcified *Gallus gallus* eggshell: implications in the calcite mineral growth pattern. *J Biol Chem*, 279, 40876–40881.

Riley, A., et al. 2014. Quantification of eggshell microstructure using X-ray micro computed tomography. *Br Poult Sci*, 55, 311–320.

Roberts, J. R., et al. 2013. Egg quality and age of laying hens: implications for product safety. *Anim Prod Sci*, 53, 1291–1297.

Roberts, J. R., et al. 2011. 16 - Avian diseases which affect egg production and quality. In: Nys, Y., et al. (eds.) *Improving the Safety and Quality of Eggs and Egg Products*. Oxford, UK: Woodhead Publishing.

Rocculi, P., et al. 2009. MAP storage of shell hen eggs, part 1: effect on physico-chemical characteristics of the fresh product. *LWT-Food Science and Technology*, 42, 758–762.

Rodriguez-Navarro, A. B., et al. 2013. Change in the chicken eggshell cuticle with hen age and egg freshness. *Poult Sci*, 92, 3026–3035.

Rodriguez-Navarro, A. B., et al. 2015. Amorphous calcium carbonate controls avian eggshell mineralization: a new paradigm for understanding rapid eggshell calcification. *J Struct Biol*, 190, 291–303.

Rose-Martel, M., et al. 2012. Proteomic analysis provides new insight into the chicken eggshell cuticle. *J Proteomics*, 75, 2697–2706.

Samiullah, et al. 2013. Effects of egg shell quality and washing on *Salmonella* Infantis penetration. *Int J Food Microbiol*, 165, 77–83.

Samiullah, et al. 2014. Effect of production system and flock age on egg quality and total bacterial load in commercial laying hens. *J Appl Poult Res*, 23, 59–70.

Samiullah, S., et al. 2017. Effect of production system and flock age on eggshell and egg internal quality measurements. *Poult Sci*, 96, 246–258.

Schäfer, A., et al. 1998. Analysis of vitelline membrane proteins of fresh and stored eggs via HPLC. *Zeitschrift für Lebensmitteluntersuchung und -Forschung A*, 206, 329–332.

Schneider, W. R. & Doetsch, R. N. 1974. Effect of viscosity on bacterial motility. *J Bacteriol*, 117, 696–701.

Schoeni, J. L., et al. 1995. Growth and penetration of *Salmonella* Enteritidis, *Salmonella* Heidelberg and *Salmonella* Typhimurium in eggs. *Int J Food Microbiol*, 24, 385–396.

Scott, T. A. & Silversides, F. G. 2000. The effect of storage and strain of hen on egg quality. *Poult Sci*, 79, 1725–1729.

Sellier, N., et al. 2007. Estimations of repeatability and heritability of egg albumen antimicrobial activity and of lysozyme and ovotransferrin concentrations. *Br Poult Sci*, 48, 559–566.

Shafee, T. M., et al. 2017. Convergent evolution of defensin sequence, structure and function. *Cell Mol Life Sci*, 74, 663–682.
Silversides, F. G. & Scott, T. A. 2001. Effect of storage and layer age on quality of eggs from two lines of hens. *Poult Sci*, 80, 1240–1245.
Silversides, F. G., et al. 2006. Effect of strain of layer and age at photostimulation on egg production, egg quality, and bone strength. *Poult Sci*, 85, 1136–1144.
Singh, R., et al. 2009. Production performance and egg quality of four strains of laying hens kept in conventional cages and floor pens. *Poult Sci*, 88, 256–264.
Sparks, N. H. & Board, R. G. 1985. Bacterial penetration of the recently oviposited shell of hens' eggs. *Aust Vet J*, 62, 169–170.
Sun, Y., et al. 2020. A comprehensive identification of chicken egg white phosphoproteomics based on a novel digestion approach. *J Agric Food Chem*, 68, 9213–9222.
Superti, F., et al. 2007. Ovotransferrin. In: Huopalahti, R., et al.. (eds.) *Bioactive Egg Compounds*. Berlin, Heidelberg: Springer Berlin Heidelberg.
Supuran, C. T., et al. 2002. Bacterial protease inhibitors. *Med Res Rev*, 22, 329–372.
Svensson, S. L., et al. 2010. Midkine and pleiotrophin have bactericidal properties: preserved antibacterial activity in a family of heparin-binding growth factors during evolution. *J Biol Chem*, 285, 16105–16115.
Swiatkiewicz, S., et al. 2015. Dietary factors improving eggshell quality: an updated review with special emphasis on microelements and feed additives. *Worlds Poult Sci J*, 71, 83–93.
Szpak, M., et al. 2014. Evaluation of the antibacterial activity of cystatin against selected strains of *Escherichia coli*. *Folia Biol (Krakow)*, 62, 187–192.
Tranter, H. S. & Board, R. G. 1984. The influence of incubation temperature and pH on the antimicrobial properties of hen egg albumen. *J Appl Bacteriol*, 56, 53–61.
Tuohimaa, P., et al. 1989. Development of progestin-specific response in the chicken oviduct. *Int J Dev Biol*, 33, 125–134.
Uyeda, A., et al. 1994. Cloning and sequencing of hen magnum cdnas encoding vitelline membrane outer layer protein I (VMO-I). *Gene*, 144, 311–312.
Van Dijk, A., et al. 2007. The beta-defensin gallinacin-6 is expressed in the chicken digestive tract and has antimicrobial activity against food-borne pathogens. *Antimicrob Agents Chemother*, 51, 912–922.
Vlckova, J., et al. 2019. Changes in the quality of eggs during storage depending on the housing system and the age of hens. *Poult Sci*, 98, 6187–6193.
Vollmer, W. 2008. Structural variation in the glycan strands of bacterial peptidoglycan. *FEMS Microbiol Rev*, 32, 287–306.
Wang, H. & Slavik, M. F. 1998. Bacterial penetration into eggs washed with various chemicals and stored at different temperatures and times. *J Food Prot*, 61, 276–279.
Wang, X. 2020. Pleiotrophin: activity and mechanism. *Adv Clin Chem*, 98, 51–89.
Wellman-Labadie, O., et al. 2008a. Antimicrobial properties of avian eggshell-specific C-type lectin-like proteins. *FEBS Lett*, 582, 699–704.
Wellman-Labadie, O., et al. 2008b. Antimicrobial activity of cuticle and outer eggshell protein extracts from three species of domestic birds. *Br Poult Sci*, 49, 133–143.
Wellman-Labadie, O., et al. 2010. Antimicrobial activity of lipophilic avian eggshell surface extracts. *J Agric Food Chem*, 58, 10156–10161.
Wesierska, E., et al. 2005. Antimicrobial activity of chicken egg white cystatin. *World J Microbiol Biotechnol*, 21, 59–64.

Whenham, N., et al. 2014. Comparative biology and expression of Tenp, an egg protein related to the bacterial permeability-increasing family of proteins. *Gene*, 538, 99–108.
Whenham, N., et al. 2015. Ovodefensins, an oviduct-specific antimicrobial gene family, have evolved in birds and reptiles to protect the egg by both sequence and intra-six-cysteine sequence motif spacing. *Biol Reprod*, 92, 154.
Wilson, P. W., et al. 2017. Understanding avian egg cuticle formation in the oviduct: a study of its origin and deposition. *Biol Reprod*, 97, 39–49.
Wong, M., et al. 1984. Collagen in the egg shell membranes of the hen. *Dev Biol*, 104, 28–36.
Woodward, S. A., et al. 1987. Influence of Florida 301 wheat in a laying hen diet on egg yolk rupture strength. *Poult Sci*, 66, 1414–1416.
Xing, J., et al. 2007. Recombinant eggshell ovocalyxin-32: expression, purification and biological activity of the glutathione S-transferase fusion protein. *Comp Biochem Physiol B Biochem Mol Biol*, 147, 172–177.
Yang, M., et al. 2016. Structure-function analysis of avian beta-defensin-6 and beta-defensin-12: role of charge and disulfide bridges. *BMC Microbiol*, 16, 210.
Yuan, C., et al. 2014. Effect of expanded cottonseed meal on laying performance, egg quality, concentrations of free gossypol in tissue, serum and egg of laying hens. *Anim Sci J*, 85, 549–554.
Zelensky, A. N. & Gready, J. E. 2005. The C-type lectin-like domain superfamily. *FEBS J*, 272, 6179–6217.
Zhao, L., et al. 2014. Expression, purification, and *in vitro* comparative characterization of avian beta-defensin-2, -6, and -12. *Avian Dis*, 58, 541–549.
Zhou, Y., et al. 2020. Quantitative comparative proteomic analysis of chicken egg vitelline membrane proteins during high-temperature storage. *J Agric Food Chem*, 68, 9816–9825.

Chapter 3

Advances in shell egg safety regulations and protocols

Vincent Guyonnet, University of Montreal, Canada

1 Introduction

While shell eggs are among the most nutritious and affordable animal-sourced foods consumed on all continents, there are now higher expectations in terms of safety and quality. In addition to safety, consumers also want a guarantee of nutritional value and wholesomeness. For egg producers, safety is also an essential element to ensure the integrity and sustainability of production. Governments will typically monitor and control the safety of food supply chains although producers have also developed their own protocols that often exceed regulatory requirements. The principle of preventing risks is widely accepted as the most effective and economic way to produce and provide safe foods to consumers.

After reviewing the main hazards that may impact the safety and quality of shell eggs, this chapter reviews regulations and protocols in place both on farms during production and throughout the shell egg value chain, including

http://dx.doi.org/10.19103/AS.2024.0150.05

increasing interactions with consumers to better inform them about egg safety measures.

2 Shell egg safety: Hazard Analysis and Critical Control Point system

The use of the Hazard Analysis and Critical Control Point (HACCP) concept is widely accepted by the food sector, both during production and processing, to identify risks and design procedures to mitigate them (Ropkins & Beck, 2000). The identification of potential risks, along with their likely origin, forms the basis for the design and implementation of a series of measures, either government-mandated or self-imposed, to eliminate the identified food safety risks or at least reduce them below an acceptable level.

The value of this HACCP approach to food safety was recognized by the Codex Alimentarius Commission, the joint group of experts from the World Health Organization (WHO) and the Food and Agriculture Organization of the United Nations (FAO), with its incorporation in 1993 within the General Principles of Food Hygiene, initially published in 1969 (Codex Alimentarius, 2011). Based on 7 principles and a series of 12 tasks, the HACCP approach towards reducing food safety risks is now applied worldwide to the production and handling of shell eggs, either through government mandates, sector-driven voluntary adoption or individual company decisions (Guyonnet, 2024).

The identification of potential hazards, the key step in this approach, relies on publications by scientists or international organizations such as the FAO (FAO, 2002), on published or internet-searchable databases (CFIA, 2008, 2024; EFSA, 2024; USDA, 2024) as well as the knowledge and experience of farm and plant operators. The potential biological, chemical and physical hazards related to the production, grading and distribution of shell eggs are outlined below with an emphasis on countries with thorough monitoring programmes within the shell egg value chain.

3 Biological, chemical and physical hazards

3.1 Biological hazards

The microbiology of shell eggs was recently reviewed by Baron, Jan and Techer (2024). Among all potential biological hazards, *Salmonella* spp. is the most important. The analysis of probable causes of foodborne outbreaks (FBO) in the EU clearly identified eggs as a major source of food-related illnesses. For each year during the 2015–2019 period, shell eggs and egg products were the food items most cited in human *Salmonella* outbreaks within the EU (Chanamé Pinedo et al., 2022). Likewise, in the recent One Health Zoonosis report from the EU (ECDC, 2023), *Salmonella* continued to be the second most common foodborne zoonosis after *Campylobacter*, with shell eggs and egg products listed as the main food vehicle for outbreaks leading to the most hospitalizations (124) and the highest number of outbreaks (86) in 2022. Interestingly, for a production of eggs slightly superior to that of the EU27 countries but about 26% less population, the USA reported for the 2018–2021 period only 27 FBO due to shell eggs and egg products, with *Salmonella* incriminated in 13 of them, causing 25 hospitalizations (CDC, 2024).

The two serotypes most incriminated in these FBO are *Salmonella enterica* serotype *enteritidis* and *typhimurium*, both able to contaminate the egg either vertically (during egg formation by colonization of the reproductive track at different locations) or horizontally (deposition at the surface of the shell followed by the penetration through the shell and shell membranes). Other Salmonella (e.g. *S. Kentucky*, *S. Infantis*, *S. corvalis* and *S. Mbandaka*) were listed in 2022 among the most frequently detected serovars from either laying hens or shell eggs in the EU27 countries (ECDC, 2023). Although several bacteria can penetrate through the eggshell and shell membranes, Gram-negative, motile and non-clustering bacteria (*Pseudomonas*, *Alcaligenes* sp., *Carnobacterium* sp. *and Acinetobacter Baumannii*) have been identified more frequently inside the egg for their ability to penetrate the eggshell (Gantois et al., 2009). An infection of the vagina at the time of oviposition or the presence of contaminated faecal materials are the most likely causes for the bacterial contamination at the surface of the eggshell. Once packed in egg cartons, the contamination of the eggshell is unlikely. However, in many countries, eggs are mostly distributed and sold in bulk, bringing new potential biological hazards depending on the environmental conditions during storage and transportation.

Although a few viruses can also be vertically transmitted, none have been implicated in food-borne diseases in humans. A recent assessment of the transmission of low pathogenic avian influenza virus via the consumption of raw shell eggs established a negligeable risk for humans (Gonzales et al., 2018). Microscopic fungi may also be identified at the surface of the eggshell, with *Alternaria sp., Penicillium sp., Chaetomium sp.,* and *Fusarium sp.* recently reported from eggs produced under different commercial housing systems (cage, barn, free-range and organic) in Poland (Tomczyk et al., 2018).

3.2 Chemical hazards

The well-documented case in 2017 of off-label use of fipronil on laying hens to control red mites in Europe highlighted the ever-present risk of chemical contamination of shell eggs (Reich & Triacchini, 2018). The presence of chemical residues in eggs may be due to intentional or unintentional events, involving veterinary medicinal products, pesticides, cleaning reagents or environmental contaminants, either at the farm level, during feed manufacturing and distribution or at the packing and grading centres.

Among veterinary medicinal products, the detection of residues of anticoccidials in shell eggs is the most frequently reported noncompliance in monitoring programmes conducting in the EU (Salvatore & Vericat Ferrer, 2024) and Australia (DAFF, 2023a). This contamination is related to the carry-over of traces of anticoccidials during the manufacturing of layers feeds or, less frequently, during delivery of feeds to layer farms. In both the EU27 countries and Australia, the rate of detection of antibiotic residues in shell eggs was extremely low to nil. The same low level of noncompliance was also reported in the USA during a survey for the 2016–2021 period (Singer, 2023). However, recent surveys conducted in China (Wang et al., 2022), Singapore (Sin et al., 2023), Ghana (Mingle et al., 2021) and Chile (Cornejo et al., 2020) demonstrated that the presence of antibiotic residues in shell eggs is not uncommon, especially when produced in backyard production systems.

In the annual monitoring programmes conducted in the EU27 countries and Australia, dioxin, dioxin-like compounds, as well as non-dioxin-like Polychlorinated Biphenyls (PCBs), were the main environmental contaminants detected in eggs, although at low levels and at a low frequency (Salvatore & Vericat Ferrer, 2024; DAFF, 2023a). However, the broader access to the outdoors by laying hens increases the potential exposure to environmental contaminants (e.g. dioxins, PCBs, pesticides and heavy metals), especially when production occurs in peri-urban zones of megacities (Piskorska-Pliszczynska et al., 2019; Samad et al., 2023).

The contamination of poultry feeds with mycotoxins (e.g. aflatoxins, ochratoxin and zearalenone) may also be a source of contamination although mycotoxins are rarely identified in eggs (Lee et al., 2016; Tomczyk et al., 2019; Wang et al., 2018). Housing systems may also impact the level of mycotoxin contamination of eggs, with a greater risk observed with eggs produced in cage-free systems and stored over several weeks (Tomczyk et al., 2018).

The use of non-food grade sanitizers and disinfectants on farms or in egg-packing plants may also result in the presence of chemical residues in the egg.

Depending on the molecular structure of chemical hazards, their deposition will preferentially occur in the egg white (water-soluble), the yolk (lipo-soluble) or on the eggshell (heavy metals and mycotoxins). Considering the physiology of the egg formation, with the simultaneous presence of several follicles at various stages of development, chemicals preferentially deposited in the yolk will accumulate in greater amount and present a greater risk to food safety (Goetting et al., 2011).

3.3 Physical hazards

The protective nature of the eggshell usually prevents the occurrence of physical hazards. However, the presence of physical hazards inside the eggs (pieces of plastic, wood, metal or dirt) is possible after the loss of integrity of the shell (broken or leaking eggs), occurring either at farm level or throughout the value chain. Although considered a physical hazard by definition, the presence of faecal materials on eggshells was previously addressed as a biological hazard.

The presence of pieces of tissues sloughed off the reproductive tract (meat spot) and of blood spots inside the egg are considered as physical hazards in some countries although their impact on food safety has not been established (Van der Fels-Klerx et al., 2017).

4 Shell egg production: on-farm general hygiene requirements

Many countries have regulations addressing specifically the conditions and requirements to produce safe eggs. However, for countries with no specific regulations related to shell egg production or no regulation at all, the Code of hygienic practice for eggs and egg products (CAC/RCP 15-1976), initially published in 1976 by the Codex Alimentarius and most recently reviewed in 2007, provides some guidance on the main principles required to produce shell eggs safe for human consumption (Codex Alimentarius, 2007). These general principles will be first reviewed before addressing specific regulations in key egg producing countries.

The code relies on three main principles:

1. From the production site to the point of retail, control measures should be in place to achieve the appropriate level of public health protection.
2. These measures should be implemented within the context of HACCP systems.
3. The control measures should be effective and validated.

Some specific aspects of good hygiene practice are discussed below.

4.1 Environmental hygiene

With the possibility of contaminating eggs directly from laying hens, indirectly from poor environmental conditions and management practices, or from contaminated feeds and water and from poor handling and storage of eggs, the farm environment presents the most risks to the safety of shell eggs. However, most biosecurity measures implemented on farms to guarantee the good health and productivity of laying hens have also a positive impact on the safety of eggs.

With hens given more frequent access to the outdoors, the potential contamination of eggs from the surrounding environment (contamination from previous land use or due to the presence of contaminants, polluted surface water, microbiological or chemical hazards from manure and other animal production in the vicinity of a farm) must be addressed carefully. The presence of wildlife should be monitored, both from a biosecurity standpoint and from a food safety standpoint, as they may contaminate the land and water supply used by laying hens. The quality of the air, both outside and inside the layer houses can also impact the level of contamination of the eggshell. Construction materials and equipment for layer houses should be selected to minimize the risks of harbouring pathogens and contaminants and be easy to clean and disinfect between flocks.

4.2 Flock health and management

Although the provision of day-old pullets from breeder flocks and hatcheries free of specific pathogens is an efficient way to reduce the risk of biological contamination of shell eggs, many countries do not have such breeder flocks, especially where egg production is dominated by smallholders. As potential sources of biological and chemical hazards, water and feeds must be sourced carefully. Municipal water should be preferred although well water is appropriate when adequately treated (e.g. filtration, UV or ozone treatment) and tested regularly.

Feeds should be purchased from feed mills using the HACCP principles as part of their good manufacturing practices (GMP). For on-farm mixing, feed ingredients should be sourced from suppliers able to guarantee their composition and quality, and finished feeds should be part of a routine testing program. Pelletization (heat) or other treatments should be applied to reduce or eliminate sources of contamination for feeds. The proper storage of feeds is also essential to limit their deterioration over time and contamination, including spillage attracting pest and wild birds.

A health management program, using a wide range of vaccines often administered during the pullet stage, along with good management practices, will reduce the risk of diseases in a flock, limiting the needs to medicate birds and decreasing the risk of residues in shell eggs. Veterinary medicinal products must be used under the supervision of a veterinarian, in compliance with local regulations in terms of usage level and withholding period. The administration of any vaccine or veterinary medicinal product to a flock should always be documented. The prompt removal of dead birds and their safe disposal will also prevent the potential transmission of pathogens to other animals or farm workers. Agricultural chemicals should be transported, stored, used and disposed of according to the instructions provided by manufacturers and local regulations.

A well designed plan prevents the bacterial contamination of shell eggs when pathogens are brought by pest. This programme will also significantly reduce the possible contamination of eggs by pest control chemicals.

4.3 Egg collection and storage on farm

The process of egg collection must be managed in a way that minimizes any damage to the eggshell, a major risk of contamination. Cracked eggs and leaking eggs should be excluded from being sold directly to consumers. Dirty eggs should also be excluded from direct consumption unless regulations allow for the cleaning of shell eggs. The collection and storage of shell eggs on farms should use tools and equipment designed for food contact and should be easy to wash and sanitize. Eggs should be transported from the farms under sanitary conditions (e.g. clean truck, absence of chemicals or other sources of contamination) to minimize any risk of contamination.

5 Shell egg production: on-farm specific country regulations and protocols

As outlined in the section on food safety hazards during egg production, the prevention of *Salmonella* contamination is the main focus of food safety regulations and protocols dedicated to the production of eggs in many countries. The presence of chemical hazards is often addressed in regulations covering all animal production systems or even any type of foods. A review of these safety regulations and protocols is presented below.

Overall, the main safety concern during egg production is related to the potential presence of *Salmonella* in eggs. A comparison of the sampling regimen adopted by the national authorities is presented in Table 1. As a country where consumers cherish the opportunity to consume raw eggs with various dishes, Japan (discussed in more detail below) has the most robust testing regiment for *Salmonella* on farms, both at the pullet stage (four sampling periods) and during production (every 12 weeks).

5.1 EU: 27 countries

The EU is the region with the most comprehensive set of regulations to control *Salmonella enteritidis* (SE) and other *Salmonella* serovars of public health significance during egg production, having set the overall target of 2% or less of positive flocks (*S. enteritidis and S. typhimurium* (ST)) of laying hens in

Table 1 Summary of testing requirements on farms for the detection of *Salmonella enteritidis* or other serovars of interest

Country/	Timing		Main samples collected (vary for cage and cage-free housing system)				
region	Pullet stage	Production stage	Faecal material	Faecal swabs	Litter swabs	Dust	Other
EU27	Day 0		x	x	x	x	Dust as substitute for litter swabs permissible
	2 weeks prior to production		x	x	x	x	
		Every 15 weeks from week 24	x	x	x	x	
USA	Week 14–16			x	x		
		Week 40–45		x	x		Dust + egg belts during investigation
		4–6 weeks post-molt		x	x		
Japan	Day 0						Birds/tissues
	Day 1–7						Tissues from dead birds
	Week 2–4			x	x	x	
	4 weeks prior to production			x	x	x	
		Week 23–30	x	x	x	x	Tissues from mortality every 3 months
		2 weeks prior to depopulation	x	x	x	x	
Australia	6 qualifying tests over 12 months			x	x		
		Every 3 months		x	x		
Argentina		18–27 weeks	x		x	x	
		70–86 weeks (9 weeks prior to depopulation)	x		x	x	
		5 weeks post-molt	x		x	x	

Colombia	Day 0		x			Tissues (spleen/livers) + swabs shipping boxes
	Week 8		x	x	x	Cloacal swabs for cage-free systems
	Week 14–16		x	x	x	
		Week 30	x	x	x	
		Week 50	x	x	x	
		Week 70	x	x	x	
		Week 85	x	x	x	

production (EU, 2011). The regulatory documents EC No. 2160/2003 on the control of *Salmonella* and other specified food-borne zoonotic agents (EU, 2003) and EU No. 517/2011 on the reduction of the prevalence of certain *Salmonella* serotypes in laying hens (EU, 2011) outline the key elements to be including in each Member State multi-year National Control programme reviewed and approved by the EU commission.

Farmers must sample pullet flocks as day-old chicks and 2 weeks prior to the start of egg production while laying hen flocks will be tested every 15 weeks during production, starting at 24 weeks of age (± 2 weeks). From their end, competent authorities will take the following samples:

- at least 1 sample per flock over 1000 birds on an annual basis;
- samples at 24 weeks (± 2 weeks) for any new flock housed in a premises previously found to be positive;
- samples from all flocks in a premises with one positive result; and
- any sample required during foodborne illness investigations.

Since the sampling procedure will affect its sensitivity greatly, the regulations provide some clear guidance:

- for flocks housed in a cage system, 2 samples of 150 grams of naturally pooled faeces will be collected from all manure belt scrapers;
- for flocks housed in cage-free systems, 2 samples using a pair of boot swabs or socks will be taken by walking throughout the house or only 1, supplemented by 1 pair of swabs from manure belts; and
- in the case of multi-tier cage-free systems, at the discretion of national competent authorities, one faecal sample or pair of boot swabs may be replaced by a sampling of dust (100 grams of dust collected from multiple locations inside the house or several moistened fabric swabs).

Along with the sampling procedure, these regulations outline the transportation and preparation of the samples, the detection methods allowed and the requirement to serotype at least one isolate from any positive sample and test for antimicrobial resistance.

In case of a positive result of any sample (faeces, dust particles or dust swab) for any relevant *Salmonella* serotype (other than vaccine strains), the flock is considered positive, and all eggs must be diverted from direct human consumption and become Class B eggs. When birds from an infected flock are slaughtered or destroyed, the proper steps must be taken to reduce any risk of spreading the disease. A flock will also be considered positive if any sample detected the presence of antimicrobials or bacterial growth inhibitors.

Due to the vertical transmission nature of *Salmonella*, these European regulations also address monitoring procedures in breeder flocks. Similar samples must be taken at 1 day of age, 4 weeks, 2 weeks prior to production and then every 2 weeks during the production of hatching eggs. In case of a positive result, all birds must be destroyed or slaughtered, all eggs in incubation must be destroyed and any non-incubated eggs from the flock destroyed or disposed of as Class B eggs. The importance of hatcheries in the control of *Salmonella* was recently reviewed (Wales and Davies, 2020), highlighting a series of hygiene and biosecurity measures broadly adopted on a voluntary basis by hatchery operators in addition to the set of rules and testing measures (sampling at least every 6 weeks) outlined in the EU 2019/2035 Rules for establishments keeping terrestrial animals and hatcheries (EU, 2019).

At the discretion of competent authorities, vaccination of pullet and laying hens against specific *Salmonella* serovars may be used within the framework of national control programmes. This vaccination would become mandatory in any Member State where the prevalence in flocks of laying hens would reach at least 10% (EU, 2006), although none of the EU Member States met this criterion in 2022 (ECDC, 2023).

A set of regulations in the EU also prevent and monitor the risks of contamination of shell eggs by veterinary medicinal products and other chemicals (EU, 2017). Apart from inspection on farms by national veterinary services (e.g. farm records, feed sampling), the core of this monitoring occurs with samples taken from packing centres and will be reviewed later.

5.2 USA

As part of the desire to reduce foodborne illnesses in the USA, the Food Safety Modernization Act (FSMA), approved in 2011, shifted the focus of food safety systems towards prevention. For the production of shell eggs, measures for the prevention of SE were adopted earlier in 2009 and are known as 'The Egg Rule'.

In addition to the procurement of day-old pullets from SE-monitored breeder flocks meeting standards equivalent to the National Poultry Improvement Plan for SE clean status, the environment of a pullet house must be sampled at 14–16 weeks of age, while the environment of the layer house must be sampled at 40–45 weeks of age and then 4–6 weeks after the end of each molt (CFR, 2023). For routine sampling, manure is the preferred substrate, collected from moistened gauze pads dragged over the manure for the entire length of each row of cages.

In cage-free systems, drag swabs of the litter and slatted area must be collected with the number of swabs varying with the width of a single-tier layer house (8.5–11 m or 28–36 feet, 6 swabs; >16.5 m or >55 feet, 12 swabs). In cage-free multi-tier systems (aviaries), drag swabs must be collected from all scratch areas or manure belts (FDA, 2011). During the investigation of a foodborne outbreak, samples from an automated egg conveyor system (egg belts and de-escalators) and fans must be taken in addition to new samples from manure (FDA, 2008). In case of a positive result at the pullet stage, the eggs must be diverted to further processing (pasteurization) during the entire life of the flock, or the Egg Testing protocol must be followed within 2 weeks of the start of production. A total of 1000 eggs, representative of 1 day of production, must be tested for SE with the process repeated 3 additional times, at 2-week intervals. If all four tests are negative for SE, the eggs from that flock may be used for direct human consumption. A positive test during production (40–45 weeks of age) will also trigger either the diversion to further processing (most likely option) or the implementation of the Egg Testing protocol (CFR, 2023).

These regulations apply to all egg producers with 3000 laying hens or more unless they are selling their eggs directly to consumers or produce eggs only for the further processing sector. This exemption for egg farms producing solely for further processing created a need for an exemption to this Egg Rule during the COVID-19 pandemic due to the shortage of shell eggs for the table market.

In addition to the SE sampling requirement, the regulations in the 21CFR Part 118 – production, storage, and transportation of shell eggs (CFR, 2023) call also for strict biosecurity protocols including a sound pest control program, the refrigeration (<45°F or 7.2°C) of shell eggs on farms starting 36 hours after laying and a written SE prevention plan along with all relevant records. However, the removal of all visible manure, dry cleaning and disinfection is mandatory only in case of a positive environmental sample.

Regarding the potential presence of chemical residues and contaminants in shell eggs, the National Residue Program (NRP) for Meat, Poultry and Egg Products in the USA combines the intervention of various US Department of Agriculture (USDA) groups, e.g. Food Safety and Inspection Service (FSIS) and Animal Plant Health and Inspection Service (APHIS), the US Environmental Protection Agency (EPA), and FDA along with State agencies. While feed samples will be collected on-farm for the detection of chemical residues, the sampling of shell eggs, as part of the FSIS sampling program, will occur more often in egg-packing plants.

5.3 China

Regulations for food products are contained in a series of National Standards (*guó biāo* in Chinese), listed by the letters GB followed by the document number and its date of publication. Technical specifications, identified by the letter T, may be related to a National Standard (GB/T) or to documents addressing the agricultural sector (*nóng yè* in Chinese) listed as NY/T (Guyonnet, 2024). There are currently no regulatory requirements for the monitoring of SE and other serovars on commercial egg farms in China. A National Standard, NY/T 1167 – 2006: Environmental Quality and Sanitary Control Code for Livestock and Poultry Farms, is addressing from a general standpoint the biosecurity measures and other means of controlling diseases in animal production systems.

Two National Standards cover the expectations for all foods, in terms of maximum residue limits for pesticides (GB 2763 – 2021) and for veterinary drugs (GB 31650 – 2019). NY/T 5032 – 2006 outlines the general requirements for 'the use of pollution-free food animal feed and feed additives' in animal farming, including egg production. A Technical Specification (NY/T 2798.11-2016: Quality and safety control – Technical production of non-pollution agricultural products: Part 11: Fresh eggs) addresses the specific requirements related to the production and grading of shell eggs within the Green Food quality scheme, a scheme comparable to organic production and very popular with affluent urban residents in China (China, 2016; Guyonnet, 2024).

5.4 Japan

In a country where shell eggs are often consumed raw on top of noodles or rice, the egg sector has been required to implement some strict protocols under careful monitoring by competent authorities. However, in addition to the implementation of strict biosecurity measures on all poultry farms, updated in 2021 by the Japanese Ministry of Agriculture, Fisheries and Forestry (MAFF), the monitoring programme for *Salmonella* on farms outlined in Notification No. 8441 is robust but not impractical to implement (MAFF, 2005).

Along with the testing of day-old chicks upon arrival and dead or culled birds during the first 7 days, the sampling procedure during the pullet stage, both at 2–4 weeks of age and 4 weeks prior to production, depends on the type of rearing system used. In cage systems, a drag swab of faeces or a gauge pad sampling of the manure scrappers will be collected in addition to 2–5 grams of dust particles from 10 different exhaust fan locations (2 composite samples). In cage-free systems (less than 6% of the housing systems used in 2022), two drag swabs across the floor surface will be performed along with the same sampling of dust particles (two composite samples).

During egg production, the sampling procedures at 23–30 weeks of age are again different for cage and cage-free systems. In cages, a drag swab will be performed on manure across each row of cages and manure will either be collected from 300 hens or faeces adhering to scrappers and manure belts sampled with a gauze pad. Dust particles from ten different exhaust fan locations are also collected. In cage-free systems, in addition to the same sampling protocol for dust particles, two drag swab samples must be collected from the floor across the entire house and 48 nest boxes are to be swabbed with a gauze pad (eight nestbox per gauze pad). In addition to the testing of dead or culled laying hens every 3 months, the sampling protocol described previously during production is repeated 2 weeks prior to the depopulation of the house. Finally, if a large number of rats are present on site (as stated in Notification No. 8441), 30 rodents should be captured and tested for *Salmonella* before implementing the depopulation of the barn. Notification No. 8441 also described the thorough sampling regimen on breeder farms and hatcheries to prevent the presence of *Salmonella* in shell eggs. A survey conducted in 2010–2011 with over 100 000 eggs collected from farms demonstrated a prevalence of 0.003% for SE, a ten-fold decrease over a survey conducted prior to the implementation of this monitoring protocol (Esaki et al., 2013).

The sampling and testing of shell eggs for residual pesticides, veterinary medicines and other chemicals is performed at the level of each prefecture, according to the directives provided by the Ministry of Health, Labour and Welfare (MHLW, 2006). Each of the 47 prefectures in Japan collects about 10 eggs on an annual basis for the analysis of potential residues, but there is no consolidated national report issued on an annual basis.

5.5 Australia

While the Australian commercial egg production has enjoyed SE-free status for many years, the detection of a few positive egg farms in the states of New South Wales (NSW) and Victoria in 2018 demonstrated the difficulty to contain the spread of this serovar. There is a legal obligation to notify state authorities in case of suspicion of SE and NSW recently instituted a mandatory SE testing programme on egg farms. At the national level, a voluntary National *Salmonella Enteritidis* Monitoring & Accreditation Program (NSEMAP) has been put in place as a cost-effective means to ensure that flocks are SE free and offers commercial farms an accreditation process required for any export.

In the year prior to receiving accreditation, a series of six sampling times (months 12, 11, 10, 9, 6 and 3 prior to certification) must take place for each flock on a farm and be negative for SE. After accreditation, each flock must be tested every 3 months and replacement pullets must either come from NSEMAP-accredited breeder flocks or from a pullet house having tested negative 4 weeks prior to transfer. A negative ELISA test for SE on 15 pullet birds, conducted 4 weeks prior to transfer, is also acceptable. Apart from this ELISA test, all other samples are taken for the detection of SE in the environment, using strict procedures to be implemented by a registered veterinarian or a farm employee trained by a registered veterinarian. In cage systems (43% of all housing systems in Australia in 2022), the swabbing of manure (manure pit or manure belt) will use the drag swab procedure, by walking across all rows of cages in the house (with a maximum of four rows swabbed per house). In cage-free systems, the litter and slatted areas must be sampled using boot swabs (two pairs or four swabs per house) or drag swabs (three per house). In addition, about 10% of the nest boxes and egg belt lines are also swabbed with pre-moistened materials. If case of a positive result, the SE isolate must be sent to a reference laboratory for further identification by whole genome sequencing (DAFF, 2023b). Currently, over 75% of the Australian layer flocks enjoy this voluntary *Salmonella*-monitoring scheme. Vaccination against SE is not allowed in Australia, but a live attenuated vaccine against ST is registered for use in poultry.

Samples of eggs are also collected for the potential detection of seven different types of chemical residues: antibiotics, anticoccidials, environmental contaminants, herbicides, insecticides and metals and results are reported as part of the national residue survey published annually by the Australian government (DAFF, 2023a).

5.6 Argentina

The control of *Salmonella* spp. is covered by Resolution 86/2016 published by the Ministry of Agro-Industry, instituting a mandatory monitoring and control programme for all commercial poultry operations (SENASA, 2016). The program, under the supervision of a veterinarian designated for each farm, requires the sampling of layer houses twice per year. Each sampling time should include birds at the beginning of production (18–27 weeks of age), towards the end of production (70–86 weeks of age and a maximum of 9 weeks before the end of the production cycle) or birds after molting (starting 5 weeks after resuming egg production). On production sites with five or more houses, the sampling will take place in two houses, while on sites with less than five houses, only one house should be sampled.

In old cage systems (deep manure pit), two 150-gram samples of faeces will be collected from 60 different locations throughout the manure pit. In automated cage system equipped with manure belts, the two 150-gram faecal samples will be collected from all rows on 60 different locations of the manure belts. In addition, a 150-gram sample of dust particle will be collected from 20 different locations representative of the house environment. In cage-free systems, two pairs of moistened boot swabs will be used to sample the environment, each covering a designated pattern for one-half of the house (SENASA, 2018).

In case of a positive result, the local veterinary authorities will inspect the farm to identify potential lapses in biosecurity measures and means to remedy them, including depopulation. The use of live vaccines or inactivated vaccines against SE and other relevant serovars is allowed to reduce the dissemination of *Salmonella* spp. and potential presence in shell eggs.

5.7 Colombia

The measures to control and reduce the prevalence of SE and ST, recognized by the Colombian government as sources of FBO, are outlined in Resolution No. 00017754 (ICA, 2019). The program, under the supervision of ICA (Colombian Agricultural Institute), is managed by egg farms that have been registered with the competent authorities and hold a valid certificate acknowledging their biosecurity plans. A series of samples must be taken both at the pullet stage and during egg production to test either birds, faecal materials or the environment, depending on the timing and the type of housing systems used.

At one day of age, tissue samples must be collected from ten chicks (one sample with livers, one sample with spleens and yolk sac residues), 50 ml of meconium collected from 20 chicks and between 20% and 25% of the chick shipping boxes should be swabbed for testing. During the pullet stage, samples must be gathered on week 8 and week 14–16, prior to transfer. In cage systems (60% of housing systems used), a 300-gram composite sample of faecal materials must be collected or 2 drag swabs must be performed over faecal materials along the length of the house. In cage-free floor systems, two cloacal swab samples, each from a pool of 10 birds, must be collected along with 2 samples of the litter collected with a pair of boot swabs. During production, samples must be gathered on Week 30, 50 and 70 (production up to 80 weeks of age) with an additional sampling time on Week 85 for longer production cycles (up to 100 weeks of age). The sampling procedures during production is similar to those described previously for the pullet stage, with different samples collected for cage and cage-free systems. A series of measures and sampling requirements are also prescribed for breeder flocks and hatcheries.

In case of a positive result for SE or ST, the eggs must be diverted from direct human consumption but may be used for further processing (pasteurization). ICA is notified and will collect on farms 5 samples of 20 eggs for testing. The same procedure will be repeated an additional 3 times at 15-day intervals. If all results from these four sampling times are negative, the SE/ST-free status of the farm is restored. Additionally, the farm must review its biosecurity and disease control programmes and initiate some changes, some at the request of ICA. Since the vaccination with live or killed SE/ST vaccines is allowed, the serotyping or other means will be required to distinguish vaccinal strains from wild strains.

The regulations related to the monitoring and control of residues in foods are outlined in Resolution No. 000770 published in 2014 (Colombia, 2014). For the egg sector, samples are collected by ICA on farms meeting any of the following criteria: farm with more than 2000 hens, farm with previous non-complaint results or farms granted specific certifications by ICA. At the national level, the aim is to collect 1 sample for every 1000 tonnes of egg produced with a minimum of 200 samples collected across the major egg-producing departments of Colombia. Although egg production in Colombia reached 812 508 tonnes in 2022, only 200 samples were collected and analyzed for the presence of prohibited products (e.g. chloramphenicol and nitrofurans), veterinary medicines (e.g. antibiotics, anticoccidials and anthelmintics) and other chemical contaminants (ICA, 2022).

5.8 South Africa

The only mandatory testing requirements for SE is in the case of shell eggs being sold after an in-shell pasteurization process with the word 'pasteurized' printed on the egg cartons. The farms supplying these eggs must be under a regular screening programme for *Salmonella* spp. and part of a validated *Salmonella* spp. management system (South Africa, 2020). This requirement comes from the fact that the in-shell pasteurization process used may not always meet the required 5-log reduction. Vaccination against SE is allowed in South Africa. Good farm practices, good feed manufacturing practices and adherence to the regulations for veterinary medicines apply to the egg sector as well as all other animal production sectors in South Africa.

6 Post-harvest shell egg handling: general hygiene requirements

The safety of shell eggs, once leaving their production site, is governed by regulations either applying broadly to any type of food or specifically enacted to address the safety of shell eggs. Most countries around the world, selling eggs in bulk, rely solely on basic food hygiene principles to safeguard the safety of shell eggs along the value chain.

While most of the global egg production is still sold as non-graded eggs, the grading of shell eggs is a mandatory step before commercialization in all Western and rich nations (Guyonnet, 2024). Packing or grading centres play a key role in ensuring the commercial availability of safe eggs. Throughout the grading and distribution process, shell eggs should be handled in a manner that avoids damages, minimizes the deposition of moisture of the eggshell (relative humidity and temperature fluctuations) and prevents contamination.

6.1 Receiving and handling of shell eggs

At the point of receipt, visual inspection will segregate shell eggs deemed unsafe or unsuitable for human consumption (broken and leaking eggs, eggs with bacterial or fungal rots). However, a number of countries have no regulatory requirement forbidding the sale of leaking or broken eggs. Where egg cleaning is allowed, the receiving of dirty eggs (within a certain percentage of the lot) is permitted and these eggs must be processed according to local regulations.

Approved dry or wet methods will be used by the grading centre to clean the eggs, taking special care to minimize damage to the shell surface. The dry method should minimize damage caused to the protective cuticle and may be followed by the spraying of food grade oil to coat the eggshell. Eggs may also be washed with water, using food grade detergents and sanitizers. Eggs should never be totally immersed in water and must be dried after washing to minimize the presence of moisture on the surface of the shell, a potential opportunity for the growth of moulds. The application of a sanitizer or food grade oil may also be allowed after washing and drying the eggs. Other methods such as in-shell pasteurization treatment used to eliminate pathogens may also be allowed as long as they respect the safety and suitability of eggs for human consumption (Codex Alimentarius, 2007).

6.2 Storage and shelf-life

Temperature fluctuations present the greatest risk to the safety and quality of shell eggs during storage, as they will contribute to the deposition of moisture at the surface of the eggs. The length of storage, the temperature and relative humidity conditions during storage are important factors that must be controlled and monitored carefully. A lower temperature of storage will minimize microbial growth, including that of SE.

The shelf-life of shell eggs depends on various factors, including:

- storage conditions (temperature, temperature variations and relative humidity);
- method for handling the eggs (segregation of cracked, leaking or dirty eggs at receiving and cleaning of eggs); and
- the type of packaging used for the eggs.

In some countries, the regulatory authorities have defined a mandatory shelf-life (best before date or other term used) for shell eggs. In other cases, a grading establishment must define the shelf-life of the eggs based on factors mentioned previously, as well as on information related to the conditions of the eggs (length between production and receiving, storage and transportation conditions) prior to receiving (Guyonnet, 2024).

6.3 Maintenance and sanitation

The proper maintenance and sanitation of the packing centre will also contribute to the availability of safe shell eggs. Therefore, a set of protocols under the general concept of GMP should be implemented: validated cleaning programme for the equipment and premises, pest control programme and waste removal practices limiting any risk of contamination. The effectiveness of these programmes and measures should be periodically monitored to guarantee the continued safety of shell eggs (Codex Alimentarius, 2007).

6.4 Distribution and commercialization

Upon leaving the packing centre, shell eggs should be transported in a way limiting breakage, damage and contamination. Vehicles used for transportation should be visibly clean and easy to clean thoroughly. Any container used during distribution should also be properly cleaned and disinfected prior to being reused. For shell eggs graded and packed, the packaging materials should provide sufficient details to identify a production lot (traceability) as well as food safety information for consumers (Codex Alimentarius, 2007).

7 Post-harvest shell egg handling: specific country regulations and protocols

In addition to these general principles of food hygiene applied to shell eggs throughout the value chain, some countries have established specific regulations for the grading, packing and distribution of shell eggs.

7.1 EU 27 countries

In the 27 countries of the EU, shell eggs are one of the agricultural products covered by EU Regulation 1308/2013, stipulating that only Class A (fresh) eggs may be sold directly to consumers after grading by weight and proper marking for traceability (EU, 2013). This evaluation of Class for shell eggs is mostly performed in packing centres/grading establishments, although it may sometimes occur immediately on farms. For instance, shell eggs from a flock identified as positive for SE are immediately downgraded to Class B and will be sent directly to an egg processing plant for transformation, thus reducing the risk of contamination of the packing centres.

For Class A shell eggs, the marketing standards, outlined in the EU Regulation 589/2008 (EU, 2008) are:

- shell and cuticle: normal shape, clean and undamaged;
- air space (dimension of the air cell) does not exceed 6 mm and 4 mm for eggs labeled as 'extra fresh';
- the yolk is visible during candling as a shadow only, without discernible outline and is slightly mobile when rotating the eggs but returning to a central position;

- the egg white is clear, translucent and the germinal disc shows no sign of development; and
- the eggs exhibit no sign of foreign matter or unexpected smell.

Class B shell eggs, not meeting these quality characteristics either immediately at receiving or over time (downgraded from Class A to Class B), will receive a different marking to clearly segregate them throughout the value chain.

As EU regulators have determined that egg washing and cleaning can cause some damage to the eggshell and cuticle, Class A eggs may not be washed unless this practice was allowed prior to 1 June 2003. In that case, egg washing may continue, providing it is in accordance with national guidelines and that the washed eggs are sold only in EU countries authorizing egg washing. Other practices such as the treatment of eggs with ultraviolet rays are allowed.

The storage and transportation of eggs will preferably be done at a constant temperature, with no refrigeration allowed before the sale to final consumers, to prevent condensation on the shell surface, a known risk of contamination of shell eggs. However, the transportation of eggs at a temperature below 5°C for not more than 24 hours or the storage at retail for not more than 72 hours is permissible. An exemption to this rule is made for the storage and distribution of shell eggs to the French overseas departments.

After sorting Class A shell eggs by weight within 10 days of being laid, they must be marked for identification and traceability. The first number indicates the type of production system used: 0 = Organic; 1 = Free-range; 2 = Barn; 4 = Cage (only enriched allowed since 2012) while the next 2 letters indicate the country of production (e.g. FR = France, ES = Spain). The rest of the marking clearly identifies a specific layer house within a farm, using different methods in the EU27 countries (e.g. France: XXX (Farm identifier) – XX (Layer house identifier); Spain: XX (Province) – XXX (Municipality) – XXX (layer house identifier)). This coding system greatly enhances traceability and provides some valuable information to both consumers and health authorities in case of FBO.

Eggs graded, marked and packed within 4 days of lay may be sold as 'extra fresh'. The 'best before' date or date of minimum durability must be indicated at packing with a maximum of 28 days from the date of lay, the date printed on the egg carton is determined by the date of lay of the oldest egg in an egg carton. Packed shell eggs must be delivered to consumers within a maximum time limit of 21 days of laying.

Egg cartons (Class A eggs) must contain the following information on their outer surface easily visible and clearly legible:

- the packing centre code;
- Quality grading ('Class A' or the letter 'A', alone or in combination with the word 'fresh';

- the weight grading (XL, L, M or S);
- the date of minimum durability ("best before date");
- the wording 'washed eggs' if applicable; and
- Some indications advising consumers to keep eggs refrigerated (chilled) after purchase.

Additionally, the packaging must indicate the type of production system (organic, free-range, barn and cage) used and the producer code must be explained on or inside the pack. The term 'extra' or 'extra fresh' may be used on egg cartons only until the ninth day after the date of lay of these eggs. Most of the information provided on eggs and egg cartons is designed to provide valuable information to consumers in terms of food safety (date of lay, best-before date, safe handling and storage, and traceability) and contribute to the overall safety of the egg value chain.

The good hygiene and manufacturing practices (GMP) of grading and packing establishments are regulated by national authorities. Establishments are allocated an identification number, indicating their suitability to perform properly the various tasks related to the inspection, grading and packing of shell eggs.

The sampling of shell eggs for residue testing usually occurs at the grading stage. The relevant standards, defined in EU regulation 2022/1646, require for each member state, as part of their multi-year national surveillance plan, to take on an annual basis a minimum of one sample for each 1000 tonnes of annual production, with a minimum of 200 samples (EU, 2022). In 2021, a total of 10 631 samples were analyzed across the EU27 countries, ranging from 0.97 sample/1000 tonnes (Spain) to 60 samples/1000 tonnes (Luxembourg), for an average of 1.8 samples/1000 tonnes. Shell eggs are testing for the presence of residues for Group A (non-approved veterinary medicinal products) and Group B (Veterinary medicinal products) products. As previously mentioned, most of the non-compliance observed is linked to Group B residues, mainly anticoccidials due a carry-over during feed processing (Salvatore & Vericat Ferrer, 2024).

7.2 USA

The regulations for shell eggs are included in two separate Acts:

- Egg Products Inspection Act (EPIA, 1970)
- Agricultural Marketing Act (AMS, 1946)

The inspection of shell eggs is mandatory for all producers and producer-packers with more than 3000 laying hens selling through retail channels.

Designed to ensure the safety of shell eggs, the USDA food safety programme consists of a random inspection of a grading facility once per quarter (EPIA, 1970). The grading of shell eggs for quality is performed on a voluntary basis and was established to boost the confidence of consumers in the safety and quality of shell eggs. If egg-packing stations or companies wish to carry the USDA grade shield on its egg cartons, they must pay for the grading service provided by USDA employees continuously present in about 220 packing plants across the USA. Local State agencies monitor the compliance to this grading process in egg-packing plants not using the USDA grading service.

The quality of an egg, determined by its grade and not its size, is based on standards for interior and exterior quality factors. Eggs sold to consumers are allocated to three different US consumer grades: AA, A or B. Some tolerances have also been defined for each grade, both at origin (packing centre) and at destination (retail stores). For instance, US consumer Grade AA at origin must consist of at least 87% AA quality eggs and a maximum of 13% A and B quality eggs (but no more than 1% B eggs due to air cell depth, blood spot dimension or yolk defects); at destination, at least 72% of the shell eggs must be AA quality grade, at least 10% A quality grade and the remaining B quality grade (but no more than 1% due to excessive air cell depth, blood spot dimension or yolk defects). The standards evaluating the quality of eggs apply to all shell eggs sold at retail, regardless of the presence or absence of the USDA shield on the egg cartons (USDA, 2000).

The washing of shell eggs is mandatory and must use equipment that is sanitarily designed, maintained in a clean and sanitary manner and thoroughly cleaned at the end of each day of operation. The wash water must be maintained at a temperature of at least 32.2°C (90°F) and always be at least 6.7°C (20°F) higher than the internal temperature of the eggs being washed. In order to prevent the bacterial contamination of shell eggs, the wash water must be changed approximately every 4 hours and more frequently if needed. Approved detergents or detergent sanitizers must be added to the wash water, while the rinse water, slightly warmer than wash water, may contain an approved sanitizer. Other chemicals (e.g. defoamers) may also be added in the wash water.

After washing, eggs must be dried, using equipment sanitarily designed and properly maintained to ensure the quick drying of the shell. Air filters must be cleaned and regularly changed to ensure that they do not constitute a potential source of contamination of the shell. The washing, rinsing and drying of shell eggs must be continuous and completed rapidly to prevent the presence of moisture (sweating) on shell eggs.

The use of oil, filtered prior to its application, is allowed to protect shell eggs and maximize the conservation of egg quality. However, eggs with excess moisture on the shell should not be oil-treated (AMS, 1946). The process of pasteurizing shell eggs (at least a 5-log reduction in total *Salmonella* count) must be approved by the FDA and shell eggs, and after treatment, the eggs will carry the statement 'USDA certified Pasteurized'.

Shell eggs must be stored and transported at the maximum ambient temperature of 7.2°C (45°F), beginning 36 hours after the time of lay (FDA, 2009). In the USA, most farms holding several thousand to a few million laying hens are directly connected through a conveyor system to a packing center. In packing centers, cooler rooms containing shell eggs to be sold at retail must also be maintained at the maximum ambient temperature of 7.2°C and with a relative humidity (70–85%) minimizing shrinkage (AMS, 1946). Accurate thermometers and hygrometers must be available to properly monitor the storage conditions. Shell eggs sold at retail, including pasteurized shell eggs, must also be distributed and maintained under refrigeration (ambient temperature ≤7.2°C or 45°F) until the final sale.

The requirements for equipment (e.g. digital-display scales, adequate candling, washing and grading equipment) and dedicated rooms for key activities (e.g. storage, candling and washing) are listed in the 7 CFR Part 56. Operational procedures, including general hygiene for employees, are also described in the regulations. In grading centres under the supervision of the USDA, the Pre-Operational inspection of the plant is conducted by USDA staff to validate the good sanitary conditions of the grading facility (AMS, 1946).

The expiry date ('Sell by' or 'Not to be sold after') is calculated from the date shell eggs are packed into retail cartons and may not exceed 30 days, including the date of pack. Another term, 'Use before', 'Use by' or 'Best before', indicates the maximum period for the expected quality and may not exceed 45 days (USDA, 2000).

There is currently no regulatory mandate to mark shell eggs individually for traceability purposes. Any marking on the shell surface is applied on a voluntary basis and often performed for marketing reasons or requested by retailers. The FDA has included shell eggs in the Food Traceability list, currently, under evaluation as part of the Food Safety Modernization Act (FSMA) designed to further protect public health. However, the inclusion of shell eggs in this list will require, once finalized, only additional traceability recordkeeping requirements and will not mandate individual egg marking (FDA, 2021).

Only shell eggs, processed under the continuous monitoring of USDA graders, may be packed in USDA-approved packaging materials bearing the USDA grade shield. Egg cartons must carry at a minimum the following informations: USDA grade (AA, A or B), weight class, date of grading expressed as Julian date, lot code and plant identifier. The placement of the expiration date (Exp. or Sell by) on egg cartons is optional.

All shell egg cartons, including pasteurized shell eggs, must carry the following safe handling statement: 'SAFE HANDLING INSTRUCTIONS: To prevent illness from bacteria: Keep eggs refrigerated, cook eggs until yolks are firm, and cook foods containing eggs thoroughly'.

7.3 China

Although more than 90% of the Chinese egg production (~29 million tonnes in 2022) is still sold as ungraded, bulk products, the rising level of income in China and higher expectations of consumers in terms of food safety and quality have created a growing demand for graded and nicely packaged shell eggs. The basic regulatory document for the egg sector, GB 2749-2015, is entitled National Food Safety Standard Eggs and Egg Products (China, 2016). The national standard, GB/T- 39438-2020 covers specifically the requirements for shell eggs packaged in egg cartons (China, 2020), while the Agricultural Standard NY/T 1551-2007 addresses aspects related to the washing and grading of eggs (China, 2007).

At reception by packing centres, shell eggs should be inspected, and visibly damaged (broken, leaking) or deformed eggs removed to limit the risk of contamination. While GB 2749-2015 covers the basic sensory requirements for shell eggs (color, odor and general conditions), the detailed quality characteristics, listed in Appendix A of GB/T 39438-2020 (China, 2020), categorized shell eggs as Premium, Class I and Class 2. Shell eggs are packed under 3 grades, AA, A and B, according to the following specifications for a given production lot:

- AA: at least 90% Premium eggs, less than 10% Class 1, no Class 2 and no more than 1% broken eggs;
- A: at least 90% Premium or Class 1 eggs, less than 10% Class 2 eggs and no more than 2% broken eggs;
- B: other combinations of Premium, Class 1 and Class 2 and no more than 3% broken eggs.

The cleaning of eggs is allowed but not mandatory. The water used must meet the requirements of GB 5749-2022 (sanitary standards for drinking water) and be maintained at a temperature between 10°C and 32°C. Detergents used should be designed for the purpose, both in terms of efficacy and safety of use on food products. After washing, the eggs may be dried with warm air (less than 45°C). Disinfectants (GB 149030.2-2012) or film coating (e.g. edible vegetable oil, silicone oil and beeswax) may be applied after drying (China, 2007).

Prior to grading, shell eggs may be stored for up to 24 hours at temperatures ranging from 4°C to 25°C (China, 2007). After grading, the use of refrigeration (rather uncommon in China) is at the discretion of the packing plant and this decision is often made by the retail chains. Packed shell eggs should be kept at a temperature lower than 25°C with a relative humidity between 70% and 88% (China, 2020).

The requirements related to grading establishments are outlined in NY/T 1551-2007, with elements related to their location (e.g. isolated, accessible via a paved road), design (e.g. building materials, ventilation, welfare areas for workers), equipment (design, cleaning and maintenance) and worker hygiene (China, 2007).

There is currently no mandatory shelf-life for shell eggs and any statement related to an 'eat before' date on egg packaging is defined by a grading station. Given the wide range of conditions used for the storage and distribution of shell eggs in China, there is currently no recognized industry standard and the 'eat before' date is rarely printed on egg cartons (China, 2020).

There is no mandatory requirement for any identification on shell eggs for traceability purposes. The marking on eggs, using a safe, food-grade ink, may identify the company (code or logo), the date of packing or other information at the discretion of companies. In practice, egg marking is mostly applied for marketing purposes and rarely from a traceability standpoint (China, 2020).

The information required on egg cartons is outlined in Appendix D of GB/T 39438-2020. Any marking should be clearly visible, easily understandable, science-based and not misleading consumers. The date of packing should be listed as YYYY-MM-DD on the packaging, along with some information related to storage conditions after purchase, and potentially provide different eat-before dates, whether stored at room temperature or under refrigeration (China, 2020).

As previously mentioned, NY/T 2798.11-2016 (Quality and safety control – Technical production of non-pollution agricultural products: Part 11: Fresh eggs) addresses the specific requirements related to the production and grading of shell eggs within the Green Food quality scheme (China, 2016). Other regulations related to the safe processing and distribution of shell eggs are:

- GB 9685 – 2016: Use of additives for food contact materials and products;
- GB 4806.1 – 2016: General safety requirements for food contact materials and products; and
- GB 14930.1 – 2015: National standard for food safety – Detergents (Guyonnet, 2024).

7.4 Japan

The regulations for shell eggs in Japan are found in two key documents:

- Hygiene management guidelines for egg grading and packaging facilities (MHLW, 1998).
- Egg standard transaction guidelines (Bulletin 12-Livestock A-3266, Ministry of Agriculture, Forestry and Fisheries (MAFF) issued in 2000), recently translated in English (Kashimori, 2017).

As a precautionary measure, shell eggs which are obviously dirty, rotten, broken or moldy should not be transported to egg grading and packaging (GP) centers. During their initial inspection at the GP center, spoiled, mouldy and broken eggs should be removed and relevant information for each lot received will be collected for traceability purposes (e.g. quantity, farm location and laying house identification, date of production and date of collections).

Shell eggs are graded for quality by evaluating, through visual inspection, candling and observations after breaking. Only shell eggs assigned to Special Grade and Grade 1 may be used for raw consumption, while Grade 2 shell eggs should always be cooked prior to consumption. Moldy, leaking, smelly, rotten, partially incubated or eggs with ruptured yolk or blood spots are considered Off-Grade and unfit for human consumption (Kashimori, 2017).

After removing shell eggs visibly broken (Off-Grade and some Grade 2), the washing of eggs must be performed using warm water (at less 30°C and no less than 5°C warmer than the internal temperature of the eggs). Sodium hypochlorite solution (>150 ppm) or a similar type of sanitizers may be added to the wash or rinse water and the rinse water must be at least 5°C warmer than that the wash water. After washing, shell eggs must be dried quickly using clean and sanitary equipment (MHLW, 1998). The pasteurization of shell eggs is allowed, although not commonly practiced.

Prior to grading, shell eggs should be stored in a cool place, not exposed directly to sunlight, and clearly identified (origin and date of production) to ensure the management of inventories using the 'First-In, First-Out' (FIFO) principle. While it is desirable to keep graded shell eggs under refrigeration (temperature ≤10°C) during storage and distribution, the law requires only to keep them at a 'temperature as low as possible'. However, attention should be made to limit large temperature fluctuations throughout the distribution channels and refrigeration, if applied, must be provided consistently until the point of sale (MHLW, 1998).

GP facilities must comply with the standards for management and operation specified in the Food Sanitation Act. For example, at the end of each day of operation, the different rooms and equipment must be cleaned and disinfected (MHLW, 1998). Although HACCP certification is not mandatory in Japan, large egg companies have obtained HACCP certification or other types of certification, including FSSC22000, ISO22000 and SQF. Since June 2021, almost all food businesses involved in production, processing, cooking and sales, including GP plants, are required to introduce hygiene practices in line with HACCP. The Japanese egg industry association issued 2019 a comprehensive HACCP handbook to facilitate the transition and compliance with these new requirements (JEIA, 2019).

There is no mandated shelf-life date for shell eggs. A Best-Before (BB) date, defined scientifically based on microbiological, physiochemical and sensory tests, indicates the date after which consumers must cook the eggs prior to consumption (Kashimori, 2017).

While a traceability coding on eggs, similar to the EU system, is under consideration, there is currently no regulatory requirement to mark eggs individually. At this time, any marking (traceability number or sticker on individual egg) is performed on a voluntary basis by GP plants and egg companies, using their own systems (Kashimori, 2017).

Due to the frequent use of raw eggs in Japanese cuisine, a known food safety risk, the requirements on packaging used by GP are different for shell eggs intended for raw consumption and all other products. In the first case, printed information, clearly visible on the outside of packaging, must indicate name and address of GP plant, BB date, storage conditions (refrigeration after purchase at a temperature ≤10°C), and some statements indicating that these eggs may be consumed raw prior to the BB date but must be cooked after the BB date. For all other egg cartons, the following information must be provided to consumers on the outside of the packaging: name and address of GP plant, date of lay, date of packing and BB date, and a statement indicating that these eggs must be cooked prior to consumption (MHLW, 1998; Kashimori, 2017).

7.5 Other countries

A recent survey of shell egg regulations in 60 countries across Asia/Oceania, Europe, North America, Latin America and Africa collected information related to the availability of national regulations specific to shell eggs, the mandate or possibility to wash shell eggs and any regulated storage conditions (Guyonnet, 2024). The results are presented in Table 2. More than 80% of the countries surveyed have some specific regulations for shell eggs, often supplemented by regulations related to food labeling, food safety or hygiene practices in food production. In addition to the USA and Japan previously reviewed, Canada is the only country mandating that shell eggs be washed prior to packing. The washing of shell eggs is allowed in 15 countries surveyed, including Australia, Brazil, Egypt, Korea and Ukraine, while it is

Table 2 Shell eggs regulations and industry practices in various countries

Country	Egg washing		Mandatory refrigeration	Published	
	Mandatory	Allowed	(graded eggs)	regulations	Other regulations
Asia/Oceania (12)					
Australia	No	Yes	Yes if washed	Yes	Various
Bangladesh	No	Yes	No	No	No
China, PR	No	Yes	No	Yes	Various
China, Taiwan	No	No	No	Yes	Feed regulations
India	No	No	No	No	Food safety and standards
Indonesia	No	Yes	No	No	No
Japan	Yes	–	No	Yes	Various
Korea	No	Yes	Yes if washed	Yes	Various
Malaysia	No	Yes	No	No	No
Philippines	No	Yes	No	No	No
Thailand	No	No	No	Yes	No
Turkey	No	No	Yes	Yes	No
Europe (32)					
EU 27	No	No	No	Yes	Various
United Kingdom	No	No	No	Yes	Various
Ukraine	No	Yes	No	Yes	No
North America (2)					
Canada	Yes	–	<10°C	Yes	Various
USA	Yes	–	<7 °C	Yes	Various

Source: Adapted from: Guyonnet (2024).

specifically prohibited in 39 countries, including, in addition to the EU27 countries, countries like Chile, Thailand or the UK. Apart from the USA and Canada, very few countries (Algeria, Libya and Tunisia) mandate a specific temperature for the storage and distribution of shell eggs, while Australia and Korea require refrigeration only if shell eggs have been washed by packing centers.

On a global basis, it is evident that the safety and quality of shell eggs are guaranteed in different ways (Guyonnet, 2018). Overall, government regulations and sector-driven practices are widely applied and contribute to the provision of safe and nutritious shell eggs to local populations.

8 Communicating safety measures to consumers

In addition to regulations requiring the placement on packaging materials of key statements related to egg safety and the proper handling of shell eggs, governments, producer organizations and individual companies provide a lot of information to consumers, mainly via dedicated websites, webpages or social media feeds.

An excellent example of this consumer outreach is given by the Egg Safety Center (ESC) in the USA (Egg Safety Center, 2024). This organization, funded by the American Egg Board (AEB), the marketing arm of the US egg sector, provides information related to the safety of shell eggs on farms, in retail stores and at home. In the case of safety on farms, the ESC combines text, graphs and videos to explain the role of egg farmers in ensuring the provision of safe eggs. For full transparency, the ESC also outlines the different food safety programmess in place and provides links to key regulations. The information provided in the section dealing with retail outlines the key considerations for the purchase of safe eggs (check for clean, uncracked shells, do not buy eggs passed their self-by date and look for the USDA shield on the egg carton), reminding consumers that shell eggs must be stored at retail under refrigeration (<45°F or <7.2°C). A section also explains in detail the various dates printed on cartons: Julian date (1-365) for the packing of eggs and the Sell-by-date (MMMDD).

Finally, the section addressing the safety of shell eggs at home covers the storage of eggs (always under refrigeration, placed in the coldest area of refrigerator, and kept in their original carton to keep track of the Julian and sell-by dates), the safe handling of shell eggs (hand washing before and after handling eggs, utensils and surfaces washed with warm, soapy water, and eggs handles separately from raw meat, seafood, and poultry to prevent cross-contamination) and cooking instructions to ensure the proper killing step to prevent any risk of contamination.

Besides the information made available by the Egg Safety Center, webpages dedicated to egg safety on the USDA and CDC websites provide additional information to consumers. While the CDC website focuses on *Salmonella*, the USDA website addresses Shell Eggs from Farm to Table with a wide range of Frequently Asked Questions, ranging from the way to read the coding on egg cartons to safe cooking methods.

Finally, individual companies will also provide additional safety information to consumers, either on packaging materials, on websites and through communications on social media. As many FBOs are due to unsafe practices at home (Cardoso et al., 2021), the use of social media to interact with consumers offer some advantages over other communication methods. Social media allows a two-way engagement with consumers and may have a bigger impact on triggering the required behavioural changes (Overbey et al., 2017).

Communication to reinforce the confidence of consumers in the national provision of safe eggs is also be achieved by the printing on egg packaging of logos (e.g. British Lion in the UK, Egg Quality Assurance or EQA in Canada) attesting to the adherence to a wide range of sector-driven standards, from production to distribution, extending well beyond national regulations.

9 Conclusion and future trends

National shell egg regulations have often been established years ago and are rather infrequently revised through lengthy processes. As a response to several major food safety events in the late 1980s–early 1990s, private food safety standards (e.g. Safe Quality Foods or SQF, British Retail Council or BRC) were developed and adopted by retailers as a guarantee of a safer food supply chain.

Major stakeholders within the global food industry established in 2000 the Global Food Safety Initiative (GFSI) with the aim to develop global food safety standards in consultation with food safety experts and the private sector (GFSI, 2024). GFSI defines the minimum requirements for food safety evaluation and certification programmes, and most often requires more than complying with local regulations. GFSI also implements a comprehensive benchmarking process to assess the applicability of current food safety private standards to the GFSI global standards.

Since 2000 GFSI has recognized various private standards including BRC, SQF, Food Safety System Certification (FSSC 22000), The International Features Standard (IFS) and Global G.A.P., with each auditing scheme covering a wide range of points to ensure that a plant is in full control of its processes and the quality, safety and legality of its products. Most often, retailers and food sector clients requesting an annual third-party audit for GP plants will accept any GFSI-recognized private food safety standard. As a business-driven initiative, GFSI aims for the continuous improvement of food safety management systems and GFSI-recognized private standards are evaluated regularly, deleting or, more often, adding new items. For instance, since the first version of the BRC Global Standard Food Safety published in 1998, this standard has been revised several times, with version 9 issued in 2022. The speed and flexibility to adapt to new situations is certainly the strength of private standards and they will become more and more relevant to the egg sector, especially in countries with weaker oversight of government agencies in terms of plant hygiene and food safety.

Regulations in some countries or regions are not exempt from criticism and revisions may be needed soon. For instance, the EU regulations for shell eggs are clearly lacking specific requirements for relative humidity and permissible fluctuations in temperature during the storage and distribution of shell eggs, well-known factors affecting the safety and quality of shell eggs along the value chain. Likewise, the EU regulations contain an inconsistent message about refrigeration, not allowed from farm to retail but strongly recommended after purchase (Fikiin et al., 2020).

A recent review by the European Food Safety Authority (EFSA) noted the use of refrigeration at retail, as a way to reduce any food safety risk, may be necessary before granting an extension to the shelf-life of shell eggs in the EU27 countries (EFSA, 2014). More countries may decide to adopt refrigeration along the egg value chain, at the very least in retail stores, to maximize the safety but also the internal quality of shell eggs.

10 Where to look for more information

The recently published *Handbook of Egg Science and Technology* (Mine et al., 2024) contains additional information regarding the quality and the microbiology of shell eggs. In terms of the evaluation of food safety risks, in addition to the websites previously mentioned, the European Centre for Disease Prevention and Control, an agency of the EU, provides an excellent and updated overview of the role of shell eggs in FBO. The EFSA publishes frequently scientific opinion on various aspects related to the safety of shell eggs, both on farms and along the value chain. Additionally, the EFSA publishes on an annual basis the reports from the EU27 countries monitoring programmes performed for *Salmonella* and other foodborne pathogens, for veterinary medicinal product residues, for pesticides and for various other undesired residues. Given the size and diversity of the EU market in terms of egg production systems and practices, these regular publications offer some valuable insight on the performance of regulations and sector-driven practices to guarantee the safety of shell eggs.

11 References

AMS (1946). Agriculture Marketing Act. 7CFR Part 56, voluntary grading of shell eggs. Code of Federal Regulations. https://www.ecfr.gov/current/title-7/subtitle-B/chapter-I/subchapter-C/part-56

Baron, F., et al. (2024). Microbiology of egg and egg products. In Y. Mine, et al. (Eds.), *1st Edition Handbook of Egg Science and Technology*, CRC Press, Boca Raton, FL, pp. 355–396.

Cardoso, M. J., et al. (2021). Salmonella in eggs: from shopping to consumption–A review providing an evidence-based analysis of risk factors. *Comprehensive Reviews in Food Science and Food Safety*, 20(3), 2716–2741. https://doi.org/10.1111/1541-4337.12753

CDC (2024). CDC dashboard – Foodborne diseases. https://wwwn.cdc.gov/norsdashboard/

CFIA (2008). Reference database for hazard identification. https://ag.umass.edu/sites/ag.umass.edu/files/projects/related-files/reference_database_for_hazard_identification.pdf

CFIA (2024, January 16). Reference database for hazard identification (RDHI). https://active.inspection.gc.ca/rdhi-bdrid/english/rdhi-bdrid/introe.aspx?i=1

CFR (2023, December 27). Code of Federal Regulations, Title 21 – Food and drugs, Chapter 1, subchapter B, - Food for Human Consumption, Part 118 Production, Storage, and Transportation of Shell Eggs (last updated 22 Dec. 2023). https://www.ecfr.gov/current/title-21/chapter-I/subchapter-B/part-118?toc=1

Chanamé Pinedo, L., et al. (2022). Sources and trends of human salmonellosis in Europe, 2015–2019: an analysis of outbreak data. *International Journal of Food Microbiology*, 379, 109850. https://doi.org/10.1016/j.ijfoodmicro.2022.109850

China (2007). Agriculture sector standard of the People's Republic of China, technical specification for sorting, cleaning and grading chicken eggs, NY/T 1551-2007 (In Chinese). http://down.foodmate.net/wap/index.php?itemid=48415&moduleid=23

China (2016). Quality and safety control – Technical production of non-pollution agricultural products: Part 11: Fresh eggs, NY/T 2798.11-2016 (In Chinese). http://down.foodmate.net/wap/index.php?itemid=48415&moduleid=23

China (2020). National standard of the People's Republic of China. Pack for shell eggs, GB/T 39438-2020 (In Chinese). http://down.foodmate.net/wap/index.php?itemid=48415&moduleid=23

Codex Alimentarius (2007). Code of hygienic practice for eggs and egg products, CAC/RCP 15-1976. https://www.fao.org/fao-who-codexalimentarius/codex-texts/codes-of-practice/en/

Codex Alimentarius (2011). General principles of Food Hygiene, CXC 1-1969. Rome, Italy. https://www.fao.org/fao-who-codexalimentarius/codex-texts/codes-of-practice/en/

Colombia (2014). Resolución 000770 Establishment of national sub sectorial plans for the monitoring and control of residues in foods (in Spanish). https://www.minsalud.gov.co/sites/rid/Lists/BibliotecaDigital/RIDE/DE/DIJ/Resolucion-0770-de-2014.pdf

Cornejo, J., et al. (2020). Assessing antibiotic residues in poultry eggs from backyard production systems in Chile, first approach to a non-addressed issue in farm animals. *Animals*, 10(6), 1056. https://doi.org/10.3390/ani10061056

DAFF (2023a). Eggs residue testing annual datasets 2021–2022. https://www.agriculture.gov.au/agriculture-land/farm-food-drought/food/nrs/nrs-results-publications

DAFF (2023b). National salmonella enteritidis monitoring and accreditation program guidelines. https://www.dpi.nsw.gov.au/__data/assets/pdf_file/0020/722081/NSEMAP-Guidelines-v2-2023.pdf

ECDC (2023). The European Union One Health 2022 Zoonoses Report. *EFSA Journal*, 21, e8442. https://doi.org/10.2903/j.efsa.2023.8442

EFSA (2014). Scientific opinion on the public health risks of table eggs due to deterioration and development of pathogens. *EFSA Journal*, 12(7), 3782.

EFSA (2024, January 16). Chemical Hazards Database (OpenFoodTox). https://www.efsa.europa.eu/en/data-report/chemical-hazards-database-openfoodtox

Egg Safety Center (2024, February 23). For egg safety, countless safeguards are in place on commercial egg farms, in processing and transport, and in the store. Retrieved February 23, 2024, https://eggsafety.org/

EPIA (1970). Egg Products Inspection Act. 7 CFR Part 57, inspection of eggs. code of federal regulations. https://www.ecfr.gov/current/title-7/subtitle-B/chapter-I/subchapter-C/part-57

Esaki, H., et al. (2013). National surveillance of Salmonella Enteritidis in commercial eggs in Japan. *Epidemiology and Infection*, 141(5), 941–943. https://doi.org/10.1017/S0950268812001355

EU (2003). Directive 2003/99/EC of the European Parliament and of the Council of 17 November 2003 on the monitoring of zoonoses and zoonotic agents. Document 32003L0099. http://data.europa.eu/eli/dir/2003/99/oj

EU (2006). Commission Regulation (EC) No 1177/2006 of 1 August 2006 implementing Regulation (EC) No. 2160/2003 as regards requirements for the use of specific control methods in the framework of the national programmes for the control of salmonella in poultry. https://eur-lex.europa.eu/legal-content/en/ALL/?uri=CELEX:32006R1177

EU (2008). Commission Regulation (EC) No 589/2008 of 23 June 2008 laying down detailed rules for implementing Council Regulation (EC) No 1234/2007 as regards marketing standards for eggs. https://eur-lex.europa.eu/legal-content/EN/TXT/?uri=celex%3A32008R0589

EU (2011). Commission regulation (EU) No. 517/2011 – Target for the reduction of the prevalence of certain *Salmonella* serotypes in laying hens of *Gallus gallus*. Document 32011R0517. http://data.europa.eu/eli/reg/2011/517/2019-03-10

EU (2013). Part VI – Eggs of hens of the Gallus gallus species. In Regulation (EU) No 1308/2013 of the European Parliament and of the Council of 17 December 2013 establishing a common organisation of the markets in agricultural products and repealing Council Regulations (EEC) No 922/72, (EEC) No 234/79, (EC) No 1037/2001 and (EC) No1234/2007 (pp. 818–819). https://eur-lex.europa.eu/legal-content/EN/TXT/?uri=CELEX:32013R1308

EU (2017). Regulation (EU) 2017/625 of the European Parliament and of the Council of 15 March 2017 on official controls and other official activities performed to ensure the application of food and feed law, rules on animal health and welfare, plant health and plant protection products. Document 32017R0625. http://data.europa.eu/eli/reg/2017/625/oj

EU (2019). Commission Delegated Regulation (EU) 2019/2035 of 28 June 2019 supplementing Regulation (EU) 2016/429 of the European Parliament and of the Council as regards rules for establishments keeping terrestrial animals and

hatcheries, and the traceability of certain kept terrestrial animals and hatching eggs. Document 32019R2035. http://data.europa.eu/eli/reg_del/2019/2035/oj
EU (2022). Commission Implementing Regulation (EU) 2022/1646 of 23 September 2022 on uniform practical arrangements for the performance of official controls as regards the use of pharmacologically active substances authorised as veterinary medicinal products or as feed additives and of prohibited or unauthorised pharmacologically active substances and residues thereof, on specific content of multi-annual national control plans and specific arrangements for their preparation. Document 32022R1646 http://data.europa.eu/eli/reg_impl/2022/1646/oj
FAO (2002). Risk assessments of Salmonella in eggs and broiler chickens: interpretative summary. Microbiological Risk Assessment Series 1. FAO, Rome, Italy. https://www.fao.org/3/Y4393E/y4393e00.htm
FDA (2008). Environmental sampling and detection of Salmonella in poultry houses. https://www.fda.gov/food/laboratory-methods-food/environmental-sampling-and-detection-salmonella-poultry-houses
FDA (2009). 21 CFR Parts 16 and 118 prevention of salmonella enteritidis in shell eggs during production, storage, and transportation; final rule. *Federal Register*, 74(130), 33030–33101. https://www.govinfo.gov/content/pkg/FR-2009-07-09/pdf/E9-16119.pdf
FDA (2011). Guidance for industry: prevention of salmonella enteritidis in shell eggs during production, storage, and transportation. https://www.fda.gov/regulatory-information/search-fda-guidance-documents/guidance-industry-prevention-salmonella-enteritidis-shell-eggs-during-production-storage-and
FDA (2021, January 12). Food traceability list. https://www.fda.gov/food/food-safety-modernization-act-fsma/food-traceability-list?utm_medium=email&utm_source=govdelivery#List
Fikiin, K., et al. (2020). Do raw eggs need to be refrigerated along the food chain?: is the current EU regulation ensuring high-quality shell eggs for the European consumers? *Trends in Food Science & Technology*, 100, 359–362. https://doi.org/10.1016/j.tifs.2020.04.003
Gantois, I., et al. (2009). Mechanisms of egg contamination by Salmonella Enteritidis. *FEMS Microbiology Reviews*, 33(4), 718–738. https://doi.org/10.1111/j.1574-6976.2008.00161.x
Goetting, V., et al. (2011). Pharmacokinetics of veterinary drugs in laying hens and residues in eggs: a review of the literature. *Journal of Veterinary Pharmacology and Therapeutics*, 34(6), 521–556. https://doi.org/10.1111/j.1365-2885.2011.01287.x
GFSI (2024, February 23). Overview – Safe food for people everywhere. Retrieved February 23, 2024, from https://mygfsi.com/who-we-are/overview/
Gonzales, J.L., et al. (2018). Assessment of low pathogenic avian influenza virus transmission via raw poultry meat and raw table eggs. *EFSA Journal*, 16(10), e05431. https://doi.org/10.2903/j.efsa.2018.5431
Guyonnet, V. (2018). Refrigerated eggs versus shelves: Is one method safer? *Poultry International*, 57(2), 14–15. https://www.wattagnet.com/articles/32905-refrigerated-eggs-versus-shelves-is-one-method-safer
Guyonnet, V. (2024). Regulations for Shell eggs in different countries. In Y. Mine, et al. (Eds.), *1st Edition Handbook of Egg Science and Technology*, CRC Press, Boca Raton, FL, pp. 323–340.

ICA (2019). Resolución No. 00017754 (05/11/2019) Establishment of a National Program for the Control and Reduction of prevalence of Salmonella paratyphoid (enteritidis and typhimurium) in Poultry production in the country (in Spanish). https://www.ica.gov.co/getattachment/0b7d2fae-f96a-4b0d-8459-2b10be4142da/20194R1775.aspx

ICA (2022). National plan for the monitoring and control of residues of veterinary medicines and other chemical substances in eggs from laying hens 2022 (in Spanish). https://www.ica.gov.co/getattachment/Areas/Pecuaria/Servicios/Inocuidad-en-las-Cadenas-Agroalimentarias/Plan-Nacional-de-Residuos/Plan-residuos-huevos-2022-ajustado.pdf.aspx?lang=es-CO

JEIA (2019). HACCP Implementation guide. Egg sorting and packaging facility / liquid egg manufacturing facility (In Japanese), Japan Egg Industry Association (Nihon Rangyo Kyokai Association). http://www.nichirankyo.or.jp/haccp-dounyu/haccp-guide.pdf

Kashimori, A. (2017). *The Illustrated Egg Handbook. Appendix* (pp. 141–149). Context Publications, Packington, Leicestershire, United Kington. https://www.contextbookshop.com/user_uploads/files/booksamples/The-Illustrated-Egg-Handbook.pdf

Lee, M., et al. (2016). Detection of foodborne pathogens and mycotoxins in eggs and chicken feeds from farms to retail markets. *Korean Journal for Food Science of Animal Resources*, 36(4), 463–468. https://doi.org/10.5851/kosfa.2016.36.4.463

MAFF (2005). Comprehensive Guidelines for Measures against Salmonella in Hen Eggs (Notification No. 8441 by the Director of Health Management Division, Food Safety and Consumer Affairs Bureau, MAFF as of January 26, 2005). https://www.maff.go.jp/e/policies/ap_health/animal/attach/pdf/index-11.pdf

MHLW (1998). Hygiene management guidelines for egg sorting and packaging facilities (Ministry of Health and Welfare Notice No. 1674 (In Japanese). https://www.jz-tamago.co.jp/wp/wp-content/uploads/2020/03/E05_3_3.pdf

MHLW (2006). Residues of pesticides, animal feed additives and veterinary drugs in foods. https://www.mhlw.go.jp/stf/seisakunitsuite/bunya/kenkou_iryou/shokuhin/zanryu/index_00016.html

Mine, Y., et al. (2024). *Handbook of Egg Science and Technology*. CRC Press, Taylor & Francis Group, Boca Raton, FL.

Mingle, C.L., et al. (2021). Veterinary Drug Residues in Beef, Chicken, and Egg from Ghana. *Chemistry Africa*, 4, 339–348. https://doi.org/10.1007/s42250-020-00225-5

Overbey, K.N., et al. (2017). A systematic review of the use of Social Media for food safety risk communication. *Journal of Food Protection*, 80(9), 1537–1549. https://doi.org/10.4315/0362-028X.JFP-16-345

Piskorska-Pliszczynska, J., et al. (2019). Contaminants in eggs: dioxins/PCBs and other toxic substances and their possible health implications. In F.J.M. Smulders, et al. (Eds.), *ECVPH Food Safety Assurance*, Volume 7, pp. 415–441. Wageningen Academic, Wageningen. https://doi.org/10.3920/978-90-8686-877-3_17

Reich, H. & Triacchini, G.A. (2018). Occurrence of residues of fipronil and other acaricides in chicken eggs and poultry muscle/fat. *EFSA Journal*, 16(5), e05164. https://doi.org/10.2903/j.efsa.2018.5164

Ropkins, K. & Beck, A.J. (2000). Evaluation of worldwide approaches to the use of HACCP to control food safety. *Trends in Food Science & Technology*, 11(1), 10–21. https://doi.org/10.1016/S0924-2244(00)00036-4.

Salvatore, S. & Vericat Ferrer, M. (2024). Report for 2022 on the results from the monitoring of veterinary medicinal product residues and other substances in live animals and animal products. *EFSA Supporting Publications*, 21(3), EN-8669 101 pp. https://doi.org/10.2903/sp.efsa.2024.EN-8669

Samad, A., et al. (2023). Intake of toxic metals through dietary eggs consumption and its potential health risk assessment on the peoples of the capital city Dhaka, Bangladesh. *Arabian Journal of Chemistry*, 16(10), 105104. https://doi.org/10.1016/j.arabjc.2023.105104

SENASA (2016). Resolución-86-2016-Ministerio de Agroindustria. Monitoring and control of Salmonella spp. on commercial Poultry farms (In Spanish). https://www.senasa.gob.ar/normativas/resolucion-86-2016-ministerio-de-agroindustria

SENASA (2018). Manual of procedures for Senasa Resolution N° 86/2016 (In Spanish). https://www.argentina.gob.ar/sites/default/files/2._manual_de_procedimientos_operativos_vigilancia_y_control_de_salmonella_spp._en_granjas_avicolas_comerciales_-res._senasa_ndeg_86.2016._version_2018_0.pdf

Sin, JE.V., et al. (2023). Surveillance of veterinary drug residues in food commonly consumed in Singapore and assessment of dietary exposure. *Heliyon*, 9(11), e21160. https://doi.org/10.1016/j.heliyon.2023.e21160

Singer, R.S. (2023). Estimates of on-farm antimicrobial usage in egg production in the United States, 2016–2021. *Frontiers in Veterinary Science*, 10, 1135377. https://doi.org/10.3389/fvets.2023.1135377

South Africa (2020). Regulations regarding the grading, packing and marking of eggs intended for sales in the Republic of South Africa, No.R. 345, Government Gazette. https://www.gov.za/sites/default/files/gcis_document/202003/43108rg11055gon345.pdf

Tomczyk, Ł., et al. (2018). Characterisation of the Mycobiota on the shell surface of table eggs acquired from different egg-laying hen breeding systems. *Toxins*, 10(7), 293. https://doi.org/10.3390/toxins10070293

Tomczyk, Ł., et al. (2019). The influence of the conditions of acquisition and storage of table eggs on changes in their quality and the presence of mycobiota and Fusarium mycotoxins. *Poultry Science*, 98(7), 2964–2971. https://doi.org/10.3382/ps/pez156

USDA (2000). Egg grading manual. https://www.ams.usda.gov/sites/default/files/media/Egg%20Grading%20Manual.pdf

USDA (2024, January 16). Food Safety Research Information Office (FSRIO). https://www.nal.usda.gov/programs/fsrio

Van der Fels-Klerx, H.J., et al. (2017). Chemical and physical hazards in the egg production chain in the Nertherlands. https://doi.org/10.18174/387371

Wales, A. & Davies, R. (2020) Review of hatchery transmission of bacteria with focus on Salmonella, chick pathogens and antimicrobial resistance. *World's Poultry Science Journal*, 76(3), 517–536. https://doi.org/10.1080/00439339.2020.1789533

Wang, L., et al. (2018). Occurrence and quantitative risk assessment of twelve mycotoxins in eggs and chicken tissues in China. *Toxins*, 10(11), 477. http://dx.doi.org/10.3390/toxins10110477

Wang, R., et al. (2022). Detection of fluoroquinolone and sulfonamide residues in poultry eggs in Kunming city, southwest China. *Poultry Science*, 101(6), 101892. https://doi.org/10.1016/j.psj.2022.101892

Chapter 4

On-farm monitoring of layer flock health using sensors and modern data analytics

Francesco Bordignon and Andrea Pezzuolo, University of Padova, Italy

1 Introduction

Animal welfare and health are pivotal to meeting ethical standards and consumer expectations and assuring the sustainability of commercial poultry production. Over recent decades, the egg industry in Europe and North America has undergone significant transformation, shifting from conventional cage housing towards cage-free alternatives, with aviary systems offering hens more space and freedom to express their natural behaviours. However, the three-dimensional (3D) structure of aviary systems adds complexity, requiring a high level of management. Simultaneously, average flock sizes have increased to improve production efficiency. Together, these changes pose significant challenges to daily farm management, including the monitoring and maintenance of flock health and welfare (van Veen et al., 2023).

http://dx.doi.org/10.19103/AS.2024.0150.06

Advancements in sensor technology and data modelling offer promising avenues for continuous, real-time monitoring of animal health and welfare. However, despite the emergence of technological applications suitable for commercial layer barns, their practical integration at farm level remains limited. This chapter reviews the latest technological and analytical advancements applicable to monitoring the health and welfare of layer flocks and explore the current issues and future perspectives on the use of sensors and advanced data analytics.

2 Sensors: state-of-the-art and current challenges

Sensor technology has advanced significantly in recent years, with notable improvements in accuracy, diversity and affordability. Wireless sensors, in particular, have found extensive use in fields such as civil engineering, environmental monitoring, emergency response and agriculture (Sassi et al., 2016). Although sensor utilisation in agriculture is a more recent development, early applications have focused on animal health and reducing the cost of management practices. In poultry farming, the current trend towards managing large flocks has driven the development and use of automatic monitoring systems and sensing technologies. These systems can complement or even replace human observations (Ramírez-Morales et al., 2017).

As sensing technologies become more economically viable and user-friendly, research has focused on exploring potential applications for assessing, managing and enhancing poultry health and welfare. Interest in this area is projected to grow over time as advancements continue. The main advantages of using sensors lie in the earlier detection of deviations from optimal conditions within poultry houses, enabling timely adjustments to management practices. In addition, sensors have significant potential to enhance our knowledge and understanding of species-specific behaviours. For health monitoring, sensors provide continuous insights into flock status, facilitating routine monitoring procedures by veterinarians.

Previous surveys involving various stakeholders, including hen farmers, poultry experts and veterinarians (Olejnik et al., 2022; van Veen et al., 2023), and recent research reports (Xin and Liu, 2017) have identified five primary areas in which sensor technologies could assist farm management. These are:

- housing and microclimate control;
- the monitoring of weight and feed intake;
- tracking locomotion and activity;
- behavioural assessments; and
- stress and disease detection.

The following sections provide a thorough overview of sensors currently being used to monitor the health and welfare of laying hen flocks. We emphasise both the advantages and critical concerns associated with various applications. Our ultimate aim is to offer valuable insights to contribute to management advancements, geared towards predictive modelling and monitoring animal health and welfare.

3 Monitoring environmental conditions

3.1 Direct monitoring approaches

Climate control is often considered a blind spot in the management of layer barns. In a recent survey of various stakeholders from the egg sector, just 20% of farmers and 33% of veterinarians prioritised climatic conditions as a key indicator of hen health and welfare (van Veen et al., 2023). In addition, climate monitoring is still primarily performed through farmers physically feeling the temperature, humidity and ventilation levels (van Veen et al., 2023). Inadequate environmental conditions and its duration of exposure significantly impact poultry health, welfare and performance (Dawkins et al., 2004; Sassi et al., 2016).

Sensors offer a potential solution, although the application of sensing technologies to monitor the environmental conditions of laying hens remains inadequate and requires future improvements, with several issues discussed below.

Current available sensors for monitoring poultry houses can be broadly classified into two categories (Qi et al., 2023):

- Those related to the climate and thermal environment (temperature, humidity and wind); and
- Those related to air quality, measuring the concentration of harmful gases such as H_2S, CO_2 and NH_3, as well as microorganisms and particulate matter.

For the thermal environment, sensors include anemometers and thermometers. Sensors monitoring gases include electrochemical sensors or spectroscopy-based sensors such as non-dispersive infrared sensors for CO_2 monitoring (Mendes et al., 2015) or tunable diode laser absorption spectroscopy for monitoring NH_3 concentrations (Tan et al., 2020).

Due to field limitations, recent studies have primarily focused on improving the portability and stability of the equipment (Qi et al., 2023). Factors that can affect the stability and efficacy of measurements include dust, the acidic environment generated by animal manure and degradation of sensor components and connections. To mitigate the challenges associated with measurements and reduce the need for material and manpower without sacrificing accuracy, wireless sensor technology has been proposed. Wireless sensor networks, for example, could find extensive application in agricultural contexts (Qi et al., 2023), including the monitoring of poultry farms.

In hen barns, temperature is often assessed based on dry-bulb air temperature alone, overlooking crucial factors such as air velocity, relative humidity and radiation. However, these variables significantly influence the birds' thermal comfort, well-being and production efficiency (Xin and Liu, 2017). To address this issue, multi-sensing systems (Fig. 1) (Bustamante et al., 2012) are used to concurrently measure and integrate data on environmental temperature, air velocity and indoor atmospheric pressure using a variety of sensors installed in poultry houses. These systems are also valuable for evaluating the design and performance of the ventilation system as well as validating and verifying computational fluid dynamic simulations. In recent years, several multi-sensor systems have shown great potential in cage-free poultry production. These sensors record data that can later be analysed to provide insights into environmental conditions as well as disease occurrence,

Figure 1 Example of a multi-sensing system from Bustamante et al. (2012) with two air velocity and temperature sensors placed at different heights (bird-level and at 1.75 m).

mortality and abnormal behaviours – all useful data to support farmers' decision-making (Ren et al., 2020).

The regulation of ventilation rates, airflow, ventilation modes (e.g. tunnel or cross ventilation) and interventions for heating or cooling are typically managed by an environmental controller. This controller adjusts environmental conditions often based solely on an average value obtained from a limited number of sensors, which may be inadequately positioned to capture comprehensive data on the birds' microenvironment. For instance, temperatures at hen level can vary by as much as 3°C compared to temperatures measured by sensors in other areas of the henhouse (Corkery et al., 2013). Overall, current monitoring systems might not comprehensively assess animal well-being, highlighting the need for further advancements and targeted applications of sensor technology in the future.

To address these challenges, the strategic placement of sensors inside the henhouse and an increase in the number of sensors within the barn are crucial for accurately monitoring bird well-being. In addition, positioning sensors at bird-level is essential for precise data collection (Corkery et al., 2013). To tackle potential temperature and humidity variations in the barn, the installation of several sensors at varying heights is suggested. However, this approach might elevate both the cost and complexity of the monitoring system (Hu et al., 2022), making optimisation of sensor placement a key factor for data management.

A recent study introduced a method that combined a genetic algorithm using dual-structure coding with the grey correlation degree technique (Hu et al., 2022). The method aimed to optimise both the quantity and positioning of temperature sensors within the henhouse. The primary objectives of this optimisation process included defining multi-layer cage system structures for laying hens, establishing initial sensor placement models, adjusting sensor quantity, determining specific sensor locations and evaluating the most effective optimisation methods (Hu et al., 2022). By calculating the correlation coefficient of temperature data from each sensor, redundant sensors were able to be removed and the number of temperature sensors decreased by 84%, with a subsequent reduction in workload. By applying a genetic algorithm of dual-structure coding and a grey correlational degree method, the number of temperature sensors further decreased by 77% (Hu et al., 2022).

3.2 Indirect monitoring approaches

Given the complexity of monitoring microclimate conditions at bird-level, research to date has focused on employing the bird as the ultimate indicator, as they incorporate all physical, health and nutritional factors (Xin and Liu, 2017). The output of animal-based recordings could include physiological measures (e.g. body temperature), behavioural changes (e.g. alterations in resting patterns) and performance measurements (e.g. reduced feed intake or egg production). Changes in these variables can indicate an unsuitable rearing environment (Xin and Liu, 2017). For example, the monitoring of a bird's facial surface temperature through infrared (IR) thermal imaging is considered a more informative indicator of a hen's welfare than measuring the ambient temperature (Giloh et al., 2012), with potential applications at a commercial scale to identify thermal stress in hens.

Knowledge of how hen physiology or behaviour responds to environmental changes remains limited, especially in real-time monitoring applications. To address this gap, image analysis and radio frequency identification (RFID) systems are used to study hen environmental preferences and behavioural changes in controlled environments (Sales et al., 2015). Correct animal identification, localisation, and time spent in that particular location are essential for assessing environmental preferences. These insights can then be amalgamated into a more comprehensive monitoring protocol implemented within poultry houses, enabling the identification of environmental conditions, including temperature and air pollution and related physiological and behavioural changes such as panic, fright and vigilance. Furthermore, sound analysis has also been proposed to conduct real-time assessments of poultry house environments by examining and categorising the sound of laying hens using a model implemented through a support vector machine (Na et al., 2019).

Environmental management in layer barns through the use of sensing technologies is often overlooked. Current practices rely on subjective evaluations, neglecting crucial factors such as temperature, humidity and ventilation. Existing sensor technologies struggle to assess the birds' microclimate accurately, although advances in multi-sensing systems, strategic sensor placement and innovative optimisation methods show great potential. Techniques such as image analysis, RFID systems and sound analysis offer opportunities to study environmental preferences and behaviours. Moreover, bridging real-time monitoring with hen responses to changes in their environment is crucial for future research.

4 Monitoring of weight and feed intake

Farmers and technicians in the egg sector rely heavily on feed and water intake as crucial indicators for identifying potential health and welfare issues in laying hen flocks (van Veen et al., 2023). Despite the attention given to monitoring these factors, current technologies have limited capabilities, which will be discussed in this section. For example, sound-based technologies are affected by environmental noise, especially when used in large bird flocks, and the efficacy of computer vision-based weighing systems may be influenced by animal genotype or growth stage.

Nevertheless, 93% of stakeholders highlighted the importance of continuous, sensor-based monitoring of production parameters, such as weight monitoring, for evaluating the health and welfare of laying hens (van Veen et al., 2023). This section explores recent advances in using acoustic sensors and cameras as promising tools to assess laying hen health and welfare through weight and feed intake measurements.

4.1 Acoustic sensors

Studies have successfully linked pecking sounds with feed intake, demonstrating high accuracy rates in broiler chickens, although there is no information available for laying hens to date. Aydin et al. (2015) were able to identify sound signals related to pecking (a sudden increase followed by a sudden decrease in amplitude) from other signals within the range 1000 Hz to 5000 kHz using a sixth-order Butterworth filter alongside sound analyses. Pecking sounds accurately identified 90% of feed intake of broiler hens within an experimental pen. Another study (Aydin et al., 2016) reported an 86% accurate detection of feed intake among broilers of various ages.

Further real-time analysis of short-term feeding behaviours in small groups of broilers (10 animals) using pecking sounds revealed promising accuracy rates – meal size was estimated at an accuracy of 90%, meal duration at 95%, daily meal count at 84% and feeding rate at 89% (Aydin and Berckmans, 2016). However, these observations were limited to small experimental groups of up to ten chickens. The practical application of these findings in commercial farms requires challenges such as filtering environmental noise and ensuring accurate detection to be addressed, especially when dealing with larger bird populations.

Monitoring hen weight offers insights into the flock's growth and feed efficiency, helping detect any deviations from the expected growth pattern among the birds. Researchers used microphones coupled with Adobe Audition™ CS6 to analyse peak frequency vocalisations emitted by chicken flocks to predict their growth, revealing an inverse relationship between these vocalisations and the age and weight of the birds (Fontana et al., 2015). Sound recording and analysis approaches could function as an early warning or continuous monitoring system, offering insight into the overall condition of the layer flock. In addition, the relationship between bird weight and sound frequency could lead to the implementation of automated tools using vocalisation to predict bird weight and growth patterns, helping farmers to monitor bird growth trends automatically. However, further research is needed in the case of laying hens and larger groups of animals. Researchers have highlighted the potential of audio and video data monitoring as a promising technique for establishing an automated growth-monitoring tool for laying hens.

4.2 Computer vision

Advances in computer vision are allowing the use of non-contact methods to estimate the weight of live animals indirectly, highlighting potential applications across various species including pigs, cows and sheep (Liu et al., 2021). In poultry, previous studies have used computer vision technologies to estimate body weight, specifically through body surface analysis in experimental conditions (De Wet et al., 2003) or through a 3D camera-based system that assesses broilers' weight (Mortensen et al., 2016).

Although camera-based systems have the advantage of weighing multiple birds simultaneously, their accuracy is often influenced by available datasets about animals and their growth performance, which vary according to genotype and production system. Moreover, accuracy can be affected by other factors, such as the ability to isolate individuals within a densely packed flock of animals with the same appearance (e.g. size, feather colour, etc.) and the potential distortion of geometric features due to changes in body posture (e.g. crowding, flapping wings) (Liu et al., 2021). Despite several ongoing research attempts, precision scales remain the preferred approach for farmers to assess bird weight.

5 Stress and disease detection

The health and productivity of laying hens are significantly affected by various stressors, with heat stress emerging as the key factor. This causes physiological discomfort, reduced performance and laying rate, as well as poor welfare and health issues, ultimately contributing to significant mortality rates (Lee et al., 2015; Saeed et al., 2019).

Rapid detection and management of stress are essential to minimising adverse impacts on health, productivity and overall well-being. Early intervention, facilitated by non-invasive, continuous monitoring tools, can effectively mitigate stress and enhance living conditions for laying hens. In the egg sector, recent advances in on-farm detection technologies offer promising opportunities for proactive stress management in laying hens. Among these, thermal imaging and sound detection have emerged as effective tools for monitoring stress levels and overall health of hens in large commercial flocks.

5.1 Infrared thermography

Infrared thermography (IRT) has emerged as a promising technology for assessing the thermal status of laying hens. Thermal imaging visualises the hens' surface temperature distribution by capturing and translating IR radiation emitted by the animals into electrical signals (Nääs et al., 2014). This non-invasive and real-time tool enables the assessment of various welfare aspects linked to thermoregulation. Bird facial temperature is typically targeted, as the surface temperature of the feather-covered body of the hens tends to reflect the ambient air temperature, regardless of the bird's thermal condition (Souza-Junior et al., 2019).

Technical parameters such as distance, image quality and camera height, along with environmental conditions such as temperature, humidity and gas levels, as well as the presence of dust, all influence the accuracy of IR detection technology (Yin et al., 2023). These challenges highlight the need for proper calibration, maintenance and consideration of environmental variables to ensure the effectiveness of IRT in hen flock monitoring. Experimental trials have successfully employed IR thermal imaging to assess various parameters associated with laying hen welfare. Several studies have utilised thermal imaging to monitor birds' thermal status, including pullet cloacal and body temperature, under various environmental conditions (Cândido et al., 2020). Moreover, the technology has been instrumental in identifying facial surface temperature variations among laying hens subjected to diverse thermal conditions (Kim et al., 2021). IRT technology has been employed to evaluate heat and cold stress, highlighting its versatility in assessing thermoregulation in laying hens (Alves et al., 2012).

IR thermal imaging extends beyond assessing thermal stress. IRT has been tested to detect signs of disease and distress within poultry production, enabling early identification of health issues among laying hens (Nääs et al., 2014). IRT is also being trialled as a potential tool for early identification and quantification of plumage damage in laying hens (Schreiter et al., 2022) and can detect featherless plumage areas at higher temperatures (Zhao et al., 2013). For instance, IRT can be used to obtain variables such as the difference between body surface temperature and the environmental temperature, as well as the proportion of featherless areas, calculated as the proportion of pixels corresponding to a temperature above 33.5°C (Pichová et al., 2017), which is above the average temperature range of feather-covered areas in chickens (24–33°C) (Cangar et al., 2008).

Recently, 3D reconstruction pipelines have been developed that integrate colour, depth and thermal information obtained through IRT to comprehensively monitor feather damage in poultry (Zhang et al., 2022). IRT has promising application potential in the detection of plumage damage, as it is more objective, accurate and repeatable compared with traditional scoring (Cook et al., 2006). However, further work is needed as the accuracy of detection varies between body regions and hen genotypes, being low for dorsal neck plumage but high for belly and back areas (Schreiter et al., 2022). Moreover, further studies are required to explore the potential implementation of plumage detection with IRT in high-density, large commercial flocks (Fig. 2).

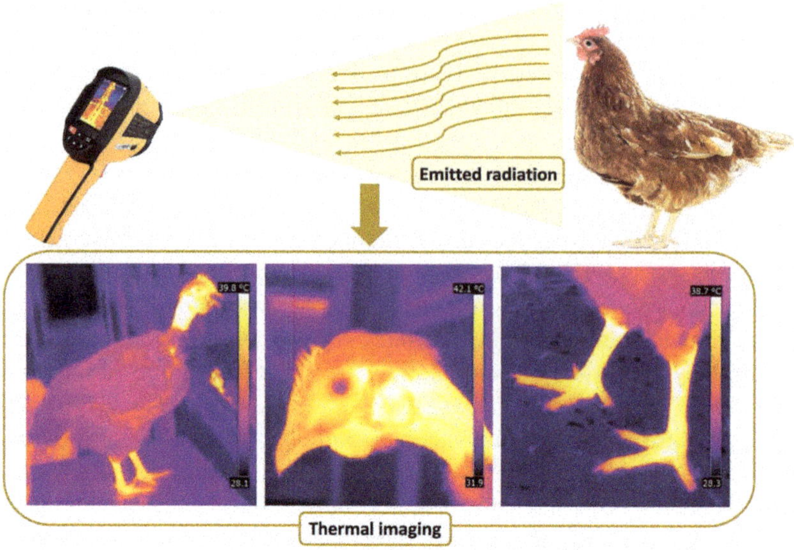

Figure 2 Thermal imaging applications in laying hens to monitor body surface temperatures. Source: Adapted from: Souza-Junior et al. (2019).

6 Tracking hen movement, localisation and behaviour

With the ban on traditional cages in the EU and the global shift towards free-range egg production, understanding how hen movement impacts egg production has become crucial. Cage-free laying hens can access a wide range of structures and spaces, including open floor areas, feeding areas, water sources, nesting boxes, perches, aviary levels, winter gardens and outdoor ranges.

In terms of health, cage-free systems offer advantages such as increased physical activity, which leads to improved leg bone health and a reduced risk of osteoporosis and mobility issues (Aguado et al., 2015). In aviary systems, uncontrolled movements increase the chances of collisions and failed landings (Ciarelli et al., 2023), often causing keel-bone damage observed at rates between 20% and 83% (Ruhnke et al., 2019).

Tracking technologies generate chronological records of location data, enabling non-invasive, long-term monitoring of the movements of multiple individuals. These technologies have become increasingly key to the early detection of health issues in farm animals, including laying hens (Montalcini et al., 2022). In cage-free poultry systems, in particular, tracking technologies have potential to identify characteristics that support animal welfare, while also providing key insights for effective flock management (Ellen et al., 2019; Li et al., 2017). Understanding individual hen movement, the influence of hen activity on specific groups within the flock, animal distribution within the aviary (Ciarelli et al., 2023) and the use of specific areas such as the floor (Sozzi et al., 2022), nests (Chien et al., 2018) or perching areas (Wang et al., 2019) is essential for monitoring large flocks in modern hen barns.

6.1 Computer vision

Computer vision technologies have found various applications in monitoring layer flocks. These include automatic counting of hens in battery cages on commercial farms to detect mortality and estimate animal density (Geffen et al., 2020), nest occupancy detection (Zaninelli et al., 2018) and the recognition (Leroy et al., 2005) and quantification (Nakarmi et al., 2014) of specific behavioural patterns. In cage-free systems, computer vision has been used to monitor pecking behaviour and related damage (Subedi et al., 2023), as well as to assess environmental preferences (Kashiha et al., 2014), such as responses to different light colours and temperature conditions (Kodaira et al., 2023). Images and videos captured through computer vision are analysed using a range of image processing tools, software and algorithms, which will be discussed later in the chapter.

6.2 Infrared technology

Sensors that are capable of capturing digital IR images could offer an alternative to help address some of the current limitations of computer vision systems. Although IR technology is primarily employed in the evaluation of bird stress and welfare (see Section 2), IR can also be applied to monitor hen localisation in specific areas, showing high sensitivity (98%) and specificity (95%) (Zaninelli et al., 2016). For instance, IR sensors can detect the presence or absence of hens to facilitate automated sanitary treatments in specific housing areas, minimising toxicity risks for both birds and workers (Zaninelli et al., 2016, 2017).

6.3 Accelerometers

Accelerometers are sensors that estimate acceleration along various axes, capturing the acceleration related to the animal's movement and gravitational acceleration. Their ability to remotely detect behavioural patterns using movement data has great potential for monitoring applications in laying hens, as observed in previous studies focusing on detecting specific hen behaviours using inertial sensors (Banerjee et al., 2012; Derakhshani et al., 2022; Fujinami et al., 2023; Nie et al., 2022) or generally describing bird activity levels (Quwaider et al., 2010).

Using body-mounted sensors in laying hens does, however, pose some challenges. The weight and size of the sensor installed on the bird's body should not induce significant changes in its behaviour and performance. The tracking device (e.g. accelerometers, tags for radio frequency identification) should weigh less than 5% of the animal's body weight (Siegford et al., 2016). Hens' smaller size and unique movement patterns result in less acceleration during their motion than larger animals. This leads to a narrower range of accelerometer values collected. Moreover, hens' quick and erratic movements create a weaker connection between the accelerometer data and their actual activity, requiring specialised analysis protocols. To identify specific behavioural activities, extracting features from hen activity data demands specialised machine learning methods designed specifically for the purpose (Siegford et al., 2016), which will be further discussed in this chapter.

6.4 Radiofrequency identification and radio signal strength

Free-range systems typically house massive groups of nearly identical hens, making individual observation by traditional monitoring almost impossible. However, health and welfare are individual characteristics and states such as suffering, pain, hunger and thirst are experienced at individual level. Moreover, these individual conditions can serve as initial warning signs to summon prompt action before they affect the entire flock. Methodologies that allow for continuous identification, tracking and study of individual hens in these large groups and complex environments are therefore essential. In this framework, radio frequency identification and radio signal strength (RSS) technologies have potential to monitor individual and flock health and welfare.

Radiofrequency identification (RFID) technology relies on passive transponders or tags that can be fastened to hens' legs or wings. When an antenna activates an external electromagnetic field, these tags emit a unique identification number. The antenna, connected to a dedicated reader, operates at 134.2 kHz or 125 kHz. The strategic placement of the antenna within the housing system facilitates the identification and tracking of individual hens in specific areas. Moreover, by employing multiple antennas, the monitoring system can extend its reach, controlling a larger number of locations. RFID tags are available in two types – active and passive. Active tags have their own power source, enabling an extended detection range, although they have a limited operational lifespan. Conversely, passive tags draw power from the antenna and have virtually limitless lifespans, although their detection range is more limited. When activated, passive tags transmit their unique identity to the reader. These tags are fastened to the monitored animal externally, typically as leg bands in laying hens.

Technology utilising RSS relies on actively transmitted radio frequency sensors equipped with RSS (>900 MHz). Essentially, this system operates by utilising devices that emit signals attached to birds via specialised harnesses

(Daigle et al., 2012). Receiving units pick up these signals, which measure the signal strength. This measurement helps estimate the approximate location of a hen.

The use of RFID and RSS technologies in laying hens has highlighted various strengths and weaknesses across different tested applications, which have been thoroughly reviewed by Siegford et al. (2016) and are summarised and briefly discussed in Table 1. With RSS technology, specific data analysis methods have been proposed to monitor the localisation of cage-free laying hens (Siegford et al., 2016). Moreover, this technology has been applied alongside accelerometers to remotely identify various behaviours (Daigle et al., 2012). Its strengths lie in precise location tracking and behavioural monitoring. However, it is imprecise for large areas due to continuous data transfer, restricted detection areas causing locational confusion, power source constraints and logistical challenges in sensor attachment without disturbing hen behaviour or signal disruptions.

Table 1 Strengths, weaknesses and applications of radio frequency identification and radio signal strength systems used in laying hen monitoring

	Radiofrequency identification (RFID)	Radio signal strength (RSS)
Key differences	Detects the presence or absence of hens at the location of the antennae	Detects movements of hens through time, space and location
Strengths	Potential coupling with other systems	Potential to be coupled with several other observations such as environmental conditions (temperature, light, humidity) and accelerometery
	Multiple individual detection at the same time	
	No need for battery or external power supply	
	Easy installation	
	Sensors can be fixed at several points of hen body (legs or wings)	
Weaknesses	Metal or water can reduce signal strength	Stationary receivers > 1 metre apart
	Fast-moving animals (like laying hens) can be difficult to detect	Potential overlapping detection fields
	Requires multiple antennae to detect the direction of movement	Short battery life
	Expensive	Metal can impair detection with consequent data loss
		For best signal capture, sensors must be placed on hen's back.
Current applications	Monitor egg laying in the nest (location of hen within a defined area)	Detect individual movement and behaviour in large groups considering the three dimensions
	Detect hen passage through a pop hole	Detect location relative to other sensors (mobile and stationary)

Source: Adapted from: Siegford et al. (2016).

In recent years, studies have focused on the use of RFID technology in group-housed hens to monitor individual feeding (Li et al., 2017) and nesting behaviour (Chien et al., 2018; Li et al., 2017), perching behaviour (Wang et al., 2019), ranging behaviour in cage-free systems (Campbell et al., 2017, 2018; Larsen et al., 2017) and environmental preferences (Sales et al., 2015). Although tracking systems have been tested in commercial laying hen barns (Gebhardt-Henrich et al., 2014; Rufener et al., 2018, 2019; Sibanda et al., 2020), their application in complex, modern multi-tier aviary systems presents various challenges, including accurately monitoring individual activity, reaching specific zones and managing the effect of varying environmental conditions.

Studies to date have primarily focused on assessing ranging behaviour by placing antennas at pop holes separating specific areas of the barn or outdoor spaces. The monitoring of hen utilisation of outdoor spaces such as verandas and free-range areas was performed through a low-frequency (125 kHz) RFID system monitoring 12 different flocks and by tracking 5-10% of the birds per flock (Gebhardt-Henrich et al., 2014). Antennas were placed on the sides of pop holes of various widths and up to 12 antennas were placed side-by-side to cover the full width. This antenna setup helped determine the hens' movement direction between areas (indoor vs outdoor), recording both entry and exit from the pop hole (Gebhardt-Henrich et al., 2014). Using RFID antennas on both sides of the pop holes allowed comprehensive and accurate tracking.

However, this approach requires a significant number of antennas and associated equipment, which increases acquisition costs. Placing just one antenna within pop holes reduces costs but limits directional information, recording only hen's presence. To address this, another study (Thurner et al., 2009) employed a high-frequency (13.54 MHz) RFID technology in a 'wide electronic pop hole' – a tunnel-like structure with adjustable width and two antennas to accommodate various hole sizes to track hen movement direction with high reliability (>97%). Several challenges persist, such as the prolonged presence of birds near the antennas, difficulty registering rapidly moving transponders (>1.5 m/s) (Gebhardt-Henrich et al., 2014), characteristics of the pop holes (i.e. presence of ramps), as well as issues relating to costs, directional information and operational reliability.

7 Data analytics to monitor hen health and welfare

As sensor technology improves and becomes more affordable, extensive datasets are collected to monitor environmental conditions, animal behaviour and health. These datasets are often broad and complex, posing traditional analysis challenges. Conventional human observation and statistics of hen behaviour might be prone to bias due to farmer experience, can be time-intensive and may overlook detections (Wang et al., 2023).

In this context, machine learning algorithms have emerged and found applications in livestock monitoring. Machine learning is an artificial intelligence field capable of handling intricate datasets with unknown relationships among measured variables, allowing future predictions or classifications. Various machine learning techniques, including decision trees, random forest, support vector machine and artificial neural networks have been employed in evaluating poultry welfare and health (Banakar et al., 2016; Zhuang et al., 2018; Pu et al., 2018; Pitesky et al., 2020), demonstrating potential for the establishment of early warning systems in poultry barns (Okinda et al., 2020). Furthermore, the transition from machine learning to deep learning applications represents an evolution in the field of artificial intelligence and animal monitoring. Deep learning is a type of machine learning that uses artificial neural networks to learn from data and are faster, more accurate and more robust than machine learning algorithms (Hao et al., 2022).

Among the array of deep learning techniques, the convolutional neural network (CNN) is one of the most widely adopted methods to analyse, combine and extract colour, geometric and textural data from images. Within the field of animal science, CNNs and similar deep learning techniques have increasingly been employed to analyse a diverse range of data related to environmental conditions, animal welfare, health and behaviour. These methods have demonstrated remarkable effectiveness in various applications, including identifying signs of injury, disease or stress by analysing images or sounds of animals, finding various applications in the field of hen monitoring. In addition, the '*You Only Look Once*' (YOLO) algorithm has attracted considerable attention in computer vision for its exceptional accuracy and speed in object detection (Yang et al., 2023).

8 Monitoring environmental conditions using computational fluid dynamics

The limitations of traditional methods for detecting and predicting environmental changes, combined with the challenge of incorporating multiple interacting factors into mathematical models, have led to the rise of modern data analytics and simulations for assessing environmental conditions in livestock settings. This modern approach combines real-time environmental data with theoretical models that account for various influencing factors. As a result, computer technologies such as computational fluid dynamics (CFD) and machine learning have become essential in enhancing the detection and management of environmental conditions within livestock systems (Qi et al., 2023). While most studies on laying hens have focused on employing CFD as a modelling tool to enhance ventilation design and management, utilising deep learning algorithms for monitoring environmental data remains largely unexplored.

CFD is a powerful tool that utilises computers and applied mathematics to simulate fluid flow in diverse scenarios, providing accurate predictions across time and space while accounting for heat and mass transport under turbulent conditions. This capability is critical because CFD can predict internal air quality, serving as a tool to continuously fine-tune ventilation to minimise emitted gases (Rojano et al., 2019). Initially, CFD applications in livestock systems focused on two-dimensional or simplified 3D models, often overlooking the influence of animals in simulation models. However, the presence of animals can influence fluid dynamics, and modelling the animal-occupied zone presents challenges due to the presence of so many animals.

Early studies on cage-reared laying hens employed CFD methods to assess how hen geometry, distribution and weight affected flow resistance (Cheng et al., 2018). These studies aimed to evaluate ventilation system performance by creating a 3D CFD model based on the actual dimensions of a laying hen house. These simulation results were then compared with field measurements of air velocity, temperature and relative humidity at 30 positions inside the building (Du et al., 2019). The findings revealed that the CFD simulation aligned closely with the experimental findings. For example, air temperatures predicted by the CFD model at the designated positions were almost identical to the experimental measurements. In terms of air velocity, the differences between measured and predicted values were less than 10%, decreasing to 5% in half of the points considered. For relative humidity, the relative errors were generally under 10%, decreasing to 5% in 18 of the 30 points. These results are significant, given the complexity of the flow. Overall, this CFD model proved to be a valuable and efficient tool for predicting the distribution and dynamic fluctuations of indoor environmental parameters in poultry houses (Du et al., 2019).

Recent studies in cage-free systems have also confirmed the effectiveness of CFD modelling in evaluating various ventilation schemes (Cheng et al., 2018) and assessing ventilation system performance at the individual bird level by incorporating animal-specific modelling (Chen et al., 2020). By using CFD simulations, researchers can evaluate the effectiveness of different ventilation schemes and make informed decisions to improve their design to ultimately improve animal health in cage-free environments.

9 Bioacoustic analysis as a welfare indicator

In recent years, advances in recording and analysis technologies have led to a growing focus on the field of bioacoustics, particularly the study of animal vocalisations (Lee et al., 2015). Vocalisations are reliable proxies of various animal responses and can be considered good welfare indicators (Sassi et al., 2016; Du et al., 2020). Specific hen vocalisations such as the gakel call, squawking and alarm calls have been linked to stressful events, highlighting potential use as stress indicators in monitoring systems for large flocks (Du et al., 2020). Previous studies have investigated vocalisation patterns in laying hens. Researchers used an automated online monitoring system to promptly identify stressful situations in real-time (Du et al., 2020). They utilised hen sounds and applied support vector machine techniques to distinguish sounds emitted by laying hens under physical stress, thermal stress and fear-induced mental stress, achieving an impressive accuracy rate (> 96%) (Du et al., 2020; Lee et al., 2015).

CNNs have been used recently to classify laying hens' sounds into eight classes, namely poly contact call, single alarm call, group alarm call, mono contact call, ordinary call, moan and threat call, high-intensity call and squawk call (Jung et al., 2021). Two types of CCN models – one based on 2D ConVnet and the other on a 1D model with long short-term memory – were developed and tested, and showed a satisfactory classification accuracy of 75.8%. Another study (Du et al., 2021) aimed to develop a sound recognition technique called 'tristimulus-formant' for continuous and automatic assessment of laying hen vocalisations at a low computational cost. Tristimulus is considered to be a timbre equivalent to the colour attributes of vision and measures the mixture of harmonics in a given sound, which are grouped into three sections according to the relative weights of the harmonics in the signal. Compared to other animal sound recognition techniques, the 'tristimulus-formant' approach showed significant potential for real-time identification and extensive data analysis. However, further research is needed to enhance our understanding of the correlation between animal behaviour recognition and animal sound recognition. For instance, a multi-modal approach integrating video and sound streaming technology could help delve deeper into this relationship.

10 Computer vision for monitoring behaviour

In recent years, computer vision has emerged as a promising non-invasive, cost-effective approach for monitoring farmed animals. Computer vision involves using mathematical principles, computer science methodologies and software programming to automatically monitor, record and identify animal behaviour, track movement, analyse welfare status and health conditions and predict weight. The computer vision process encompasses three main phases – observation, diagnostics and prediction. This analysis is generally run by machine learning or deep learning models tailored to extract object characteristics (e.g. the profile and physical characteristics of a hen) and forecast the intended category, ultimately assessing the precision of classification (Yang et al., 2022).

10.1 Image analysis for behaviour and health monitoring

Several studies have recently been carried out to apply machine learning algorithms to monitor behaviour, health and localisation through computer vision in laying hens. For instance, 2D image analysis through deep learning algorithms using CNNs have been tested to detect and classify a wide range of hen behaviours such as mating, standing, feeding, drinking, wing spreading and fighting (Wang et al., 2019; Yang et al., 2023) and to detect specific behaviours such as feeding (Hao et al., 2022) and drinking (Li et al., 2020).

After the introduction of YOLO, studies focused on its application to classify hen behaviours such as feeding, mating, standing and fighting (YOLO v3 – Wang et al., 2020; YOLO v4 – Wang et al., 2023) or to effectively recognise various behaviours in hens from 1 to 50 weeks of age, including feeding, drinking, walking, perching, dust bathing and nesting (Yang et al., 2023), or measuring comfort behaviours such as dust bathing (YOLO v4; Sozzi et al., 2022). YOLO (v5 and v6) has also been extensively applied to detect hen localisation and spatial distribution in the aviary and on the floor (Bist et al., 2023a; Guo et al. 2023; Li et al., 2023; Sozzi et al., 2022; Yang et al., 2022), to detect dead (Bist et al., 2023b) or sick (Wang et al., 2019) hens and to identify hen piling (Bist et al., 2023c). Interestingly, a specific YOLO algorithm (*YOLOv5x-hens*) was developed and tested to monitor hen behaviours in cage-free systems under varying light conditions, observational angles and hen ages, with an overall accuracy rate of 95% (Yang et al., 2022).

Overall, the performance of these algorithms is generally very high, showing an accuracy rate ranging from 60% to as high as 95.3% in behaviour classification and up to 99.17% in behaviour detection, depending on the specific behaviour or variable considered and the deep learning model used. On average, the segmentation accuracy of behaviours is about 87% (Yang et al., 2023).

Implementing deep learning algorithms linked to computer vision for analysing hen behaviour in field conditions can be challenging due to their complexity, often requiring expertise to manage them effectively. As an alternative, various software options for video and image analysis are available for farmers to monitor and evaluate broiler or hen behaviour. These include Pocket Observer Version 3 (Noldus Information Technology; Wageningen, The Netherlands) (Hunniford et al., 2017), eYeNamic™ system (Fancom BV, Panningen, The Netherlands) (Peña Fernández et al., 2018) and EthoVision XT 10 (Noldus, Leesburg, VA, USA) (Fraess et al., 2016).

Various challenges still need to be addressed to further optimise computer vision analysis in monitoring hen health. These challenges include technological issues related to environmental conditions and practical implementation challenges. In fact, computer vision systems based on visible light-based sensors remain susceptible to environmental conditions, including light intensity (Hao et al., 2022), dust, aviary layout and flock density (Yang et al., 2023), camera height (Bist et al., 2023a), the colour of floor area and hen feathers (Yang et al., 2023) and hen size (Yang et al., 2022). Furthermore, the real-time monitoring of birds is particularly challenging when identifying lameness-associated alterations and stress-related behaviours among rapid movements and illumination fluctuations.

10.2 Infrared thermography analysis and animal body reconstruction

As discussed above, numerous studies have utilised machine vision technology to monitor the health and behaviour of laying hens. However, these studies have primarily relied on 2D RGB imaging, typically susceptible to lighting variations. In this context, employing IRT might offer a feasible solution to counteract fluctuations in light intensity. Utilising IRT is crucial for monitoring feather damage in hens.

Enhancing this process involves creating 3D images of the hen's body, enabling a detailed analysis of the depth of the damage. A recent study (Zhang et al., 2022) proposed a 3D reconstruction pipeline for a comprehensive hen monitoring system, incorporating colour, depth and thermal data derived from optical and thermal/IR camera systems. The process involved isolating the hen image from its background through a deep learning-based network (adaptive aggregation method; AANet) using colour and depth images. Four image types – showing colour, depth, thermal and mask – were employed as inputs to create a 3D hen model utilising RGB–Depth–-Thermal maps. Results showed that this system outperformed the assessment of feather damage solely through 2D colour or thermal IR images. The 3D model allows for precise evaluation of the depth of chicken feather damage. This approach paves the way for automated and intelligent feather damage and weight monitoring in poultry farming.

Thermal images can also be analysed to predict heat stress in laying hens. Artificial neural networks and automatic linear modelling procedures have been proposed to analyse the data. In addition, the predictive ability of these algorithms to forecast heat stress index from the age of birds, housing system and various thermo-physiological traits such as pulse rate and respiration rate was analysed (Yakubu et al., 2018). Both artificial neural networks and automatic linear modelling had advantages over multiple linear regression when predicting heat stress index from hen respiration rate, offering an opportunity for continuous monitoring of the thermal comfort and overall wellbeing of the hens.

Thermal image analysis has found other applications, such as utilising deep learning methods to identify illness in laying hens (Li et al., 2021). In particular, a YOLOv3-based neural network structure was utilised to identify distinctive features of laying hens, with the primary objective of rapidly identifying specific features of hen bodies within a vast dataset. Following this, a comparative examination of the temperatures between healthy and sick laying hens was conducted to identify any potential pathological conditions present among the hens.

11 Analysing movement data from body-mounted sensors

11.1 Data from inertial sensors

Combining wearable inertia sensors with machine learning offers a practical approach to monitoring laying-hen behaviours. This approach could help support farmers to efficiently manage hens within cage-free housing systems (Derakhshani et al., 2022). While accelerometers are widely used, a consistent method for processing the data they collect is lacking, creating discrepancies that hinder result comparability across studies. Moreover, behavioural studies employing inertial sensors are frequently restricted to smaller-scale systems with limited animal samples due to practical constraints, potentially undermining their ability to accurately reflect behaviours observed in larger commercial systems. Nevertheless, in a study by Derakhshani et al. (2022), machine learning models demonstrated remarkable success in predicting three activity levels of laying hens (static, semi-dynamic and highly dynamic). Notably, the bagged tree model emerged as the top performer, showing an accuracy rate of 90%. These results highlight the proficiency of machine learning in accurately discerning diverse activity levels within laying hen behaviour.

While these analytic methods are becoming increasingly operable within low-density environments, employing them for individual hen tracking in today's cage-free housing systems remains a challenge. Nevertheless, the effectiveness of these methods is improving due to rapid advancements in hardware, such as lightweight sensors facilitating extensive data acquisition and powerful processors enabling deep network training, as well as software enhancements such as advanced frameworks for optimisation and parallel computations (Shahbazi et al., 2023). Inspired by the excellent performance of deep learning models in tasks such as image classification and object detection, numerous studies have explored the feasibility of transforming signals from inertial sensors into the format of an image. This technique, often called 'signal imaging' (Shahbazi et al., 2023), aims to render sensor data recognisable by widely used deep learning-based models employed in computer vision (see Section 7).

11.2 Data from radio frequency identification

Recognising 'routine' behaviour patterns is essential for effective flock management. Moreover, using temporal change detection algorithms to signal shifts in hen dynamics can enable timely interventions, preventing incidents like smothering by addressing uncoordinated mass movements (Barrett et al., 2014).

Employing individual tracking technology (RFID) and agglomerative and K-means cluster analysis allows for identifying flock sub-population patterns (Sibanda et al., 2020). This automated classification method could significantly help farmers in decision-making processes, optimising egg production by considering the behaviour and specific needs of various flock sub-populations (clusters), rather than relying on conventional information based on monitoring the flock's average performance. For instance, by utilising unsupervised learning and machine learning algorithms alongside data visualisation, Sibanda et al. (2020) identified specific hen clusters associated with time budgets. Notably, hens displaying more consistent time spent at each location tended to stay longer than hens with erratic movement patterns.

RFID technology and machine learning have also been tested to measure the average durations spent within the aviary, aiming to predict individual hen health status (Welch et al., 2023). A random forest-based machine learning workflow showed promising but moderate performance, correctly predicting an average of 28% of cases for Spotty Liver Disease, 33% for roundworm and 33% for tapeworm infections by the end of the production cycle. These results could form the basis of future research in this area and guide the development of more precise and effective health monitoring methods for laying hens.

12 Conclusion and future trends

The monitoring of farm animals has traditionally relied heavily on farmers' observations, judgement and experience. Nowadays, the global consolidation of egg production poses fresh challenges, making it increasingly difficult for farmers to maintain an adequate level of detailed monitoring and management. Technological advancements have made it feasible for cameras, microphones and sensors to complement farmers' observations and oversight within animal housing systems. This technology offers significant advantages by enabling continuous monitoring of animals and improving flock management by providing comprehensive and continuous data.

Further optimisation is required, particularly in regard to the frequency of data recording by sensors and tools. Although this can help provide significant quantities of raw data, processes such as saving, sending, analysing and modelling the data require energy and adequate computing capacity. Performance data are often not available at the farm level.

The availability of real-time information to support farmer decisions is also a strategic advantage for the sustainability of these technologies. This represents an important tool for assessing the quality of the data collection and analysis processes to provide the farmer with useful information and correct operational timing.

In conclusion, current technological developments allow important data to be obtained in various livestock areas, but future objectives must consider the quality and sustainability of the information (Pezzuolo et al., 2021) offered to the farmer through dedicated digital footprint systems.

13 References

Aguado, E., et al. (2015) 'Bone mass and bone quality are altered by hypoactivity in the chicken', *PLoS One*, 10(1), e0116763. https://doi.org/10.1371/journal.pone.0116763.

Alves, F., et al. (2012) 'Impact of exposure to cold on layer production', *Revista Brasileira de Ciência Avícola*, 14(3), 223–226. https://doi.org/10.1590/S1516-635X2012000300010.

Aydin, A., et al. (2015) 'A real-time monitoring tool to automatically measure the feed intakes of multiple broiler chickens by sound analysis', *Computers and Electronics in Agriculture*, 114, 1–6. https://doi.org/10.1016/j.compag.2015.03.010.

Aydin, A. and Berckmans, D. (2016) 'Using sound technology to automatically detect the short-term feeding behaviours of broiler chickens', *Computers and Electronics in Agriculture*, 121, 25–31. https://doi.org/10.1016/j.compag.2015.11.010.

Banakar, A., et al. (2016) 'An intelligent device for diagnosing avian diseases: Newcastle, infectious bronchitis, avian influenza', *Computers and Electronics in Agriculture*, 127, 744–753. https://doi.org/10.1016/j.compag.2016.08.006.

Banerjee, D., et al. (2012) 'Remote Activity Classification of Hens Using Wireless Body Mounted Sensors', In *Proceedings of the 2012 Ninth International Conference on Wearable and Implantable Body Sensor Networks*, London, UK, 9–12 May; IEEE: New York, NY, pp. 107–112.

Barrett, J., et al. (2014) 'Smothering in UK free-range flocks. Part 1: incidence, location, timing and management', *Veterinary Record*, 175(1), 19–19. https://doi.org/10.1136/vr.102327.

Sassi, N., et al. (2016) 'Technology and poultry welfare', *Animals*, 6(10), e62. https://doi.org/10.3390/ani6100062.

Bist, R.B., et al. (2023a) 'A novel YOLOv6 object detector for monitoring piling behavior of cage-free laying hens', *AgriEngineering*, 5(2), 905–923. https://doi.org/10.3390/agriengineering5020056.

Bist, R.B., et al. (2023b) 'Automatic detection of cage-free dead hens with deep learning methods', *AgriEngineering*, 5(2), 1020–1038. https://doi.org/10.3390/agriengineering5020064.

Bist, R.B., et al. (2023c) 'A novel YOLOv6 object detector for monitoring piling behavior of cage-free laying hens', *AgriEngineering*, 5, 905–923. https://doi.org/10.3390/agriengineering5020056.

Bustamante, E., et al. (2012) 'Multisensor system for isotemporal measurements to assess indoor climatic conditions in poultry farms', *Sensors*, 12(5), 5752–5774. https://doi.org/10.3390/s120505752.

Campbell, D.L.M., et al. (2017) 'Outdoor stocking density in free-range laying hens: radio-frequency identification of impacts on range use', *Animal*, 11(1), 121–130. https://doi.org/10.1017/S1751731116001154.

Campbell, D.L.M., et al. (2018) 'Using radio-frequency identification technology to measure synchronized ranging of free-range laying hens', *Animals*, 8(11), 210. https://doi.org/10.3390/ani8110210.

Cândido, M.G.L., et al. (2020) 'Effects of heat stress on pullet cloacal and body temperature', *Poultry Science*, 99(5), 2469–2477. https://doi.org/10.1016/j.psj.2019.11.062.

Cangar, Ö., et al. (2008) 'Quantification of the spatial distribution of surface temperature of broilers', *Poultry Science*, 87, 2493–2499. https://doi.org/10.3382/ps.2007-00326.

Chen, L., et al. (2020) 'Computational fluid dynamics modeling of ventilation and hen environment in cage-free egg facility', *Animals*, 10(6), 1067. https://doi.org/10.3390/ani10061067.

Cheng, Q., et al. (2018) 'CFD study of the influence of laying hen geometry, distribution and weight on airflow resistance', *Computers and Electronics in Agriculture*, 144, 181–189. https://doi.org/10.1016/j.compag.2017.12.003.

Chien, Y.-R. and Chen, Y.-X. (2018) 'An RFID-based smart nest box: an experimental study of laying performance and behavior of individual hens', *Sensors*, 18(3), 859. https://doi.org/10.3390/s18030859.

Ciarelli, C., et al. (2023) 'Space use and navigation ability of hens at housing in the aviary for the laying phase: effect of enrichment with additional perches and genotype', *Poultry Science*, 102, 102962. https://doi.org/10.1016/j.psj.2023.102962.

Cook, N.J., et al. (2006) 'Assessing feather cover of laying hens by infrared thermography', *Journal of Applied Poultry Research*, 15(2), 274–279. https://doi.org/10.1093/japr/15.2.274.

Corkery, G., et al. (2013) 'Incorporating smart sensing technologies into the poultry industry', *Journal of World's Poultry Research*, 3, 106–128.

Daigle, C.L., et al. (2012) 'Noncaged laying hens remain unflappable while wearing body-mounted sensors: levels of agonistic behaviors remain unchanged and resource use is not reduced after habituation', *Poultry Science*, 91(10), 2415–2423. https://doi.org/10.3382/ps.2012-02300.

Dawkins, M., et al. (2004) 'Chicken welfare is influenced more by housing conditions than by stocking density', *Nature*, 427(6972), 342–344. https://doi.org/10.1038/nature02226.

De Wet, L., et al. (2003) 'Computer-assisted image analysis to quantify daily growth rates of broiler chickens', *British Poultry Science*, 44(4), 524–532. https://doi.org/10.1080/0007166031000161619 2.

Derakhshani, S.M., et al. (2022) 'Implementation of inertia sensor and machine learning technologies for analyzing the behavior of individual laying hens', *Animals*, 12(5), 536. https://doi.org/10.3390/ani12050536.

Du, L., et al. (2019) 'Computational fluid dynamics aided investigation and optimization of a tunnel-ventilated poultry house in China', *Computers and Electronics in Agriculture*, 159, 1–15. https://doi.org/10.1016/j.compag.2019.02.020.

Du, X., et al. (2020) 'Assessment of laying hens' thermal comfort using sound technology', *Sensors*, 20(2), 473. https://doi.org/10.3390/s20020473.

Du, X., et al. (2021) 'A tristimulus-formant model for automatic recognition of call types of laying hens', *Computers and Electronics in Agriculture*, 187, 106221. https://doi.org/10.1016/j.compag.2021.106221.

Ellen, E., et al. (2019) 'Review of sensor technologies in animal breeding: phenotyping behaviors of laying hens to select against feather pecking', *Animals*, 9(3), 108. https://doi.org/10.3390/ani9030108.

Fontana, I., et al. (2015) 'An innovative approach to predict the growth in intensive poultry farming', *Computers and Electronics in Agriculture*, 119, 178–183. https://doi.org/10.1016/j.compag.2015.10.001.

Fraess, G.A., et al. (2016) 'Automated behavioural response assessment to a feeding event in two heritage chicken breeds', *Applied Animal Behaviour Science*, 179, 74–81. https://doi.org/10.1016/j.applanim.2016.03.002.

Fujinami, K., et al. (2023) 'Evaluating behavior recognition pipeline of laying hens using wearable inertial sensors', *Sensors*, 23(11), 5077. https://doi.org/10.3390/s23115077.

Gebhardt-Henrich, S.G., et al. (2014) 'Individual tracking of laying hens with an RFID-System', *Landtechnik*, 69, 301–307.

Geffen, O., et al. (2020) 'A machine vision system to detect and count laying hens in battery cages', *Animal*, 14(12), 2628–2634. https://doi.org/10.1017/S1751731120001676.

Giloh, M., et al. (2012) 'Skin surface temperature of broiler chickens is correlated to body core temperature and is indicative of their thermoregulatory status', *Poultry Science*, 91(1), 175–188. https://doi.org/10.3382/ps.2011-01497.

Guo, Y., et al. (2023) 'Automatic detection of brown hens in cage-free houses with deep learning methods', *Poultry Science*, 102(8), 102784. https://doi.org/10.1016/j.psj.2023.102784.

Hao, H., et al. (2022) 'Research on laying hens feeding behavior detection and model visualization based on convolutional neural network', *Agriculture*, 12(12), 2141. https://doi.org/10.3390/agriculture12122141.

Hu, X., et al. (2022) 'Optimal placement of laying hen house temperature sensors based on genetic algorithm', *IEEE Access*, 10, 7234–7244. https://doi.org/10.1109/ACCESS.2022.3141090.

Hunniford, M.E. and Widowski, T.M. (2017) 'Nest alternatives: adding a wire partition to the scratch area affects nest use and nesting behaviour of laying hens in furnished cages', *Applied Animal Behaviour Science*, 186, 29–34. https://doi.org/10.1016/j.applanim.2016.10.018.

Jung, D.-H., et al. (2021) 'Classification of vocalization recordings of laying hens and cattle using convolutional neural network models', *Journal of Biosystems Engineering*, 46(3), 217–224. https://doi.org/10.1007/s42853-021-00101-1.

Kashiha, M.A., et al. (2014) 'Performance of an image analysis processing system for hen tracking in an environmental preference chamber', *Poultry Science*, 93(10), 2439–2448. https://doi.org/10.3382/ps.2014-04078.

Kim, N.Y., et al. (2021) 'Changes in facial surface temperature of laying hens under different thermal conditions', *Animal Bioscience*, 34(7), 1235–1242. https://doi.org/10.5713/ab.20.0647.

Kodaira, V., et al. (2023) 'Assessment of preference behavior of layer hens under different light colors and temperature environments in long-time footage using a computer vision system', *Animals*, 13(15), 2426. https://doi.org/10.3390/ani13152426.

Larsen, H., et al. (2017) 'Individual ranging behaviour patterns in commercial free-range layers as observed through RFID Tracking', *Animals*, 7(12), 21. https://doi.org/10.3390/ani7030021.

Lee, J., et al. (2015) 'Stress detection and classification of laying hens by sound analysis', *Asian-Australasian Journal of Animal Sciences*, 28(4), 592–598. https://doi.org/10.5713/ajas.14.0654.

Leroy, T., et al. (2005) 'Computer vision based recognition of behavior phenotypes of laying hens', *ASAE Annual Meeting. American Society of Agricultural and Biological Engineers*, 1. https://doi.org/10.13031/2013.19471.

Li, G., et al. (2023) 'Interindividual distances and orientations of laying hens under 8 stocking densities measured by integrative deep learning techniques', *Poultry Science*, 102(11), 103076. https://doi.org/10.1016/j.psj.2023.103076.

Li, L., et al. (2017) 'A UHF RFID system for studying individual feeding and nesting behaviors of group-housed laying hens', *Transactions of the ASABE*, 60(4), 1337–1347. https://doi.org/10.13031/trans.12202.

Li, N., et al. (2020) 'Review: automated techniques for monitoring the behaviour and welfare of broilers and laying hens: towards the goal of precision livestock farming', *Animal*, 14(3), 617–625. https://doi.org/10.1017/S1751731119002155.

Li, P., et al. (2021) 'Detection of sick laying hens by infrared thermal imaging and deep learning', *Journal of Physics: Conference Series*, 2025(1), 012008. https://doi.org/10.1088/1742-6596/2025/1/012008.

Liu, D., et al. (2021) 'Separate weighing of male and female broiler breeders by electronic platform weigher using camera technologies', *Computers and Electronics in Agriculture*, 182, 106009. https://doi.org/10.1016/j.compag.2021.106009.

Quwaider, C. L., et al. (2010) 'Development of a wireless body-mounted sensor to monitor location and activity of laying hens in a non-cage housing system', *Transactions of the ASABE*, 53(5), 1705–1713. https://doi.org/10.13031/2013.34890.

Montalcini, C.M., et al. (2022) 'Evaluation of an active LF tracking system and data processing methods for livestock precision farming in the poultry sector', *Sensors*, 22(2), 659. https://doi.org/10.3390/s22020659.

Mortensen, A.K., et al. (2016) 'Weight prediction of broiler chickens using 3D computer vision', *Computers and Electronics in Agriculture*, 123, 319–326. https://doi.org/10.1016/j.compag.2016.03.011.

Na, D., et al. (2019) 'Environment analysis of laying hens based on detailed sound classification using SVM', *Journal of Theoretical and Applied Information Technology*, 97, 4886–4896.

Nääs, I.A., et al. (2014) 'Infrared thermal image for assessing animal health and welfare', *Journal of Animal Behaviour and Biometeorology*, 2(3), 66–72. https://doi.org/10.14269/2318-1265/jabb.v2n3p66-72.

Nakarmi, A.D., et al. (2014) 'Automated tracking and behavior quantification of laying hens using 3D computer vision and radio frequency identification technologies', *Transactions of the ASABE*, 57(5), 1455–1472.

Nie, L., et al. (2022) 'Effects of body-mounted Inertial Measurement Unit (IMU) backpacks on space use and behaviors of laying hens in a perchery system', *Agriculture*, 12(11), 1898. https://doi.org/10.3390/agriculture12111898.

Okinda, C., et al. (2020) 'A review on computer vision systems in monitoring of poultry: a welfare perspective' *Artificial Intelligence in Agriculture*, 4, 184–208. https://doi.org/10.1016/j.aiia.2020.09.002.

Mendes, L.B., et al. (2015) 'NDIR gas sensor for spatial monitoring of carbon dioxide concentrations in naturally ventilated livestock buildings', *Sensors*, 15, 11239–11257. https://doi.org/10.3390/s150511239.

Olejnik, K., et al. (2022) 'Emerging precision management methods in poultry sector', *Agriculture*, 12(5), 718. https://doi.org/10.3390/agriculture12050718.

Peña Fernández, A., et al. (2018) 'Real-time monitoring of broiler flock's welfare status using camera-based technology', *Biosystems Engineering*, 173, 103–114.

Pezzuolo, A., et al. (2021). Digital technologies and automation in livestock production systems: a digital footprint from multisource data. In *2021 IEEE International Workshop on Metrology for Agriculture and Forestry (MetroAgriFor)*, 258–262. https://doi.org/10.1109/MetroAgriFor52389.2021.9628544.

Pichová, K., et al. (2017) 'Assessment of the effect of housing on feather damage in laying hens using IR thermography', *Animal*, 11(4), 661–669. https://doi.org/10.1017/S1751731116001981.

Pitesky, M., et al. (2020) 'Data challenges and practical aspects of machine learning-based statistical methods for the analyses of poultry data to improve food safety and production efficiency', *CABI Reviews*, PAVSNNR202015049. https://doi.org/10.1079/PAVSNNR202015049.

Pu, H., et al. (2018) 'Automatic recognition of flock behavior of chickens with convolutional neural network and kinect sensor', *International Journal of Pattern Recognition and Artificial Intelligence*, 32(7), 1850023. https://doi.org/10.1142/S0218001418500234.

Qi, F., et al. (2023) 'Environmental factor detection and analysis technologies in livestock and poultry houses: a review', *Agriculture*, 13, 1489. https://doi.org/10.3390/agriculture13081489.

Ramírez-Morales, I., et al. (2017) 'Automated early detection of drops in commercial egg production using neural networks', *British Poultry Science*, 58(6), 739–747. https://doi.org/10.1080/00071668.2017.1379051.

Ren, G., et al. (2020) 'Agricultural robotics research applicable to poultry production: a review', *Computers and Electronics in Agriculture*, 169, 105216. https://doi.org/10.1016/j.compag.2020.105216.

Rojano, F., et al. (2019) 'Modelling the impact of air discharges caused by natural ventilation in a poultry house', *Biosystems Engineering*, 180, 168–181. https://doi.org/10.1016/j.biosystemseng.2019.02.001.

Rufener, C., et al. (2019) 'Keel bone fractures are associated with individual mobility of laying hens in an aviary system', *Applied Animal Behaviour Science*, 217, 48–56. https://doi.org/10.1016/j.applanim.2019.05.007.

Rufener, C., et al. (2018) 'Finding hens in a haystack: consistency of movement patterns within and across individual laying hens maintained in large groups', *Scientific Reports*, 8(1), 12303. https://doi.org/10.1038/s41598-018-29962-x.

Ruhnke, I., et al. (2019) 'Free-range laying hens: using technology to show the dynamics and impact of hen movement', *Animal Production Science*, 59(11), 2046. https://doi.org/10.1071/AN19256.

Saeed, M., et al. (2019) 'Heat stress management in poultry farms: a comprehensive overview', *Journal of Thermal Biology*, 84, 414–425. https://doi.org/10.1016/j.jtherbio.2019.07.025.

Sales, G.T., et al. (2015) 'Quantifying detection performance of a passive low-frequency RFID system in an environmental preference chamber for laying hens', *Computers* and *Electronics in Agriculture*, 114, 261–268. https://doi.org/10.1016/j.compag.2015.03.008.

Schreiter, R. and Freick, M. (2022) 'Research Note: Is infrared thermography an appropriate method for early detection and objective quantification of plumage damage in white and brown feathered laying hens?', *Poultry Science*, 101(9), 102022. https://doi.org/10.1016/j.psj.2022.102022.

Shahbazi, M., et al. (2023) 'Deep learning for laying hen activity recognition using wearable sensors', *Agriculture*, 13(3), 738. https://doi.org/10.3390/agriculture13030738.

Sibanda, T.Z., et al. (2020) 'Characterising free-range layer flocks using unsupervised cluster analysis', *Animals*, 10(5), 855. https://doi.org/10.3390/ani10050855.

Siegford, J., et al. (2016) 'Assessing activity and location of individual laying hens in large groups using modern technology', *Animals*, 6(2), 10. https://doi.org/10.3390/ani6020010.

Souza-Junior, J.B.F., et al. (2019) 'Estimating sensible heat loss in laying hens through thermal imaging', *Computers and Electronics in Agriculture*, 166, 105038. https://doi.org/10.1016/j.compag.2019.105038.

Sozzi, M., et al. (2022) 'Measuring comfort behaviours in laying hens using deep-learning tools', *Animals*, 13(1), 33. https://doi.org/10.3390/ani13010033.

Subedi, S., et al. (2023) 'Tracking pecking behaviors and damages of cage-free laying hens with machine vision technologies', *Computers and Electronics in Agriculture*, 204, 107545. https://doi.org/10.1016/j.compag.2022.107545.

Tan, H., et al. (2020) 'Detection of ammonia concentration in livestock poultry houses based on tunable diode laser absorption spectroscopy', *Transactions of the Chinese Society of Agricultural Engineering*, 36, 186–194. https://doi.org/10.11975/j.issn.1002-6819.2020.13.022.

Thurner, S., et al. (2009) 'Using a wide electronic pop hole based on RFID-technology with high-frequency transponders to monitor the ranging behaviour of laying hens in alternative housing systems', *Precision Livestock Farming*, 9, 243–249. https://doi.org/10.3920/9789086866632_030.

Van Veen, L.A., et al. (2023) 'Perception of laying hen farmers, poultry veterinarians, and poultry experts regarding sensor-based continuous monitoring of laying hen health and welfare', *Poultry Science*, 102(5), 102581. https://doi.org/10.1016/j.psj.2023.102581.

Wang, F., et al. (2023) 'Application of deep learning methods in behavior recognition of laying hens', *Frontiers in Physics*, 11, 1139976. https://doi.org/10.3389/fphy.2023.1139976.

Wang, J., et al. (2020) 'Real-time behavior detection and judgment of egg breeders based on YOLO v3', *Neural Computing and Applications*, 32(10), 5471–5481. https://doi.org/10.1007/s00521-019-04645-4.

Wang, K., et al. (2019) 'An RFID-based automated individual perching monitoring system for group-housed poultry', *Transactions of the ASABE*, 62(3), 695–704. https://doi.org/10.13031/trans.13105.

Welch, M., et al. (2023) 'An initial study on the use of machine learning and radio frequency identification data for predicting health outcomes in free-range laying hens', *Animals*, 13(7), 1202. https://doi.org/10.3390/ani13071202.

Xin, H. and Liu, K. (2017) 'Precision livestock farming in egg production', *Animal Frontiers*, 7(1), 24–31. https://doi.org/10.2527/af.2017.0105.

Yakubu, A., et al. (2018) 'Predicting heat stress index in Sasso hens using automatic linear modeling and artificial neural network', *International Journal of Biometeorology*, 62(7), 1181–1186. https://doi.org/10.1007/s00484-018-1521-7.

Yang, X., et al. (2022) 'A deep learning model for detecting cage-free hens on the litter floor', *Animals*, 12(15), 1983. https://doi.org/10.3390/ani12151983.

Yang, X., et al. (2023) 'An automatic classifier for monitoring applied behaviors of cage-free laying hens with deep learning', *Engineering Applications of Artificial Intelligence*, 123, 106377. https://doi.org/10.1016/j.engappai.2023.106377.

Yin, M., et al. (2023) 'Non-contact sensing technology enables precision livestock farming in smart farms', *Computers and Electronics in Agriculture*, 212, 108171. https://doi.org/10.1016/j.compag.2023.108171.

Zaninelli, M., et al. (2017) 'A monitoring system for laying hens that uses a detection sensor based on infrared technology and image pattern recognition', *Sensors*, 17(6), 1195. https://doi.org/10.3390/s17061195.

Zaninelli, M., et al. (2018) 'Development of a machine vision method for the monitoring of laying hens and detection of multiple nest occupations', *Sensors*, 18(2), 132. https://doi.org/10.3390/s18010132.

Zaninelli, M., et al. (2016) 'First results of a detection sensor for the monitoring of laying hens reared in a commercial organic egg production farm based on the use of infrared technology', *Sensors*, 16(10), 1757. https://doi.org/10.3390/s16101757.

Zhang, X., et al. (2022) 'Feather damage monitoring system using RGP-depth-thermal model for chickens', *Animals*, 13(1), 126. https://doi.org/10.3390/ani13010126.

Zhao, Y., et al. (2013) 'Use of infrared thermography to assess laying-hen feather coverage', *Poultry Science*, 92(2), 295–302. https://doi.org/10.3382/ps.2012-02629.

Zhuang, X., et al. (2018) 'Development of an early warning algorithm to detect sick broilers', *Computers and Electronics in Agriculture*, 144, 102–113. https://doi.org/10.1016/j.compag.2017.11.032.

Part 2

Post-harvest safety

Chapter 5

The use of edible coatings to improve the safety and extend the shelf life of eggs

Paula Gabriela da Silva Pires, IFC, Brazil

1 Introduction

Eggs are nutritionally interesting for humans as they are a source of proteins, vitamins, and minerals and are widely used in the food industry as they combine nutritional, organoleptic, and functional properties, such as clarifying agent, binder, gelling, and emulsifying. There are several factors that can affect egg quality: genetics and age hen, feeding program, diseases, housing system, temperature, humidity, handling, and storage. The longer the storage time, the greater the deterioration of the internal quality, due to the higher CO_2 movement through the eggshell, especially at room temperature.

Eggs lose quality quickly and to prolong the life of eggs, it would be necessary to keep the eggs refrigerated from production to consumption, which would result in an increase in electric energy expenditure, production costs, and, consequently, an increase in the price paid by consumers.

http://dx.doi.org/10.19103/AS.2024.0150.09

Jenkins (1919) reported the positive effects of the use of refrigeration on egg conservation for the first time more than a century ago. In the 1970s, the number of outbreaks related to *Salmonella enteritidis* (SE) contamination in the US increased dramatically and eggs were the main vehicle for SE infection in humans. Egg refrigeration was adopted as part of US interventions aimed at preventing SE contamination and growth in eggs. Currently the US interventions aimed at preventing the contamination and growth of SE in eggs have included early and sustained refrigeration of shell eggs (Braden, 2006). However, in many countries – including the EU – the eggs should not be refrigerated before sale to the final consumer, according to Regulation (EC) No 589/2008 (CEC, 2008). In Latin American countries, the refrigeration of eggs at points of sale is not mandatory by law. Fikiin et al. (2020) suggested that efforts should be made to harmonize the huge discrepancies between relevant codes and practices in Europe, the US, and the rest of the world, thereby overcoming trade barriers.

Washing eggshells can also accelerate the loss of egg quality. Egg washing is a process carried out to remove dirt and fecal material from the eggshell and consequently reduce microbial contamination. The egg washing process can affect the thickness of the individual layers of the shell and reduce or completely remove the cuticle present on the eggshell (Samiullah et al. 2013), accelerating the process of deterioration in egg quality. Liu et al. (2016) and Jones et al. (2018) reported the negative impact of washing in the eggshell. The practical recommendation for washing eggs varies in different countries. For example, the US, Japan, and Australia adopt egg-washing procedures routinely. However, many countries – including the Latin America and EU – still have resisted the practice.

Coatings are a low-cost tool used in various foods to reduce moisture loss, gas exchange, and oxidation processes (Dhall, 2013). The use of coatings to extend the shelf life of eggs has been consistently studied over the years. Spamer (1931) reported that Dutch farmers used oil to preserve eggs as early as 1807. Research over the past 200 years has led to the use of certain colorless, odorless, and less-flavored paraffin base. Rosser (1942) and Romanoff and Yushok (1948) were among the first researchers to document the positive effects of using mineral oil to coat eggs. Currently, research is focused on the use of sustainable materials to create new coatings. More sustainable coatings made from renewable sources such as agro-industrial waste can be an alternative to the use of raw materials of fossil origin, e.g. mineral oil of petroleum origin. Furthermore, there is an increase in consumer interest in waste-free and more sustainable products.

The chapter demonstrates the main changes observed in egg quality during storage and presents the materials commonly used in egg coatings, in addition to discussing gaps, trends and future opportunities for the development of more complex coatings with greater capacity to preserve quality and safety of stored eggs.

2 Egg quality and shelf life

The eggshell has thousands of pores that allow for the movement of moisture and CO_2 to the outside. The porosity of the eggshell favors the reduction of the egg's useful life due to the loss of water and degradation of the proteins contained in the albumen. During the egg storage period, continuous physical-chemical changes occur that can result in changes in the freshness, odor, and flavor of the egg. The longer the storage period and temperature, the greater the deterioration in egg quality. Several responses can be evaluated to estimate the decrease in egg quality during storage (Table 1).

2.1 Egg weight loss

The decrease in egg weight occurs gradually throughout the storage period and is an easy metric to perform, as there is no need to break the eggshell. Water loss depends mainly on temperature and relative humidity during storage. In places with milder temperatures, egg weight loss will occur more slowly. Weight loss assessment is a parameter widely used by researchers and is generally carried out at weekly intervals starting from day zero (fresh egg). The higher the ambient temperature, the faster the eggs will lose weight. Egg weight loss occurs due to the probable loss of ammonia, nitrogen and hydrogen sulfide, products of the chemical degradation of organic constituents (Solomon, 1997).

Table 1 Main quality factors affected during storage

Parameter	Cause
Weight egg loss	Loss of ammonia, nitrogen, and hydrogen sulfide that are products of the chemical degradation of their organic constituents.
Haugh unit decrease	Fluidization and loss of viscosity of dense albumen occur because of the hydrolysis of amino acid chains, which, when degraded, release water linked to large protein molecules.
Albumen and yolk pH increase	Biochemical changes in the yolk and the transfer of water from the albumen.
Yolk index decrease	The water released during the hydrolysis reaction of the albumen's amino acids is transferred to the yolk, which consequently increases in weight becoming decentralized and less dense.
Yolk color	Carotenoids can be degraded by the oxidative process, changing pigmentation during storage.
S-Ovoalbumin	Transformation of albumin into S-ovoalbumin and dissociation of the ovomucin–lysozyme complex, with destruction of the ovomucin gel.

Weight loss of around 2–3% during storage is acceptable (FAO, 2003). However, studies have shown that egg weight loss is quite pronounced in eggs stored at room temperature, especially in countries with higher average temperatures and in the hottest seasons of the year. Musa et al. (2011) carried out a study in Iraq and observed a weight loss of 25.27% in eggs stored at 40°C for 4 weeks, while Pires et al. (2018) observed a weight loss of 8.28% in eggs stored for 8 weeks in Brazil (20°C). Suryani et al. (2017) reported 7.34% egg weight loss after 5 weeks of storage at ambient conditions (26–32°C, 60–72% RH). Suppakul et al. (2010) observed that after 28 days of storage at room temperature in Thailand, eggs show an average weight loss of 8.83%.

2.2 Haugh unit

The Haugh unit (HU) is the most used metric to express albumen quality. Pires et al. (2020) demonstrated that around 75% of studies use HU as a measure to evaluate the freshness of eggs. The Haugh unit is a logarithmic equation proposed by Haugh in 1937 to evaluate egg quality through the ratio of egg weight to albumen height. Thus, the liquefaction of dense albumen is evidenced by a decrease in albumen height and a reduction in HU values. However, egg weight can be affected by factors such as the breed and age of the laying hens. In this sense, HU would not be the most suitable metric to evaluate the freshness of eggs.

Several studies have demonstrated a decrease in HU values with increasing egg storage time (Caner, 2005; Vlĉková et al., 2019). As the storage period increases, there is a decrease in the dense portion of albumen and an increase in the proportion of liquid albumen. The fluidization and loss of viscosity of dense albumen occur due to the hydrolysis of amino acid chains, which, when degraded, release water linked to large protein molecules. Eggs can be classified based on US standards for individual shell egg quality (USDA, 2020); AA (company) = Haugh unit value above 72; A (fairly firm) = 71 to 60; B (weak and watery) = less than 60. Yuceer and Caner (2014) demonstrated that eggs changed from AA to A and B after 4 and 2 weeks of storage at 25°C, while Ryu et al. (2011) verified a change from grade AA to A in the first week and to grade B after 2 weeks of storage at 25°C. Pujols et al. (2014) documented that fresh eggs had HU values of 84.84 (grade AA) and after 5 weeks of storage, the HU values decreased significantly (29.82, grade B). The moment at which the 'AA' grade of eggs changes varies according to each study. This change appears to be associated with storage conditions (temperature and humidity).

2.3 Albumen pH

Biochemical changes occur in albumen during storage. The loss of gases through the eggshell, such as CO_2, during storage is mainly responsible for the changes observed in albumen pH. Furthermore, the water present in the albumen can migrate to the yolk during storage, which can also cause changes in the pH of the albumin and yolk. pH is a metric that efficiently describes the changes that occur in the albumen during storage, demonstrating the loss of egg quality and freshness. Furthermore, it is important to highlight that this analysis is not influenced by other factors such as the age of the laying hen.

Normally the pH of the albumen of a fresh egg ranges from 7.6 to 7.9. However, with increasing storage time, albumen pH can reach 9.5 (Alleoni and Antunes, 2001). Loss of CO_2 can result in altered egg flavor due to increased alkalinity. Ryu et al. (2011) have observed an increase in albumen pH from 8.81 to 9.37 after 6 weeks of storage at 25°C. Pires et al. (2019b) showed that the average initial albumen pH of the eggs was 8.05 and this value increased to 9.48 at the end of 6 weeks of storage at 20°C.

2.4 Measurements in egg yolk

The yolk may also undergo changes during the storage period but the increase in pH occurs more slowly in yolk than albumen. Pissinati et al. (2014) has documented a maximum increase in yolk pH from 6.13 to 6.26 after storage. Samli et al. (2005) observed an increase in yolk pH from 5.75 to 6.08 after 10 days of storage at 29°C. The change in the pH of the yolk is less pronounced when compared to the change in the pH of the albumen, and therefore the pH of the yolk is less used as a metric to evaluate the freshness of eggs.

During the hydrolysis reaction of albumen amino acids, water is released, which is transferred to the yolk. Because of the degradation of the albumen, the yolk absorbs the water derived from the protein decomposition, becoming flatter. Consequently, there is an increase in weight of the yolk, weakening of the vitelline membrane, making the yolk decentralized and less dense. The yolk index (YI) is often used to determine the freshness of the yolk and can be evaluated through the height and width of the yolk. The fresh egg YI can vary from 0.30 to 0.50 and the higher the YI, the better the quality of the yolk. Pissinati et al. (2014) observed that after 6 weeks, the YI decreased from 0.37 to 0.23. Decreased YI values indicate fragility of the yolk membrane, which can easily result in bud breakage, indicating a lower quality yolk.

3 Use of edible coatings in eggs

The process of coating eggshells for long-term storage has been widely used for many decades and has recently received renewed attention as a method of quality retention. Many studies have been conducted evaluating techniques that allow prolonging the shelf life of eggs (Pires et al., 2020). The use of coatings prevents the loss of volatile compounds and slows the rate of food spoilage by controlling the transfer of moisture and oxygen through the eggshell. Research reports improvements in maintaining egg quality and reduced eggshell breakage after coatings are applied. Finding a substance capable of minimizing the deterioration of the interior quality of eggs and at the same time adding enough strength to the shells to reduce the likelihood of cracking would be a considerable advantage for the poultry industry. Eggshells coated with wheat gluten, soybean protein, whey protein, or carboxymethyl cellulose showed greater puncture resistance than those of uncoated eggs (Xie et al., 2002). Biladeau and Keener (2009) demonstrated that wax-coated eggs had 4–10% greater shell strength than uncoated eggs. Increasing eggshell strength will potentially decrease the number of broken eggs, thus decreasing the number of discarded eggs, which will result in significant savings for the egg industry.

Furthermore, increasing the storage time of eggs results in a reduction in food waste, in addition to allowing eggs to reach locations further away from production points with preserved quality. Egg coatings are made using lipids, polysaccharides, and proteins and in some cases substances with antimicrobial properties are included. The frequently used substances, their particularities, and results of their applications will be discussed in the following sections.

4 Choice of coating raw materials

The raw material used in the production of coatings must provide adequate adhesion and coverage to the egg, capable of forming a barrier to limit gas exchange between the egg and the external environment. The efficiency of using a coating depends on the ability of the substance to form a homogeneous and good quality coating. Many studies have investigated different substances for the development of coatings capable of delaying the degradation of eggs after oviposition (Morsy et al., 2015; Pires et al., 2020, 2022). However, there is still a great opportunity for the development of coatings based mainly on agricultural by-products.

Different materials that are abundant around the world are being used in the development of coatings. The choice of raw material to produce the coating may be conditioned by the availability of resources in each region. Cereals produced on a large scale and available in several countries, such as corn, rice, wheat, and soybeans, as well as fruits such as apples (pectin), grapes, bananas, and oranges, deserve attention and are already used in the manufacture of edible coatings.

For example, Brazil is a large rice producer and therefore, Pires et al. (2018) used rice protein as an alternative for coating eggs stored at room temperature. The results obtained in the study demonstrated that the use of rice protein concentrate coatings can be an effective alternative to prolong the life of stored eggs. Figure 1 shows the quality of eggs coated or not coated with rice protein after 21 days of storage. There is an estimate that for every 100 kg of rice processed, an average of 14 kg of broken rice is obtained, with a lower commercial value, around 20% of the value compared to whole grains (Oliveira, 2014). Broken rice can be used to develop coatings. Furthermore, it is possible to develop more complex coatings by combining rice with other substances such as propolis (Pires et al., 2019b) and essential oils (Pires et al., 2019a). Oliveira et al. (2023) used rice flour coating in association with rosemary essential oil to preserve the internal, microbiological, and sensory quality of quail eggs. Akarca et al. (2021) carried out a study in Turkey and used apricot, almond, and cherry resins to coat eggs and demonstrated that these substances had great barrier properties, reducing the speed of egg deterioration. However, in other countries, these resins would probably be considered expensive products.

Some native species that are still little studied may have important characteristics for the development of egg coatings. The Brazilian Amazon Forest is home to numerous native species with economic and technological potential for the development of edible coatings. For example, Copaiba oil (*Copaifera langsdorffii*), plant from the Brazilian Amazon, has characteristics such as flexible coverage and antimicrobial potential that allow its use in covering eggs (Mendonça and Onofre, 2009). Brasil et al. (2019) evaluated the use of copaiba oil to coat eggs and observed that coated eggs had better internal quality after storage compared to uncoated eggs. Pires et al. (2019a) evaluated the use of copaiba essential oil associated with rice protein in coating eggs and demonstrated that the coating could preserve albumen quality for 2–3 weeks longer compared to eggs that did not receive the same treatment.

Rufino et al. (2023) observed that the use of a coating made from a bioproduct from andiroba oil (tree of Amazonian origin) presented satisfactory results in preserving the physical, bacteriological, and nutritional aspects of stored eggs. However, the authors highlighted the implications of using andiroba oil on the sensory attributes of eggs, since higher concentrations

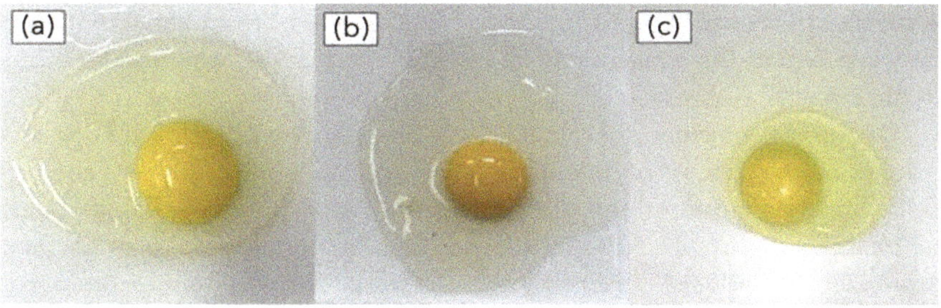

Figure 1 Quality of eggs coated or not coated with rice protein after 21 days of storage.

and prolonged exposure times can lead to decreased consumer acceptance due to changes in sensory parameters. Mendonça et al. (2023) developed an edible coating containing babassu starch, babassu oil, and plasticizers to maintain egg quality and found that coatings containing starch and/or babassu oil reduced egg weight loss and preserved the albumen of eggs stored for 28 days. Babassu (*Attalea speciosa*) is a species of great economic and cultural importance for the north and northeast regions of Brazil.

Aloe vera grows all over the world. The gel extracted from Aloe vera leaves has attracted great attention as an active coating material (Maan et al., 2021) as it contains various bioactive compounds, phenolic content, and minerals. Mudannayaka et al. (2019) reported that the use of a coating made with Aloe vera gel presented similar results to the use of mineral oil for preserving eggs. However, the inclusion levels of Aloe vera gel for coating manufacturing need to be investigated. Furthermore, combining Aloe vera gel with other substances seems a promising approach for developing new coatings.

5 Mineral oil

In many countries, the egg industry uses mineral oil to coat eggs. Mineral oil demonstrates excellent sealing properties, preventing the evaporation of moisture and gases. The efficiency of using mineral oil to preserve the quality of stored eggs has already been extensively demonstrated by several studies over the years (Rosser, 1942; Romanoff and Yushok, 1948; Wong et al., 1996; Jones et al., 2018).

Jirangrat et al. (2010) and Jones et al. (2018) observed less weight loss in eggs coated with mineral oil than in uncoated eggs after 15 weeks of storage. Wardy et al. (2010) found that the use of mineral oil extended the shelf life of eggs by 4 weeks. Torrico et al. (2010) observed the use of mineral oil was efficient in maintaining the quality of eggs in storage. The authors demonstrated that the pH of albumen in eggs coated with mineral oil was lower than uncoated eggs up to 5 weeks. Pires et al. (2018) observed that eggs coated with mineral oil showed an average weight loss of 0.87% after 60 days of storage, while uncoated eggs showed 8.28% egg weight loss after the same period. Mineral oil demonstrates excellent properties for sealing the pores present in the egg shell, preventing the evaporation of moisture and gases, which reduces egg weight loss.

Despite the efficiency of mineral oil in blocking the pores of the egg shell and slowing the process of deterioration in egg quality, mineral oil is a chemical compound derived from petroleum. Coatings made from renewable sources can be a conscious alternative to the use of raw materials of fossil origin (such as petroleum products). This fact is even more interesting from a commercial point of view, as consumers will probably be more concerned about issues involving the sustainability of agricultural production and interested in waste-free products.

6 Lipid coatings

Lipids have been used in egg conservation since 1807 (Spamer, 1931). Recently, Pires et al. (2020) summarize studies on the effects of using coatings and found that lipids are the most used materials in preparing egg coatings (Table 2). Lipids offer good protection against water due to their hydrophobic nature. Lipid-based coatings have a low affinity for water, which explains the low permeability to water vapor.

Ryu et al. (2011) reported that coating eggs with mineral oil, corn oil, sunflower oil, canola oil, grape seed oil, and soybean oil demonstrated equally satisfactory results in preserving eggs stored for 5 weeks at room temperature. Wardy et al. (2010) reported that the use of soybean oil extended the shelf life of eggs by 4 weeks. Nongtaodum et al. (2013) used different oils (coconut, palm, rice bran, and soy) to preserve the internal quality of eggs. The study demonstrated that the use of oils helped maintain the quality of the albumen (maintaining grade A) for at least 4 weeks longer than eggs without any coating. Furthermore, the use of oils provided a shinier shell than uncoated eggs.

Waxes are a good choice for creating coatings as they have an excellent barrier to water vapor. However, they create brittle coatings due to their hardness. Biladeau and Keener (2009) demonstrated that wax-coated eggs showed negligible water loss through the shell compared to uncoated eggs. Furthermore, the use of wax was able to maintain the quality of the albumen (grade AA) for at least 8 weeks. The authors reported that the wax-coated eggs had negligible water loss through the shell compared with 5% for the uncoated eggs. In addition, the wax-coated eggs had a 4–10% higher shell strength than all others. Eyng et al. (2021) evaluated the quality and lipid oxidation of eggs coated with a carnauba wax-based product and stored for up to 28 days. Carnauba (*Copernicia cerifera*) is a tropical palm tree of Brazilian origin

Table 2 Summary of different lipids used for egg coating

Material	Storage condition	Effects	Reference
Wax	12 weeks at 7°C	Shelf-life extension	Biladeau and Keener (2009)
Vegetable oil	5 weeks at 25°C	Shelf-life extension	Nongtaodum et al. (2013)
Vegetable oil	5 weeks at 25°C	Preserve internal quality	Ryu et al. (2011)
Soybean oil	7–15 weeks at 4–25°C	Shelf-life extension	Wardy et al. (2010)
Carnauba wax	4 weeks at 25°C	Shelf-life extension	Eyng et al. (2021)

from which the wax is extracted. The authors reported that the weight loss of eggs stored at 25°C during storage was, on average, 37.3% lower for eggs coated with carnauba wax compared to uncoated eggs. However, there was no difference in the lipid oxidation of coated or uncoated eggs.

Resins have also been used as coatings. Shellac is a natural resin that originates from insects. Shellac coatings provide excellent sealing of eggshell pores, preventing evaporation of moisture and gases. Caner and Yuceer (2015) found that shellac is more effective in preventing weight loss than coatings based on whey protein or zein. Musa et al. (2011) concluded that shellac-based coatings can preserve the internal quality of chicken eggs stored at 40°C.

7 Polysaccharide coatings

Polysaccharides are large natural polymers formed by chains of monosaccharides linked together by glycosidic bonds. Polysaccharides are insoluble in water, and they are carbohydrates, also known as glycans. Polysaccharides are widely used in the production of edible coatings as they form more homogeneous films. Starches, pectin, and chitosan are the polysaccharides most frequently used to coat eggs (Table 3).

Starch is a carbohydrate made up mainly of glucose with glycosidic bonds. It is the most common carbohydrate in human nutrition and is found in large quantities of foods, such as potatoes, rice, and wheat. Several studies have demonstrated positive results from using starch to coat eggs (Sharaf Eddin and Tahergorabi, 2019). The cassava starch coating has been shown to be efficient against water loss and gas transport, preserving eggs for longer during storage at 25°C (Lima et al., 2020).

Chitosan has been widely used as coating due to its well-documented antimicrobial properties. Chitosan is a natural, nonallergen, and biocompatible polymer. The main industrial source of chitosan is seafood waste from the

Table 3 Summary of different polysaccharides used for egg coating

Material	Storage condition	Effects	Reference
Cassava, yam starch	4 weeks at 25°C	Shelf-life extension	De Brito Mota et al. (2017)
Sweet potato starch + thyme	4 weeks at 25°C	Preserve internal quality	Eddin and Tahergorabi (2019)
Pectin-alginate, pectin-alginate-laurolyl	6 weeks	Control of *S. enteritidis*	De Leo et al. (2018)
Chitosan	5 weeks at 25°C	Shelf-life extension	Bhale et al. (2003)
Chitosan	3 weeks at 4°C or 23°C	Maintaining egg quality	Jo et al. (2011)
Chitosan	6 weeks at 24°C	Shelf-life extension	Yuceer and Caner (2020)

processing of foods such as crab, squid, and shrimp. The positive effects of using chitosan-based coatings on egg preservation are well documented (Jo et al., 2011; Pujols et al., 2014). Chitosan coating can extend the shelf life of eggs by at least 3 weeks during storage at 25°C compared to uncoated eggs (Kim et al., 2009). Suresh et al. (2015) demonstrated that coating eggs with shrimp α-chitosan increased egg shelf life by almost 4 weeks at 22 ± 1°C and 3 weeks at 32 ± 1°C compared to eggs without the coating. Other studies have shown that chitosan coating can increase the shelf life of eggs by 2–3 weeks when compared to uncoated eggs (Lee et al., 1996; Bhale et al., 2003; No et al., 2005).

Pectin is a polysaccharide frequently used for the manufacture of films and coatings. Pectin is a type of fiber, it dissolves easily in water, found naturally in fruits and vegetables, such as apples, beets, and citrus fruits. Oliveira et al. (2020) demonstrated that a pectin-based coating effectively maintained egg quality during 35 days of storage. Considering the large supply of fruits, vegetables, and their residues, the use of pectin for the preparation of coatings seems an interesting opportunity that deserves to be explored.

Cellulose is the most abundant polysaccharide in nature. Cellulose stands out for being an insoluble carbohydrate and resistant to various chemical reactions. Cellulose-based coatings can reduce the weight loss of stored eggs and maintain the Grade A classification for a longer period compared to uncoated eggs (Suppakul et al., 2010). Xie et al. (2002) demonstrated that carboxylmethylcellulose coating can improve the mechanical properties and minimize microbial contamination of eggshell.

Coatings produced from polysaccharides are sensitive to moisture, permeable to water vapor, and have poor mechanical properties compared to those made from other materials. In this case, the combination of polysaccharides with other materials is advantageous to improve the performance of the coatings, such as the water vapor barrier property. Rachtanapun et al. (2021) used cassava starch as the main polysaccharide for the formulation of coatings due to its high purity, low cost, chemical modification capacity, and abundance. Starch was used in combination with gelling agents and waxes to prepare new edible materials for coating eggs. The coating developed in the study extended the shelf life and quality of grade AA eggs for 4 weeks at 25°C. The use of wax (paraffin) provided advantages such as flexibility and water resistance, which facilitated total coverage of the shell pores, maintenance of the HU value and reduced egg weight loss. Paraffin is recommended as a good coating material to increase the shelf life of eggs (Farag et al., 1994).

8 Protein coatings

Proteins are biological macromolecules formed by one or more polypeptides (amino acid polymers), which are arranged in a unique way. Proteins can be easily modified to achieve desirable film properties (Han and, 2005). Proteins present a great barrier to oxygen, CO_2, and lipids, especially at low relative humidity and are therefore good film-formers (Lacroix, 2014). Various proteins (e.g. gelatin, keratin, and vegetable proteins) can be used in the production of edible coatings (Table 4). Animal proteins already investigated for egg coat development include gelatin (Pissinati et al., 2014) and whey protein (Caner, 2005; Caner and Yuceer, 2015), and of vegetable origin include wheat gluten (Xie et al., 2002), soy protein (Biladeau and Keener, 2009), zein (Caner and Yuceer, 2015), and rice protein (Pires et al., 2018).

Pissinati et al. (2014) demonstrated interesting results in preserving stored eggs when using a gelatin-based coating. Gelatin can be obtained from different sources (pigs, cattle, and fish). However, fish gelatin-based coatings have yet to be tested on eggs (Pires et al., 2022). Gelatin from fish shows good potential for use in coatings due to its abundance, film-forming capacity, and excellent barrier properties against O_2 and CO_2 (Etxabide et al., 2021). Caner and Yuceer (2015) tested the effectiveness of protein coatings (whey protein isolate (WPI) or concentrate and zein) on the interior quality, functional properties, and resistance to eggshell cracking. The authors reported that protein coatings also improved the shell strength and functional properties of eggs and may be a viable technology for maintaining the internal quality of eggs during long-term storage.

Protein coatings are fragile and have low adhesion. Plasticizers are non-volatile substances that, when added to a material, change its mechanical and/or physical properties (Alleoni, 2006). Plasticizers are added to formulations to induce flexibility in the coatings. Furthermore, the use of plasticizers improves the adhesion of coatings to foods (Baldwin et al., 1995).

Table 4 Summary of different protein used for egg coating

Material	Storage condition	Effects	Reference
Whey protein isolate	4 weeks at 25°C	Shelf-life extension	Caner (2005)
Rice protein concentrate	6 weeks at 20°C	Shelf-life extension	Pires et al. (2018)
Egg albumen	4 weeks	Improve eggshell strength and preserve egg quality	Wong et al. (1996)
Wheat gluten	4 weeks	Shelf-life extension	Wong et al. (1996)
Lysozyme	6 weeks at 24°C	Shelf-life extension	Yuceer and Caner (2020)
Soy protein isolate	12 weeks at 7°C	Shelf-life extension	Biladeau and Keener (2009)
Zein	4 weeks at 24°C	Shelf-life extension	Caner and Yuceer (2015)
Gelatin	6 weeks at 25°C	Preserve internal quality	Pissinati et al. (2014)

Glycerol is the most used plasticizer in egg coatings due to its abundance and lower cost. However, other substances have demonstrated better results when added to coatings to extend the shelf life of eggs. Pires et al. (2021b) tested the use of glycerol, sorbitol, and propylene glycol (not associated with proteins) as a coating for eggs and found that eggs coated only with these substances had a wet appearance to the touch and the formation of mold on the surface after a week of storage. However, when these substances were associated with whey protein (Pires et al., 2021b) or rice protein (Pires et al., 2018), they were effective in preserving the internal quality of the eggs. These results are probably related to the level of use of plasticizers. According to Yang and Paulson (2000), coatings with glycerol concentrations above 75% result in sticky behavior. In other words, when used in a solution with other substances, plasticizers are incorporated at lower levels. Furthermore, the use of plasticizers isolated from other substances can be tested as long as lower concentrations are used, in which case there would be a need to dilute the plasticizer.

9 Bioactive coatings to improve egg safety

The egg is generally sterile shortly after laying when produced under healthy breeding conditions. However, contamination of eggs can occur by a microbiota containing microorganisms that spoil food and, sometimes, pathogenic bacteria that cause food outbreaks in humans (Baron and Jan, 2011). Contaminated eggs or egg products are common vehicles for outbreaks involving bacteria such as *Salmonella*. Many substances can be added to coating formulations to add additional protective features, such as antimicrobial action.

Currently, essential oils have received great attention from researchers, and several studies have been published in recent years. Essential oils are naturally occurring volatile aromatic compounds found in the seeds, bark, stems, roots, flowers, and other parts of plants and can have both pleasant and potent fragrances. Essential oils give plants their distinctive scents, as well as offering them protection and playing an important role in their pollination. In addition to the intrinsic benefits of plants, essential oils have long been used in various applications due to their antimicrobial and antioxidant properties.

The use of essential oils in egg coatings has proven to be an effective and innovative way of maintaining quality and preventing bacterial contamination. Essential oil coatings are a mixture of essential oils and biopolymers such as proteins, starches, lipids, etc. Vale et al. (2023) developed a coating based on WPI and garlic essential oil (GEO). Quail eggs coated with WPI/GEO had fewer aerobic mesophilic bacteria, fewer *Enterobacteriaceae*, and fewer molds and yeasts when compared to uncoated eggs. Araújo et al. (2023) demonstrated that corn starch (CS)-based coating enriched with basil essential oil (BEO) reduced the growth of total aerobic mesophilic bacteria, *Enterobacteriaceae*, molds and yeasts on the shell surface of quail eggs when compared to uncoated eggs. Furthermore, the coating preserved the internal quality of the quail eggs during 4 weeks of storage.

Sharaf Eddin and Tahergorabi (2019) reported that a sweet potato starch-based coating containing thyme essential oil reduced the growth of *Salmonella* throughout the storage. Upadhyaya et al. (2016) observed that coatings based on pectin and gum arabic with the addition of phytochemicals (carvacrol and eugenol) were able to reduce SE in eggshells. Sun et al. (2021) developed a coating of chitosan and beeswax (BW) or BEO and concluded that the addition of 0.5 BW or 1% BEO extended the shelf life of eggs and presented antibacterial effects against the main bacteria present in the eggshell.

Propolis is a resin produced by bees that contains a wide variety of substances and demonstrates satisfactory results when used to coat eggs (Pires et al., 2021a). Pires et al. (2019b) reported that coatings based on rice protein and propolis can increase the shelf life of eggs. Pires et al. (2021a) demonstrated that Brazilian green propolis (which is obtained from a plant native to Brazil, *Baccharis dracunculifolia*) can be used to prevent egg contamination and to increase the shelf life of eggs stored at 20°C. Ezazi et al. (2021) developed a coating based on propolis extract and chitosan that was effective against *S. enteritidis*. The authors described that the solution has unique and important properties for creating an egg coating, including good film formation, good barrier property, and antimicrobial activity.

Other materials are being investigated in the development of antimicrobial coatings for eggs. Morsy et al. (2015) demonstrated that a pullulan coating increased the shelf life of eggs by at least 2 weeks compared to that of uncoated eggs. Furthermore, the use of pullulan with nisin as a coating material was advantageous in reducing microbial load over a 10-week storage period. Yuceer and Caner (2020) reported that lysozyme-chitosan coating can improve the antibacterial characteristics of eggs while extending the storage period at room temperature.

10 Limitations in use of egg coatings

Although the terms are often used interchangeably, edible films and coatings have different presentations. Films are preformed separately from the food and later applied over it. Edible coatings or coatings can be applied directly to food in dipping or spraying methods, with the formation of a thin film on the food after drying (Gennadios and Weller, 1990). In eggs, coatings are used. In most studies carried out to verify the efficiency of new coatings, the application is carried out by immersing the eggs in the coating solution. According to the legislation of some countries, the washing operation must be continuous and completed as soon as possible, not allowing immersion-type

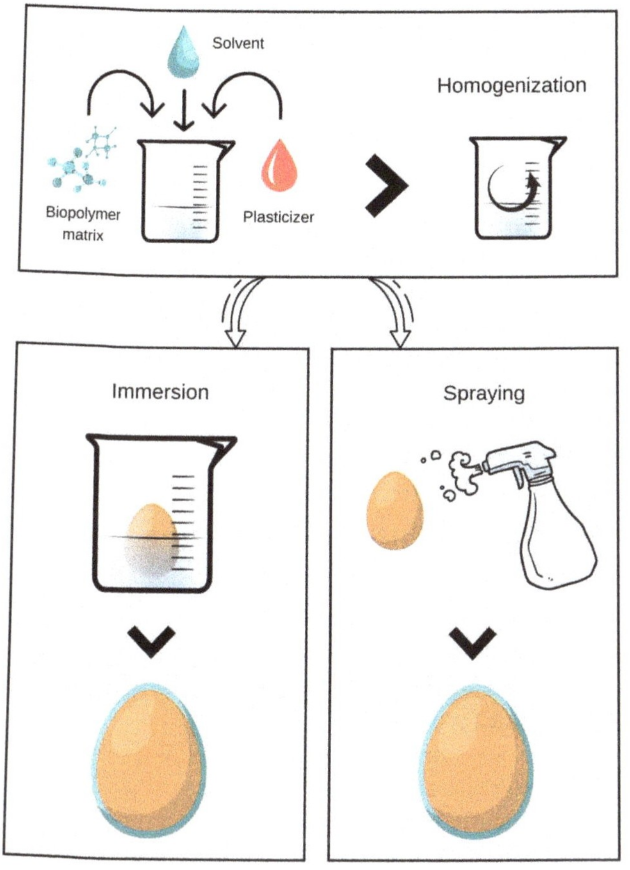

Figure 2 Experimental application scheme of coatings on eggs.

egg washing equipment. The same applies to the use of coatings. This avoids any contamination of the solution that can occur at the time of immersion, which would result in the possible contamination of the next eggs that would be immersed in the coating later. The experimental application scheme of coatings on eggs is shown in Fig. 2. Future studies may focus on the feasibility of using the coating on a commercial scale and on the analysis of the final cost for its production. Finally, the real applicability of the developed coatings will be achieved through research on a commercial scale.

11 Future trends in coating development

Despite the increased publications in recent years, there is still a dearth of information on the effect of edible coatings on the sensory characteristics of the eggs. The development of new products must take into account sensory studies, consumer acceptability, and the economic and environmental impact. Also, future studies are needed to verify the correlation between the use of coatings and the presence of active antimicrobial agents in order to minimize contamination by microorganisms.

There is a striking difference between the number of published articles and patents generated on the development of egg coverings. This is probably due to the need for fewer tests and repetitions used for the elaboration of an article, lack of large industrial scale tests, more generic formulations, and lack of successful commercial applications. In addition, there is still a lack of wide communication about their real advantages of applying egg coverings in the poultry industry. In the future, additional evaluations may be carried out, expanding the perimeter of the study, including other considerations, e.g. economic and regulatory, to assess the feasibility of applying coatings on a commercial scale.

12 Where to look for further information

Further reading

- A comprehensive review on coatings and egg safety is 'Egg quality and safety with an overview of edible coating application for egg preservation', by Sharaf Eddin et al. (2019).
- A systematic review summarizing research on the use of egg coatings is 'The use of coatings in eggs: A systematic review' by Pires et al. (2020).
- A good review on trends for the development of new egg coatings is 'Egg coatings: trends and future opportunities for new coatings development' by Pires et al. (2022).

Key journals/conferences

- PS (Poultry Science Journal) is well attended by members of the egg community.
- PSA (Poultry Science Association) holds regular meetings, including Egg Quality Section.
- WPC (World's Poultry Congress) holds meetings every 4 years on Poultry Science.

Key research in this area can be found at the following organizations

- United States Department of Agriculture (USDA) (www.usda.gov).
- Brazilian Agricultural Research Company (www.embrapa.br/suinos-e-aves).

13 References

Alleoni, A. C. C. and Antunes, A. J. (2001), 'Haugh unit as a measure of the quality of hen eggs stored under refrigeration', *Sci. Agric.*, 58, 681–685.

Alleoni, A. C. C., et al. (2006), 'Recobrimento de laranja 'Pêra' com filme de concentrado protéico de soro de leite associado a plastificantes', *Pesqui. Agropecu. Bras.*, 41, 1221–1226.

Akarca, G., et al. (2021), 'The effect of resin coating on the quality characteristics of chicken eggs during storage', *J. Food. Sci.*, 86(4), 1243–1257.

Araújo, M. V., et al. (2023), 'Preserving the internal quality of quail eggs using a corn starch-based coating combined with basil essential oil', *Process.*, 11, 1612.

Baldwin, E., et al. (1995), 'Edible coatings for lightly processed fruits and vegetables', *HortSci.*, 30, 35–38.

Baron, F. and Jan, S. (2011), 'Egg and egg product microbiology'. In Nys, Y., et al. (Eds), *Improving the Safety and Quality of Eggs and Egg Products*. Egg Chemistry, Production and Consumption, Woodhead Publishing Limited, Cambridge, UK, pp. 330–350.

Bhale, S., et al. (2003), 'Chitosan coating improves shelf life of eggs', *J. Food. Sci.*, 68, 2378–2383.

Biladeau, A. M. and K. Keener. (2009), 'The effects of edible coatings on chicken egg quality under refrigerated storage', *Poult. Sci.*, 6, 1266–1274.

Brasil, R., et al. (2019), 'Physical-chemical and sensorial quality of eggs coated with copaiba oil biofilm and stored at room temperature for different periods', *Brazil. J. Poult. Sci.*, 21(4), 001-006eRBCA-2019-0930 .

Braden, C. R. (2006), '*Salmonella enterica* serotype Enteritidis and eggs: a national epidemic in the United States', *Clin. Infec. Dis.*, 43, 512–517.

Caner, C. (2005), 'Whey protein isolate coating and concentration effects on egg shelf life', *J. Sci. Food. Agric.*, 85, 2143–2148.

Caner, C. and Yuceer, M. (2015), 'Efficacy of various protein-based coating on enhancing the shelf life of fresh eggs during storage', *Poult. Sci.*, 94, 1665–1677.

Commission, E. (2008), Commission Regulation (EC) No 889/2008 of 5 September 2008 laying down detailed rules for the implementation of Council Regulation (EC), No 834/2007 on organic production and labelling of organic products with regard to organic production, labelling and control. *Official J. L.*, 250, 1–84.

De Brito Mota, A. S., et al. (2017), 'Internal quality of eggs coated with cassava and yam starches', *Rev. Bras. Cienc. Agr.*, 12(1), 47–50.

De Leo, R., et al. (2018), 'Application of pectin-alginate and pectin-alginate-laurolyl arginate ethyl coatings to eliminate *Salmonella enteritidis* cross contamination in egg shells', *J. Food Safet.*, 38, e12567.

Dhall, R. (2013), 'Advances in edible coatings for fresh fruits and vegetables: a review', *Crit. Rev. Food Sci. Nutr.*, 53, 435–450.

Etxabide, A., et al. (2021), 'Effect of curcumin, betanin and anthocyanin containing colourants addition on gelatin films properties for intelligent films development', *Food Hydrocoll.*, 115, 106593.

Eyng, C., et al. (2021), 'Carnauba wax coating preserves the internal quality of commercial eggs during storage', *Semin. Cienc. Agrar.*, 42(3), 1229–1243.

Ezazi, A., et al. (2021), 'Development of a chitosan-propolis extract edible coating formulation based on physico-chemical attributes of hens eggs: optimization and

characteristics edible coating of egg using chitosan and propolis', *Food Biosci.*, 40, 100894.
Farag, R., et al. (1994), 'Biochemical and microbial studies on the efficiency of some coating materials for egg preservation', *Int. J. Food Sci. Nutr.*, 45, 263–273.
Fikiin, K., et al. (2020), 'Do raw eggs need to be refrigerated along the food chain? Is the current EU regulation ensuring high-quality shell eggs for the European consumers?', *Trends Food Sci. Technol.*, 100, 359–362.
Food and Nations, (2003), 'Egg marketing: a guide for the production and sale of eggs', *FAO Agricultural Services Bulletin*, 150.
Gennadios A. and C. L. Weller. (1990), 'Edible films and coatings from wheat and corn proteins', *Food Technol.*, 44(10), 63–69.
Han, J. H. and Gennadios, A. (2005), 'Edible films and coatings: a review. In: Han, J. H. (Ed), *Innovations in Food Packaging;* Elsevier Academic, Oxford, England, pp. 239–262.
Haugh, R. (1937), 'The Haugh unit for measuring egg quality', *US Egg Poult. Mag.*, 43, 522–255.
Jenkins, M. K. (1919), Commercial preservation of eggs by cold storage. US Department of Agriculture.
Jirangrat, W., et al. (2010), 'Effects of mineral oil coating on internal quality of chicken eggs under refrigerated storage', *Int. J. Food Sci. Techn.*, 45, 490–495.
Jo, C., et al. (2011), 'Effects of chitosan coating and storage with dry ice on the freshness and quality of eggs', *Poult Sci.*, 90, 467–472.
Jones, D., et al. (2018), 'Impact of egg handling and conditions during extended storage on egg quality', *Poult. Sci.*, 97., 716–723.
Kim, S. H., et al. (2009), 'Effects of chitosan coating and storage position on quality and shelf life of eggs', *Int. J. Food Sci. Techn.*, 44, 1351–1359.
Lacroix, M. and Vu K. D. (2014), 'Edible coatings and films materials: proteins. In: Han, J. H. (Ed). *Innovations in Food Packaging,* 2nd. Elsevier Academic, Oxford, England, pp. 277–304.
Lee, S. H., et al. (1996), 'Effect of chitosan coating on quality of egg during storage'. *J. Korean Soc. Food Nutr.*, 25, 288–293.
Lima, V. V. C., et al. (2020), 'Coberturas de amido de mandioca e inhame na conservação de ovos', *Res. Soc. Dev.*, 9, e23984949-e23984949.
Liu, Y.-C., et al. (2016), 'Effects of egg washing and storage temperature on the quality of eggshell cuticle and eggs', *Food Chem.*, 211, 687–693.
Maan, A. A., et al. (2021), 'Aloe vera gel, an excellent base material for edible films and coatings'. *Trends Food Sci. Techn.*, 116, 329–341.
Mendonça, D. E. and S. B. Onofre. (2009), 'Atividade antimicrobiana do óleo-resina produzido pela copaíba - Copaifera multijuga Hayne (Leguminosae)', *Rev. Bras. Farm.*, 19, 577–581.
Mendonça, G. R., et al. (2023), 'Effectiveness of babassu starch and oil coating with plasticizers to improve shelf life of eggs', *J. Coat. Techn. Res.* (in press). doi:10.1007/s11998-023-00815-1
Morsy, M. K., et al. (2015), 'Efficacy of antimicrobial pullulan-based coating to improve internal quality and shelf-life of chicken eggs during storage', *J. Food Sci.*, 80, M1066–M1074.
Mudannayaka, A., et al. (2019), 'Effect of beeswax, gelatin and Aloe vera gel coatings on functional properties and visual sensory attributes of chicken eggs stored under room temperature', *Int. J. Sci. Res. Public.*, 9(12), 96107.

Musa, T. N., et al. (2011), 'The effect of shellac as coating material on the internal quality of chicken eggs', *Int. J. Poult. Sci.*, 10, 38–41.

No, H. K., et al. (2005), 'Comparison of shelf life of eggs coated with chitosans prepared under various deproteinization and demineralization times', *J. Food Sci.*, 70, s377–s382.

Nongtaodum, S., et al. (2013). 'Oil coating affects internal quality and sensory acceptance of selected attributes of raw eggs during storage', *J. Food Sci.*, 78, S329–S335.

Oliveira, C. A. O., et al. (2014), 'Rice flour and derived products as alternatives to rice production chain', Brazil. *J. Agri-Ind Produc.*, 16, 61–67.

Oliveira, G., et al. (2022), 'Combination of cassava starch biopolymer and essential oils for coating table eggs', *Front. Sustain. Food Syst.*, 6, 987229.

Oliveira, G. S., et al. (2020), 'Conservation of the internal quality of eggs using a biodegradable coating', *Poult Sci.*, 99, 7207–7213.

Pires, P. G. S., et al. (2018), 'Rice protein coating in extending the shelf-life of conventional eggs', *Poult Sci.*, 98, 1918–1924.

Pires, P. G. S., et al. (2019a), 'Effects of rice protein coating enriched with essential oils on internal quality and shelf life of eggs during room temperature storage', *Poult Sci.*, 99, 604–611.

Pires, P. G. S., et al. (2019b), 'Effects of rice protein coatings combined or not with propolis on shelf life of eggs', *Poult Sci.*, 98, 4196–4203.

Pires, P. G. S., et al. (2020), 'The use of coatings in eggs: a systematic review', *Trends Food Sci. Techn.*, 106: 312–321.

Pires, P. G. S., et al. (2021a), 'Development of an innovative green coating to reduce egg losses', *Clean. Eng. Techn.*, 2, 100065.

Pires, P. G. S., et al. (2021b), 'Plasticizer types and whey protein coatings on internal quality and shelf life of eggs stored for 42 days', *Sci. Agric.*, 78(suppl 1), e20200271.

Pires, P. G. S., et al. (2022), 'Egg coatings: trends and future opportunities for new coatings development', *World. Poult. Sci. J.*, 78(3), 751–763.

Pissinati, A., et al. (2014), 'Internal quality of eggs subjected to different types of coating and stored for 35 days at 25° C', *Semin Cienc. Agrar.*, 35, 531–540.

Pujols, K. D., et al. (2014), 'Comparing effects of α-vs. β-chitosan coating and emulsion coatings on egg quality during room temperature storage', *Int. J. Food Sci. Techn.*, 49, 1383–1390.

Rachtanapun, P., et al. (2021), 'Effect of egg-coating material properties by blending cassava starch with methyl celluloses and waxes on egg quality', *Polymers.*, 13, 3787.

Romanoff, A. L. and Yushok, W. (1948), 'Preservation of intact eggs by sealing with chemical agents'. *J. Food Sci.*, 13, 331–335.

Rosser, F. (1942), 'Preservation of eggs: ii. Surface contamination on eggshell in relation to spoilage', *Can. J. Res.*, 20, 291–296.

Rufino, J. P. F., et al. (2023), 'Preserving eggs through coating applications using a bioproduct based on andiroba oil'. *Soc. Sci. Res. Net.* (in press). doi:10.2139/ssrn.4571546

Ryu, K. N., et al. (2011), 'Internal quality and shelf life of eggs coated with oils from different sources', *J. Food Sci.*, 76, S325–S29.

Samiullah, A., et al. (2013), 'Effects of egg shell quality and washing on salmonella infantis penetration', *Int. J. Food Microb.*, 165(2), 77–83.

Samli, H. E., et al. (2005), 'Effects of storage time and temperature on egg quality in old laying hens', *J. Appl. Poultry Res.*, 14, 548–553.

Sharaf Eddin, A. and Tahergorabi, R. (2019), 'Efficacy of sweet potato starch-based coating to improve quality and safety of hen eggs during storage', *Coat.*, 9, 205.

Solomon, S. E. (1997), *Egg and Eggshell Quality*. Iowa State University Press, Ames, Iowa pp. 1–149.

Spamer, C. (1931), 'Historic methods and present practices of preserving eggs in Holland', *US Egg Poult Mag.*, 37, 48.

Sun, R., et al. (2021), 'Effect of basil essential oil and beeswax incorporation on the physical, structural, and antibacterial properties of chitosan emulsion based coating for eggs preservation'. *LWT.*, 150(8), 112020.

Suppakul, P., et al. (2010), 'Efficacy of cellulose-based coating on enhancing the shelf life of fresh eggs', *J. Food Eng.*, 98, 207–213.

Suresh, P. V., et al. (2015), 'Application of chitosan for improvement of quality and shelf life of table eggs under tropical room conditions'. *J. Food Sci. Techn.*, 52(10), 6345–6354.

Suryani, Y., et al. (2017), 'Effect of propolis coating on the quality of eggs: microbial contamination and haugh unit', *Res J. Pharm. Biol. Chem. Sci.*, 8, 1776–1784.

Torrico, D. D., et al. (2010), 'A novel emulsion coating and its effects on internal quality and shelf life of eggs during room temperature storage', *Int. J. Food Sci. Techn.*, 45, 2241–2249.

Upadhyaya, I., et al. (2016), 'Inactivation of *Salmonella enteritidis* on shell eggs by coating with phytochemicals'. *Poult Sci.*, 95(9), 2106–2211.

USDA. (2020), 'Department of Agriculture. United States standards, grades, and weight classes for shell eggs', *AMS* 56, 210.

Vale, I. R. R., et al. (2023), 'Whey protein isolate and garlic essential oil as an antimicrobial coating to preserve the internal quality of quail eggs', *Coat.*, 13, 1369.

Vlčková, J., et al. (2019), 'Changes in the quality of eggs during storage depending on the housing system and the age of hens', *Poult Sci.*, 98, 6187–6193.

Wardy, W., et al. (2010), 'Edible coating affects physico-functional properties and shelf life of chicken eggs during refrigerated and room temperature storage', *Int. J. Food Sci. Techn.*, 45, 2659–2668.

Wong, Y., et al. (1996), 'Evaluation of mechanical and barrier properties of protein coatings on shell eggs', *Poult. Sci.*, 75, 417–422.

Xie, L., et al. (2002), 'Edible film coating to minimize eggshell breakage and reduce post-wash bacterial contamination measured by dye penetration in eggs', *J. Food Sci.*, 67(1), 280–284.

Yang, L. and Paulson, A. T. (2000), 'Mechanical and water vapour barrier properties of edible gellan films', *Food Res. Int.*, 33, 563–570.

Yuceer, M. and Caner, C. (2014), 'Antimicrobial lysozyme–chitosan coatings affect functional properties and shelf life of chicken eggs during storage', *J. Sci. Food Agric.*, 94, 153–162.

Yuceer, M. and Caner, C. (2020), 'The effects of ozone, ultrasound and coating with shellac and lysozyme–chitosan on fresh egg during storage at ambient temperature. Part II: microbial quality, eggshell breaking strength and FT-NIR spectral analysis', *Int. J. Food Sci. Techn.*, 55, 1629–1636.

Chapter 6

Pasteurization treatments of whole and liquid egg products: optimizing safety and quality parameters

Valérie Lechevalier, INRAE, L'Institut Agro Rennes Angers, STLO (Science and Technology of Milk and Eggs), Rennes, France

1 Introduction

Eggs are an important part of the human diet and are consumed all over the world as shell eggs (Table eggs) or as egg products (e.g. liquid, frozen or dried egg white, egg yolk or whole egg) used as ingredient in many foodstuffs. The safety of eggs is thus a major concern for both egg industry and consumers. Generally, the inside of a chicken egg is considered as sterile immediately after laying (Qi et al., 2020), but the surface of the egg shell is contaminated with a thousand to one million bacteria as the intestinal bacteria of the hens can adhere to the eggshell surface. Moreover, soil bacteria from the poultry farm environment can also adhere to the egg's surface after laying and becoming dominant with time (Hatta & Kubo, 2024). Fortunately, eggs have a remarkable arsenal of antimicrobial defenses (cuticle, egg shell, shell membranes, alkaline pH in egg white, antimicrobial molecules in egg white, vitelline membrane) that can, however, be affected by the conditions during egg production, collection, packaging, storage, and manufacturing in case of egg products (Baron et al., 2024). Significant advances in egg safety

http://dx.doi.org/10.19103/AS.2024.0150.10

have improved our understanding between egg quality and safety and the risk of foodborne pathogens as well as mitigation strategies. However, eggs and egg products continue to be implicated in food poisoning resulting from bacterial contamination: in the European union, 45.6% of reported *Salmonella* outbreaks in 2018 were caused by the consumption of eggs or egg products and in the USA, eggs are also the main vector for *Salmonella* foodborne illnesses. Moreover, eggs and egg products can also be spoiled by non-pathogenic bacteria that lead to high economic losses for the egg industry (Baron et al., 2024). Guaranteeing egg and egg product safety is thus the major concern of the egg industry and up to now, it is mainly ensured by pasteurization treatments.

For egg product manufacturers, the second objective is to provide their customers with functional products for their applications (foaming, gelling, and binding properties). However, these properties rely on preserving the structure of proteins, which unfortunately are particularly sensitive to denaturation by heat. Balancing the objectives of microorganism destruction while preserving the functional properties constitutes a challenging issue, which has been the focus of research over recent decades. Emerging alternative technologies for processing eggs and egg products have been recently reviewed by Bhat et al. (2023). Some of them (pulsed light, cold plasma, ohmic heating, microfluidization, radiofrequency, and hyperbaric storage) can at least partly substitute pasteurization while preserving functional properties.

More recently, the emergence of energy transitions challenges the sector by considering the energy-intensive aspect of the applied processes. Alternative technologies developed for preserving functional properties can thus also be considered under this point of view.

2 Pasteurization principles

In the egg product industry, microbiological safety of egg products is mainly achieved by heat treatment. The destruction of microorganisms at constant temperature follows a first-order kinetics:

$$\text{Log } N0/N = t/D_\theta \qquad \text{(Eq. 1)}$$

Where $N0$ and N are the initial and final (after treatment) viable population, respectively, t is the treatment time, and D_θ is the decimal reduction time at the temperature θ, i.e. the time required to divide by ten the initial population. According to Eq. 1, complete destruction can never be reached and the number of surviving microorganisms depends directly on the initial number of microorganisms.

Increasing the temperature decreases the decimal reduction time D_θ. The relationship between D_θ and the temperature is given by the following equation:

$$\text{Log } D_\theta/D_{\theta ref} = (\theta_{ref}-\theta)/z \quad \text{(Eq. 2)}$$

where $D_{\theta ref}$ is the value of D_θ at the reference temperature θ_{ref}, and z is the temperature increase required to divide by ten the decimal reduction time.

The thermal resistance characteristics of a microorganism are fully described if D_θ and z are known. These values may vary depending on the type (species, strain) and the physiological state of the microorganism (spore or vegetative form, potential adaptation to a previous stress) and the nature and composition, the pH and the water activity of the egg product (Baron et al., 2024).

3 Contamination issues in eggs and egg products

3.1 Preventing foodborne illnesses due to eggs and egg products

Current performance levels require that pasteurization treatment aims for a 5–6 decimal reductions to be considered as effective. However, egg pasteurization requirements vary according to the legislation of the countries: for example, in Australia, New Zealand, UK, or China, authorities require that liquid whole egg be heated to a temperature no lower than 64°C for at least 2.5 min. In the USA, a lower temperature is required, 60°C for 3.5 min, that appeared sufficient to ensure satisfactory pasteurization of liquid whole egg and egg yolk. In France, there is no statutory heat treatment; only microbiological results are determined by regulations. To achieve this, the heat treatments generally applied to egg products varied from 65°C to 68°C for 5–6 min for whole egg and egg yolk and from 55°C to 57°C for 2–5 min in liquid egg white. Based on these ranges of time–temperature values applied worldwide, the reduction levels for *Salmonella enteritidis* in egg products can be calculated, as it is a pathogenic microorganism of importance for foodborne illnesses in egg products (Table 1).

Table 1. Efficacy on *Salmonella enteritidis* destruction for the worldwide used pasteurization treatments of liquid egg products

	Whole egg	Egg yolk	Egg white
D-values (min)	$D_{60°C}$ = 0.22–0.44	$D_{63°C}$ = 0.27–0.35	$D_{57°C}$ =1.7--2.1
z-values (°C)	3.3	4.6–6.6	4.3–5
Treatments	60–68°C for 23.5–6 min	60–68°C for 23.5–6 min	55–57°C for 2–5 min
Number of decimal reductions	8–151	45–500	2.4–2.9
Treatment efficacy	Sufficient	Sufficient	Insufficient

Source: Adapted from: Baron et al. (2024).

The number of decimal reductions reached in egg white thus appeared as insufficient as it did not achieve a 5-log reduction of *Salmonella*. However, it should be highlighted that egg white is not a favorable environment for bacterial development (alkaline pH, antimicrobial molecules).

For many microorganisms, the heat treatment that does not result in destruction (sublethal treatment) may result in an increase into tolerance to subsequent heat stress through the induction of a set of specific proteins called 'heat shock proteins'. The time–temperature applied must then be rigorously monitored. In the specific case of *Salmonella* in egg products, an increase in the thermal resistance of the bacteria is observed when (1) a sublethal treatment is applied beforehand, allowing the induction of such heat shock proteins (Shah et al., 1991), (2) the pH of egg white is closer to 8 than 9 (fresh eggs) (Palumbo et al., 1996), or (3) salt or sugar is added (Palumbo et al., 1995) as a decrease of aw generally increased the thermal resistance of microorganisms (Denis et al., 1995).

After pasteurization, a significant decrease is obtained for a number of enterobacteria, such as *E. coli*, and bacteria of the genera Brochotrix and *Pseudomonas* in egg products (Protais et al., 2006). The effectiveness of pasteurization also appears to be good for *Campylobacter*, *Salmonella* (Protais et al., 2007b), and *Listeria* (Moore & Madden, 1993). The characteristics of the latter (pathogenicity, ability to create biofilms and growth at low temperature) require manufacturers to be very vigilant in respecting pasteurization rates, cleaning and disinfection practices, and maintaining cold chain temperatures. It is nevertheless important to highlight that the incidence of *Salmonella* and *Listeria* still be low in egg products: 0.16% of the 1246 egg products samples analyzed in 2019 in the EU and 0.14% of the 15 258 pasteurized liquid egg products analyzed between 2008 and 2017 in the USA were *Salmonella* positive and none of the 26 samples analyzed in the EU were positive to *Listeria*, while 0.09% *Listeria monocytogenes* positive pasteurized egg product samples were reported in the USA in 2016–2017 (EFSA and ECDC, 2021; FSIS, 2021). Pasteurized products can then be stored at 4°C from between a few days to a few weeks.

3.2 Preventing egg product spoilage

Streptococci, *Enterocci*, and *Bacillus* spores are less affected by pasteurization than other bacteria (Protais et al., 2006). However, particular attention has to be paid to the presence and potential issues linked to *Enterococcus* and *Bacillus* as their involvement in sanitary and/or spoilage issues must not be underestimated.

Bacteria of the genus *Enterococcus* are among the most thermotolerant of the non-spore-forming bacteria. Due to their ubiquitous nature and especially

Table 2. Prevalence of *Enterococcus* in raw and pasteurized liquid whole egg (Protais et al., 2007a) and egg white (Techer et al., 2015) and main species identified

	Whole egg	Egg white
Raw	92.4%	38.8%
Pasteurized	20.1%	93.4%
Number of samples	144 (raw/pasteurized)	18 (raw)/15 (pasteurized)
Incidence of species in positive samples	*E. faecalis* (64.6%) *E. durans* (12.1%) *E. faecium* (8.8%) *E. casseliflavus* (7.4%) *E. hirae* (5.5%) *E. gallinarum* (1.4%) *E. avium* (0.5%)	*E. faecalis* (33%) *E. rivorum* (6.1%) *E. avium* (6.1%) *E. hirae* (3.1%) *E. faecium* (3.1%) *E. pseudoavium* (3.1%)

the presence in the commensal flora of the chicken digestive tract, it is not surprising to find them on the eggshell and subsequently in the egg products. The prevalence of *Enterococcus* species has been quantified in raw and pasteurized whole egg and egg white (Table 2).

The distribution of the species was not the same for the raw and pasteurized whole egg, the heat treatment being more effective on *E. faecalis* than *E. faecium* or *E. durans*, for example.

The results on egg white show that the prevalence is much higher in pasteurized egg white than in raw egg white. This illustrates the selection pressure of pasteurization: by eliminating the heat-sensitive microflora, pasteurization allows *Enterococcus* to become the dominant flora.

Due to the ubiquitous nature and the ability of *Bacillus* to sporulate, it is also difficult to avoid contamination of the raw and pasteurized egg products. The problem of elimination is particularly acute since egg products can only be subjected to moderate heat treatment due to the heat sensitivity of the constituents and especially proteins. Byrne et al. (2006) reported *D*-values for *B. cereus* spores ranging from 2 min at 95°C to 32.1 min at 85°C and a *z*-value of 8.5°C. Compared to the treatments generally applied to egg products (Table 1), it can be concluded that these treatments are ineffective on spore-forming microorganisms. In addition, these treatments can trigger the germination of *Bacillus* spores thanks to the destruction of competing microorganisms potentially present in a vegetative form.

4 Pasteurization of shell eggs

Shell surface contamination is an obvious risk for the internal contamination of eggs, either by penetration through the shell or by contact when the shell is intentionally broken. This is why washing grade A eggs ('fresh' or 'table' eggs) is carried out in many countries such as the USA, Canada, Australia, and Japan (Hierro et al., 2009). However, the EU does not consider washing of shell eggs intended for retail sale sufficient (OJEU, 2003). Additional techniques for decontamination of shell surfaces such as those presented hereafter are thus considered.

4.1 Hot water bath

Immersion in a hot water bath at 57°C for 25 min is an efficient decontamination technique against *S. enteritidis* (−3 log cfu mL^{-1}), with no detrimental effect on the major egg properties (Hou et al., 1996). Complete inactivation can be even obtained with more drastic treatments (50–57 min at 58°C), but this is too long to be cost-effective and damages egg white quality (Schuman et al., 1997). Higher temperature (100°C, 3 s) causes numerous shell cracks (Himathongkham et al., 1999).

4.2 Hot air

Hou et al. (1996) developed a low-temperature-extended time (180 min at 55°C in a hot air oven) pasteurization process for intact shell eggs which can achieve a 5-log reduction of *S. enteritidis*. This treatment did not significantly impact the in vitro digestibility of unpasteurized and pasteurized albumen measured by the AOAC (Association of Official Analytical Chemists) method (Hank et al., 2001). However, more recently, Wang et al. (2018b) observed that egg white in vitro digestibility was significantly higher after an in-shell pasteurization treatment of 32 min at 56°C.

James et al. (2002) identified 180°C for 8 s as the treatment enabling the highest surface temperature that can be achieved without detrimental changes to internal egg quality. Pasquali et al. (2010) neither observe any quality changes after exposing eggs twice to a temperature of 600°C for 8 s with an interval of 30 s of cold air, obtaining a 1.9 log cfu/eggshell reduction of *S. enteritidis*.

4.3 Ultrasonication

Ultrasonication is a clean non-thermal technology that uses sound waves above 20 kHz. It has been reported to reduce microbial populations of eggs: maximum reductions of 1.61, 2.52, and 2.05 log CFU/egg of *Salmonella* spp., *Staphylococcus* spp., and yeast/molds, respectively, were achieved for the shell eggs treated at 35 kHz for 15 min (Aygun & Sert, 2012).

5 Pasteurization of liquid egg products

Pasteurization of egg products is a well-established technology today. Several technologies are available for the pasteurization of liquid egg products. They are still under research to improve thermal efficiency, reduce fouling, increase cycle times, and minimize biofilm formation. Egg white, for which the applicable heat treatments do not always achieve a satisfactory level of decimal reductions of *S. enteritidis* is of particular concern.

5.1 Batch pasteurization

Batch pasteurization is sometimes used in small units where continuous pasteurization is not economic because of the small volumes treated. The equipment used is an airtight, pressure-resistant tank equipped with a heating and cooling device. The tank is pressurized with sterile air to prevent contamination from the outer air during cooling. In the biggest tanks (3000–8000 L), a recirculation system equipped with a heat exchanger allows for more efficient heating and quicker pasteurization cycles.

For whole eggs, it is not possible to apply a heat transfer fluid temperature exceeding 65°C at the end of heating. In practice, temperatures between 60°C and 62°C are targeted for both the egg white and yolk, with holding times of at least 15 min. These standards also help maintaining satisfactory functional properties. However, it is challenging to find operational conditions for egg white that reconcile bacterial destruction and the maintenance of foaming properties. In all cases, particular attention must be paid to the presence of foam, whose thermal conductivity is significantly lower than that of liquid egg products (Dickinson et al., 1970). The cleanability of such a system, with significant heating surfaces, complex agitation, and pressurization, must be carefully qualified. Despite its seemingly simple design, this technology actually requires expert management for effective implementation.

The thermal balance of batch pasteurization is very poor, as the entire product and installation (tank and heat transfer circuits) are heated and cooled without any heat recovery possibility. However, material losses are low since there is no dilution front. This technique is still used because it offers complementary solutions to continuous pasteurizers: treatment of viscous products, pasteurization of small batches, pasteurization of products rich in dry extract without a dilution front, pasteurization with long holding times, and pasteurization with degassing in the heating phase.

5.2 Hot-filling

Hot-filling consists of maintaining a high temperature during filling and cooling after packaging closure. In the case of egg products, a wide variety of thermal standards are possible, as reported in numerous national, European, and American patents. Some consider hot-filling with lower temperatures around 54–56°C, followed by holding in a warm chamber for 12–72 h. Others propose hot-filling around 60°C with a short holding time, similar to what is described in the case of batch pasteurization. The heating or pasteurization prior to hot-filling is done quite simply in a batch process or with a plate heat exchanger. The thermal efficiency is better than batch pasteurization but does not allow for heat recovery. The ability to use containers of various shapes, for which aseptic packaging can sometimes be challenging, is one of the main attractions of this technology. However, the long hot and cold holding times complicate the mechanization of this process and limit achieving high production rates. From a practical standpoint, controlling the foam that forms during hot-filling and developing a cooling system to quickly reach temperatures below 4°C at the core are the specific technical challenges that need to be addressed.

This method has been developed for packaging of less than 5 kg, whether in whole egg or egg yolk, as products produced in this way have good hygiene quality. However, applying such treatment to egg white at the typical temperatures of 55–57°C remains challenging because extended cooling times lead to excessive exposure of the egg white to temperatures that result in a degradation of functional properties (Wang & Wang, 2009). Nevertheless, there is an innovative and effective temperature range for this product, utilizing the egg white's intrinsic bactericidal properties (Liot & Anza, 1996). According to the claims of this patent, a treatment at 42–45°C for 1–5 days can sterilize the egg white and produce a shelf-stable product at room temperature.

5.3 Continuous pasteurization

However, batch pasteurization and hot-filling, as previously presented, remain marginal practices. Indeed, manufacturers generally prefer to implement continuous processes with heat recovery systems and satisfactory levels of productivity. These technologies are widely used for the production of fresh, frozen, and dried liquid egg products. For fresh liquid products, ultra-clean packaging, positioned immediately after pasteurization, is required to achieve minimum shelf life limits (SLL) of at least 4 weeks at 4°C without breaking the cold chain. Thermal treatments exceeding 70°C with short holding times (known as HTST or high temperature-short time treatments) have also been proposed. Beyond the undeniable pasteurizing efficiency of these treatments, other factors must be considered in achieving a long shelf life. The quality of the raw material, the hygienic design of equipment, the design and parameters of the cleaning-in-place (CIP) station, and the packaging process organization are also decisive. It is worth noting that the determination of the SLL is the responsibility of the egg product manufacturer, based on a validation protocol they impose on themselves. The choice of target bacteria and the decision to break the cold chain or not are determined, in part, by the intended use of the products and the quality of the distribution network.

5.4 Standard pasteurization equipment with heat recovery

Pasteurizers used in egg products factories are mostly plate heat exchangers (Fig. 1) or tubular (Fig. 2) exchangers with an initial heat recovery section in which the cold product is preheated by the warm pasteurized product. The following section heats the egg product to the pasteurization temperature; in this section, the temperature difference between the hot water in the exchanger and the egg product has to be less than 1°C. After holding (3–10 min) and precooling in the recovery section, the last section cools the product below 4°C.

Increasing the number of plates in the heat exchanger section improves the thermal efficiency of the system but also increases pressure drop, which affects flow, depending on the viscosity of the product being processed. The

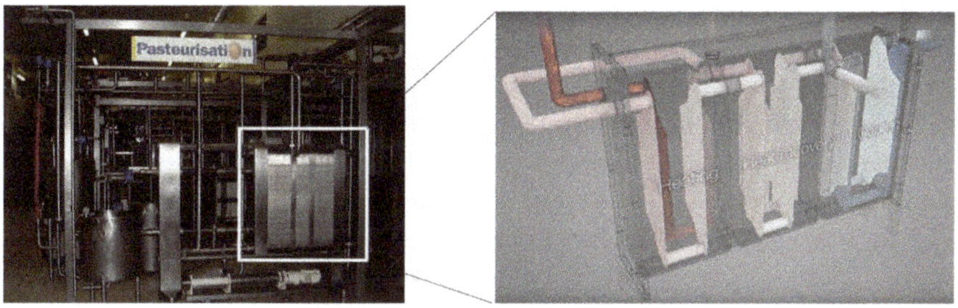

Figure 1 Multi-section pasteurizer gasketed plate-and-frame heat exchanger (Alfa Laval).

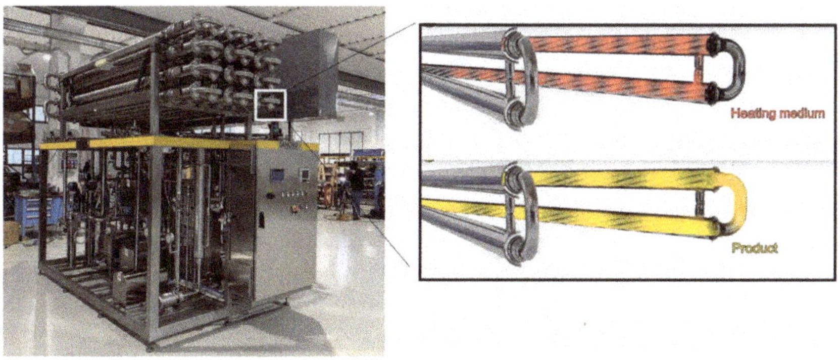

Figure 2 Multi-section tubular heat exchanger pasteurizer (SANOVO).

optimal configurations (sizing of the exchanger) differ then for egg white and egg yolk pasteurization (Gut et al., 2005a, 2005b).

Plate heat exchangers are commonly used for egg yolk pasteurization, although this design is difficult to operate for highly viscous products. Generally, whole egg and egg white can be pasteurized on equipment designed for egg yolk; in such cases, the flow rate is increased to limit fouling in the hot section, which is naturally higher than in egg yolk. Conversely, equipment designed for whole egg or egg white pasteurization cannot be satisfactorily used for yolk treatment, since its viscosity is higher.

As for other food products, a tubular heat exchanger is a more appropriate design for increasing viscosity, as it usually enables users to reach higher Reynolds numbers (i.e. turbulent flow) with an acceptable pressure drop. Thus, tubular heat exchangers can be used to pasteurize whole eggs. This equipment enables users to reach higher temperatures (67–68°C for 1–3 min for whole egg), compared to classical plate heat exchangers (64°C for 2–3 min).

At the outlet of the heating section, egg product pasteurizers are equipped with relatively large chambers to maintain the product temperature for 3–10 min depending on the configurations. Caution should be exercised in controlling the residence times inside the holding tubes, as they can vary by about 20% around the average residence time (Landfeld et al., 2001). It is tempting to increase the tube diameter to reduce the total length, space, and installation cost. However, this can result in the risk of laminar flow, with low flow velocities in the tubes, leading to increased residence time dispersion and fouling. Additionally, the flow rates required for cleaning, which should be performed under turbulent conditions, significantly impact investments in CIP equipment.

Egg yolk stands out due to its significantly higher viscosity at low temperatures compared to egg white and whole egg. This phenomenon is further accentuated with salted egg. This results in high pressure drops in the cooling zone of heat exchangers, requiring adjustments. Using a cooling fluid below 2°C indeed leads to an increase in the viscosity of the yolk in contact with the exchange surface, which hinders thermal transfer efficiency. Therefore, it is preferable to control the cooling fluid temperature between 2°C and 3°C to limit the exchange surface required for cooling. It is often necessary to add a booster pump at the entrance to the cold section to limit excessive pressures upstream of this cold zone. However, this solution is less attractive than optimizing the heat exchange and flow of egg yolk from the pasteurized circuit through the selection of plates with a suitable design.

Plate heat exchangers are the most common technology for natural or salted egg yolk. The viscosity of the products does not easily allow for achieving a turbulent flow in a tubular configuration, which may lead to insufficient exchange surface. The choice of plates, defined by the spacing between two plates and the channel geometry, must be adapted to the product's viscosity. An optimization of the design has been proposed for natural egg yolk (Gut et al., 2005a). It is generally possible to process egg white and whole egg on equipment designed for egg yolk, with flow rates often 2–3 times higher than for egg yolk, to limit fouling in the hot section that would be naturally more important than for egg yolk. Conversely, an installation optimized for egg white or whole egg cannot process egg yolk under satisfactory conditions because of its higher viscosity.

5.5 Continuous pasteurization: high-temperature/short-time equipment

The high-temperature/short-time (HTST) concept was introduced for egg products in the USA in the mid-1980s in order to lower protein denaturation. It aimed to reduce the necessary holding times by applying higher temperatures. The effect of thermal treatments ranging from 63.7°C to 72.2°C for 192.2 s to 2.7 s on whole egg was studied in the laboratory in terms of soluble protein losses and increased product viscosity (Ball et al., 1987; Hamid-Samimi et al., 1984). It was established that the thermal treatment should not result in a loss of more than 5% of soluble proteins to maintain satisfactory functional properties.

The operating limit of plate heat exchangers is determined by the relatively low pressure drops they can withstand, leading to lower flow rates and increased residence times. In practice, it is difficult with this geometry to exceed 67°C for whole egg and 56°C for egg white without encountering fouling in the heating zone, which is incompatible with satisfactory operational conditions. This is why HTST equipment uses tubular exchangers, which are easier to clean, can tolerate higher pressure drops (up to 30 bar), and thus reduce the risk of contamination between compartments.

5.6 Ohmic heating

Ohmic heating occurs when an alternating electrical current is passed through a food, resulting in the internal generation of heat due to the electrical resistance of the food. This technique has many advantages: temperature is quickly achieved, there are low heat losses since thermal energy is generated directly within the food itself, and liquids are uniformly heated (Pereira & Vicente, 2010). This technology is particularly suitable for the processing of viscous liquid foods such as egg products (Icier & Ilicali, 2005). Reznik (1993, 1996) registered two patents for the application of ohmic heating in liquid egg product pasteurization.

A device, called the 'tube-in-passage' or Actijoule® (Fig. 3), involves applying an electrical potential difference at the ends of a tube, heating the

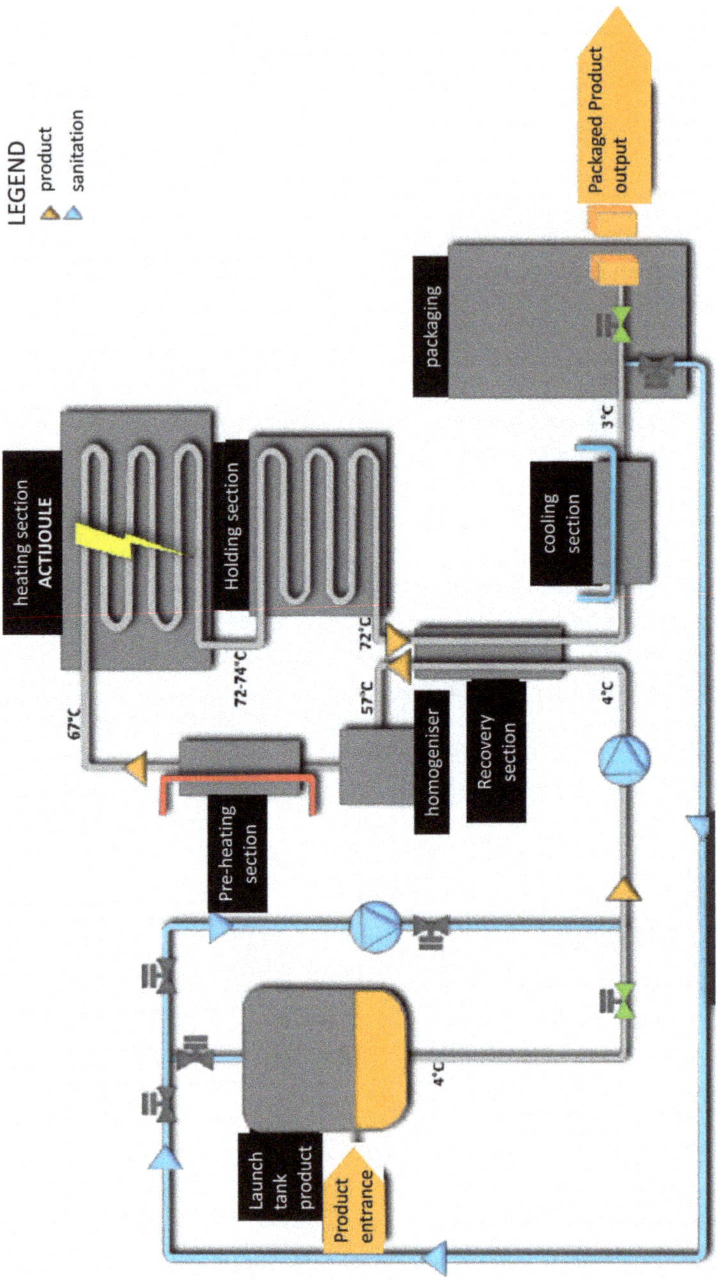

Figure 3 Schematic diagram of a current passage tube pasteurizer or Actijoule® (Actini).

product through Joule's effect and ensuring a constant temperature difference between the wall and the product over the entire exchange surface. This makes it possible to finely control temperatures at the tube contact points (within 0.2°C), reducing fouling and increasing the cycle times between cleanings. For the most viscous products (yolk and sweet and/or salted products), thermal recovery is achieved with tri- or quadrilobed tubes (Fig. 4), which, on the one hand, increase the exchange surface with constant dimensions and, on the other hand, achieve better temperature homogeneity in the section by reducing the critical distance between the wall and center. The Actijoule® system, due to the strong turbulence generated in the product flow (Reynolds number of 5000–7000), allows pushing the treatment temperature limits of whole egg from 66°C to 70°C compared to classical tubular configurations; the product is then held for 10–120 s depending on the treatment. Actiflash®, a technology derived from Actijoule®, is characterized by a small tube diameter and thus a highly turbulent flow; the Reynolds number can reach values up to 15 000 for low viscosity products like whole egg or egg white. The overall heat transfer coefficient can then reach 5000 $W \cdot m^{-2} \cdot °C^{-1}$, instead of 1400 $W \cdot m^{-2} \cdot °C^{-1}$ with Actijoule®. This technology, applied in addition to conventional heating for a passage from 67°C to 74°C in 14 s for whole egg, allows thermal treatments ranging from 71°C for 60–90 s to 74°C for 10–20 s. This type of treatment achieves a higher reduction in all bacterial populations than a conventional treatment of 65°C for 6 min while limiting fouling and denaturation of constituents.

Residual contaminations observed in egg products after pasteurization come from thermoresistant bacteria released from biofilms established in the hot zone and downstream of pasteurization. Reducing the contact surfaces with the product is therefore an identified area of progress to improve the hygienic quality of egg products. Research is focused on developing systems that reduce

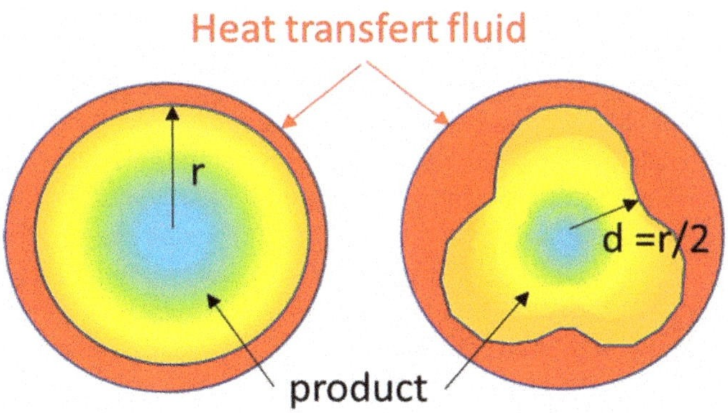

Figure 4 Classic tubular geometry (a) versus 'tri-lobe tube' geometry (b) developed to increase the exchange surfaces (Actini).

heating surfaces. For example, the company Diamond markets a cavitation-friction heating device; the egg product is fed into a high-speed rotor-stator system that provides the final heating of the product. To our knowledge, this device has not yet been the subject of published studies describing its cleanability and the functional quality of the products obtained.

5.7 Continuous pasteurization: specific aspects of egg white pasteurization

Raw egg white is not a favorable medium for bacterial growth (Baron et al., 2024), and microbiological contaminations observed in egg processing plants remain low. However, pasteurization in liquid form is a mandatory step for the commercialization of liquid egg white, with the exception of egg white powder, for which bacterial decontamination is achieved by powder heating (see Section 6).

Although, thermo-bacteriology follows well-established laws, allowing access to characteristic parameters such as the decimal reduction time at a given temperature (D_θ) and the temperature increase (z) necessary to reduce this characteristic time by a factor of 10 (see Section 2), in practice, several factors make it very difficult to calculate the absolute efficiency of a thermal treatment in industrial conditions in advance. First, thermo-sensitivity varies greatly from one bacterial strain to another, and the bacterial populations contaminating products in an industrial setting are always extremely complex. Secondly, the few D_θ and z values available in the literature for the bacterial strains of interest have generally been determined from laboratory experiments, which more or less verify the assumption that the temperature is instantaneously raised to a value θ, i.e. that the influence of temperature rise and fall times is negligible. However, in cases where these times are not negligible, the destruction attributed to a treatment (constant temperature θ – time t) is underestimated, which results in overestimating the D_θ values. This is somewhat analogous to the industrial challenge of accounting for the dispersion of residence times observed in thermal treatment facilities, which determines the local intensity distribution of treatment experienced by the fluid. Finally, the physicochemical characteristics of the products to be treated, especially pH, significantly modify bacterial thermo-sensitivity. Froning et al. (2002) report, for a mixture of five strains of *Salmonella* and treatment of egg white in immersed capillary tubes at 56.7°C, a decrease in the D_θ value as pH increases: from 2.47 min at pH 7.8, $D_{56.7°C}$ decreases to 2.25 min at pH 8.2, 1.99 min at pH 8.8, and 0.85 min at pH 9.3. However, even under optimal pH conditions and with these D_θ values, which are among the lowest reported in the literature, it appears that standard pasteurization of liquid egg white at 57°C for 2–5 min results in a decimal reduction of the *Salmonella* population of approximately 2–6 log. Considering the usually expressed requirements for microbiological safety and considering the numerous limitations associated with these theoretical calculations, it appears that the pasteurization of liquid egg white remains a delicate operation to undertake, and the results cannot be guaranteed systematically. Therefore, the initial microbiological quality of the product is of particular importance. In all cases, pasteurized liquid egg white must undergo release testing before being placed on the market.

5.8 Impact of egg products pasteurization on their qualities

5.8.1 Impact on protein structure

The effect of pasteurization on protein structure has been quite well characterized; the tertiary structure of proteins begins to change between 50°C and 60°C and depending on their molecular weight, stability, and size, the secondary structure is affected thereafter when temperature reaches over 65°C (Vanga et al., 2017). Pasteurization treatment decreases protein solubility as early as 1 min at 60°C for whole egg (Herald & Smith, 1988; Uysal et al., 2017), 2.5 min at 72°C for egg yolk (Le Denmat et al., 1999), and 20 min at 55°C for egg white (Van der Plancken et al., 2006).

Egg yolk viscosity, solubility, and electrophoresis profile are not affected when heated under 69°C. Above 69°C, protein solubility decreased strongly and apparent viscosity increased sharply because of aggregation of the proteins (Le Denmat et al., 1999). Blume et al. (2015) confirmed by Fourier transform infra-red spectroscopy that no secondary structure change was observed in egg yolk under 75°C. Egg yolk proteins are thus well preserved by industrial pasteurization treatments that stay under these temperatures.

Using capillary electrophoresis, Uysal et al. (2017) showed that heat treatments at 60°C for 1–5 min did not significantly modify the whole egg protein profile in electropherograms. When treated at 64°C, few proteins attributed to supernatant protein of yolk (γ-livetin) or undefined egg white protein decreased or disappeared. At 68°C, the electropherograms revealed that proteins were substantially affected; the bands for ovotransferrin, lysozyme, and yolk supernatant proteins were strongly decreased. However, ovalbumin and ovomucoid were not affected by these heating conditions. These results agreed with Lechevalier et al. (2017a) who observed that pasteurization treatment of whole egg at temperatures below 66°C does not induce protein aggregation.

Pasteurization treatment of liquid egg white causes an increase in protein surface hydrophobicity and exposed sulfhydryl content and a decrease in protein residual enthalpy, indicating a partial loss of protein structure. Ovotransferrin and lysozyme signals in differential scanning calorimetry disappeared after 20 min at 60°C, whereas ovalbumin signal remained as it is less heat sensitive (Van der Plancken et al., 2006). Lechevalier et al. (2005) reported that an industrial pasteurization treatment (56°C for 6 min) induced the aggregation of ovotransferrin and caused the unfolding of ovalbumin. Egg white proteins are thus sensitive to pasteurization treatment, the impact on their structure depends on the protein under consideration. Liquid egg white pasteurization is thus a tricky process because egg white proteins are very heat sensitive.

5.8.2 Consequences on protein digestibility

Pasteurization-induced changes in secondary and tertiary structures of egg proteins lead to the exposure of enzyme cleavage sites and consequently may increase protein digestibility (Dupont & Nau, 2019). Most of the studies have been performed on egg white proteins since they are the most heat-sensitive egg proteins. Van der Plancken et al. (2002) and Jimenez-Saiz et al. (2011) showed that susceptibility of ovalbumin to hydrolysis was increased after a heat treatment higher than 60°C for 10–30 min, whereas the ovomucoid susceptibility was not changed after 30 min at 60°C. Van der Plancken et al. (2003) showed that heat-induced changes in the susceptibility of egg white proteins to enzymatic hydrolysis can be described by a first-order fractional conversion model. The rate constant of the model was higher for egg white proteins than for pure ovalbumin solution probably because some egg white proteins have lower thermal stability than ovalbumin. Van der Plancken et al. (2004) reported that when egg whites were heated between 50°C and 85°C for 20 min, two temperature ranges (67.5°C and 77°C) could be distinguished in terms of susceptibility to enzymatic hydrolysis. They attributed the last one to ovalbumin heat denaturation which was responsible for the strong increase in susceptibility to enzymatic hydrolysis, whereas the lower temperature range (between 55°C and 60°C) was associated with the denaturation of the more sensitive egg white proteins such as ovotransferrin or lysozyme. Liu et al. (2017) observed that heat treatment of egg white at 60°C for 10 min at pH over 5 significantly reduced ovotransferrin and lysozyme bands intensity on SDS PAGE analysis. However, they did not notice any significant change in *in vitro* peptic digestibility of egg white proteins, maybe because of the shorter treatment time than the one performed by Van der Plancken et al. (2004).

Nevertheless, when increasing the temperature up to 80°C, Liu et al. (2017) reported an increase in ovalbumin and lysozyme susceptibility to pepsinolysis irrespective of the pH. This agreed with Jimenez-Saiz et al. (2011) who observed an increase in the susceptibility of ovalbumin to pepsinolysis after 15 min of treatment at 90°C and Martos et al. (2011) and Takagi et al. (2003) who obtained complete hydrolysis of ovalbumin after 5 min or 30 min at 100°C, respectively. Van der Plancken et al. (2004) reported that an increase in susceptibility to enzymatic hydrolysis was faster with increasing temperature at atmospheric pressure, whereas under high pressure, an antagonistic effect was observed since the association of the treatments enhanced protein aggregation. It was therefore recommended to work at lower temperatures during pressure treatments to optimize the susceptibility of enzymatic hydrolysis of egg white proteins.

Pasteurization temperature of 60–90°C has been reported to decrease the efficacy of protease inhibitors (Damodaran, 2017) and egg white contains several protease inhibitors such as ovomucoid with antitryptic activity, ovostatin with inhibitory activity against several metalloproteases and serine proteases, and ovoinhibitor inhibiting trypsin, chymotrypsin, and some other proteases (Wu et al., 2010). This inactivation of protease inhibitors during pasteurization treatment is also an explanation to the increase in protein digestibility.

Lechevalier et al. (2017a) evaluated the effect of pasteurization rates on *in vitro* digestibility of whole egg proteins. They observed that pasteurization time and its interaction with pasteurization temperature showed a significant effect on the protein degree of hydrolysis at the end of both gastric and intestinal stages. It also affected the *in vitro* digestion kinetics as indicated by SDS PAGE. Less intense pasteurization treatments (up to 10 min at 60°C or 62°C and 4 min at 64°C) showed a higher rate of protein hydrolysis, whereas more intense treatments (more than 4 min at 64°C or 66°C for 4–10 min) induced protein aggregation that reduced the digestibility.

5.8.3 Consequences on protein techno-functional properties

Heat treatment also damages egg white foaming properties. Industrial pasteurization (56°C – 6 min in plate heat exchangers) can decrease foaming capacity and stability by 10% (Kline et al., 1966; Lechevalier et al., 2005); up to a threefold increase in drainage was observed by Ferreira et al. (1995). Heating during egg white drying also decreases egg white foaming capacity between 10% and 60%, depending on the drying pH (Hill et al., 1965), and foam stability by 20%. However, these results are contradicted as Hammershøj et al. (2004) observed an increase of foaming capacity by 10% and no change in foam stability.

Egg white gelling properties are less sensitive to heat treatments, in the range classically used for pasteurization (i.e. 3–6 min at 56°C). However, as soon as the temperature of 58°C is reached, dense and irregular protein aggregates are formed, increasing egg white turbidity (Seideman et al., 1963). This the reason why the pasteurization temperature of egg white is usually still under 57°C.

Ohmic heating, as an alternative technology to conventional pasteurization, can improve some whole egg properties such as foam overrun and gel hardness

(Alemprese et al., 2019). However, depending on the voltage (70V or 110V) and the target temperature (57–63°C), the color, the rheology, the coagulation capacity and the generation of volatile compounds can be altered (Rosa et al., 2023). 3D computer simulation models were developed by Wang et al. (2021) to visualize the temperature and distribution of degrees of pasteurization at several cross sections. Although, ohmic heating treatment is well known to offer better temperature uniformity than traditional pasteurization methods, they highlighted some non-uniformities in temperature distribution that they managed to reduce by combining ohmic heating treatment with an external heating approach.

6 Pasteurization of egg white powder

Egg powders are very convenient ingredients for food manufacturers. Especially, egg white is widely dried since it considerably decreases its transportation and storage costs as 8.5 kg of liquid egg white yields 1 kg of powder. This is less true for egg yolk that is rich in dry matter and for which the difference between liquid and powder volume is less (2.3 kg of liquid yolk is needed to yield 1 kg of powder) (Lechevalier et al., 2013). The main technology used for egg product drying is spray drying.

Egg yolk and whole egg are pasteurized in liquid form before drying. On the contrary, egg white is first dried and then pasteurized in its dry form.

Egg white powder heating (or dry-heating) is a commonplace process in the egg product industry, allowing higher pasteurization rates than those for liquid egg white (Baron et al., 2003). In spite of these treatments, the survival of *Salmonella* in spray-dried egg powders has been reported in several countries. Perez-Reyes et al. (2020, 2021a) showed that the aw of egg powders increased with the rising temperature of dry heating. They attributed this result to the denaturation process at temperatures above 60°C that may generate chemical and structural changes in the ovalbumin and ovotransferrin, the main proteins of egg white, since their respective denaturation temperatures are of 84 and 61°C. When temperature is increased, the hydrogen bonding and electrostatic interactions are weakened. As a result, ovalbumin and ovotransferrin become more flexible, resulting in more side groups becoming exposed to the surrounding solvent. When these bonds are broken, the water surrounding the protein forms new hydrogen bonds with the newly exposed groups thus increasing aw. This increase of aw allows lowering $D_{80°C}$-values of *Salmonella*. However, the higher aw and longer dry heating time had a negative impact on egg white powder color, tap and bulk densities (Perez-Reyes et al., 2021b).

First developed to pasteurize egg white powders (55–65°C for 3–5 days), it has since been used to overcome the negative effects of spray drying on egg white powders' foaming and gelling properties, after Kato et al. (1989) demonstrated the benefit of higher temperature and time conditions (80°C for 5–10 days).

6.1 Dry heating technologies

The most common technology consists of heating and holding the temperature of packaged egg white powder in a room with control of temperature and, if possible, relative humidity. The position of pallets/boxes/bags in the room, air flow and air moisture regulation are designed to maximize the efficiency and uniformity of heating without drying out the powder (as it could occurred depending on the tightness of packaging used). Indeed, such excessive drying would generate an obvious yield loss. Moreover, residual moisture elimination would have an energy cost higher than the one for powder heating, since 72.8 kJ kg^{-1} is needed to increase egg powder temperature from 20–80°C, against 317 kJ kg^{-1} if a decrease of residual moisture from 8% to 0.5% simultaneously occurs (Landfeld et al., 2007); this would induce insufficient heat treatment of the powder. It is thus of particular interest to control air relative humidity by providing water vapor during heating, in order to limit the water transfer from the powder to ambient air. It is also necessary to provide dried cold air during cooling to reduce condensation on the packaging, which could otherwise induce caking of the powder.

The physical and thermal properties of egg white powder, including densities, specific heat capacity (cp), thermal conductivity (k), thermal diffusivity (α), and effusivity (e) are useful for estimating the amount of energy and the heat transfer rates needed in dry heating treatment (Boreddy & Subbiah, 2015). These authors modelized cp and k as a function of powder moisture content, dry heating temperature, and powder bulk density (Eqs. 3 and 4, respectively):

$cp = 1.11462 - 0.00229T + 0.0146M + 0.00189TM$ (Eq. 3)

$k = 0.005659 + 0.000112\rho + 0.002404M + 0.000108T$ (Eq. 4)

where M is moisture content ($5.1\% \leq M \leq 10.2\%$), T is the temperature ($20°C \leq T \leq 100°C$), and ρ is the powder bulk density (394 kg $m^{-3} \leq \rho \leq 594$ kg m^{-3}).

In contrast to k, α of egg white powder decreased with an increase in powder bulk density, moisture content, and temperature (Boreddy & Subbiah, 2015). Szafner et al. (2014) compared the effusivity of fresh hen eggs with rehydrated egg powders. The effusivity of rehydrated egg white powders was practically the same as that of the fresh egg white whereas the difference in effusivity between fresh egg yolk and rehydrated egg yolk, and between the blend of fresh egg and the rehydrated whole egg powder was significant.

Dry heating can also be carried out in a conical double-jacketed tank, with a rotating mixing screw to increase heat transfer (Fig. 5). This 'dynamic

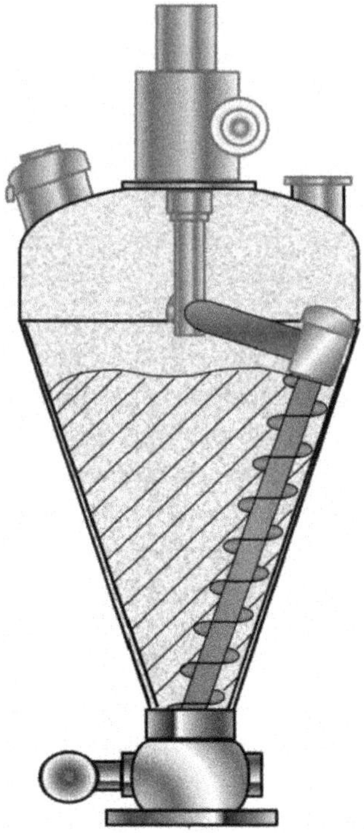

Figure 5 Conical double-jacketed tank with rotating mixing screw (hatched area shows product).

dry heating' allows shorter thermal treatments at higher temperatures (85–100°C), which are inaccessible in hot rooms, because of the increased tendency toward caking in this temperature range, close to powder's glass transition temperature. However, such equipment requires good control of cooling to avoid condensation in the powder. Moreover, the feeding area of non-pasteurized powder has to be well separated from the discharge area of pasteurized powder.

With such devices, heating at 90°C over 21 h is enough to obtain superior powder foaming and gelling properties than those in corresponding liquid product (Hammershøj et al., 2004). Both methods are time-consuming, have high energy costs and require large storage facilities. A promising alternative method could be continuous dry-heating in a fluidized bed. Powder particles are suspended in a fluidized manner by forcing hot air/gas through a punched sheet metal at a sufficiently high gas velocity to overcome the gravity effect. The close gas-to-particle contact ensures a high heat transfer rate, which gives a uniform and fast heat treatment (Hammershøj et al., 2006a).

Continuous dry-heating in a fluidized bed has been compared to heating in a conical tank (Hammershøj et al., 2006a, 2006b). This process requires a minimum treatment of 1 h at 115°C but is limited by the difficulty in reconciling pasteurization efficiency and good functional properties in the heated powder.

However, this last approach as well as others such as high-pressure processing (Van der Plancken et al., 2007a) and irradiation (Froehlich et al., 2015; Kim et al., 2016; Song et al., 2009) are not adopted by the egg processing industry for commercial treatment of egg white powders due to limitations on mass production, cost or consumer resistance (Boreddy et al., 2014).

Radiofrequency (RF) heating methods are known for rapid and volumetric heating of food products as it generates heat within the food product by dipolar and ionic movements (Boreddy et al., 2016). RF heating is advantageous in terms of large-scale processing of low-moisture food products due to greater penetration depth and uniform heating. RF unit is primarily used to preheat the food product to the desired temperature and then the food product is held at that desired temperature for a desired time in a hot air oven or hot room. Thus, RF-assisted thermal processing reduces the come-up time and the overall heating time of the food products. Boreddy et al. (2016) showed that the optimal RF-assisted treatment conditions to produce standard egg white powders with similar functional properties as the traditionally processed (hot room processed) standard egg white powders were 90°C for 8 h, compared to 14 days of traditional hot room processing at 58°C. To design continuous RF-treatment systems, Boreddy and Subbiah (2016) measured the dielectric properties of egg white powder as a function of powder moisture content and temperature and proposed second-order models to fit the experimental data. Chen et al. (2019) completed the study by modeling the RF heating of egg white powder in movement on a conveyor belt and Wei et al. (2020) validated the RF-assisted thermal process in a microbial challenge on *Salmonella* and *Enterococcus faecium*.

Microwave technology was also used to improve the foaming properties of egg white powder (Wang et al., 2020). They obtained an optimal treatment of 6 min at 440W, increasing powder solubility, solution viscosity, protein flexibility and thus decreasing their surface tension, which was beneficial to the protein adsorption at the interface and enhance foam ability of 15% compared to conventional dry heat treatment. These effects were reinforced when microwave treatment was performed in the presence of maltodextrin (Chang et al., 2021).

Recently, Wen et al. (2022) evaluated the effect of superheated steam, an emerging and efficient heating technology, on the molecular structure of ovomucoid, an allergenic protein of egg white. Ovomucoid powder was heated for 2–10 min between 120°C and 200°C that increased its oxidation degree, free SH content, and surface hydrophobicity. Dimers and trimers appeared after the treatment. Some changes were also observed in the primary structure since nitro, phospho, carboxymethyl, and sulfo modifications were reported.

6.2 Impact on protein structure and consequences on protein techno-functional properties

Dry-heating treatment varies from 65°C to 85°C over 5–14 days (with the time of heating increasing at lower temperatures), depending on the powder functionalities required (minimal treatment being enough for bacterial destruction) (Lechevalier et al., 2013). Egg powder technological properties are determined by variations in time, temperature, water content and pH (Kato et al., 1989; Mine, 1996, 1997; Baron et al., 2003). A usual industrial treatment can improve foam stability and foaming capacity by 20% and 30%, respectively (Berardinelli et al., 2006). However, Van der Plancken et al. (2007b) reported that the water content significantly affects the properties of egg white powder during dry heating and must remain lower than 6.8%. Wang et al. (2018a) observed that dry heating up to 15 days at 75°C decreased protein enthalpy and denaturation temperature and increased hydrophobicity and tryptophan exposure. Chang et al. (2020) showed that protein particle size and zeta potential decreased with dry heating time at 75°C, while protein hydrophobicity increased. This indicated that dry heat treatment may induce the exposure of hydrophobic chains and charged groups from intramolecular to protein surface. An increase in solution viscosity was also observed with increase in dry heating time and was attributed to enhanced hydrophobic interactions between protein molecules (Chang et al., 2020). Wang et al. (2018a) observed the formation of aggregates as early as 3 days of dry heating at 75°C, implying specifically to lysozyme. Desfougères et al. (2011) have shown that dry heating of egg white lysozyme induced some protein structural modifications, such as 2–5 succinimidyl residue formation, without any change in the secondary or tertiary structure of the protein and led to a strong improvement in powder functionality. Both Baron et al. (2003) and Talansier et al. (2009) obtained better foaming properties when dry heating time and temperature increased (up to 75°C for 15 days and 80°C for 8 days, respectively). However, mild treatments (70°C for 3 days, 75°C for 3 or 5 days, and 80°C for 1 day) produced the best egg white foaming properties while saving energy (Talansier et al., 2009). Chang et al. (2020) also obtained the best foaming capacity and stability after a dry heating treatment of 5 days at 75°C. Intermediate dry heating treatments (2–5 days at 70°C or 1–2 days at 80°C or 90°C) were found to be a good compromise to improve interfacial properties of egg white without enhancing too much protein resistance to digestion or protein antigenicity (Lechevalier et al., 2017b; Wang et al., 2018a).

However, Wei et al. (2013) suggested higher treatments (10 days at 85°C) to optimize the gelling properties of egg white powder. Mine (1996) has obtained the highest gel strength and elasticity in the pH region between 8.5 and 9.5. Gels from egg white powder dry heated 15 days at 80°C had excellent gel properties with the highest transparency and water holding capacity as well as a fine-stranded and orderly network structure compared to the spray-dried egg white powder and current heat-dried egg white powder (9 days at 80°C) (Cheng et al., 2020). But Jiang and Kim (2013) obtained higher gel hardness after dry heating for 10.3 days at 84.8°C and 10.3% of relative humidity. RF dry heat treatment can also significantly increase the gel firmness from egg white powder since a huge enhancement of gel firmness (48.6%) was obtained after less than 1 day at 80°C, the highest firmness being obtained after 10 days at 70°C using this technology (Kar et al., 2020).

Using 'dynamic dry heating', treatment such as 21 h at 90°C resulted in a two-fold increase in the gelling properties. However, gels were significantly darker, greener, and less yellow than after dry heating in a hot room (Hammershøj et al., 2004). Continuous dry heating in a fluidized bed has been compared to heating in a conical mixer (Hammershøj et al., 2006a, 2006b). This process requires a minimum treatment of 1 h at 115°C but is limited by the difficulty in reconciling pasteurization efficiency and good technological properties in the heated powder.

Egg white powder dry heating can also be used to graft different molecules onto proteins. In particular, the co-drying of egg white with polymers such as guar gum hydrolysates has been studied. Powder heat treatment generates, through Maillard reactions, protein–polysaccharides complexes with better emulsifying and gelling properties than standard egg white powder (Kato et al., 1993; Matsudomi et al., 2002). Linked sugar residues improved the organization of the gel network by reducing the contact area between soluble polymers; it also improves the water retention capacity by binding large quantities of water (Handa & Kuroda, 1999; Matsudomi et al., 2002). In the same way, the co-drying of egg white with phosphate enables protein phosphorylation during dry heating, which improves its foaming properties (Hayashi et al., 2009). The gelling properties of egg white powders are also improved by increasing protein net charge through conjugation with glucuronic acid or phosphorylation during dry heating at neutral pH (Aoki et al., 1999; Li et al., 2003, 2004).

Ma et al. (2019) compared the efficiency of heating and the Maillard reactions applied at the dry state on egg white powders. They showed that the storage stability of egg white powders modified by the Maillard reactions (MDEW) was better than that of dry heated egg white (DHEW) during the entire storage period, although MDEW developed a brown color more easily than DHEW after 6 months of storage.

6.3 Consequences on protein digestibility

The extent of conformational alterations and aggregation of proteins during dry heating process depends on several factors, including temperature and time of the process, moisture content, and pH (Mine, 1996). The literature is scarce on the effect of spray drying on egg protein digestibility. Wang et al. (2018a) reported a digestibility of 86.9% for egg white powder and only 51% for raw liquid egg white. However, no significant difference was observed between the digestibility of alkaline and neutral egg white powders (Rao et al., 2020). Bodwell et al. (1980) estimated the fecal digestibility of spray-dried whole egg protein and reported a range of 92% to 97%. Because nitrogen is intensely metabolized, absorbed, and secreted in the colon, fecal digestibility values are not necessarily equal to ileal ones (Evenpoel et al., 1998).

More studies have been performed on the effect of dry heating on egg white proteins (Lechevalier et al., 2015; 2017b; Wang et al., 2018a; Wen et al., 2022). Lechevalier et al. (2015) observed that protein electrophoresis profile changed with dry heating rate as it enhanced the protein unfolding and aggregation. The change in degree of protein unfolding and/or aggregation affected the progression of *in vitro* digestion. At the end of digestion, samples dry heated for 1 day at 70°C were more hydrolyzed than the control, samples dry heated for more than 2 days at 80°C or 90°C were less hydrolyzed than the control, and samples dry heated for more than 2 days at 70°C or 1 day at 80°C or 90°C were as hydrolyzed as the control. Egg white protein digestibility was thus improved by low-temperature dry heat treatment that favored protein unfolding and flexibility, whereas intense treatments decreased protein digestibility due to aggregation that reduces the enzyme accessibility to their cleavage sites. Similar results were reported by Wang et al. (2018a) who found that protein digestibility of egg white powders increased significantly from 86.9% for unheated samples to 89.01% for samples dry heated for 15 days at 75°C. Over 15 days heating time, protein digestibility decreased. The results showed a good resistance of ovalbumin to pepsin hydrolysis at all heating treatments and its corresponding SDS PAGE band slightly decreased after 9 days at 75°C, whereas lysozyme band decreased as early as 3 days at 75°C. The ovotransferrin band was absent even in the nonheated samples due to its high susceptibility to pepsin. The aggregates formed during the first 15 days of dry heating were digested by pepsin. No significant difference was observed in the amino acid composition of the digests suggesting that the dry heating process did not damage the nutritional value of egg white proteins.

Superheated steam treatment has been reported to increase the protein susceptibility to trypsin hydrolysis of ovomucoid (Wen et al., 2022) and showed an increase from 86.7% for the control to 99% and 100% for the samples treated for 10 min at 120°C and 200°C, respectively. Although ovomucoid is a trypsin inhibitor, superheated steam treatment increased its susceptibility to hydrolysis as only few structural domains attribute resistance to trypsin digestion.

7 Conclusion

The primary challenge in the egg and egg product industry is to provide consumers with microbiologically safe products. This objective is mostly achieved today through thermal stabilization treatments applied to egg products, notably through the pasteurization of whole eggs, liquid egg whites, and yolks, as well as the dry-heating of egg white powders. However, these treatments alter the functional and nutritional quality of the products, primarily due to the high heat sensitivity of proteins. Hence, the second challenge in the egg industry is to optimize the microbiological stabilization treatments to minimize their impact on product functionalities. This challenge has been the subject of extensive research in recent decades.

Today, the issue of the sustainability and environmental impact of processes questions the existing solutions, incorporating the energy dimension of the processes. The use of less energy-intensive alternative technologies is therefore becoming a focal point in the industry. Given the thermal sensitivity of egg products, these matters have already been researched, particularly concerning the combination of various thermal or non-thermal treatments (hurdle technologies), aiming to achieve the desired bacterial destruction while preserving constituents, especially the egg proteins (Khan et al., 2017; Agregan et al., 2023). These strategies could be now reconsidered by including the variable of energy consumption. The new issue of egg sector is now how to provide egg and egg products, safe, functional and healthy while saving energy and water during the processes.

8 Where to look for further information

The following books provide a good overview of the subject:

- Bhandari, B., et al. (Eds). (2023) *Handbook of food powders: processes and properties*, 2nd edition, Woodhead Publishing, Sawston.
- Bhat, Z.F., et al. (Eds). (2023) *Processing technologies and food protein digestion*, Elsevier Academic Press, London.
- Mine, Y., et al. (Eds). (2024) *Handbook of egg science and technology*, CRC Press, Boca Raton. http://doi.org/10.1201/9781003254430

Key research in this area can be found at the following organizations:

- UMR 1253 INRAE/L'institut Agro Rennes-Angers Science and technology of milk and eggs (France) (www6.rennes.inrae.fr/stlo_eng).

- Department of Food Science – Food Technology (Denmark) (www.food.au.dk).
- Actini (France) (www.actini.com).
- SANOVO Technology group (Italy) (www.sanovogroup.com).

9 References

Agregan, R., et al. (2023) 'The use of novel technologies in egg processing', *Food Rev. Int.*, 39, 2854–2874. https://10.1080/87559129.2021.1980887

Alemprese, C., et al. (2019) 'Effects of ohmic heating on technological properties of whole egg', *Innov. Food Sci. Emerg. Technol.*, 58, 102244. https://doi.org/10.1016/j.ifset.2019.102244

Aoki, T., et al. (1999), 'Improvement of heat stability and emulsifying activity of ovalbumin by conjugation with glucuronic acid through the Maillard reaction', *Food Res. Int.*, 32, 129–133. https://dx.doi.org/10.1016/S0963-9969(00)00140-X

Aygun, A. & Sert, D. (2012), 'Effects of ultrasonic treatment on eggshell microbial activity, hatchability, tibia mineral content, and chick performance in Japanese quail (Coturnix coturnix japonica) eggs', *Poultry Sci.*, 91, 732–738. https://doi.org/10.3382/ps.2011-01901

Ball, H.R. Jr., et al. (1987), 'Functionality and microbial stability of ultra-pasteurized aseptically refrigerated whole egg', *J. Food Sci.*, 52(5), 1212–1218. https://doi.org/10.1111/j.1365-2621.1987.tb14046.x

Baron, F., et al. (2003), 'Effect of dry heating on the microbiological quality, functional properties, and natural bacteriostatic ability of egg white after reconstitution', *J. Food Protect.*, 66, 825–832. https://doi.org/10.4315/0362-028x-66.5.825

Baron, F., et al. (2024), 'Microbiology of egg and egg products', in Mine, Y., et al. (Eds), *Handbook of egg science and technology*, CRC Press, Boca Raton, pp. 355–396.

Berardinelli, A., et al. (2006), 'Interfacial properties of egg white powders before and after dry pasteurisation', *XII European poultry conference*, Verone, Italy, 10–14 September.

Bhat, Z.F., et al. (2023), 'Emerging technologies for processing egg and egg products', in Bhat, Z.F., et al. (Eds), *Processing technologies and food protein digestion*, Elsevier Academic Press, London, pp. 135–174.

Blume, K., et al. (2015), 'Exploring the relationship between protein secondary structures, temperature-dependent viscosities and technological treatments in egg yolk and LDL by FTIR and rheology', *Food Chem.*, 173, 584–593. https://doi.org/10.1016/j.foodchem.2014.10.084

Bodwell, C.E., et al. (1980), ''Protein digestibility of the same protein preparations by human and rat assays and by *in vitro* enzymic digestion assays', *Am. J. Clin. Nutr.*, 33, 677–686. https://doi.org/10.1093/ajcn/33.3.677

Boreddy, S.R. & Subbiah, J. (2015), 'Physical and thermal properties of spray-dried egg white powder', *Trans. ASABE*, 58(5), 1409–1416.

Boreddy, S.R. & Subbiah, J. (2016), 'Temperature and moisture bib0015 dependent dielectric properties of egg white powder', *J. Food Eng.*, 168, 60–67. http://dx.doi.org/10.1016/j.jfoodeng.2015.07.023

Boreddy, S.R., et al. (2014), 'Effect of radio frequency assisted thermal processing on quality and functional properties of egg white powder', *Trans. ASABE*, 57(6), 1761–1770. http://dx.doi.org/10.13031/trans.57.10781

Boreddy, S.R., et al. (2016), 'Novel radiofrequency-assisted thermal processing improves the gelling properties of standard egg white powder', *J. Food Sci.*, 81(3), E665–E671. https://doi.org/10.1111/1750-3841.13239

Byrne, B., et al. (2006), 'Thermal inactivation of *Bacillus* cereus and clostridium perfringens vegetative cells and spores in pork luncheon roll', *Food Microbiol.*, 23(8), 803–808. https://doi.org/10.1016/j.fm.2006.02.002

Chang, C., et al. (2020), 'Effect of dry-heat and guar gum on properties of egg white powder: analysis of foaming capacity and baking performance', *Food Hydro.*, 99, 105333. http://dx.doi.org/10.1016/j.foodhyd.2019.105333

Chang, C., et al. (2021), 'Microwave induced glycosylation of egg white protein: study on physicochemical properties and baking performance', *Food Hydro.*, 118, 106569. https://doi.org/10.1016/j.foodhyd.2020.106569

Chen, J., et al. (2019), 'Modeling of radio frequency heating of egg white powder continuously moving on a conveyor belt', *J. Food Eng.*, 262, 109–120. https://doi.org/10.1016/j.jfoodeng.2019.05.029

Cheng, Y., et al. (2020), 'Effect of dry heating on egg white powder influencing water mobility and intermolecular interactions of its gels', *J. Sci. Food Agric.*, 101, 433–440. https://doi.org/10.1002/jsfa.10652

Damodaran, S. (2017), 'Amino acids, peptides and proteins', In Damodaran, S. & Parkin, K.L. (Eds), *Fennema's food chemistry*, CRC Press, Boca Raton, pp. 235–356. https://doi.org/10.1201/9781315372914

Denis, C., et al. (1995), 'Thermoresistance of *Salmonella* enteritidis inoculated in egg products. Influence on water activity on thermoresistance', *Industries alimentaires et agricoles*, 112, 659–663.

Desfougères, Y., et al. (2011), 'Succinimidyl residue formation in hen egg-white lysozyme favors the formation of intermolecular covalent bonds without affecting its tertiary structure', *Biomacromolecules*, 201(12), 156–166. https://doi.org/10.1021/bm101089g

Dickinson, R.W., et al. (1970), 'Temperatures in egg product foams during vat pasteurisation', *J. Food Sci.*, 35(6), 866–871.

Dupont, D. & Nau, F. (2019), 'The importance of length scales', In Gouseti, O., et al. (Eds), *From bite to nutrient interdisciplinary approaches to food digestion*, Springer, ville, pp. 129–143. http://dx.doi.org/10.1007/978-3-030-03901-1_6

EFSA & ECDC (European Food Safety Authority and European Center for Disease prevention and Control) (2021), 'The european union one health 2019 zoonoses report', *EFSA J.*, 19(2), 6406. https://doi.org/10.2903/j.efsa.2021.6406

Evenpoel, P., et al. (1998), 'Digestibility of cooked and raw egg protein in humans as assesses by stable isotope techniques', *J. Nutr.*, 128, 1716–1722. https://doi.org/10.1093/jn/128.10.1716

Ferreira, M., et al. (1995), 'Instrumental method for characterizing protein foams', *J. Food Sci.*, 60, 90–93. https://doi.org/10.1111/j.1365-2621.1995.tb05613.x

Froning, G.W., et al. (2002), 'International egg pasteurization manual'. https://www.incredibleegg.org/wp-content/uploads/2023/07/Pasteurization_Manual.pdf

FSIS (2021), 'Food safety and Inspection Service; U.S. Department of Agriculture. Tables & results: Microbiological testing program for pasteurized egg products, calendar years 1995–2017', retrieved on October 24, 2023 from https://www.fsis.usda.gov

Froehlich, A., et al. (2015), 'Sensory aspects and reduction of *Salmonella* in irradiated egg powder', *Cienc. Agrotec.*, 39(5), 506–513. http://dx.doi.org/10.1590/S1413-70542015000500009

Gut, J.A.W., et al. (2005a), 'Continuous pasteurization of egg yolk: thermophysical properties and process simulation', *J. Food Proc. Eng.*, 28, 181–203. https://doi.org/10.1111/j.1745-4530.2005.00416.x

Gut, J.A.W., et al. (2005b), 'Continuous pasteurization of egg white: thermophysical and flow properties and process design', *In: V Congreso Iberoamericano de Ingeniería de Alimentos, Puerto Vallarta.*

Hammershøj, M., et al. (2004), 'The significance of critical processing steps in the production of dried egg albumen powder on gel textural and foaming properties', *J. Sci. Food Agric.*, 84(9), 1039–1048. http://dx.doi.org/10.1002/jsfa.1772

Hammershøj, M., et al. (2006a), 'Dry pasteurization of egg albumen powder in a fluidized bed. I. Effect on microbiology, physical and chemical parameters', *Int. J. Food Sci. Technol.*, 41(3), 249–261. http://dx.doi.org/10.1111/j.1365-2621.2005.01058.x

Hammershøj, M., et al. (2006b), 'Dry-pasteurization of egg albumen powder in a fluidized bed. II. Effect on functional properties: gelation and foaming', *Int. J. Food Sci. Technol.*, 41(3), 263–274. https://doi.org/10.1111/j.1365-2621.2005.01059.x

Hamid-Samimi, M., et al. (1984), 'Flow behavior of liquid whole egg during thermal treatments', *J. Food Sci.*, 49(1), 132–136. https://doi.org/10.1111/j.1365-2621.1984.tb13689.x

Handa, A. & Kuroda, N. (1999), 'Functional improvements in dried egg white through the Maillard reaction', *J. Agric. Food Chem.*, 47, 1845–1850. https://doi.org/10.1021/jf9811018

Hank, C.R., et al. (2001), 'The effect of shell egg pasteurization on the protein quality of albumen', *Poultry Sci.*, 80, 821–824. https://doi.org/10.1093/ps/80.6.821

Hatta, H. & Kubo, N. (2024), 'Shelf life of eggs', in Mine, Y., et al. (Eds), *Handbook of egg science and technology*, CRC Press, Boca Raton, pp. 313–321.

Hayashi, Y., et al. (2009), 'Improvement of foaming property of egg white protein by phosphorylation through dry-heating in the presence of pyrophosphate', *J. Food Sci.*, 74(1), C68–C72. https://doi.org/10.1111/j.1750-3841.2008.01019.x

Herald, T.J. & Smith, D.M. (1988), 'Functional properties and composition of liquid whole egg proteins as influenced by pasteurization and frozen storage', *Poultry Sci.*, 68, 1461–1469. https://doi.org/10.3382/ps.0681461

Hierro, E., et al. (2009), 'Inactivation of *Salmonella enterica* serovar Enteritidis on shell eggs by pulsed light technology', *Int. J. Food Microbiol.*, 135 (2), 125–130. https://doi.org/10.1016/j.ijfoodmicro.2009.07.034

Hill, W.M., et al. (1965), 'Spray-drying egg white at various pH levels', *Poultry Sci.*, 44, 1155–1163. https://doi.org/10.3382/ps.0441155

Himathongkham, S., et al. (1999), 'Efficacy of disinfection of shell eggs externally contaminated with *Salmonella enteritidis*: Implications for egg testing', *Int. J. Food Microbiol.*, 49(3), 161–167. https://doi.org/10.1016/S0168-1605(99)00092-6

Hou, H., et al. (1996), 'Pasteurization of intact shell eggs', *Food Microbiol.*, 13, 93–101. https://doi.org/10.1006/fmic.1996.0012

Icier, F. & Ilicali, C. (2005), 'Temperature dependent electrical conductivities of fruit purees during ohmic heating', *Food Res. Int.*, 38, 1135–1142. https://doi.org/10.1016/j.foodres.2005.04.003

James, C., et al. (2002), 'Surface pasteurisation of shell eggs', *J. Food Eng.*, 53(2), 193–197. https://doi.org/10.1016/S0260-8774(01)00156-X

Jiang, Z.Y. & Kim, Y.H. (2013), 'Optimization of the heating conditions in the production process of high gel strength egg white powder', *Adv. Mat. Res.*, 662, 409–412. http://dx.doi.org/10.4028/www.scientific.net/AMR.662.409

Jimenez-Saiz, R., et al. (2011), 'Human immunoglobulin E binding to heated and glycated ovalbumin and ovomucoid before and after *in vitro* digestion', *J. Agric. Food Chem.*, 59, 10044–10051. https://doi.org/10.1016/j.foodres.2017.11.040

Kar, A., et al. (2020), 'Effect of traditional and radiofrequency assisted thermal processing on the gel firmness of egg white powder', *LWT-Food Sci. Technol.*, 133, 110091. https://doi.org/10.1016/j.lwt.2020.110091

Kato, A., et al. (1989), 'New approach to improve the gelling and surface functional properties of dried egg white by heating in dry state', *J. Agric. Food Chem.*, 37, 433–437. https://doi.org/10.1021/jf00086a036

Kato, A., et al. (1993), 'Improvement of emulsifying properties of egg white proteins by the attachment of polysaccharide through Maillard reaction in a dry state', *J. Agric. Food Chem.*, 41(4), 540–543. https://doi.org/10.1021/jf00028a006

Khan, I., et al. (2017) 'Hurdle technology: a novel approach for enhanced food quality and safety – a review', *Food Control*, 73, 1426–1444. https://10.1016/j.foodcont.2016.11.010

Kim, H.J., et al. (2016), 'Microbial safety and physicochemical characteristics of electron beam irradiated whole egg powder', *Food Sci. Biotechnol.*, 25, 637–642. https://doi.org/10.1007/s10068-016-0089-4

Kline, L., et al. (1966), 'Further studies on heat pasteurisation of raw liquid egg white', *Food Technol.*, 20, 114–116.

Landfeld, A., et al. (2001), 'Residence time distribution during egg yolk pasteurisation', *Czech. J. Food Sci.*, 20, 193–201. http://dx.doi.org/10.17221/3531-CJFS

Landfeld, A., et al. (2007), 'Sorption and thermal properties of dried egg white', *J. Food Eng.*, 87(1), 59–63. http://dx.doi.org/10.1016/j.jfoodeng.2007.09.030

Le Denmat, M., et al. (1999), 'Protein denaturation and emulsifying properties of plasma and granules of egg yolk as related to heat treatment', *J. Food Sci.*, 64, 194–197. https://doi.org/10.1111/j.1365-2621.1999.tb15863.x

Lechevalier, V., et al. (2013), 'Powdered egg', In Bhandari, B., et al. (Eds), *Handbook of food powders*, Woodhead Publishing, Cambridge, pp. 484–512.

Lechevalier, V., et al. (2015), 'Effects of dry heating on the progression of *in vitro* digestion of egg white proteins: contribution of multifactorial data analysis', *Food Funct.*, 6, 1578. https://doi.org/10.1039/c4fo01156b

Lechevalier, V., et al. (2017a), 'Pasteurisation of liquid whole egg: optimal heat treatments in relation to its functional, nutritional and allergenic properties', *J. Food Eng.*, 195, 137–149. https://doi.org/10.1016/j.jfoodeng.2016.10.007

Lechevalier, V., et al. (2017b), 'Effect of dry heat treatment of egg white powder on its functional, nutritional and allergenic properties', *J. Food Eng.*, 195, 40–51. http://dx.doi.org/10.1016/j.jfoodeng.2016.09.022

Lechevalier, V., et al. (2005), 'Statistical analysis of effects of industrial processing steps on functional properties of pasteurized liquid egg white', *J. Sci. Food Agric.*, 85, 757–769. https://doi.org/10.1002/jsfa.2042
Li, C.P., et al. (2003), 'Phosphorylation of egg white proteins by dry-heating in the presence of phosphate', *J. Agric. Food Chem.*, 51, 6808–6815. https://doi.org/10.1021/jf030043+
Li, C.P., et al. (2004), 'Improvement of functional properties of egg white proteins through phosphorylation by dry-heating in the presence of pyrophosphate', *J. Agric. Food Chem.*, 52, 5752–5758. https://doi.org/10.1021/jf0498259
Liot, R. & Anza, L. (1996), 'Procédé de traitement de blanc d'œuf liquide', *Patent FR*, n° 96, 08356.
Liu, Y.F., et al. (2017), '*In vitro* peptic digestion of ovomucin-depleted egg white affected by pH, temperature and pulsed electric fields', *Food Chem.*, 231, 165–174. https://doi.org/10.1016/j.foodchem.2017.03.136
Ma, Y., et al. (2019), 'Changes in the gel characteristics of two hen egg white powders modified by dry heating and the Maillard reaction during long-term storage', *LWT Food Sci. Technol.*, 109, 123–129. http://dx.doi.org/10.1016/j.lwt.2019.04.003
Martos, G., et al. (2011), 'Mechanisms underlying differential food allergy response to heated egg', *J. Allergy Clin. Immunol.*, 127(4), 990–997.e992. https://doi.org/10.1016/j.jaci.2011.01.057
Matsudomi, N., et al. (2002), 'Improvement of gel properties of dried egg white by modification with galactomannan through the Maillard reaction', *J. Agric. Food Chem.*, 50(14), 4113–4118. https://doi.org/10.1021/jf0114334
Mine, Y. (1996), 'Effect of pH during the dry heating on the gelling properties of egg white proteins', *Food Res. Int.*, 29(2), 155–161. https://doi.org/10.1016/0963-9969(96)00008-7
Mine, Y. (1997), 'Effect of dry heat and mild alkaline treatment on functional properties of egg white proteins', *J. Agric. Food Chem.*, 45, 2924–2928. https://doi.org/10.1021/JF970158B
Moore, J. & Madden, R.H. (1993), 'Detection incidence of *Listeria* species in blended raw egg', *J. Food Protec.*, 56(8), 652–654. https://doi.org/10.4315/0362-028X-56.8.652
OJEU (Official Journal of the European Union) (2003), 'Commission regulation (ES) No 2295/2203 introducing detailed rules for implementing council regulation No 1907/90 on certain marketing standards for eggs'.
Palumbo, M.S., et al. (1995), 'Thermal resistance of *Salmonella* spp. and *Listeria* monocytogenes in liquid egg yolk and egg yolk products', *J. Food Protec.*, 58(9), 960–966. https://doi.org/10.4315/0362-028X-58.9.960
Palumbo, M.S., et al. (1996), 'Thermal resistance of *Listeria* monocytogenes and *Salmonella* spp. in liquid egg white', *J. Food Protec.*, 59(11), 1182–1186. https://doi.org/10.4315/0362-028X-59.11.1182
Pasquali, F., et al. (2010), 'Hot air treatment for surface decontamination of table eggs', *Food Control*, 21(4), 431–435. https://doi.org/10.1016/j.foodcont.2009.07.003
Pereira, R.N. & Vicente, A.A. (2010), 'Environmental impact of novel thermal and non-thermal technologies in food processing', *Food Res. Int.*, 43(7), 1936–1943. https://doi.org/10.1016/j.foodres.2009.09.013
Perez-Reyes, M.E., et al. (2020), 'Influence of low water activity on the thermal resistance of *Salmonella* Enteritidis PT 30 and *Enterococcus faecium* as its surrogate in

egg powders', *Food Sci. Technol. Int.*, 27(2), 184–193. https://doi.org/10.1177/1082013220937872

Perez-Reyes, M.E., et al. (2021a), 'The influence of elevated temperatures and composition on the water activity of egg powders', *J. Food Process Preserv.*, 45, e15269. http://dx.doi.org/10.1111/jfpp.15269

Perez-Reyes, M.E., et al. (2021b), 'Influence of water activity and dry-heating time on egg white powders quality', *LWT Food Sci. Technol.*, 140, 110717. http://dx.doi.org/10.1016/j.lwt.2020.110717

Protais, J., et al. (2006), 'Identification et comportement des bactéries d'altération dans la matrice œuf entier liquide', *Sciences et techniques avicoles*, 57, 4–13.

Protais, J., et al. (2007a), 'Identification of the enterococcis genus in the matrix of liquid whole eggs', *7ième Journées de la recherche avicole*, 28–29 March 2007, pp. 528–531.

Protais, J., et al. (2007b), '*Campylobacter* sp. et *Listeria* monocytogenes dans l'œuf entier liquide', *7ième Journées de la recherche avicole*, 28–29 March 2007, pp. 532–535.

Qi, L., et al. (2020), 'Non-destructive testing technology for raw eggs freshness: a review', *SN App. Sci.*, 2, 1113. https://doi.org.10.107/s42452-020-2906-x

Rao, P.S., et al. (2020), 'Effect of pH and heat treatment on the antioxidant activity of egg white protein derived peptides after simulated in-vitro gastrointestinal digestion', *Antioxidants*, 9, 1114. https://doi.org/10.3390/antiox9111114

Reznik, D. (1993), 'Liquid egg continuous pasteurization for high microbial kill rate by electro-heating, with AC frequency to prevent electrolysis, and cooling, for highly functional liquid egg with improved storage stability, avoiding product coagulation', GB Patent GB-2265538.

Reznik, D. (1996), 'Electro-heating food products, especially liquid egg products for pasteurization or sterilisation – by applying alternative current to food as it passes through electroheating zone defined between pairs of electrodes spaced from barrier contacting food', US Patent 96US624651, GB Patent GB2282052 A.

Rosa, D.A., et al. (2023) 'Effect of ohmic heating temperature and voltage on liquid whole egg processing', *Innov. Food Sci. Emerg. Technol.*, 89, 103490. https://doi.org/10.1016/j.ifset.2023.103490

Seideman, W.E., et al. (1963) 'Factors affecting heat coagulation of egg white' *Poultry Sci.*, 42, 406–417. https://doi.org/10.3382/ps.0420406

Schuman, J.D., et al. (1997), 'Immersion heat treatments for inactivation of *Salmonella enteritidis* with intact eggs', *J. Appl Microbiol.*, 83(4), 438–444. https://doi.org/10.1046/j.1365-2672.1997.00253.x

Shah, D.B., et al. (1991), 'Thermal resistance of egg-associated epidemic strains of *Salmonella enteritidis*', *J. Food Sci.*, 56 (2), 391–393. https://doi.org/10.1111/j.1365-2621.1991.tb05287.x

Song, H.P., et al. (2009), 'Improvement of foaming ability of egg white product by irradiation and its application', *Radiat. Phys. Chem.*, 78(3), 217–221. http://dx.doi.org/10.1016/j.radphyschem.2008.10.001

Szafner, G., et al. (2014), 'Photopyroelectric assessment of the thermal effusivity of fresh hen egg and of rehydrated egg powders', *J. Therm. Anal. Calorim.*, 120(1), 363–368. http://dx.doi.org/10.1007/s10973-014-3927-8

Takagi, K., et al. (2003), 'Comparative study of in vitro digestibility of food proteins and effect of preheating on the digestion', *Biol. Pharm. Bull.*, 26(7), 969–973. https://doi.org/10.1248/bpb.26.969

Talansier, E., et al. (2009), 'Optimization of dry heat treatment of egg white in relation to foam and interfacial properties', *LWT Food Sci. Technol.*, 42, 496–503. *http://dx.doi.org/10.1016/j.lwt.2008.09.013*

Techer, C., et al. (2015), 'Microbial quality of industrial liquid egg white: assumptions on spoiling issues in egg-based chilled desserts', *J. Food Sci.*, 80(2), M389–M398. https://doi.org/10.1111/1750-3841.12764

Uysal, R.S., et al. (2017), 'Effects of heat treatment parameters on liquid whole egg proteins', *Food Chem.*, 216, 201–208. https://doi.org/10.3390/foods12071474

Van der Plancken, I., et al. (2002), 'Effect of thermal treatment on the susceptibility of egg white proteins to enzymatic hydrolysis', *Commun. Agric. Appl. Biol. Sci.*, 67, 309–312. https://doi.org/10.1021/jf026019y

Van der Plancken, I., et al. (2003), 'Heat-induced changes in the susceptibility of egg white proteins to enzymatic hydrolysis: a kinetic study', *J. Agric. Food Chem.*, 51, 3819–3823. https://doi.org/10.1021/jf026019y

Van der Plancken, I., et al. (2004), 'Kinetic study on the changes in the susceptibility of egg white proteins to enzymatic hydrolysis induced by heat and high hydrostatic pressure pretreatment', *J. Agric. Food Chem.*, 52, 5621–5626. https://doi.org/10.1021/jf049716u

Van der Plancken, I., et al. (2006), 'Effect of heat treatment on the physicochemical properties of egg white proteins: a kinetic study', *J. Food Eng.*, 75, 316–326. https://doi.org/10.1016/j.jfoodeng.2005.04.019

Van der Plancken, I., et al. (2007a), 'Foaming properties of egg white proteins affected by heat or high-pressure treatment', *J. Food Eng.*, 78, 1410–1426. http://dx.doi.org/10.1016/j.jfoodeng.2006.01.013

Van der Plancken, I., et al. (2007b), 'Effect of moisture content during dry-heating on selected physicochemical and functional properties of dried egg white', *J. Agric. Food Chem.*, 55, 135–137. https://doi.org/10.1021/jf062370y

Vanga, S.K., et al. (2017), 'Review of conventional and novel food processing methods on food allergens', *Crit. Rev. Food Sci. Nutr.*, 57, 2077–2094. https://doi.org/10.1080/10408398.2015.1045965

Wang, C.S., et al. (2021) 'Analysis of thermal processing of liquid eggs using high frequency ohmic heating: experimental and computer simulation approaches', *Innov. Food Sci. Emerg. Technol.*, 73, 102792. https://doi.org/10.1016/j.ifset.2021.102792

Wang, G. & Wang, T. (2009), 'Effect of yolk contamination, shearing and heating on foaming properties of fresh egg white', *J. Food Sci.*, 74(2), C147–C156. https://doir.org/10.1111/j.1750-3841.2009.01054.x

Wang, J., et al. (2018a), 'Physicochemical properties, *in vitro* digestibility and antioxidant activity of dry-heated egg white protein', *Food Chem.*, 246, 18–25. https://doi.org/10.1016/j.foodchem.2017.10.128

Wang, X., et al. (2018b), 'Effect of different heat treatments on *in vitro* digestion of egg white proteins and identification of bioactive peptides in digested products', *J. Food Sci.*, 83, 1140–1148. https://doi.org/10.1111/1750-3841.14107

Wang, X., et al. (2020), 'Microwave technology as a new strategy to induce structural transition and foaming properties improvement of egg white powder'. *Food Hydro.*, 101, 105530. https://doi.org/10.1016/j.foodhyd.2019.105530

Wei, X., et al. (2013), 'Optimization of the heating conditions in the production process of high gel strength egg white powder', *Adv. Mat. Res.*, 662, 409–412. http://dx.doi.org/10.4028/www.scientific.net/AMR.662.409

Wei, X., et al. (2020), 'A microbial challenge study for validating continuous radio-frequency assisted thermal processing pasteurization of egg white powder', *Food Microbiol.*, 85, 103306. http://dx.doi.org/10.1016/j.fm.2019.103306

Wen, P.W., et al. (2022), 'Effects of superheated steam treatment on the allergenicity and structure of chicken egg ovomucoid', *Foods*, 11, 238. https://doi.org/10.3390/foods11020238

Wu, J., et al. (2010), 'Bioactive proteins and peptides from egg proteins', In Mine, Y., et al. (Eds), *Bioactive proteins and peptides as functional foods and nutraceuticals*, Wiley-Blackwell, Ames, pp. 247–264. https://doi.org/10.1002/9780813811048.ch17

Chapter 7

Advances in innovative postharvest egg pasteurization and disinfection technologies to optimize egg safety and quality

Daniela Bermudez-Aguirre, Joshua Carter, Joseph Sites, and Brendan A. Niemira, USDA-ARS Eastern Regional Research Center, USA[1]

1 Introduction

Eggs and egg products include several products that are commercialized worldwide. These terms include but are not limited to whole eggs, liquid whole egg (LWE), liquid egg white, liquid egg yolk, yolk powder, white powder, and frozen egg, among others. The composition of these products also varies widely depending on the country. In the United States, the Food Safety and Inspection Service (FSIS) defines the composition and physicochemical characteristics of the most common egg products (2020). These changes in the composition of the egg products are important to consider because the inactivation or survival of microorganisms will be affected by the increase or decrease in acidity and the presence of solutes like salt or sugar, among others.

1 The mention of trade names or commercial products in this article is solely to provide specific information. It does not imply a recommendation or endorsement by the U.S. Department of Agriculture. USDA is an equal opportunity provider and employer.

http://dx.doi.org/10.19103/AS.2024.0150.11
Published by Burleigh Dodds Science Publishing Limited, 2025.

The main pathogen associated with egg and egg products is *Salmonella* spp. (Ribeiro et al., 2023; Keerthirathne et al., 2016; Pasquali et al., 2016; Monfort et al., 2011; Alvarez et al., 2007). Recently, it has been established that the strains associated with most poultry outbreaks in the United States are Enteritidis, Typhimurium, I:4,5,12:i:-, Infantis, and Heidelberg, besides some serogroups (O:4, O:7, O:9) as reported by the National Advisory Committee on Microbiological Criteria in Foods (NACMCF, 2024). However, the presence of other microorganisms, such as *Listeria monocytogenes*, was found in a foodborne outbreak in 2019 involving hard-boiled eggs (CDC, 2024). Additional pathogens linked to outbreaks involving eggs are *Campylobacter* spp., *Escherichia* spp., and *Staphylococcus* spp. (Aygun, 2017).

Most of the research has been focused on the inactivation of *Salmonella* spp. in eggs and egg products. Although it is believed that this pathogen can be present only on the exterior eggshell surface, the truth is that *Salmonella* spp. can also be present inside the eggshell. It is important to differentiate these two scenarios; the presence of microbial cells outside the eggshell on the egg's surface usually results in cross-contamination. Once the egg is laid and placed, transferred, or stored in different locations, the surface can become contaminated if the pathogen is present on any contact surface such as boxes, processing lines, egg cartons, employee gloves, or hands. However, eggs can also be contaminated if any hatchery area is positive for *Salmonella*, such as nests, feathers, feces, or cages. This is known as horizontal transmission. This kind of contamination also applies to liquid egg products that become contaminated by pathogens when they come into contact with non-sanitary surfaces. However, when the hens are positive for *Salmonella* (usually do not present any symptoms), the microbial cells infect the egg inside the reproductive system of the bird, and the eggs are already contaminated with the pathogen. This is known as vertical transmission (Bermudez-Aguirre and Niemira, 2022).

Then, there is a real need to evaluate food processing technologies that can ensure the microbial safety of the eggs not only for the possible cross-contamination of the eggshell but also for the probability of internal *Salmonella* infection that could lead to foodborne outbreaks. It is important that these technologies do not affect the quality of the egg and egg products and maintain the functionality of the egg albumin, for example. This chapter presents some of the innovations in egg processing technologies to minimize the risk of vertical and horizontal contamination in the egg industry, highlighting not only the research advances but also the research needs.

2 Pasteurization and disinfection of eggs

Thermal treatment is usually applied to food to inactivate microorganisms and decrease enzymatic activity, providing the food with some stability during storage. Depending on the processing conditions during thermal processing, the food can be pasteurized when the vegetative cells (i.e. pathogenic microorganisms) are inactivated or sterilized when the spores are killed by heat. Both processes vary in the intensity of the heat applied (i.e. temperature) and the processing time but also in the final stability of the product. Pasteurized foods are usually kept refrigerated to extend the shelf life and delay microbial growth and enzymatic activity, usually for a few days. In pasteurization, the goal is to achieve at least a 5 log reduction (99.999%) of the target microorganism. Sterilized food can be stored at room temperature, and the shelf life is usually longer (several months). In sterilized food, microbial activity is not detected during the storage (if the process was conducted correctly), but the nutritional and sensory quality of the product decreases over time.

Disinfection, decontamination, and sanitation are terms that refer to milder treatments or activities than pasteurization or sterilization. Disinfection and decontamination are related to a process in which some mild treatments using sanitizers (i.e. chemical agents such as chlorine; or natural agents such as organic acids) are applied to the food surface to reduce the microbial counts, usually in the order of 2–3 log reductions (99–99.9%). Sanitation is a term related to the hygienic conditions that are followed during the production of food; in the specific case of eggs, it applies to the conditions observed from the poultry hatcheries, the egg processing lines until the eggs reach the consumer (Bermudez-Aguirre and Niemira, 2022).

These terms in the poultry industry, specifically related to the egg production chain, are targeted for *Salmonella* spp. The pasteurization of the liquid egg products (white, yolk, and whole egg) is intended to achieve at least 5 log reduction of the pathogen without affecting the quality of the product. Eggs contain thermosensitive components, such as the albumen proteins. Ovotransferrin represents 12% of the protein content in albumen, and its denaturation temperature is 61°C (Froning et al., 2002). In the case of intact shell eggs, most of the research has been conducted to achieve decontamination of the eggshell, 2–3 log reductions of different pathogens. However, eggs represent a unique surface with high variability between each unit. The eggshell surface has a very irregular surface with a high porosity, as observed in Fig. 1. Minor cracks that are not observable by sight are also part of this unique exterior layer. This surface represents the perfect niche for the microorganism to hide, protect, and grow. Although in some countries like the United States, washing the eggs with water containing sanitizers (such as chlorine) is a common practice to reduce the microbial flora of eggs, in other countries this practice for eggs is banned or optional. A unique technology has been recently developed to pasteurize in-shell eggs and ensure the egg content is *Salmonella*-free, and the quality attributes are like a raw egg. That will be discussed in further sections of this chapter.

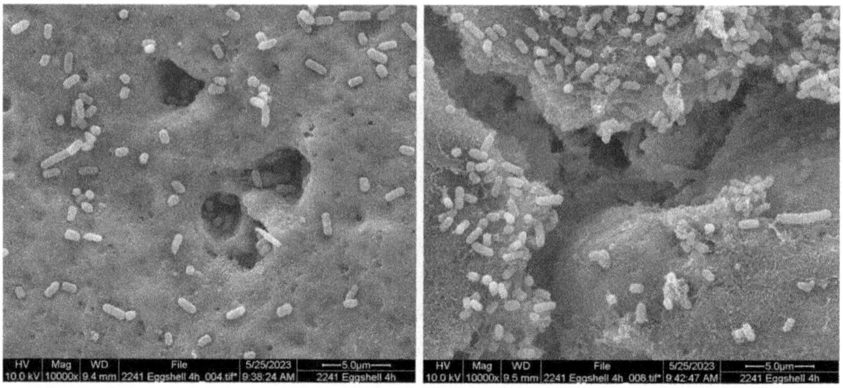

Figure 1 Eggshell surface inoculated with cells of *Salmonella Typhimurium* ATCC53647 showing the natural presence of pores and cracks in the uneven shell surface (magnification 10 000x).

Some of the egg quality attributes that are often evaluated and that represent the freshness of the egg are the Haugh unit (HU), the yolk index (YI), the yolk color, the albumen and yolk pH, the eggshell breakage strength, the vitelline membrane strength, the albumen turbidity, and its functional properties (foaming ability and stability), among others. The definition and methodology to calculate these properties have been previously discussed by several authors (Illera et al., 2023; Bermudez-Aguirre and Niemira, 2022; Yang et al., 2020; DSM, 2018). So, the main challenge for the innovative technologies for pasteurization and decontamination is to successfully inactivate *Salmonella* spp. but to preserve the egg quality at the same time.

3 Innovative food technologies

Through the search for some alternatives to decrease the thermal damage to food products, such as the loss of sensory attributes or the decrease in the nutrient content, some technologies emerged in the food science arena to overcome some of these problems but ensure at the same time the food safety of the product. Some of these technologies have been researched for many years until they have successfully been installed in the industry and launched a wide variety of products with outstanding quality. Some of these innovative technologies use preservation factors different from heat, and microbial inactivation occurs because of exposure to pressure, electricity, sound waves, radiation, light, or gases. These are known as nonthermal technologies, and the list has been growing in the last two decades. Some of these technologies are high hydrostatic pressure, pulsed electric fields, ultrasound, irradiation, ultraviolet light, pulsed light, cold plasma, ozone, and super-critical carbon dioxide, to mention a few. On the side of innovative thermal processing technologies, these options offer fast, effective, and volumetric heating in food, reducing the processing time or temperature and decreasing the negative effects of heat on the product. The most common innovative thermal interventions are microwave (MW) and radio frequency (RF).

4 Nonthermal technologies

Several of the nonthermal technologies have been tested in egg and egg products. Some of them have been used as processing aids to develop egg products like high hydrostatic pressure with successful results in terms of quality (Juliano et al., 2006a, 2006b) or combined with high temperature to sterilize egg products (Juliano et al., 2012). Other innovative technologies like pulsed electric fields were intensively tested to pasteurize LWE a few decades ago (Monfort et al., 2010a, 2010b; Hermawan et al., 2004), but further research has been conducted in a slow pace mainly due to the lack of adequate processing equipment for egg products to keep intact the quality. Meanwhile, some other nonthermal technologies are currently tested not only for pasteurization of liquid products but also for the microbial decontamination of the eggshell. Some of these technologies are discussed in the following paragraphs.

4.1 Irradiation

Irradiation technology has been in the food science arena for several decades. Egg and egg products have been widely tested under different types of treatments, assessing the quality of the products and the microbial inactivation. Gamma irradiation, or gamma rays, was the first type of irradiation tested a few years ago. Currently, most irradiation studies rely on electron beams and low-energy X-rays. Gamma irradiation uses as energy sources radioactive forms of Cobalt 60 or Cesium 137; meanwhile, electron beam (also known as e-beam) is composed of high-energy electrons boosted from an accelerator. Low-energy X-rays are produced by reflecting a flow of high-energy electrons off a substance (i.e. heavy metal) (FDA, 2024).

In general, irradiation has been applied to eggshells, frozen and liquid egg products, and egg powder products (Fig. 2). The use of gamma irradiation to pasteurize in-shell eggs has been approved by the FDA with a dose of 3 kGy. The World Health Organization (WHO), Centers for Disease Control and Prevention (CDC), and the Department of Agriculture confirm the safety of irradiated products (FDA, 2024; Perry et al., 2011).

Irradiation is considered a nonthermal or cold pasteurization technology, so many of the quality properties of eggs are preserved during the processing (Song et al., 2009). The main preservation factor that is applied during irradiation is the energy coming from the sources and damaging the DNA of

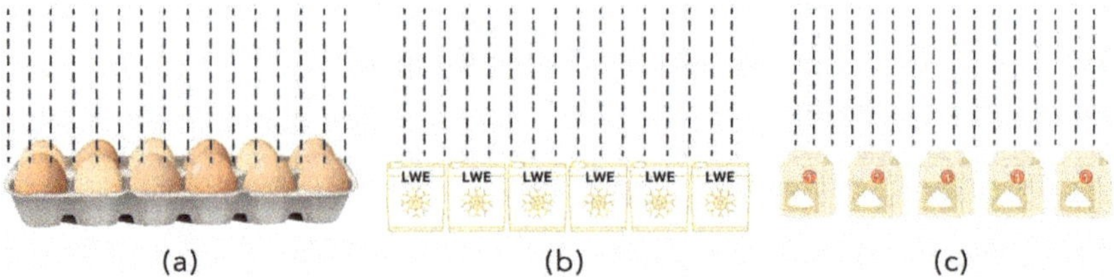

Figure 2 Representation of irradiation applied to (a) raw shell eggs, (b) frozen whole egg, and (c) powder egg products. Irradiation can be applied with gamma rays, low-energy X-rays, or electron beam (e-beam).

viruses and bacteria, carbohydrates, lipids, and breaking down the covalent bonds (Bisht et al., 2021). This technology has been applied to conduct the microbial decontamination of the eggshell (Table 1) and also when the microorganisms have been inoculated inside the egg (Table 2). There are few studies using irradiation to decontaminate the eggshell, and most of them are from several years ago. The studies reported in Table 1 show the inactivation of *Salmonella* cells using Cobalt 60 as the main source of gamma rays. When the dose is about 2 kGy and 3 kGy, cells were not detected on the eggshell (Tellez et al., 1995). In a further study, the use of chitosan together with irradiation also enhanced microbial decontamination (Yun et al., 2012).

In Table 2, the studies deal with *Salmonella* cells inoculated inside the egg or in liquid egg products. The research presented by Rodrigues et al. (2011) tested 20 different strains of *S. enteritidis* inoculated inside the egg. Samples were processed with several doses of gamma irradiation, with the highest being 5 kGy. After this strong treatment, there were two *Salmonella* strains that survived the treatment. Opposite results were found by the team of Liu et al. (2009b) using a similar inoculation level but finding no survivors after only 2 kGy. In a more comprehensive study, Alvarez et al. (2007) studied the effect of irradiation followed by heat in two strains of *Salmonella* inoculated LWE and using several GRAS (generally recognized as safe) compounds (nisin, EDTA, sorbic acid, and carvacrol). The research was conducted using a device operating with Cs 137, and it was found to have a synergistic effect between irradiation, heat, and additives. Nisin was the best additive to reduce *Salmonella* counts when used together with irradiation and heat, suggesting a possible reduction in time or temperature to minimize quality changes.

Regarding the egg quality after the irradiation processing, there is a vast number of reports dealing with gamma irradiation, and electron beam applied to liquid egg white, powder egg white, frozen egg white, liquid egg yolk, frozen yolk, powder yolk, and powder whole egg, as presented in Table 3. Most of the results after the irradiation of egg products presented no changes or improved properties, such as the foaming capacity and stability. The most affected property after irradiation is the viscosity. In contrast, Table 4 lists changes in

Table 1 Microbial decontamination of eggshells through innovative technologies

Technology	Microorganism	Processing conditions	Results	Reference
Gamma irradiation				
	S. enteritidis	Cobalt 60, 2 h, 1 kGy, 2 kGy, 3 kGy. Initial inoculum on eggshell 10^8 cfu/egg	3.9 log reduction (1 kGy) ND (2 kGy and 3 kGy)	Tellez et al., 1995
Plus chitosan	*S. typhimurium*	Cobalt 60, >1 kGy, 1% chitosan, 10^5 cfu/ml	ND	Yun et al., 2012
UV light				
	S. typhimurium	254 nm, 620 mW/cm², 1 min, 3 min, 5 min, 7 min. Initial inoculum 6.4 log/egg shell	> 6 log reduction	Kuo et al., 1997
	S. enteritidis, *Listeria monocytogenes*, enterohemorrhagic *Escherichia coli*	UV-C continuous (0.05–3 J/cm²; 10 mW/cm², 5–300 s) UV pulsed light (1.25–18 J/cm²)	1.6–3.8 log reduction Better continuous treatment than pulsed	Holck et al., 2018
	S. enteritidis	Pulsed UV-light system, 20 s, 9.5 cm distance from lamp, 23.6 J/cm². Initial inoculum 7–8 log CFU/cm²	5.3 CFU/cm² log reduction	Keklik et al., 2010
	S. typhimurium, *S. enteritidis*, *S. senftenberg*	UV-C light, 254 nm, 15 s. Initial inoculum 10^5 CFU/egg	2 log reduction	Mattioli et al., 2020
	E. coli and *E. faecium*	Pulsed UV-light system, 3 pulses (360 ss/pulse), 4.9 J/cm². Initial inoculum 10^4–10^5 log CFU/cm²	4.62 and 3.82 log reduction, respectively	Cassar et al., 2021

(Continued)

Table 1 *(Continued)*

Technology	Microorganism	Processing conditions	Results	Reference
Pulsed light				
	S. enteritidis	Pulsed light, 100 ss per pulse and 600 J/pulse. Max fluence 12 J/cm². Initial inoculum: 4.5 log CFU/unwashed egg; 6.3 log CFU/washed egg	3.6 log reduction for unwashed eggs; 1.8 log reduction for washed eggs	Hierro et al., 2009
	S. typhimurium	Pulsed light, 325 ss per pulse, room temperature (20–23°C), 5.5 cm distance from lamp, 2.1 J/cm². Initial inoculum 6 log CFU/egg	5.0 and 4.9 log CFU reduction for washed and unwashed egg, respectively	Lasagabaster et al., 2011
	E. coli ATCC 8879	Pulsed light, 15 cm from the lamp, 1.32 J/cm². Initial inoculum 6.5 log CFU/egg	3.77 log reduction CFU/egg	Wang et al., 2022
Cold plasma				
	S. enteritidis	HVACP 85 kV, 60 Hz, 15 min, dry air. Initial inoculum 10^7 CFU/egg	5.53 log reduction	Wan et al., 2017
	S. enteritidis	NTP, 400 W, 30 slm Argon, 5 eggs/min, 4 cm distance from plasma jet. Initial inoculum 10^8 CFU/egg	3.57 log reduction	Lin et al., 2021
	S. enteritidis	CAAP-J 800W, 120 s, 20 mm distance from the jet, air. Initial inoculum 10^7 CFU/egg	ND	Abdoli et al., 2024
	S. enteritidis	RBD plasma 15 kV, 12.7 kHz, 90 min, low moisture air (35% RH) and high moisture air (65% RH). Initial inoculum 5.5 log CFU/egg	2.5 log and 4.5 log reduction, according to the moisture content of air	Ragni et al., 2010

S. enteritidis	PAA gliding arc discharge, 50 W and 6 min OR 100 W and 3 min. Initial inoculum ~ 6 log CFU/egg	>4 log reduction (below the detection limit)	Guo et al., (2023)
S. enteritidis	DBD plasma, 25–30 kV, 10 –12 kHz, direct treatment (He/O_2 mixtures, 10 min) and indirect treatment (air, 25 min)	<10^2 cells/egg	Georgescu et al., (2017)
S. typhimurium	CNP, 600 W, 2 min, *Salmonella* biofilm created on shell at 25°C – 3 days	3.13 log CFU/cm^2	Lin et al., 2020
S. typhimurium	PAW, plasma jet 195V, 22 kHz, 200 ml water, 5 cm distance from the jet, 10 min generation. Contact time with PAW 2.5 min. Initial inoculum 7.7 log CFU/egg	5.1 log reduction	Wang et al., 2023
Klebsiella michiganensis	PAW generated with atmospheric pressure cold plasma system, 295 V, 22.5 kHz, 300 ml water, 5 min generation, pH 2.95. Washing time 3 min with massaging. Initial inoculum 7.9 log CFU/egg	5.28 log reduction	Narasimhan et al., 2023

ND: not detected; HVACP: high voltage atmospheric cold plasma; RBD: resistance barrier discharge; RH relative humidity; DBD: dielectric barrier discharge; CNP: cold nitrogen plasma; NTP: nonthermal plasma; slm: standard liter per minute; PAA: plasma activated air; CAAP-J: cold atmospheric pressure air plasma – jet; PAW: plasma activated water.

Table 2 Microbial inactivation on eggs and egg products using innovative technologies

Technology	Product	Microorganism	Processing conditions	Results	References
Gamma irradiation					
	In-shell egg	*S. enteritidis* (20 strains)	Cobalt 60, 0.49–5 kGy, 53.7 Gy/min, 10^4 CFU/egg initial inoculum	2 strains survived after 5 kGy	Rodrigues et al., 2011
	In-shell egg	*S. typhimurium*	Cobalt 60, 0.5 kGy, 1 kGy, 1.5 kGy, and 2 kGy, 3.5 log CFU/g initial inoculum	No viable cells after 2 kGy	Liu et al., 2009b
Low-energy X- rays					
	In-shell egg	*S. enteritidis* (monoculture biofilm)	X-ray, 3 kW, 160 keV, 1 Gy/s (air), 225 Gy, 25°C, 50–60% humidity. Initial inoculum 10^7 cfu/egg	3.5 log reduction	Zhang et al., 2021
Electron beam					
	Liquid egg white	*S. typhimurium*	Linear accelerator, 10 MeV, 8.1 kW, 75–105 kGy/min, 2.5–3.3 kGy. Initial inoculum 10^7 cfu/ml	No cells detected after processing	Wong et al., 1996
UV light					
	Liquid egg white	*E. coli* K-12	Flow-through pulsed UV-light, 40 ml/min, 5 cm distance from lamp, 3 passes. Initial inoculum 6 log CFU/ml	1.57 log reduction	Ouyang et al., 2020
	Liquid egg white	*E. coli* K-12	Centrifugal UV irradiator, 10°C, 3.9 J/ml. Initial inoculum 8 log CFU/ml	5 log reduction	Geveke and Torres, 2013
	Liquid egg white	*E. coli* K-12, *E. coli* O157:H7, *Listeria monocytogenes*	UV fluence 1.3 mW/cm^2, 20 min, 0.153 cm sample depth. Initial inoculum 10^7–10^8 CFU/ml	0.89, 1.4, 0.96 log reduction, respectively	Unluturk et al., 2010

Table 3 Quality parameters of egg products treated with innovative technologies

Technology: gamma irradiation				
Egg product	Egg property	Processing conditions	Results	Reference
White (liquid)	Emulsifying properties, particle size, protein content, rheological properties	Cobalt 60, 3.4 kGy/h, 1 kGy, 3 kGy, and 5 kGy	Dispersion of droplets improved, decrease on particle size, α-helix and β-sheet protein content decreased, decrease on elasticity and viscosity decreased	Liu et al., 2020
White (liquid)	pH	Cobalt 60, 10°C, 20 kGy/h, 1 kGy, 2 kGy, and 5 kGy	No changes	Song et al., 2009
	Viscosity		Decreased	
	Foaming ability		Improved	
White (frozen)	pH	Cobalt 60, −35°C, 1 kGy, 2.5,kGy, and 4 kGy	No changes	Ma et al., 1993
	% solids		No changes	
	% protein		No changes	
	Functional properties (foaming, emulsifying, gelling)		No changes	
White (spray-dried powder)	Foam stability	Cobalt 60, 18 −22°C, 2–16 kGy, 6.5 kGy/h	Better stability after treatment	Clark et al., 1992
White (powder)	pH	Cobalt 60, 10°C, 20 kGy/h, 1 kGy, 2 kGy, and 5 kGy	No changes	Song et al., 2009
	Foaming ability		Improved	
White (powder)	Viscosity	Cobalt 60; 5 kGy, 15 kGy, and 25 kGy	No changes	Ferreira and Del Mastro, 1998

(Continued)

Table 3 (*Continued*)

Technology: gamma irradiation				
Egg product	Egg property	Processing conditions	Results	Reference
White protein (powder)	Foaming characteristics, turbidity, and protein oxidation	Cobalt 60; 20 kGy/h, 12°C, 2.5 kGy and 5 kGy	Increase foaming ability, decrease foaming stability. Increase on turbidity and protein oxidation	Liu et al., 2009a
Yolk (powder)	Rheological properties and color	Cobalt 60; 5 kGy, 15 kGy, and 25 kGy	Changes on both properties	Ferreira and Del Mastro, 1998
Yolk (frozen)	pH	Cobalt 60, −35°C, 1 kGy, 2.5 kGy, and 4 kGy	No changes	Ma et al., 1993
	% solids		No changes	
	% protein		Significant changes	
	Apparent viscosity		Significant changes	
Whole egg (powder)	Viscosity	Cobalt 60; 5 kGy, 15 kGy, and 25 kGy	Changes	Ferreira and Del Mastro, 1998
Technology: Electron beam irradiation				
Egg product	**Egg property**	**Processing conditions**	**Results**	**Reference**
White (liquid)	Color and viscosity	Electron beam irradiation, 10 MeV, 8.1 kW, 75–105 kGy/min, 2.5–3.3 kGy	No changes on color; viscosity more stable during storage	Wong et al., 1996
White (powder)	Degree of hydrolysis (DH), bound water content (BWC), denaturation temperature (DT)	Electron beam irradiation, 10 MeV, 15 kW, 25°C, 1.08, 3.24 kGy and 5.40 kGy	Increase on DH, decrease on BWC and DT	Jin et al., 2017
White protein (powder)	$ABTS^+$ free radical scavenging rate, particle size, surface hydrophobicity	Electron beam irradiation, 10 MeV, 15 kW, 1 kGy, 1.5 kGy, 2 kGy, 2.5 kGy; 1 h, 1.5 h, 2 h	Increase on $ABTS^+$ (from 30.16% to 49.48%), particle size and surface hydrophobicity	Liu et al., 2019

White protein (powder)	Solubility, $ABTS^+$, emulsifying activity index (EAI)	Electron beam irradiation, 10 kGy/pass, 5 kGy, 10 kGy, 20 kGy, 30 kGy, 50 kGy, 70 kGy, and 100 kGy	Better solubility, increase on $ABTS^+$ and EAI	Liu et al., 2023
Yolk (liquid)	Rheological properties, color, protein content	Electron beam irradiation, 10 MeV, 8.1 kW, 2.3–3 kGy	Changes on the denaturation temperature, storage modulus (G) delayed during storage, retained more soluble protein, yolk was less bright	Huang et al., 1997
Technology: UV light				
Egg product	**Egg property**	**Processing conditions**	**Results**	**Reference**
White (liquid)	pH, lipid oxidation, turbidity, color, and foaming properties	Flow-through pulsed UV-light, 40–80 mL/min, 1–3 passes, distance from lamp 5–13 cm	Changes on pH, lipid oxidation, turbidity, and color. No changes on foaming capacity and stability	Ouyang et al., 2020
White, whole egg, yolk (all liquid)	Vitamin A, B_2, B_5, C, E; mineral content (P, Cl, K, Na, Ca, Mg, Fe, Zn) and carotenoids (lutein and zeaxanthin)	UV light, 20.1 J/l (egg white), 32.3 J/l (whole egg), 115.6 J/l (egg yolk)	Reduction on vitamin A and C and carotenoids, by 80%, 66%, and 61%, respectively	Mendes de Souza et al., 2015
White, whole egg, yolk (all liquid)	Foaming properties and emulsifying activity index (EAI)	UV light, 20.1 J/l (egg white), 32.3 J/l (whole egg), 115.6 J/l (egg yolk)	Better foaming stability and ability, increase on EAI	Mendes de Souza et al., 2015

Table 4 Quality parameters of shell egg processed under innovative technologies

Technology				
Gamma irradiation	Egg property	Processing conditions	Results	Reference
	HU	Cobalt 60, 2 h, 1 kGy, 2 kGy, 3 kGy	Decrease on 50%	Tellez et al., 1995
	HU	Cobalt 60, 0.5 kGy	Decreased	Liu et al., 2009b
	HU	Cobalt 60, 0.87–1 kGy/h, 0.97 kGy, 2.37 kGy, and 2.98 kGy	Decreased	Ma et al., 1990
	Yolk color	Cobalt 60, 2 h, 1 kGy, 2 kGy, 3 kGy	Decrease on color when 2-3 kGy used	Tellez et al., 1995
	Yolk color		Decreased	Liu et al., 2009b
	Lipid oxidation	Cobalt 60, 2.0 kGy	Increased	Liu et al., 2009b
	Viscosity	Cobalt 60, 0.5kGy , 1 kGy, 1.5 kGy, 2 kGy	Decreased	Liu et al., 2009b
	Foaming ability		Improved	Liu et al., 2009b
	Apparent viscosity		Decreased for egg white, increased for yolk	
Nanosecond electron beam	% Ca on eggshell	500 keV, 50 ns, 200 pps	Calcium content was increased with the higher doses (25–40 kGy)	Sokovnin et al., 2019
Low-energy x-ray	Puncture force	X-ray, 3 kW, 160 keV, 1 Gy/s (air), 225 Gy, 25°C, 50–60% humidity. Initial inoculum 10^7 cfu/egg	Unchanged	Zhang et al., 2021
	Yolk color, viscosity, foaming capacities	Linear accelerator, 10 MeV, 5.6 kW, 61.3 kGy/min, 1–2 kGy	Yolk color was not affected, white viscosity decreased significantly, foaming capacity and stability decreased	Min et al., 2005

	Turbidity, viscosity, foaming properties	10 MeV, 5.6 kW, 61.3 kGy/min, room temperature, 2.5 kGy, 5 kGy, and 10 kGy	Increase turbidity and decrease viscosity of egg white. Foaming capacity and stability not affected at lower dose	Min et al., 2012
UV light				
	Albumen height, eggshell strength, cuticle presence	Pulsed UV-light system, 1–30 s, 9.5 cm and 14.5 cm distance from lamp	No changes on the evaluated parameters	Keklik et al., 2010
	Weight loss, HU, YI, albumen pH during storage	Slightly acidic electrolyzed water (SAEW) 4 min, 30 mg/l, pH 6.37, 980 mV plus UV treatment (4 min, 10.2 W/cm^2)	Improved quality parameters during storage (6 weeks, 25°C)	Sheng et al., 2020
Pulsed light				
	HU, YI, albumen pH, weight loss	Pulsed light, 15 cm from the lamp, 0.66 J/cm^2/pulse × 2 pulses (1.32 J/cm^2)	No changes after processing	Wang et al., 2022
	Rheological properties of egg white	Pulsed light, 5.5 cm from the lamp, 325 ss per pulse, room temperature (20–23°C), 10.5 J/cm^2	No changes on complex modulus, storage and loss moduli	Lasagabaster et al., 2011
	White absorbance, immunoreactivity, viscosity, gelling temperature, and foaming properties	Pulsed light, 0–18 pulses, 0.50 ss per pulse, 3 cm distance from lamp, 25°C, 1.75–31.5 J/cm^2	Changes on absorbance, increase on immunoreactivity, no changes on viscosity, decrease on gelling temperature, higher foaming stability	Manzocco et al., 2013
Cold plasma				
	HU, Yolk color, pH albumen/yolk, vitelline membrane strength, egg weight	HVACP 85 kV, 60 Hz, 15 min, dry air. Initial inoculum 10^7 CFU/egg	No changes in quality	Wan et al., 2017

(Continued)

Table 4 (*Continued*)

	Technology			
Gamma irradiation	Egg property	Processing conditions	Results	Reference
	Eggshell color, albumen pH, yolk index	CNP, 600 W, 2 min	No changes on the product	Lin et al., 2020
	Specific gravity, weight loss, yolk and albumen index, pH, HU	NTP, 400 W, 30 slm Argon, 5 eggs/min, 4 cm distance from plasma jet. Initial inoculum 10^8 CFU/egg	No changes on the product after processing and during storage	Lin et al., 2021
	Eggshell color and strength	PAW was generated with atmospheric pressure cold plasma system, 295V, 22.5 kHz, 300 ml water, 5 min generation, pH 2.95. Washing time 3 min with massaging.	Minor changes in color, no changes on shell strength	Narasimhan et al., 2023
	pH, yolk color, HU, eggshell breaking strength	CAAP-J 800W, 120 s, 20 mm distance from the jet, air.	Minor changes in color	Abdoli et al., 2024
	Color, albumen pH, weight loss, yolk index	RBD plasma 15 kV, 12.7 kHz, 90 min, low moisture air (35% RH) and high moisture air (65% RH).	No changes	Ragni et al., 2010
	HU, albumen and yolk pH, color, vitelline membrane strength, shell strength	HVACP, 100 kV RMS, 60% HR air, 1 min of treatment and 6 h post-treatment storage (4°C) OR 5 min of treatment and 4 h post treatment storage (4°C)	No changes	Illera et al., 2022

HVACP, high voltage atmospheric cold plasma; RBD, resistance barrier discharge; RH, relative humidity; CNP, cold nitrogen plasma; NTP, nonthermal plasma; slm, standard liter per minute; CAAP-J, cold atmospheric pressure air plasma – jet; PAW, plasma activated water; HU, Haugh unit.

egg quality when the treatment was applied to in-shell eggs. For example, the HU decreased in all the studies reported in Table 4; additional changes were observed in the yolk color and lipid oxidation.

Gamma irradiation has also been tested to reduce the allergenicity of some albumen proteins. In a study conducted by Seo et al. (2004) the egg white was treated with gamma rays coming from Cobalt 60 (10°C, 20 kGy/h, 10–20 kGy). Results showed a reduction in ovalbumin (OVA) from the control sample (432.88 sg/g) to the treated sample (14.27 sg/g), meaning a reduction in the allergenicity.

4.2 Ultraviolet and pulsed light

Ultraviolet light known as UV – C is the most common wavelength used in food applications. This wavelength goes from 100 nm to 280 nm, and the effect of this radiation on the microorganism is directly on the DNA. UV photons dimerize with the pyrimidine inside the microbial DNA, leading to cellular death (Kaavya et al., 2021).

The use of UV – C for microbial decontamination and pasteurization has been studied in several products. In the specific case of eggs, UV – C has been tested mainly as a batch process for decontaminating the eggshell and as a continuous treatment of liquid egg products (Figure 3). In the first instance, most of the reported studies present a good level of decontamination on the eggshell surface, as shown in Table 1. The main drawbacks during decontamination and limiting the effectivity of the treatment are related to the presence of different microflora or additional pathogens on the surface, the presence of shadows or pores hiding the cells and protecting them from the UV – C radiation, and the incorrect distance of the UV – C lamp from the eggshell. This technology has limitations in food products when the surface is irregular. As shown previously in Figure 1, the eggshell presents a very irregular surface, making the UV – C radiation difficult to reach those cells inside the porosity or cracks. When the treatment is applied to a liquid product such as liquid egg white the efficiency of UV – C is limited because of the turbidity of the product. Microbial inactivation of several pathogens in liquid egg white is reported in Table 2 and, in most cases, is limited to 2 log reduction. This fact is well-known for UV – C technology in other products and represents a hurdle for microbial inactivation in using ultraviolet for those products that are not as clear as water. It is often required to increase the turbulence of the liquid product inside the UV – C reactor to increase the contact between the radiation and all the points of the product (Bermudez-Aguirre and Niemira, 2022).

Regarding the quality of eggs after UV – C processing, in Table 3, there are some results for LWE, liquid egg white, and liquid egg yolk. In the study

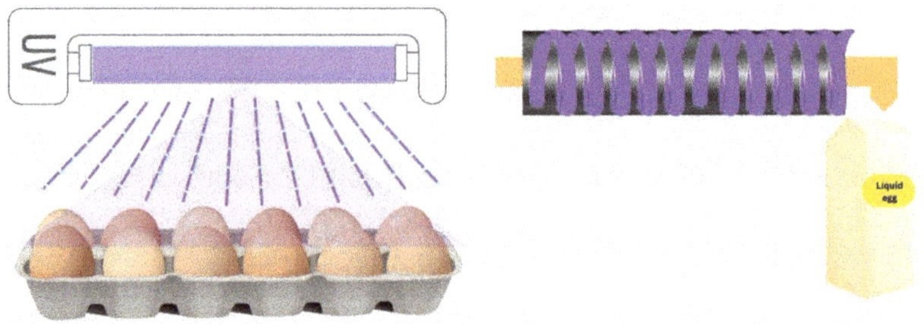

Figure 3 Representation of ultraviolet treatment applied to (a) raw shell eggs and (b) liquid whole egg using a coil-UV – C reactor

reported by Ouyang et al. (2020) in egg white there were changes in pH, lipid oxidation, turbidity, and color. However, there were no changes in the foaming capacity and stability. Similar results were observed by Mendes de Souza et al. (2015) in egg white and an increase in the emulsifying activity index of egg yolk. Some reductions in the vitamin content and carotenoids have been reported by Mendes de Souza et al. (2015) because of the effect of the UV photons on the stability of these biomolecules. A couple of studies about the use of UV – C on in-shell eggs are reported in Table 4 regarding the quality. In general, the effect of the light treatment did not affect the quality of the eggshell such as the strength or the cuticle presence (Keklik et al., 2010); UV – C improved the quality of eggs when it was applied together with slightly acidic electrolyzed water and stored for 6 weeks at 25°C (Sheng et al., 2020).

Pulsed light (PL) is a technology that has been officially approved by the FDA, and it is based on a series of short-time and high-power pulses emitting radiation on microbial cells (Fig. 4). Microbial inactivation occurs because photochemical and photothermal effects affect the DNA and avoid its multiplication (Wang et al., 2022). The inactivation of pathogens such as *Salmonella* spp. and *Escherichia coli* on the eggshell surface has been reported by several researchers as shown in Table 1. The work conducted by Hierro et al. (2009) presented interesting results when PL was applied to unwashed and washed eggs. The maximum log reduction was observed for unwashed eggs (3.6 log reduction) contrasting with a 1.8 log reduction in washed eggs. The authors suggested a possible photoreaction of the pathogen after treatment but also discussed the eggshell surface and porosity as one of the main factors affecting the inactivation. The work presented by Lasagabaster et al. (2011) in Table 1 is similar, and the inactivation was almost the same between washed and unwashed eggs. Regarding the quality, most of the research conducted with PL for shell decontamination presented no changes in the main quality parameters as observed in Table 4. One of the main problems when comparing the research done with novel technologies is not only the lack of uniformity

Figure 4 Representation of pulsed light technology applied to raw shell eggs.

and consistency in reporting processing conditions but also the very diverse equipment used for microbial inactivation. A global harmonization initiative for food science is underway to try to standardize the way the processing conditions are reported in order to compare the research between different groups and reduce the duplication of experiments. This initiative is an international network of scientists and organizations looking to harmonize food safety regulations (GHI, 2024).

4.3 Cold plasma

Cold plasma is one of the newest interventions tested for microbial inactivation in food applications. This technology uses an electrical discharge applied to a gas (commonly air) to produce free radicals, active species, ions, protons, electrons, and UV radiation. Additional working gases include Argon, Helium, and mixtures of oxygen and nitrogen, to mention a few. All these active species contained in a cloud known as plasma glow have an antimicrobial effect on the microbial cells, spores, or even viruses (Bermudez-Aguirre, 2020; Hertrich et al., 2017; Han et al., 2016; Hong et al., 2009; Kayes et al., 2007). Cold plasma has proved to have minimal to no effect on the quality of several food products (Pankaj et al., 2018). As a novel, green, inexpensive, and easy-to-operate technology, together with the emerging need to control the presence of pathogens in food products and reduce foodborne outbreaks, cold plasma has been extensively tested in eggs for decontamination, as shown in Table 1. Again, it is difficult to compare between treatments because of the different configurations of the plasma devices, but the level of decontamination is constantly high (~ >3 log reduction) in the tested pathogens. Some of the different cold plasma configurations include corona discharge (Fig. 5a), dielectric barrier discharge or DBD (Fig. 5b), or plasma jet (Fig. 5c). Comparing cold plasma to UV – C light for shell decontamination, the first technology offers the possibility to reach those cells hiding inside the natural porosity of the eggshell and inactivate them because of the presence of free radicals in the plasma glow. Furthermore, it is possible to enhance microbial inactivation through cold plasma when moisture is added to the working gas (air). Since the water molecules present in the relative humidity of the air will be broken down by the electrical discharge that generates the plasma glow, additional free radicals coming from the water will act on the microbial cells, as shown in work reported by Ragni et al. (2010).

Recently, to improve the efficiency of cold plasma for microbial inactivation, a new term has been introduced: plasma-activated water (PAW). In this technology, the electrical discharge to generate the plasma glow is applied to the water surface or inside the water to provide antimicrobial properties to the water and act as an antimicrobial wash for food surfaces such as produce, meat, or eggs. The use of PAW has been successfully reported in eggs by several researchers (Narasimhan et al., 2023; Wang et al., 2023; Shah et al., 2023). In the study reported by Wang et al. (2023), PAW was used as a sanitizer for the eggshell surface and was able to reach pasteurization standards (> 5 log reduction) of *Salmonella typhimurium* in about 5 min.

In terms of quality, cold plasma offers a convenient solution for microbial decontamination with minimal to no changes in quality as presented in Table 4. The main minor change reported in plasma-treated eggs is related to the yolk color. Abdoli et al. (2024) and Narasimhan et al. (2023) observed the change

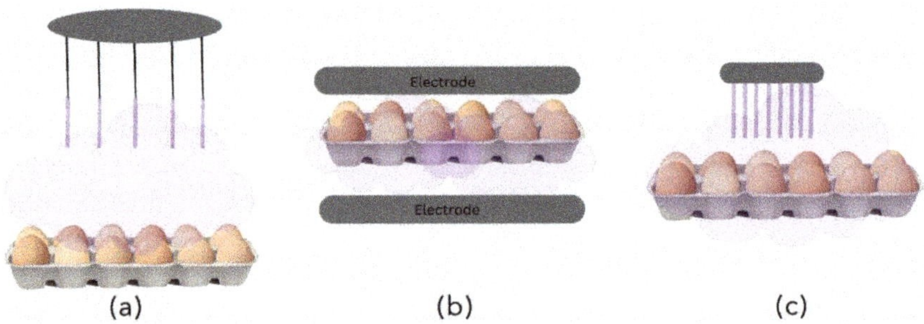

Figure 5 Representation of cold plasma treatment applied to raw shell eggs using different plasma devices: (a) corona discharge, (b) dielectric barrier discharge (DBD), and (c) plasma jet.

in this quality parameter. The reason behind the color change is the possible oxidation of the carotenoids present in the yolk. Cold plasma generates some free radicals and ozone (when air is used as working gas) that are potent oxidizing agents. However, this fact can be minimized when a different working gas is used (i.e. argon) or the system has been optimized for microbial inactivation and egg quality. Several studies conducted with cold plasma in eggs studied the quality during storage and results presented the successful application of this technology to preserve the quality (Illera et al., 2022; Lin et al., 2021).

5 Case study: radio frequency to pasteurize in-shell eggs

All the previous technologies discussed in this chapter have presented the positive effects on microbial inactivation mainly on the eggshell and the possibility to preserve the egg quality during the processing. However, the possibility of inactivating *Salmonella* cells inside the egg (vertical transmission) has been only addressed by irradiation, as shown in the examples presented in Table 2. However, an innovative technology that is considered a thermal treatment has been explored in the last few years, as an option to pasteurize in-shell eggs and eliminate any microbial risk. This technology is known as RF and shares a similar working principle with MW. The main difference between RF and MW is the working frequencies. According to the Federal Communications Commission (FCC), specific frequencies are allowed for Industrial, Science, and Medical (ISM) uses that do not interfere with other communication systems. The allowed frequencies for RF for food applications are 13.56 MHz, 27.12 MHz, and 40.68 MHz (Bermudez-Aguirre and Niemira, 2023a); meanwhile, MW uses frequencies at 915 MHz and 2450 MHz (Chandrasekaran et al., 2013).

The working principle of RF relies on the dielectric and thermal properties of the product. Food contains water, which is a polar molecule. When food is placed in contact with an electric field (RF or MW), the water molecules rotate, aligning with the field. This rotation produces friction between the water molecules and this microscopic phenomenon generates heat from inside the product; this is known as volumetric heating (Bermudez-Aguirre and Niemira, 2023a). The heat intensity depends on several factors, such as the dielectric and thermal properties of the food. The dielectric properties of food represent how the product can reflect, store, or utilize energy. The dielectric properties that are important during RF processing are the relative permittivity (ε), the dielectric constant (ε'), and the dielectric factor (ε''). The relative permittivity (ε) is related to the ability of the material to interact with the magnetic field. The dielectric constant (ε') represents a measurement of the food to absorb, reflect, or transmit energy. Meanwhile, the loss factor (ε″) measures the ability of the food to dissipate heat during the heating (Bermudez-Aguirre and Niemira, 2023a). These properties are also affected by the presence of other food components such as solutes (salts, sugar), fat, and water content, among others. The thermal properties of food, such as the specific heat (Cp), which measures the required energy that needs to be applied to the food to increase the temperature, and the thermal conductivity (*K*), which measures the ability of the food to conduct heat, are also important during RF processing. These properties and the specific characteristics of the RF equipment allow the food to warm up and start volumetric heating. During this novel technology applied to in-shell eggs, heat is produced inside the egg, and the contents (i.e. white and yolk) are heated volumetrically.

The USDA-ARS developed and patented an egg pasteurizer based on RF energy. The system was initially reported by Geveke et al. (2017, 2018). The system has been improved over the years with important modifications in the electronic part but also in the main processing unit. The system can pasteurize in-shell eggs in only 20 min, inactivating 5 log reduction of *S. typhimurium* inoculated inside the egg. The quality after the RF processing is the same as a raw egg without the risk of *Salmonella* contamination, and it was preserved through storage (Geveke et al., 2017, 2018; Yang et al., 2020; Yang and Geveke, 2020; Bermudez-Aguirre and Niemira, 2023b; Bermudez-Aguirre and Niemira, 2023c). The conventional thermal pasteurization of in-shell eggs is a lengthy process (>60 min) in which eggs are submerged in water and held for a specific time. The temperature, usually 57°C, must be carefully controlled to avoid the thermal damage of the egg proteins, adding costs to the final price. Comparing the RF processing to the conventional thermal pasteurization of eggs, RF offers an excellent option to ensure microbial safety and maintain egg quality during processing, mainly for those special markets that require safe eggs, such as nursing homes, hospitals, schools, and those applications that raw eggs are required.

The current RF system is shown in Fig. 6 with several improvements. The new system works at 40.68 MHz and has a new matching network that minimizes the loss of energy, the current reflected power is less than 1%. The eggs are placed on top of rollers that allow the egg's movement during processing for a more uniform treatment. The internal temperature of the egg components (i.e. yolk and albumen) are recorded during processing using fiber optic sensors. The electrodes for the RF system are attached to 3D-custom-printed paddles. The building materials were chosen based on durability to make this system an option for small and large egg processors. Eggs are in contact with water spray during RF processing to regulate the temperature and avoid albumen coagulation. This new version of the patented system allows the inactivation of *S. typhimurium* in in-shell eggs, achieving pasteurization standards (5 log

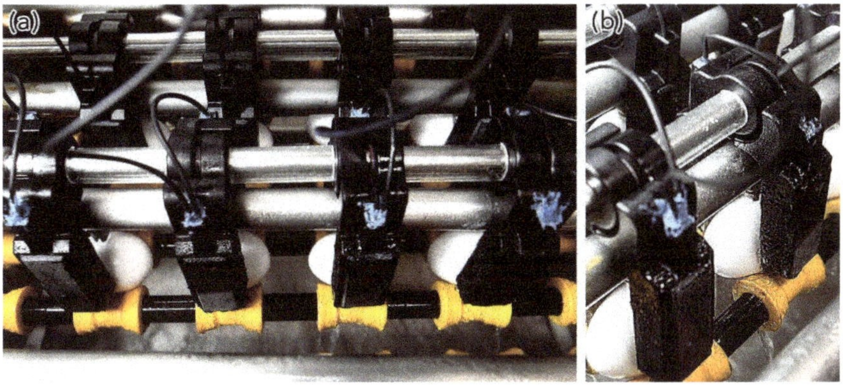

Figure 6 Radiofrequency (RF) egg pasteurizer patented by USDA – ARS. (a) The current configuration allows the pasteurization of 24 eggs in a semi-continuous mode. (b) The eggs are held by 3D-custom-printed paddles that contain electrodes. The RF system operates at 40–68 MHz.

reduction) and preserving the quality of the eggs during storage. RF-processed eggs have been tested for sub-lethally injured cells or cell survivors during refrigerated storage (7°C), and none of them were found, ensuring the microbial quality of the eggs (unpublished data). This unit is in the process of scaling up for full commercialization in the near future, offering affordable, quick, and easy technology to ensure the microbial safety of the eggs and preserve the overall quality.

6 Conclusion and future trends

Eggs represent a unique food system with very diverse characteristics inside and outside. The egg contents, such as white (albumen) and yolk, have very different characteristics, composition, and behavior under the innovative interventions described in this chapter. The eggshell is also an interesting irregular system that poses a challenge for scientists during microbial decontamination. Eggs are a target for different regulatory agencies, national and international, because of the constant presence of eggs and egg products in several foodborne outbreaks dealing with *Salmonella* spp. The options for decontamination and pasteurization have been discussed in this chapter with more benefits than drawbacks. However, an essential part of the research into eggs is related to the lack of regulations worldwide about the processing conditions that eggs must have before they reach the final market such as washing. Eggs are also controversial when storage conditions are discussed universally, finding discrepancies between the storage temperature in the market and at home. Perhaps more processing technologies are expected to be tested for eggs and egg products in the near future, and some will become a commercial reality such as RF, but the research for the best technology should go together with the seek for international regulations during the whole egg production chain.

7 Where to look for further information

Further reading

- Zhang, H.Q., et al. (2025). *Nonthermal processing technologies for food* (2nd ed.). IFT Press. Wiley-Blackwell, Hoboken, NJ. Forthcoming.
- Koutchma, T. (2021). *Validation of food preservation processes based on novel technologies*. Academic Press, Oxford, UK. 264 pp.

Key journals/conferences

- *Innovative Food Science and Emerging Technologies (IFSET)* publishes the latest trends in innovative food processing and preservation technologies.
- The Institute of Food Technologists (IFT) Annual Meeting is held during the summer, including Nonthermal Processing meetings.
- The International Association of Food Protection (IAFP) Annual Meeting is held during the summer.

8 References

Abdoli, B., et al. (2024). Cold atmospheric pressure air plasma jet disinfection of table eggs: Inactivation of *Salmonella enterica*, cuticle integrity and egg quality. *International Journal of Food Microbiology*, 410, 110474. https://doi.org/10.1016/j.ijfoodmicro.2023.110474

Alvarez, I., et al. (2007). Inactivation of *Salmonella* Enteritidis and *Salmonella* Senftenberg in liquid whole egg using generally recognized as safe additives, ionizing radiation, and heat. *Journal of Food Protection*, 70(6), 1402–1409. https://doi.org/10.4315/0362-028X-70.6.1402

Aygun, A. (2017). The eggshell microbial activity. In P.Y. Hester (Ed.), *Egg Innovations and strategies for improvements*. Elsevier, Academic Press, Oxford, UK, pp. 165–175. https://doi.org/10.1016/B978-0-12-800879.9.00013-5

Bermudez-Aguirre, D. and Niemira, B. (2023a). Radio frequency treatment of food: a review on pasteurization and disinfestation. *Foods*, 12, 3057. https://doi.org/10.3390/foods12163057

Bermudez-Aguirre, D. and Niemira, B. (2023b). Microbial inactivation models of *Salmonella* Typhimurium of radio frequency treated eggs. *Food Control*, 148, 109634. https://doi.org/10.1016/j.foodcont.2023.109634

Bermudez-Aguirre, D. and Niemira, B. (2023c). Modelling quality changes and *Salmonella* Typhimurium growth in storage for egg pasteurized by radio frequency treatments. *Food Control*, 148, 109638. https://doi.org/10.1016/j.foodcont.2023.109638

Bermudez-Aguirre, D. and Niemira, B. (2022). A review on egg pasteurization and disinfection: Traditional and novel processing technologies. *Comprehensive Reviews in Food Science and Food Safety*, 1–29. https://doi.org/10.1111/1541-4337.13088

Bermudez-Aguirre, D. (2020). Advances in the inactivation of microorganisms and viruses in food and model systems using cold plasma. In D. Bermudez-Aguirre (Ed.), *Advances in cold plasma applications for food safety and preservation*. Academic Press, Oxford, UK.

Bisht, B., et al. (2021). Food irradiation: Effect of ionizing and non-ionizing radiations on preservation of fruits and vegetables: a review. *Trends in Food Science and Technology*, 114, 372–385. https://doi.org/10.1016/j.tifs.2021.06.002

Cassar, J.R., et al. (2021). The efficacy of pulsed ultraviolet light processing for table and hatching eggs. *Poultry Science*, 100, 100923. https://doi.org/10.1016/j.psj.2020.12.021

CDC (2024). Centers for Disease Control and Prevention. https://www.cdc.gov, Accessed December 2023.

Clark, D.C., et al. (1992). The effect of irradiation on the functional properties of spray-dried egg white protein. *Food Hydrocolloids*, 5(6), 541–548.

Chandrasekaran, S., et al. (2013). Microwave food processing: a review. *Food Research International*, 52, 243–261. https://dx.doi.org/10.1016/j.foodres.2013.02.033

DSM (2018). *DSM egg quality manual*. DSM Nutritional Products Ltd, Basel.

Ferreira, L.F.S. and Del Mastro, N.L. (1998). Rheological changes in irradiated chicken eggs. *Radiation Physics and Chemistry*, 52(1–6), 59–62.

Froning, G.W., et al. (2002). *International egg pasteurization manual*. Published in collaboration with the United Egg Association and the American Egg Board.

Food and Drug Administration (FDA) (2024). Food irradiation: what you need to know. https://www.fda.gov, Accessed January 2024.

Food Safety and Inspection Service (FSIS) (2020). Food safety guideline for egg products. September 9, 2020. USDA FSIS.

Georgescu, N., et al. (2017). Inactivation of *Salmonella* enterica serovar Typhimurium on egg surface, by direct and indirect treatments with cold atmospheric plasma. *Food Control*, 76, 52–61. https://dx.doi.org/10.1016/j.foodcont.2017.01.005

Geveke, D.J., et al. (2018). Improvement in the radio frequency method to pasteurize shell eggs by automation and cost reduction. *International Journal of Food Science and Technology*, 53, 2500–2508. https://doi.org/10.1111/iffs.13843.

Geveke, D.J., et al. (2017). Pasteurization of shell eggs using radio frequency heating. *Journal of Food Engineering*, 193, 53–57. https://doi.org/10.1016/j.jfoodeng.2016.08.009.

Geveke, D.J. and Torres, D. (2013). Liquid egg white pasteurization using a centrifugal UV irradiator. *International Journal of Food Microbiology*, 162, 43–47. https://dx.doi.org/10.1016/j.ijfoodmicro.2012.12.005

GHI (2024). Global Harmonization Initiative. https://www.globalharmonization.net

Guo, L., et al. (2023). Inactivation of *Salmonella enteritidis* on the surface of eggs by air activated with gliding arc discharge plasma. *Food Control*, 148, 109662. https://doi.org/10.1016/j.foodcont.2023.109662

Han, L., et al. (2016). Mechanisms of inactivation by high-voltage atmospheric cold plasma differ for *Escherichia coli* and *Staphylococcus aureus*. *Applied and*

Environmental Microbiology, 82(2), 450–458. https://doi.org/10.1128/AEM.02660-15

Hermawan, N., et al. (2004). Pulsed electric field treatment of liquid whole egg inoculated with *Salmonella Enteritidis*. *Journal of Food Safety*, 24, 71–85.

Hertrich, S., et al. (2017). Cold plasma inactivation of *Salmonella* in pre-packed mixed salads is influenced by cross-contamination sequence. *Journal of Food Protection*, 80, 2132–2136. https://doi.org/10.4315/0362-028x.JFP-17-242

Hierro, E., et al. (2009). Inactivation of *Salmonella enterica* serovar Enteritidis on shell eggs by pulsed light technology. *International Journal of Food Microbiology*, 135, 125–130. https://doi.org/10.1016/j.ijfoodmicro.2009.07.034

Holck, A.L., et al. (2018). Comparison of UV-C and pulsed UV light treatments for reduction of *Salmonella, Listeria monocytogenes* and enterohemorrhagic *Escherichia coli* on eggs. *Journal of Food Protection*, 81(1), 6–16. https://doi.org/10.4315/0362-028X.JFP-17-128

Hong, Y.F., et al. (2009). Sterilization effect of atmospheric plasma on *Escherichia coli* and *Bacillus subtilis* endospores. *Letters in Applied Microbiology*, 48(1), 33–37. https://doi.org/10.1111/j.1472-765x.2008.02480.x

Huang, S., et al. (1997). Effect of electron beam irradiation on physical, physicochemical, and functional properties of liquid egg yolk during frozen storage. *Poultry Science*, 76, 1607–1615.

Illera, A.E., et al. (2023). Effect of high voltage atmospheric cold plasma on chicken eggs quality during refrigerated storage. *Food Bioscience*, 53, 102754. https://doi.org/10.1016/j.fbio.2023.102754

Illera, A., et al. (2022). High voltage atmospheric cold plasma decontamination of *Salmonella* enteritidis on chicken eggs. *Innovative Food Science and Emerging Technologies*, 82, 103210. https://doi.org/10.1016/j.ifset.2022.103210

Jin, Y., et al. (2017). Effects of structure changes on hydrolysis degree, moisture state, and thermal denaturation of egg white protein treated by electron beam irradiation. *LWT – Food Science and Technology*, 77, 134–141. https://dx.doi.org/10.1016/j.lwt.2016.11.025

Juliano, P., et al. (2012). Shelf-stable egg-based products processed by high pressure thermal sterilization. *Food Engineering Reviews*, 4, 55–67. https://doi.org/10.1007/s12393-011-9046-4

Juliano, P., et al. (2006a). Descriptive analysis of precooked egg products after high-pressure processing combined with low and high temperature. *Journal of Food Quality*, 29, 505–530.

Juliano, P., et al. (2006b). Texture and water retention improvement in high-pressure thermally treated scrambled egg patties. *Journal of Food Science*, 71(2), E52–E61.

Kaavya, R., et al. (2021). Emerging non-thermal technologies for decontamination of *Salmonella* in food. *Trends in Food Science and Technology*, 112, 400–418. https://doi.org/10.1016/j.tifs.2021.04.011

Kayes, M.M., et al. (2007). Inactivation of foodborne pathogens using a one atmosphere uniform glow discharge plasma. *Foodborne Pathogens and Disease*, 4(1), 50–59. https://doi.org/10.1089/fpd.2006.62

Keerthirathne, T.P., et al. (2016). A review of temperature, pH, and other factors that influence the survival of *Salmonella* in mayonnaise and other raw egg products. *Pathogens*, 5, 63. https://doi.org/10.3390/pathigens5040063

Keklik, N.M., et al. (2010). Pulsed UV light inactivation of *Salmonella* Enteritidis on eggshells and its effects on egg quality. *Journal of Food Protection*, 73(8), 1408–1415.

Kuo, F.L., et al. (1997). UV irradiation of shell eggs: Effects on populations of aerobes, molds, and inoculated *Salmonella typhimurium*. *Journal of Food Protection*, 60(6), 639–643.

Lasagabaster, A., et al. (2011). Pulsed light technology for surface decontamination of eggs: Impact on *salmonella* inactivation and egg quality. *Innovative Food Science and Emerging Technologies*, 12, 124–128. https://doi.org/10.1016/j.ifset.2011.01.007

Lin, L., et al. (2020). Inhibitory effect of cold nitrogen plasma on *Salmonella* Typhimurium biofilm and its application on poultry egg preservation. *LWT – Food Science and Technology*, 126, 109340. https://doi.org/10.1016/j.lwt.2020.109340

Lin, C.M., et al. (2021). Applying a large-scale device using non-thermal plasma for microbial decontamination on shell eggs and its effects on sensory characteristics. *LWT – Food Science and Technology*, 142, 111067. https://doi.org/10.1016/j.lwt.2021.111067

Liu, Y., et al. (2023). Structural characterization and functional properties of egg white protein treated by electron beam irradiation. *Innovative Food Science and Emerging Technologies*, 84, 103262. https://doi.org/10.1016/j.ifset/2022.103262

Liu, X., et al. (2020). Effects of different doses of Co-60 gamma-ray irradiation treatment on the microrheological and emulsifying properties of liquid white egg. *Colloids and Surfaces A*, 589, 124461. https://doi.org/10.1016/j.colsurfa.2020.124461

Liu, X., et al. (2019). Electron beam irradiation-induced structural changes increase the antioxidant activities of egg white protein. *LWT – Food Research International*, 111, 846–852. https://doi.org/10.1016/j.lwt.2019.05.066

Liu, X.D., et al. (2009a). Effect of irradiation of foaming properties of egg white proteins. *Poultry Science*, 88, 2435–2441. https://doi.org/10.3382/ps.2009.00063

Liu, X.D., et al. (2009b). Effect of combination of chitosan coating and irradiation on physicochemical and functional properties of chicken egg during room-temperature storage. *Radiation Physics and Chemistry*, 589–591. https://doi.org/10.1016/j.radphyschem.2009.03.015

Ma, C.Y., et al. (1993). Effect of gamma irradiation on the physicochemical and functional properties of frozen liquid egg products. *Food Research International*, 26, 247–254.

Ma, C.Y., et al. (1990). Gamma irradiation of shell eggs. Internal and sensory quality, physicochemical characteristics, and functional properties. *Canadian Institute of Food Science and Technology Journal*, 23(4–5), 226–232.

Manzocco, L., et al. (2013). Effect of pulsed light on selected properties of egg white. *Innovative Food Science and Emerging Technologies*, 18, 183–189. https://dx.doi.org/10.1016/j.ifset.2013.02.008

Mattioli, S., et al. (2020). Impact of ozone and UV irradiation sanitation treatments on the survival of *Salmonella* and the physical-chemical characteristics of hen eggs. *Journal of Applied Poultry Research*, 29, 409–419. https://doi.org/10.1016/j.japr.2020.01.004

Mendes de Souza, P., et al. (2015). Functional properties and nutritional composition of liquid egg products treated in a coiled tube UV-C reactor. *Innovative Food Science and Emerging Technologies*, 32, 156–164. https://dx.doi.org/10.1016/j.ifset.2015.09.004

Min, B., et al. (2012). Irradiation of shell egg on the physicochemical and functional properties of liquid egg white. *Poultry Science*, 91, 2649–2657. https://dx.doi.org/10.3382/ps.2012-02345

Min, B.R., et al. (2005). Effect of irradiating shell eggs on quality attributes and functional properties of yolk and white. *Poultry Science*, 84, 1791–1796.

Monfort, S., et al. (2011). Design of a combined process for the inactivation of *Salmonella* Enteritidis in liquid whole egg at 55°C. *International Journal of Food Microbiology*, 145, 476–482. https://doi.org/10.1016/j.ijfoodmicro.2011.01.027

Monfort, S., et al. (2010a). Inactivation of *Salmonella Typhimurium* and *Staphylococcus aureus* by pulsed electric fields in liquid whole egg. *Innovative Food Science and Emerging Technologies*, 11, 306–313. https://doi.org/10.1016/j.ifset.2009.11.007

Monfort, S., et al. (2010b). Evaluation of pulsed electric fields technology for liquid whole egg pasteurization. *Food Microbiology*, 27, 845–852. https://doi.org/j.fm.2010.05.011

NACMCF (2024). Response to questions posed by the food safety and inspection service: enhancing *Salmonella* control in poultry products. *Journal of Food Protection*, 87, 100168. https://doi.org/10.1016/j.jfp.2023.100168

Narasimhan, S.L., et al. (2023). Effect of cold plasma-activated water as an environmentally friendly sanitizer in egg washing. *Poultry Science*, 102, 102893. https://doi.org/10.1016/j.psj.2023.102893

Ouyang, B., et al. (2020). Inactivation of *Escherichia coli* K-12 in liquid egg white by a flow-through pulsed UV light treatment system. *Journal of Food Protection*, 83(3), 418–425. https://doi.org/10.4315/0362-028X.JFP-19-386

Pankaj, S.K., et al. (2018). Effects of cold plasma on food quality: a review. *Foods*, 7, 4. https://doi.org/10.3390/foods7010004

Pasquali, F., et al. (2016). Modelling survival behavior of *Salmonella enterica* ser. Enteritidis, Typhimurium and Tennessee on table eggs during storage at different temperatures. *Food Control*, 59, 314–319. https://dx.doi.org/10.1016/j.foodcont.2015.05.037

Perry, J.J., et al. (2011). Quality of shell eggs pasteurized with heat or heat -ozone combination during extended storage. *Journal of Food Science*, 76(7), S437–S444. https://doi.org/10.111/j.1750-3841.2011.02294.x

Ragni, L., et al. (2010). Non-thermal atmospheric gas plasma device for surface decontamination of shell eggs. *Journal of Food Engineering*, 100, 125–132. https://doi.org/j.jfoodeng.2010.03.036

Ribeiro, A.C., et al. (2023). Influence of pressure pre-treatments on liquid whole egg thermal pasteurization – microbiological, physicochemical and functional properties. *Food Chemistry Advances*, 2, 100293. https://doi.org/10.1016/j.focha.2023.100293

Rodrigues, E.C.P., et al. (2011). Effects of gamma irradiation on the viability and phenotypic characteristics of *Salmonella Enteritidis* inoculated into specific pathogen free eggs. *Journal of Food Protection*, 74(12), 2031–2038. https://doi.org/10.4315/0362-028X.JFP-11-086

Seo, J.H., et al. (2004). Change of an egg allergen in a white layer cake containing gamma-irradiated egg white. *Journal of Food Protection*, 67(8), 1725–1730.

Shah, U., et al. (2023). Optimization of plasma-activated water and validation of a potential surrogate for *Salmonella* for future egg washing processes. *Journal of Food Protection*, 86, 100029. https://doi.org/10.1016/j.jfp.2022.100029

Sheng, X.W., et al. (2020). A combined approach using slightly acidic electrolyzed water and UV exposure to improve internal egg quality during storage. *Poultry Science*, 99, 6007–6012. https://doi.org/10.1016/j.psj.2020.07.034

Sokovnin, S.Y., et al. (2019). Properties of hen's eggs after surface irradiation by nanosecond electron beam. *Radiation Physics and Chemistry*, 165, 108398. https://doi.org/10.1016/j.radphyschem.2019.108398

Song, H.P., et al. (2009). Improvement of foaming ability of egg white product by irradiation and its application. *Radiation Physics and Chemistry*, 78, 217–221. https://doi.org/10.1016/j.radphyschem.2008.10.001

Tellez, Z., et al. (1995). Effects of gamma irradiation on commercial eggs experimentally inoculated with *Salmonella Enteritidis. Radiation Physics and Chemistry*, 46(4–6), 789–792.

Unluturk, S., et al. (2010). Modeling inactivation kinetics of liquid egg white exposed to UV-C irradiation. *International Journal of Food Microbiology*, 142, 341–347. https://doi.org/j.ijfoodmicro.2010.07.013

Wan, Z., et al. (2017). High voltage atmospheric cold plasma treatment of refrigerated chicken eggs for control of *Salmonella* Enteritidis contamination on egg shell. *LWT – Food Science and Technology*, 76, 124–130. https://dx.doi.org/10.1016/j.lwt.2016.10.051

Wang, B., et al. (2022). Surface decontamination of whole eggs using pulsed light technology and shelf life study of combined pulsed light and vaseline coating during room temperature storage. *Food Control*, 137, 108411. https://doi.org/10.1016/j.foodcont.2021.108411

Wang, Q., et al. (2023). Plasma-activated water for inactivation of *Salmonella* Typhimurium avirulent surrogate: Applications on produce and shell egg and understanding the mode of action. *LWT – Food Science and Technology*, 187, 115331. https://doi.org/10.1016/j.lwt.2023.115331

Wong, Y.C., et al. (1996). Comparison between irradiated and thermally pasteurized liquid white egg on functional, physical and microbiological properties. *Poultry Science*, 75, 803–808.

Yang, Y., et al. (2020). Quality of radio frequency pasteurized shell eggs during extended storage under normal and moderate abuse conditions. *Food Control*, 116, 107330. https://doi.org/10.1016/j.foodcont.2020.107330

Yang, Y. and Geveke, D.J. (2020). Shell egg pasteurization using radio frequency in combination with hot air or hot water. *Food Microbiology*, 85, 103281. https://doi.org/10.1016/j.fm.2019.103281.

Yun, H., et al. (2012). Predicting optimal conditions to minimize quality deterioration while maximizing safety and functional properties of irradiated egg. *Radiation Physics and Chemistry*, 81, 1163–1165. https://doi.org/10.1016/j.radphyschem.2011.11.027

Zhang, H., et al. (2021). Low-energy x-ray inactivation of *Salmonella* Enteritidis on shell eggs in mono-/co-culture biofilms with *Pseudomonas fluorescens. Food Control*, 123, 107742. https://doi.org/10.1016/j.foodcont.2020.107742

Part 3

Quality

Chapter 8

Developing and implementing shell egg quality standards

Takehiko Hayakawa, DSM Nutritional Products Asia Pacific, Japan

1 Introduction

The importance of egg quality in poultry science extends beyond mere animal farming concerns, encompassing critical aspects of food safety, nutrition, and commercial interests (Scanes, 2007). Eggs are a vital component of the global food supply and an affordable food source. Their quality significantly impacts consumer health and the economic viability of the poultry industry. This chapter delves into the multifaceted nature of egg quality, elucidating the complex interplay of various factors affecting it.

Egg quality is broadly classified into external and internal qualities. External egg quality is evaluated by indicators such as egg size, weight, eggshell strength, thickness, colour, and cleanliness. These indicators are crucial for ensuring the egg's structural integrity, making it suitable for consumption or further processing. Eggshell composition and characteristics are influenced by genetic, nutritional, age, and environmental factors, reflecting the health and condition of laying hens. Internal egg quality, including the yolk and egg white, is equally important. The colour of the egg yolk, the roundness of the yolk, and the quality of the albumen, including the height of thick albumen, are indicators of the egg's nutritional value and freshness. These qualities also vary depending on the diet, age, and health of the hen. Additionally, post-collection

http://dx.doi.org/10.19103/AS.2024.0150.12

storage conditions on the farm greatly affect the preservation of these internal qualities.

This chapter provides a comprehensive overview of current knowledge on egg quality, presenting these advances in a consolidated manner. It explores the latest scientific findings and methodologies for egg quality evaluation and improvement, compares international standards, and discusses the practical implications of these standards in various markets. The goal is to provide a resource that is scientifically rigorous and directly applicable to the needs of poultry science researchers, educators, and professionals.

2 Understanding egg quality

2.1 Definition and historical perspective

Egg quality in poultry science is a multifaceted concept that encompasses various physical, biological, and chemical attributes of eggs (Wilson, 2017). This concept is not only pivotal for the poultry industry but also for consumer health and nutrition. Historically, the understanding of egg quality has evolved from basic observations to a more complex scientific inquiry. Initially, egg quality definitions focused primarily on external characteristics such as eggshell strength, size, and shape. However, as the hen egg industry grew, and consumer demands for egg quality and hygiene increased, the need for a more detailed and scientific definition emerged. As a result, internal egg quality, i.e. yolk and egg white quality, nutritional composition, and other internal factors, were incorporated into the definition. Technological advances have further subdivided these parameters, and measurement methods have moved from manual to digital for more accurate and comprehensive evaluation.

2.1.1 Early research and understanding of egg quality

The historical evolution of egg quality evaluation also reflects changing consumer demands and regulatory standards. In the early days, the primary concern was the physical integrity of the egg, which is essential for transportation and handling. However, as scientific knowledge expanded, the focus shifted to aspects such as nutritional value, safety, and even the welfare of the laying hen. Today, egg quality is a dynamic field of research that integrates aspects of genetics, nutrition, health, and environmental factors. Ongoing research in this area is constantly changing our understanding and standards of what constitutes a high-quality egg, reflecting advances in poultry science and consumer expectations. The concept of egg quality has evolved from an assessment of simple physical characteristics to a sophisticated evaluation that includes multiple scientific disciplines. This evolution is a testament to advances in food and poultry science and its responsiveness to changing market demands and scientific insights.

The journey to understanding egg quality in poultry science began more than a century ago with groundbreaking research by pioneers such as von Natusius (Palmer, 1967) and Romanoff (1929). Their early work laid the foundation for understanding the structure and composition of the eggshell, focusing primarily on the gross structure, which is known to be predominantly calcium carbonate. This was important research because the integrity of the eggshell is critical to the overall quality of the egg and the health of the hen. By the late 1960s, research by Scott et al. (1971) at Cornell University brought attention to the dietary needs of laying hens, particularly calcium sources like finely ground limestone and oyster shells. Their findings were crucial in improving eggshell quality and understanding the hen's calcium metabolism. Subsequent studies by researchers such as Roland and Bryant (1994) focused on optimizing calcium levels and particle size in the diet to improve eggshell quality. These studies provided important insights into the management of laying hen diets to improve eggshell formation.

Our understanding of eggshell strength and quality was further advanced by the exploration of various measurement methods developed to predict eggshell problems. However, they also faced the challenge of how to translate the knowledge gained in the laboratory to commercial production.

The use of scanning electron microscopy by King and Robinson (1972) led to a better understanding of eggshell microstructure. It revealed complexities beyond the simple calcium carbonate layer and highlighted the protein matrix basis of the eggshell membrane. These studies have revealed a variety of eggshell anomalies and their importance. The 1990s marked significant progress in protein research related to eggshells. Hincke et al. (1995) opened new avenues for understanding the eggshell's molecular composition and its impact on eggshell quality. These historical perspectives set the stage for contemporary research, emphasizing the complex interplay of nutrition, genetics, and environmental factors in determining egg quality.

2.1.2 Advances in egg quality assessment

Advances in egg quality assessment in the last century have transformed our understanding of egg properties and their implications. Early studies were limited to basic physical attributes such as egg size, eggshell thickness, and weight. However, as poultry science progressed, the focus shifted to detailed analysis of egg components such as egg white, yolk, and eggshell. The advent of molecular biological methods has ushered in a new era in egg quality assessment. Researchers began to explore the genetic factors that influence egg quality and gained a more nuanced understanding of how genetics interact with environmental factors to influence egg characteristics. This approach has made it possible to identify specific genes associated with desirable traits such as laying performance and egg size. Additionally, technological advancements in analytical chemistry have allowed for more precise measurements of egg components. Techniques such as spectroscopy and chromatography have been employed to quantify vitamins, minerals, and other nutrients in eggs, providing a clearer picture of their nutritional value.

2.1.3 Impact of genetics research

The genetics of laying hens have profoundly influenced our understanding of egg quality. Research in these areas has elucidated how specific genetic traits and nutritional interventions can affect the quality of eggs produced by laying hens. Genetic research has significantly deepened our understanding of egg quality traits in poultry science. This research, integral to modern breeding programs, has elucidated how genetic factors influence traits such as eggshell strength, egg weight, and shape. For example, Blanco et al. (2014) measured genetic parameters of various quality traits such as egg weight, shell breaking strength, egg shape index, shell thickness, and albumen height and reported that traits like eggshell strength depend not only on physical attributes such as shell thickness but also on the uniformity and quality of the eggshell structure. This insight is crucial, as it moves the focus of breeding programs beyond mere physical measurements to the underlying genetic influences that dictate these traits. The genetic basis of egg quality extends beyond eggshell characteristics. Other traits, such as egg size and shape, have also been found to be genetically influenced. This understanding has ushered in a new era of precision in poultry breeding, where selection is based not just on observable traits but also on genetic markers linked to desirable qualities.

2.1.4 Impact of nutrition research

In poultry science, nutritional research has explored the impact of various nutrients on egg characteristics and quality, along with their roles. For example, calcium and phosphorus have been identified as vital for the development of strong, healthy eggshells (Bar et al., 2002). The balance and availability of these minerals in a hen's diet directly affect eggshell quality. The progression of nutritional research in poultry science has revealed the significant impact of diet on various aspects of egg quality (Kodani et al., 2022). Studies have examined how different feed compositions, including the balance of vitamins, minerals, and other nutrients, affect not only eggshell quality but also the internal quality of eggs, such as the composition of the albumen and yolk (Zang et al., 2011; Ghasemi et al., 2022). Understanding the specific nutritional needs of laying hens has led to the development of targeted feeding strategies. These strategies are designed to optimize egg quality by ensuring that hens receive a balanced diet tailored to their specific physiological needs. This approach is crucial for maintaining the health and productivity of the hens and directly influences the quality of eggs they produce. Recent studies have expanded to explore the impact of novel feed additives, such as probiotics and prebiotics, on egg quality (Wan et al., 2023). These additives have shown potential in enhancing eggshell quality and overall egg production. Additionally, research has delved into the effects of feed on the sensory attributes of eggs, such as taste and colour, which are important for consumer preference (Liu et al., 2017).

2.2 Importance of egg quality in different contexts

2.2.1 Nutritional value and consumer health

Eggs are a crucial source of high-quality protein, vitamins, and minerals, contributing significantly to a balanced diet. The nutritional value of eggs, directly impacted by their quality, influences the health benefits they offer. As noted by Ruxton (2010), eggs are especially rich in essential nutrients like vitamin D, vitamin B12, and selenium, which are important for bone development, immunity, and so on. Surai and Sparks (2001) reported that nutritional quality enhancement in eggs deliver what kinds of benefits to consumers. Therefore, ensuring the highest quality of eggs is essential not only for taste and appearance but also for maximizing their nutritional benefits and contributing to human nutrition.

2.2.2 Economic significance for the poultry industry

The economic value of egg quality is substantial in the poultry industry. Egg quality attributes such as eggshell strength, yolk colour, and albumen quality significantly impact marketability and consumer preference, thus influencing the profitability of poultry farms. Generally, high-quality eggs such as branded eggs, specialty eggs, and nutrients-enriched eggs have higher market prices compared to regular eggs, reflecting consumer demand. This market preference underscores the importance of maintaining superior egg quality as a key factor in the economic success and competitiveness of the poultry industry. Hence, the focus on egg quality is not only a matter of product excellence but also a strategic economic decision for poultry businesses.

2.2.3 Food safety and public health

Egg quality is pivotal for food safety and public health. As De Reu et al. (2006) point out, lower quality eggs are more prone to microbial contamination, increasing the risk of foodborne illnesses such as salmonellosis. Regulatory standards, like those set by the FDA in the United States and EFSA in Europe, play a crucial role in ensuring egg safety. These standards are designed to prevent contamination and protect public health. Consistent monitoring and adherence to these quality standards are essential in maintaining consumer trust and preventing health risks associated with egg consumption.

2.2.4 Global food security

Eggs play an important role in global food security, particularly in regions struggling with malnutrition. Eggs are an affordable and nutritious source of protein and are essential in diets where food diversity is limited. The nutritional value of eggs should therefore be effective in combating human malnutrition. Ensuring the production of high-quality eggs is therefore critical to global efforts to combat hunger and improve nutritional status. This is particularly important in developing countries, where eggs can be an important dietary component to alleviate malnutrition.

2.2.5 Cultural and regional variations

Perceptions and standards of egg quality vary widely across cultures and regions, reflecting diverse food preferences and culinary traditions. Evaluations of attributes such as egg size, eggshell colour, and yolk colour differ in different parts of the world. For example, consumers in many European countries prefer brown-shell eggs, while white-shell eggs are more popular in India (IEC, 2024). Eggs fortified with certain nutrients may be more favoured in regions where certain nutrients are deficient. Such regional preferences have a direct impact on the poultry industry, influencing everything from breeding choices to feed composition. Understanding these differences is crucial for egg producers aiming for global markets. This requires a flexible and responsive approach to hen egg production that is tailored to the cultural and dietary needs of different people.

2.2.6 Environmental impact

The environmental impact of egg production is a critical aspect that intersects with egg quality. Research by Xin et al. (2011) has highlighted various environmental aspects of egg production systems. Key findings include the impact of housing systems on air quality and emissions, with high-rise cage houses generally having poorer air quality and emitting more ammonia compared to manure belt cage houses. The study also notes that non-cage houses tend to have lower air quality and may lead to higher feed and energy use due to colder conditions in cold weather. Furthermore, the environmental footprint of different production systems, including emissions and resource utilization, is an area of ongoing research. This emphasizes the need for sustainable practices in egg production, balancing quality with environmental stewardship.

3 External quality: eggshell characteristics

3.1 Factors affecting eggshell quality

3.1.1 Genetic factors

Eggshell quality, a paramount aspect of poultry production, is influenced by a complex interplay of genetic factors. Recognizing the heritability of traits like eggshell thickness, strength, and structure, early research by Hunton (1982) reported that the work focused on these characteristics. The genetic diversity within breeds plays a crucial role, with selective breeding aiming to enhance desirable traits while maintaining overall breed health. Advances in poultry genetics have significantly contributed to understanding the genetic architecture of eggshell traits. Specific genetic markers associated with eggshell quality have been identified, offering a nuanced understanding of the factors influencing traits like calcium metabolism, which is critical for eggshell formation. The advent of genome-wide association studies (GWAS) and quantitative trait locus (QTL) mapping, as shown in the work of Rowland et al. (2019) and Liu et al. (2018) has enabled breeders to identify genetic variation directly related to eggshell quality, and by understanding specific gene sequences quality could be targeted for improvement. Despite these advancements, the genetic improvement of egg quality faces challenges due to low heritability and intensive selection focused on egg production. The balance between improving specific egg quality traits and maintaining overall productivity and health of the flock remains a key challenge. While significant progress has been made in understanding and manipulating the genetic factors influencing eggshell quality, this field continues to evolve. Future advancements will likely bring even more sophisticated tools and techniques, offering breeders new opportunities to enhance egg quality while adhering to ethical and welfare standards.

3.1.2 Nutritional factors

Eggshell quality in poultry is significantly influenced by nutrition, involving a delicate balance of minerals, vitamins, and other dietary components. Calcium is the major constituent of eggshells (Chang, 2021), and its effective utilization requires the synergistic action of phosphorus and Vitamin D3.

These elements facilitate calcium absorption and metabolism, influencing eggshell formation and strength . To prevent eggshell quality issues such as thin shells, soft shells, and cracked eggs, it is essential to maintain a sufficient calcium level of 4–5 g per bird per day. However, the intake balance of calcium and phosphorus is important for the health of eggshells and bones. Vitamin D3, specifically its metabolized form 25 hydroxycholecalciferol (25(OH)D3), is vital for calcium and phosphorus metabolism. It facilitates their absorption and utilization, directly influencing eggshell quality. As hens age, reduced levels of 25(OH)D3 can impair calcium absorption, weakening the eggshell. Manganese, copper, and zinc are vital trace minerals for eggshell quality. They contribute to biochemical processes essential for eggshell formation. It is recommended to adopt well-defined mineral sources, with some benefits observed in supplying these minerals from organic sources. Electrolytes, specifically sodium (Na), potassium (K), and chloride (Cl), are involved in maintaining acid-base balance, a critical factor in eggshell formation. Ensuring an electrolyte balance of around 200 mEq/kg of feed is necessary for optimal eggshell quality (Chang, 2021). In heat-stressed birds, which often lay eggs with thinner eggshells, maintaining this balance becomes even more crucial. Adjusting the electrolyte levels in feed, along with the addition of vitamins C and E, can improve eggshell strength and thickness under prolonged heat stress conditions. Recent research has focused on dietary microelements and feed additives, exploring the impact of pre- and probiotics, organic acids, and herb extracts on intestinal health and mineral availability. While the efficacy of these supplements varies, several studies indicate that eggshell quality can be positively affected by optimizing specific feed additives (Wiaętkiewicz et al., 2015; Świątkiewicz et al., 2018). Implementing effective nutritional strategies is integral to poultry management. This involves regular monitoring and adjustments of dietary components, particularly as hens age. Nutritionists must balance economic viability with the specific nutritional needs of the hens, ensuring diets that support optimal eggshell quality. Nutrition is a critical factor in determining eggshell quality in poultry. A well-balanced diet, rich in essential minerals, vitamins, and other nutrients, is key to producing strong and viable eggshells. As poultry nutrition science advances, new dietary formulations and additives offer promising avenues for further enhancing eggshell quality.

3.1.3 Environmental and management factors

Environmental and management factors play a pivotal role in determining the quality of eggshells, a key concern in the realm of poultry science. This comprehensive discussion aims to explore how various environmental and managerial conditions critically influence egg production and eggshell quality, based on extant literature and studies in the field. Such knowledge is not only academically intriguing but also of immense practical value to poultry farmers and researchers seeking to optimize egg production.

3.1.4 Temperature

The optimal temperature range for laying hens is typically between 18 and 24°C (65 to 75°F). Deviations from these ranges, either too high or too low, can induce stress in hens, with consequent adverse effects on egg production and eggshell integrity. Optimal temperature ranges are essential for sustaining hen health and productivity. Higher temperatures, especially above 30°C (86°F), can cause heat stress, while lower temperatures can lead to a reduction in feed intake and egg production. A seminal study by Badran et al. (2020) highlighted that heat stress significantly diminishes egg production and compromises eggshell quality in laying hens. Therefore, maintaining temperature within an ideal spectrum is crucial for optimal egg production.

3.1.5 Humidity

Ideal humidity levels for egg-laying hens are usually in the range of 40–60%. High humidity, especially when combined with high temperatures, can exacerbate heat stress, and is also associated with an increased risk of diseases and can impair the hen's ability to produce high-quality eggs (Kim et al., 2022). On the other hand, very low humidity can lead to respiratory issues and dust problems. Equally significant is the role of humidity in the poultry environment.

3.1.6 Lighting

The lighting regime, including the duration and intensity of light exposure, profoundly impacts the laying cycle of hens. Research by Farghly et al. (2019) has shown that manipulating light patterns can enhance egg production and potentially improve eggshell quality. Adjusting lighting schedules can synchronize laying cycles and increase overall egg output, making lighting an important variable in poultry management.

3.1.7 Housing systems

The housing system employed in poultry farming, such as cage versus free-range systems, also impacts eggshell quality. Different systems provide varying levels of environmental control, space, and social interaction, influencing egg quality. It evaluates eggshell quality, microbial contamination of eggshells, and penetration of microorganisms into the egg content in different housing systems (enriched cage vs free range) and at different hen ages (Vlčková et al., 2018). It found that the housing system and age significantly affected eggshell quality and microbial contamination.

3.2 Measurement and standards for eggshell quality

3.2.1 Eggshell strength

Eggshell strength and eggshell thickness are important indicators of egg quality. Eggshell strength is usually measured by applying pressure until the egg cracks and using a device to quantify the force required. This method is known as the destructive trial method and provides a direct measure of the eggshell's resistance to breakage. The value also varies depending on whether the egg is measured vertically (small end and large end) or horizontally (equator), but is often measured by applying pressure to the vertical part of the egg with the small end part down. The eggshell strength can range widely from 1 to 7.5 kgf, but a value of 3 kgf or more is desirable for distribution as packed eggs (DSM, 2022). If the eggshell strength is less than 3 kgf, hairline cracks or cracks are likely to occur or have already done so in many cases.

3.2.2 Eggshell thickness

Eggshell thickness is not as important as eggshell strength in terms of quality standards in egg distribution, but, as with eggshell strength, there are differences depending on which part of the eggshell is measured, although eggshell callipers and thickness gauges are often used in poultry and egg research. It is recommended to always measure the same part of the eggshell, i.e. the part of the eggshell on the Equator that is flattest and excluding the eggshell membrane. The range of eggshell thickness ranges from 0.20 to 0.57 mm, but a minimum of 0.3 mm is preferred (DSM, 2022). Khaliduzzaman et al. (2020) introduced a non-destructive method for measuring eggshell thickness using terahertz (THz) reflectance spectroscopy.

3.2.3 Eggshell colour

Eggshell colour varies greatly depending on the genetic background of the hen breed, and the quality of the eggshell colour is often judged subjectively at the distribution stage. Generally, the key is whether or not the

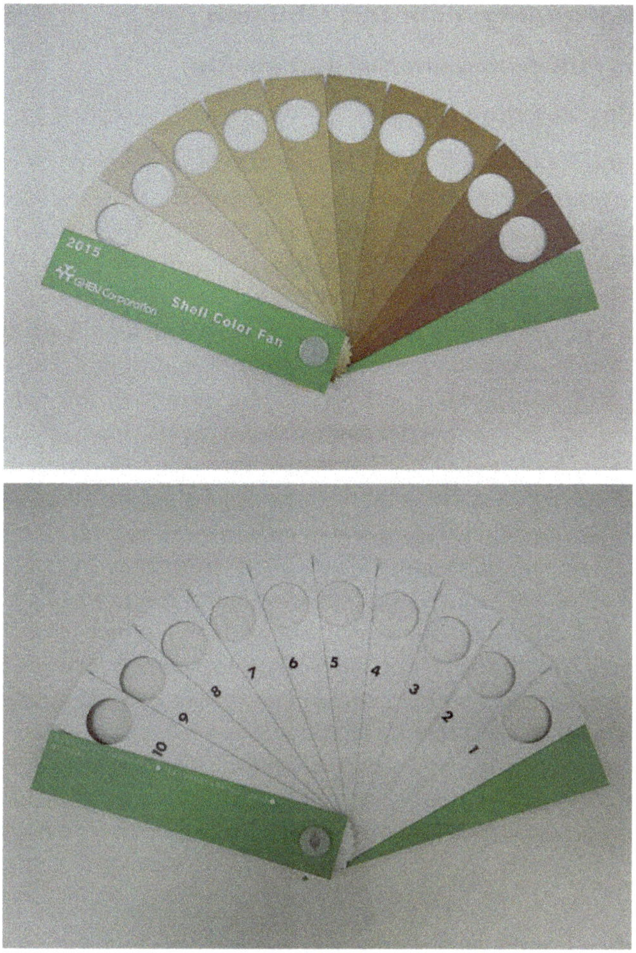

Figure 1 Shell Colour Fan. Courtesy of GHEN corporation, Japan.

eggshell colour is uniform across the entire eggshell and whether or not the colour is extremely different from other individuals. In the Japanese market, the GHEN corporation offers a Shell Colour Fan for brown eggs, which is an index from 1 to 10 (Fig. 1). Some egg producers control the quality of eggshell colour by visually checking the corresponding colour by aligning the eggshell with the round hole in the fan blade and then referring to the number on the back.

In some research applications, CIE L*a*b* values are used to observe detailed colour changes.

4 Internal quality: yolk and albumen

4.1 Factors influencing internal egg quality

4.1.1 Poultry nutrition impact

The quality of a hen's egg, particularly its yolk and albumen, is significantly influenced by poultry nutrition, which directly impacts the composition, quality, and healthfulness of the eggs produced. A diet enriched with essential nutrients ensures superior nutritional quality of eggs and enhances characteristics like yolk color and albumen consistency. Proteins, especially amino acids such as methionine and lysine, are crucial for albumen formation, with deficiencies leading to thinner albumen and reduced egg size (Roberts, 2004). Essential fatty acids, found in flaxseed and fish oil, are necessary for yolk formation, enriching the yolk with omega-3 beneficial for human consumption (Lopez-Bote et al., 1998). Vitamins A, D, E, K, and B-complex, along with minerals like calcium and phosphorus, are vital for maintaining egg quality and structure, with deficiencies leading to issues like weak eggshells or poor yolk quality (Nys et al., 2001). Feed composition, including the incorporation of plant extracts and probiotics, directly affects egg quality, enhancing egg white's antibacterial properties and shelf life. Biological antioxidants reduce oxidative stress, improving yolk quality and colour (Surai & Sparks, 2001). Supplementation with canthaxanthin improves yolk colour and increases yolk vitreous membrane strength and antioxidant capacity. Nutritional deficiencies can lead to diseases like encephalomalacia from vitamin E deficiency, affecting the nervous system and egg quality, and muscular dystrophy from selenium deficiency, further impacting egg quality, highlighting the critical role of balanced nutrition in egg production and quality.

4.1.2 Poultry health and disease impact

The health and disease status of poultry is a crucial determinant of the internal quality of eggs. Diseases in poultry not only affect the well-being of the birds but also have a direct impact on the quality of eggs they produce. Infectious diseases, respiratory problems, parasitic infections, and stress all negatively impact egg quality. Ensuring optimal health through proper management, nutrition, and disease prevention is essential for producing high-quality eggs. Diseases like Avian Influenza and Newcastle Disease can severely impact egg production and quality. Infected birds often produce eggs with thin eggshells, abnormal shapes, and poor internal quality. The albumen may become watery, and the yolk may be misshapen or discoloured (Capua & Alexander, 2004). Respiratory diseases, such as Infectious Bronchitis, can lead to a decrease in egg production and the production of eggs with wrinkled eggshells and watery albumen. These changes are attributed to the damage caused to the oviduct and respiratory system of the hens (Najimudeen et al., 2021). Parasites like mites and lice can stress the birds, leading to a reduction in egg quality.

4.1.3 Impact of vaccination and medication

Vaccination and medication play a crucial role in maintaining poultry health, but they can also influence egg quality. Medications, such as antibiotics, can lead to residues in eggs, raising food safety concerns and potentially affecting the egg's internal quality. The use of antibiotics during the laying period is therefore restricted and regulated. Therefore, the use of organic acids as an alternative to antibiotic growth promoters has been explored due to their potential benefits in maintaining gut health and improving egg quality (Ebeid & Al-Homidan, 2022).

4.1.4 Stress and its effects

Stress, whether due to environmental factors, handling, or disease, can significantly impact egg quality. Stressful conditions can lead to a decrease in egg production and affect the albumen quality, making it thinner and less viscous (Purswell & Branton, 2015). Stress can also cause blood spots and meat spots in eggs, affecting their internal quality and consumer acceptability. The study by Purswell and Branton highlights the effects of environmental stress, specifically air velocity, on laying hen performance and egg quality, demonstrating the sensitivity of poultry to their environmental conditions.

4.1.5 Egg storage environment impact

The storage environment of eggs post-laying significantly influences their internal quality, with optimal temperature, humidity, and storage duration being critical for preserving the yolk and albumen's quality. Proper packaging and handling are also paramount to ensuring eggs maintain their freshness and nutritional value. Storing eggs at low temperatures, typically below 7°C (45°F), is vital to slow the aging process, minimize microorganism growth, and prevent the deterioration of albumen and yolk, although extremely low temperatures can freeze the contents, altering texture and potentially cracking the shell. Conversely, high temperatures accelerate aging, thin the albumen by increasing its pH, weaken the yolk membrane–raising the risk of breakage and mixing with the albumen–and enhance microbial growth, which poses food safety risks (Bell & Weaver, 2002). Adequate humidity prevents excessive moisture loss, crucial for maintaining egg weight and internal quality (Yimenu et al., 2018), while good air quality and ventilation prevent odour and gas absorption, thereby preserving egg taste and quality. Thus, temperature, relative humidity, and air flow velocity are essential environmental factors affecting the freshness indices of hen eggs.

4.1.6 Storage duration

The duration of storage significantly affects egg quality (Okonkwo et al., 2021). Over time, the quality of the albumen deteriorates; it becomes thinner and less able to hold the yolk in place, and the pH of the albumen increases, which can affect its functional properties, such as foaming (Sokołowicz et al., 2022). The yolk absorbs water from the albumen, becoming larger and more prone to rupture. Prolonged storage can also lead to the development of off-flavours and reduce the overall freshness of the egg.

4.1.7 Egg handling

Proper handling during storage is essential to maintain egg quality (Jones et al., 2018). Packaging materials should protect the eggs from external contamination and physical damage. Handling should minimize vibrations and shocks to prevent the rupture of yolk membranes and the mixing of yolk and albumen. The quality of eggs, especially the yolk, can be significantly affected by storage conditions, as changes in yolk lipid composition can occur under extreme storage conditions, potentially leading to a deterioration in edible value (Meng et al., 2022).

4.2 Measurement and standards for internal quality

The assessment of the internal quality of eggs is paramount, not only because it influences consumer preferences and the nutritional value of the eggs but also because it directly impacts the freshness and safety of the eggs. This evaluation can be categorically divided into two main components: the quality of the egg white (albumen) and the quality of the egg yolk. However, the assessment encompasses additional parameters, such as the detection of meat spots and blood spots. These factors are critical in determining the overall quality and acceptability of eggs for consumption.

4.2.1 Haugh unit

The Haugh unit (HU) is a widely recognized measure for evaluating the internal quality of hen's eggs, particularly the quality of the egg white (albumen). Developed by Raymond Haugh in 1937 (Haugh, 1937), the Haugh unit is an empirical index that has become a standard in the poultry industry for assessing egg freshness and quality. The principle behind the Haugh unit is that fresher eggs have thicker and more viscous albumen, resulting in a higher albumen height when the egg is broken onto a flat surface. The Haugh unit score correlates this height with the egg's weight to provide a standardized measure of quality. A higher Haugh unit indicates a fresher egg with better quality albumen. However, it should be noted that the challenge with the Haugh unit is that the value over time can vary significantly due to the rapid loss of concentrated egg white under room temperature conditions, and the variation in the value can be significant depending on where the centre of the concentrated egg white is taken. Therefore, it is advisable to evaluate eggs under refrigerated storage.

The Haugh unit is calculated using the formula:

$$HU = 100 \times log\left(H - 1.7W^{0.37} + 7.6\right)$$

Where H is the height of the albumen (in millimetres), and W is the weight of the egg (in grams).

The USDA grading system classifies eggs based on Haugh unit values as follows:

AA quality: Haugh unit value of 72 or higher.
A quality: Haugh unit value from 60 up to, but not including, 72.
B quality: Haugh unit value lower than 60.
C quality: Haugh unit value of 30 or less.

4.2.2 Air cell

The air cell is also referred to as an air chamber and air space, but here we use the term air cell, which is the most commonly accepted term in academic contexts. The air cell is widely recognised for referring to the space within an egg that expands as the egg ages. This expansion is a direct result of moisture loss and content contraction, making the air cell a critical indicator of an egg's freshness and overall quality. The air cell is located at the broader end of the egg, between the outer and inner shell membranes. Its size is considered a rough indicator of the egg's age and is closely associated with internal quality. The depth of the air cell, particularly when exceeding 6 mm, can signify the egg's age and is utilised in grading and standardisation practices. For more information, see the European Union and USDA egg standard manuals described below. This process includes the classification of eggs into uniform categories based on quality and physical characteristics, which is essential for maintaining consumer confidence and ensuring product reliability (FAO, 2003).

4.2.3 Yolk Index

The Yolk Index is a measure used to assess the quality of the egg yolk in hens' eggs. This index is particularly important as it provides insights into the freshness and the nutritional value of the egg, which are critical factors for both consumers and the poultry industry. The Yolk Index is a simple yet effective method for assessing egg quality. It is based on the principle that as eggs age, the yolk absorbs water from the albumen, causing it to swell and become less dense. This change in physical properties results in a flatter yolk with a lower height-to-diameter ratio. Therefore, a higher Yolk Index is indicative of a fresher egg with a yolk that has maintained its shape and density.

The Yolk Index is calculated by measuring the height and diameter of the egg yolk and then using the formula:

$$Yolk\,index = \frac{Hx2}{(D1 + D2)\,of\,the\,Yolk}$$

Where H is the height of the egg yolk (in centimetres or millimetres), and D1 and D2 are the diameters of the egg yolk (in centimetres or millimetres). D2 measures the diameter at a location 90° rotated from the diameter measured by D1.

As shown in Fig. 2, the Yolk Index values varied with storage temperature and storage period and were classified as extra fresh for values above 0.38, fresh for values between 0.28 and 0.38, and regular for values below 0.28, which is recommended for use in commercial egg applications (Kashimori, 2021).

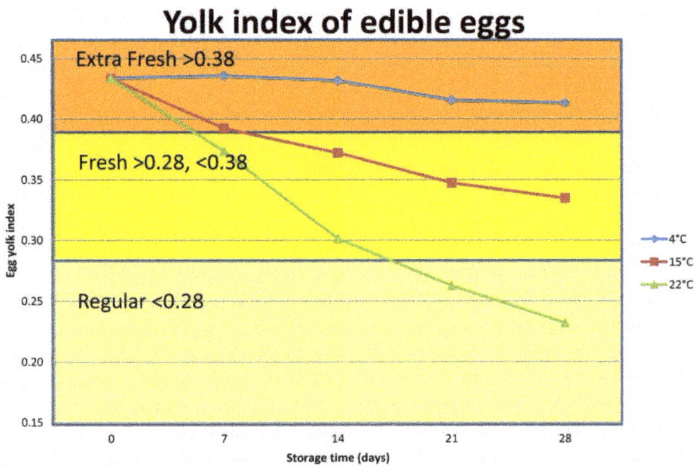

Figure 2 Freshness of edible eggs as determined by yolk index. Courtesy of DSM Nutritional Products, Switzerland.

Figure 3 DSM YolkFan™. Courtesy of DSM Nutritional Products, Switzerland.

4.2.3 Yolk colour

The traditional method for measuring egg yolk colour involves visual comparison against a colour chart, the DSM YolkFan™ (ex Roche Yolk Colour Fan), as Fig. 3 which has a series of numbered colour shades ranging from 1 (pale) to 16 (dark orange). The yolk is assigned a number from the fan that best matches its colour. This method, while subjective, is widely used in the industry as a global standard evaluation tool. In recent years, NABEL's DET6500 (NABEL: https://digitaleggtester.com/en/; accessed 26 November 2023) and DSM's digital YolkFan (DSM: https://www.dsm.com/anh/products-and-services/tools/digital-yolkfan.html; accessed 26 November 2023) have made it possible to objectively measure egg yolk colour with digitised measuring instruments, which were previously judged subjectively by the human eye.

5 Egg quality standards by market

5.1 United States

The United States places a paramount emphasis on maintaining high standards of egg quality, ensuring that consumers have access to products that are both safe and of excellent quality. This commitment is underscored by the extensive information available on the websites of the U.S. Department of Agriculture (USDA) and the Food and Drug Administration (FDA), which jointly oversee the regulatory framework for egg quality standards (USDA, 2000; FDA, 2023). These standards are not only vital for consumer safety but also play an integral role in the domestic and international trade of egg products.

The USDA's Agricultural Marketing Service (AMS) is a key player in this process, providing detailed information about the grades and standards for shell eggs. Eggs in the United States are classified into three main consumer grades based on their quality: Grade AA, Grade A, and Grade B. Grade AA eggs, representing the highest quality, are characterized by their freshness and minimal defects. In contrast, Grade B eggs are often used in processed food products due to their lower grade, characterized by a greater number of allowable defects, such as larger air cells or the presence of blood spots. The grading process is meticulous, taking into account various factors including eggshell integrity, air cell size, and yolk quality. For instance, U.S. Consumer Grade AA eggs at origin must consist of at least 87% AA quality eggs, with specific tolerances set for lower quality eggs and various defects.

In support of this grading system, the USDA also offers an Egg Grading Manual (USDA, n.d.-a), which serves as an invaluable resource for both novice and experienced egg graders. This manual ensures the correct interpretation and application of U.S. standards, grades, and weight classes for eggs, fostering consistency and reliability in egg quality assessment.

The FDA's role complements that of the USDA by focusing on the safety aspects of egg production and consumption. It shares regulatory authority with the USDA over egg safety in the United States, playing a critical role in ensuring the safety, purity, and accurate labelling of food products, including shell eggs. The FDA's responsibilities extend to regulating the production, transportation, and storage of shell eggs, emphasizing the prevention of contamination by bacteria such as Salmonella. This is achieved through stringent guidelines and regulations, including guidance documents, compliance programs, and egg safety final rules, which provide comprehensive information on these standards (FDA, 2022).

The historical development of U.S. egg quality standards highlights a journey of significant evolution, adapting to technological advancements and changing consumer expectations. A landmark in this journey was the enactment of the Egg Products Inspection Act in 1970, which established quality grades based on internal and external qualities of eggs (USDA, n.d.-b). This act marked a departure from the earlier practice where egg quality was largely determined by local practices, leading to significant regional variations. Over the years, these standards have been refined to include detailed criteria for eggshell strength, yolk consistency, and albumen clarity, reflecting a commitment to enhancing food safety and aligning with international quality benchmarks.

The current comprehensive egg quality standards in the United States categorize eggs into the aforementioned grades. These grades are determined by factors such as eggshell cleanliness and strength, air cell size, yolk positioning, and the clarity and viscosity of the albumen. Size categorization, based on weight, is independent of these quality grades and includes classifications such as jumbo, large, and medium. The meticulous enforcement of these standards is critical in maintaining high standards of egg quality, safeguarding public health, and sustaining consumer trust in egg products.

Consumer preferences and industry trends significantly influence these egg quality standards. The growing demand for specialty eggs, such as organic, free-range, or omega-3 enriched, has led to a diversification in quality criteria. This shift reflects a broader consumer movement towards ethically produced and nutritionally enhanced products, driving innovation in production and processing technologies.

However, the U.S. egg industry faces several challenges in maintaining these high standards. Disease outbreaks in poultry, such as avian influenza, and logistical challenges pose threats to egg quality and supply chain stability. Future standard revisions are likely to integrate sustainable practices and enhance animal welfare measures, driven by regulatory compliance and increasing consumer demand for ethically produced food products.

5.2 Europe

The classification and grading of eggs within the European Union are governed by strict standards to ensure quality and uniformity (European Union, 2018). Eggs are classified as either Class A or Class B, with specific criteria for each class. Class A eggs must have a normal-shaped eggshell, a clean and undamaged surface, an air space inside no bigger than 6 mm, a slightly mobile yolk without a clearly visible outline, clear and translucent white, no foreign matter or smells, and no sign of germ development. They are also graded by weight into four categories: XL (more than 72 g), L (63–72 g), M (53–62 g), and S (less than 53 g). These eggs must not be washed or cleaned before or after grading, and they must not be treated for preservation or chilled below 5°C. Class B Eggs are eggs that do not meet the quality thresholds of Class A eggs or are Class A eggs that have been downgraded. They are typically used for industrial or other non-retail purposes. The grading process must be carried out in packing centres that are equipped with appropriate facilities. Only these centres can grade and pack eggs, ensuring a standardized process across the EU. This classification system plays a crucial role in maintaining egg quality across the EU, providing transparency and reliability for consumers, and ensuring that egg producers adhere to high standards of quality.

The European Union has established detailed regulations for the traceability of eggs to ensure high quality, consumer protection, and consistency in the EU market. These regulations specify how eggs and their packaging should be marked, including the method of farming, country of origin, and a unique producer identification code. This system enables consumers to know the origin and method of production of the eggs they purchase. Each egg is marked with a number indicating the farming method used. The numbering system is as follows:

- 0 for organic eggs
- 1 for free-range eggs
- 2 for barn eggs
- 3 for enriched cage eggs

The eggs also include a country code, which follows the ISO 3166 standard. For instance, "FR" indicates France, and other EU countries have their respective codes. Finally, the eggs carry a unique code for the individual farm where they were produced. This enhances traceability back to the source, which is critical for addressing food safety or animal welfare issues.

Recent updates to the EU's egg marketing standards, effective from November 2023 and November 2024, focus on improving traceability and aligning with EU sustainability goals (AGRINFO, 2023). These updates include mandatory marking of eggs at the production site for EU producers and adjustments to the standards for free-range and organic eggs, especially concerning avian flu precautions.

5.3 Japan

In Japan, egg quality standards are a sign of close attention to food safety and consumer health. These standards, administered by the Ministry of Agriculture, Forestry and Fisheries (MAFF), cover not only the physical characteristics of eggs, but also aspects such as hen health and feed quality. The Japanese system is characterised by its emphasis on egg quality, not only in terms of items such as the colour and size of the eggshell, but also in terms of yolk, egg white, and air cell, reflecting the cultural importance of eggs in the Japanese diet. For example, a traditional Japanese meal is called "Tamagokake Gohan" in Japanese, which consists of rice topped with a raw egg and a small amount of soy sauce. In particular, a major difference from overseas standards is the distinction between whether eggs can be eaten raw or must be heat-processed. These standards are intended to ensure that eggs are not only safe but also meet the high expectations of Japanese consumers.

The development of egg quality standards in Japan has been influenced by the country's rich food culture and rapid technological progress. Traditional practices initially governed egg production and grading, with an emphasis on local preferences and regional distinctions. However, as the egg industry expanded and public awareness of food safety increased, the country adopted a more standardised, science-based approach. This shift was crucial in meeting the growing urban population and demand for consistent, high-quality egg products. The historical progression of these standards reflects the evolution of Japanese food safety priorities and the integration of modern quality control techniques.

The regulatory framework for egg quality in Japan is robust and comprehensive. The Ministry of Agriculture, Forestry and Fisheries (MAFF) plays a key role in setting and enforcing these standards, with regular inspections and quality control measures in place throughout the egg production supply chain. This regulatory system ensures that all aspects of egg production, from hen welfare and feed quality to egg processing and packaging, comply with the highest standards. Compliance with these regulations is not just about meeting quality standards, but also about maintaining Japan's reputation for food safety and agricultural excellence.

Details of Japanese egg quality standards can be found on the MAFF website (https://www.maff.go.jp/e/index.html; accessed 26 December 2023) and the Japan Poultry Association website (https://www.jpa.or.jp/; accessed 26 December 2023), but unfortunately, these are only available in Japanese and there are few sources of information in English. However, please refer to *The Illustrated Egg Handbook* (Kashimori, 2021), pp. 157–167, for details on egg standards.

6 Future trends in egg quality research

The poultry industry is witnessing a remarkable era of technological innovation, profoundly influencing egg production and quality. These advancements are not just enhancing efficiency and productivity but also revolutionizing the ways in which egg quality is assessed and ensured. From precision agriculture to sophisticated quality assessment tools, technology is at the forefront of shaping the future of egg production, addressing both industry needs and consumer expectations.

6.1 Emerging technologies in egg quality assessment

Emerging technologies in egg quality assessment are significantly advancing the egg industry, particularly with the advent of Egg Industry 4.0 (EI 4.0). Innovations like the Internet of Things (IoT), various optical sensors, robotics, artificial intelligence (AI), big data, and cloud computing are transforming traditional practices into a smart and sustainable egg industry. These technologies enhance automation, improve biosecurity, and promote animal welfare. They also improve intelligent grading and quality inspection, thereby increasing efficiency. Non-destructive optical sensing technologies, a key component of EI 4.0, facilitate non-invasive and accurate egg quality assessment. This integration of technologies optimizes productivity, quality control, and resource management, contributing to the sustainable development of the egg industry (Ahmed et al., 2023).

6.2 Genetic advances in poultry breeding

Recent advancements in genetics have significantly impacted the field of poultry science, particularly in understanding and improving egg quality in laying hens. The integration of genomic tools has enabled a deeper exploration of the genetic factors influencing egg traits, leading to more targeted breeding strategies. The application of genomic prediction techniques, as demonstrated by Wolc et al. (2023), compared the accuracy of genomic predictions using medium-density SNP chips and low-pass sequencing, identifying significant SNPs across various egg quality traits. This approach has opened new avenues for enhancing egg quality through genetic selection. Furthermore, the research by Dunn et al. (2021) challenges the previously held belief that increased egg production compromises bone quality in laying hens. Their findings suggest no significant genetic correlation between egg production persistency and bone quality, indicating that breeding for higher egg production may not adversely affect the hens' skeletal health. Lastly, the study by Rowland et al. (2019) highlights the genetic control of egg quality traits under environmental stress, such as heat. This research underscores the importance of considering environmental factors in breeding programs to ensure the resilience and productivity of laying hens. These genetic advancements offer promising directions for future research in egg quality, focusing on sustainable and ethical breeding practices that enhance both the welfare of laying hens and the quality of their eggs.

6.3 Enhancements in feed and nutrition

Enhancements in feed and nutrition are pivotal in the poultry industry, especially in improving egg quality. The incorporation of nutritional additives such as vitamins, carotenoids, omega-3 fatty acids, and minerals into poultry diets has become a strategic approach to enhance the nutritional profile of eggs. These additives play a crucial role in improving various egg quality parameters. Vitamins are essential for maintaining the health and productivity of laying hens. Vitamin E, for instance, is known for its antioxidant properties, which help in improving the shelf life and quality of eggs. Vitamin D is crucial for calcium metabolism, directly impacting eggshell quality. Carotenoids, naturally occurring pigments found in plants, are added to feed to enhance yolk colour, a key quality attribute for consumer acceptance. Lutein and zeaxanthin, specific types of carotenoids, also contribute to the nutritional value of eggs, as they are beneficial for human eye health. Omega-3 fatty acids, particularly those found in fish oil and flaxseed, are added to hen diets to enrich eggs with these beneficial fats. Omega-3 enriched eggs are increasingly popular among health-conscious consumers due to their cardiovascular and cognitive health benefits. Minerals like selenium and zinc are important for the immune health of hens and the quality of eggs. Selenium, for example, is known for its role in antioxidant defence systems in the body.

These nutritional enhancements not only benefit the consumer by improving the nutritional quality of eggs but also support the health and productivity of the hens. However, the precise formulation of these additives in poultry diets requires careful consideration to balance the nutritional needs of the hens, the desired quality of the eggs, and economic viability. The ongoing advancements in poultry nutrition research continually refine the understanding and application of these feed additives to optimize egg production and quality.

6.4 Improvements in housing and welfare practices

Improvements in housing and welfare practices in the poultry industry have been increasingly focused on enhancing the well-being of birds while maintaining or improving egg quality. There has been a notable shift towards free-range systems and enriched cage environments, reflecting growing attention to animal welfare and consumer preferences (Alig et al., 2023; He et al., 2022). Free-range systems offer hens more natural living conditions, allowing them to exhibit natural behaviours such as foraging and dust bathing. This freedom not only impacts their physical health positively but also their psychological well-being (He et al., 2022). However, it's important to note that free-range environments aren't without their challenges, such as exposure to predators and environmental factors. Enriched cage environments, on the other hand, aim to improve the standard cage systems by providing additional features like perches, nesting areas, and more space per bird. These enhancements are designed to cater to the natural instincts and needs of the hens, leading to improved welfare outcomes. Studies have shown that these environments can positively impact the birds' health and egg quality, although the specific impacts can vary depending on the exact nature of the enrichment (Alig et al., 2023). The industry's move towards these improved housing and welfare practices is a response to a combination of ethical considerations, consumer demands, and scientific understanding of poultry behaviour. To put these into practice, we will not only have to meet rising standards for animal welfare, but also look at the sustainable production of high-quality poultry products.

6.5 Automation and data analytics in egg production

The integration of automation and data analytics in egg production is indeed a growing and transformative trend within the agricultural sector, including poultry farming (Anand et al., 2023). These technologies are increasingly being employed to improve efficiency, monitor various parameters, and enhance overall productivity in egg production. Automation in the egg industry has seen significant advancements, particularly in egg collection. In recent years, the use of mechanical robots has become more common, especially in larger-scale operations. These robots have automated tasks that were traditionally labour-intensive, like egg collection, sorting, and packaging (FarmingUK Team, 2021). This shift towards automation is not only improving efficiency but also reducing the physical strain on workers and increasing precision in handling eggs. Data analytics play a crucial role in optimizing farm operations. Through the use of various digital technologies, including flock management apps and smart sensors, farmers can now capture and analyse data related to bird health, egg production, and environmental conditions. This real-time monitoring and analysis will enable more informed decision-making, leading to better management of resources and improved animal welfare. The integration of these technologies also extends to sustainability practices in egg production. Eco-friendly packaging is increasingly becoming a part of the modern egg production process. The focus on sustainability is not just about meeting environmental standards but also about ensuring the long-term viability of egg production as a business. The adoption of these technological innovations is expected to continue growing, driven by the need for efficiency, sustainability, and improved quality in egg production. As these technologies evolve, they are likely to bring even more significant changes to the poultry industry, shaping the future of egg production towards a more automated and data-driven approach.

6.6 Sustainability and environmental practices

The focus on sustainability and environmental practices in egg production aligns with global trends in agriculture, emphasizing the reduction of environmental impacts. Innovations in this sector are directed towards creating more energy-efficient systems and utilizing sustainable feed sources (Dierenfeld, 2023). One significant aspect is the low environmental impact of egg production compared to other protein sources. Eggs require less water compared to other popular protein sources, like fruits and other animal protein sources (Mekonnen, 2012), and produce fewer greenhouse gas emissions per gram of protein (UN, no date). This makes eggs a sustainable choice in diets, contributing to environmental sustainability. Additionally, egg production is adaptable to various climates and seasons, making it a sustainable food source globally. The eggshells, which are the only inedible part for humans, are compostable and contribute to nutrient-dense soil for plants. Modern sustainable poultry farming practices include using alternative energy sources, implementing water-saving systems, and reducing waste through composting. These practices not only reduce the environmental footprint but can also lead to cost savings. Precision livestock farming technologies, such as automated feeders and computerized monitoring systems, optimize feed consumption, reducing waste and ensuring hens receive the precise nutrients needed for optimal egg production (Tedeschi & Mendes, 2023). Additionally, improving production efficiency and egg quality traits is essential for sustainability. A significant portion of the carbon footprint in hen layer production systems comes from feed. Therefore, innovations in feed utilization are crucial. For example, the use of feed enzymes such as phytase and carbohydrase in laying hen diets has been shown to increase productivity and profitability and reduce environmental impact. In conclusion, the egg industry's commitment to sustainability is evident in its adoption of eco-friendly practices, precision feeding technologies, and innovations in feed formulations. These efforts contribute to the production of a nutritious food source while minimizing the environmental impact, aligning with the industry's goal of sustainable egg production.

6.7 Challenges and considerations

Addressing the challenges and considerations in future directions of egg quality research involves acknowledging several critical factors. The implementation of new technologies in egg production, while promising, comes with its own set of challenges. Firstly, high investment costs pose a significant challenge. The development and integration of advanced technologies such as automation, data analytics, and genetic engineering require substantial financial investment. For many egg producers, especially smaller operations, these costs can be prohibitive. This financial barrier can slow down the adoption of innovative practices and technologies that could otherwise enhance efficiency and sustainability. Secondly, the requirement for skilled personnel is another critical consideration. The integration of new technologies in egg production demands a workforce with specialized skills and knowledge. Training existing employees or hiring new, skilled workers can be a challenging and costly process. There's also the challenge of keeping pace with continuously evolving technology, which requires ongoing education and adaptability. Ethical considerations are also paramount. As genetic advances and automation become more prevalent, ethical concerns regarding animal welfare and environmental impact come to the forefront. Balancing technological advancement with ethical farming practices is essential. This involves ensuring that animal welfare is not compromised and that environmental sustainability is maintained. There's also the aspect of public perception and consumer acceptance of products derived from technologically advanced egg production systems. These challenges underscore the need for a holistic approach that considers economic viability, workforce development, ethical farming practices, and environmental sustainability. Collaborative efforts between industry stakeholders, government bodies, and educational institutions are crucial to address these challenges effectively. By fostering an environment that supports innovation while considering these key factors, the egg production industry can move forward sustainably and ethically.

7 Conclusion

Egg quality is a critical element in the sphere of poultry science, profoundly impacting consumer health, industry standards, and global food security. This chapter has explored the intricate dynamics of both external and internal egg qualities, focusing on how these aspects are influenced by genetic, nutritional, and environmental factors. As we have seen, the external qualities, including eggshell strength, size, and colour, play a significant role in determining the egg's suitability for consumption and processing. Internally, the nutritional value and freshness of the egg are paramount, marked by the quality of the albumen and the yolk.

The importance of egg quality extends beyond agricultural concerns, encompassing significant areas of food safety, nutrition, and commercial viability. High-quality eggs contribute substantially to a balanced diet, offering essential nutrients crucial for human health. Moreover, egg quality significantly influences the poultry industry's profitability, dictating marketability and consumer preferences. This aspect underlines the economic relevance of maintaining superior egg quality as a strategic focus for poultry businesses.

The chapter has provided a comprehensive overview of the standards and methodologies employed in assessing and improving egg quality. It compares international standards, emphasizing the need for uniformity in quality assessment, and discusses the practical implications of these standards in different markets. By examining the latest scientific findings and technological advancements, the chapter offers a holistic perspective on egg quality, aligning academic insights with practical applications.

Looking ahead, the future of egg quality research is promising, with emerging technologies and genetic advancements paving the way for more precise and sustainable approaches. However, the chapter also underscores the challenges and considerations that accompany these advancements, particularly in the context of ethical and environmental concerns.

In conclusion, this chapter serves as a pivotal resource for researchers, educators, and professionals in poultry science. It emphasizes the need for ongoing research and development in egg quality, advocating for a balanced approach that harmonizes technological innovation with ethical and sustainable practices. As the demand for high-quality eggs continues to grow, driven by consumer awareness and regulatory standards, the poultry industry must adapt and evolve, ensuring that egg quality remains at the forefront of its strategic objectives.

8 References

AGRINFO. (2023). Revised egg maketing standards. AGRINFO. (https://agrinfo.eu/book-of-reports/revised-egg-marketing-standards/); accessed February 14th 2024.

Ahmed, M. W., et al. (2023). Non-destructive optical sensing technologies for advancing the egg industry toward Industry 4.0: a review. *Comprehensive Reviews in Food Science and Food Safety*, 22(6), 4378–4403. https://doi.org/10.1111/1541-4337.13227.

Alig, B., et al. (2023). The effect of housing environment on commercial brown egg layer production, USDA grade and USDA. *Animals*, 13(4), 694. https://doi.org/10.3390/ani13040694.

Anand, J., et al. (2023). Technological applications in smart farming: a bibliometric analysis. *Advanced and Sustainable Technologies*, 2(2), 30–40. https://doi.org/10.58915/aset.v2i2.334.

Badran, A. M. and Abd-Elaal, S. A. (2020). Effect of acute heat stress conditions on egg production, eggshell quality, intestinal calcium transport and calbindin of the laying hens. *Egyptian Poultry Science Journal, Article 18*, 40(1), 291–303. https://doi.org/10.21608/EPSJ.2020.81755.

Bar, A., et al. (2002). Re-evaluation of calcium and phosphorus requirements in aged laying hens. *British Poultry Science*, 43(2), 261–269. https://doi.org/10.1080/00071660120121481.

Bell, D. D. and Weaver, W. D. (2002). *Commercial Chicken Meat and Egg Production*. Springer Science + Business Media, New York, NY.

Blanco, A. E., et al. (2014). Genetic parameters of egg quality traits on different pedigree layers with special focus on dynamic stiffness. *Poultry Science*, 93(10), 2457–2463. https://doi.org/10.3382/ps.2014-04132.

Capua I. and Alexander, D. J. (2004). Avian influenza: recent developments. *Avian Pathology*, 33(4), 393–404. https://doi.org/10.1080/03079450410001724085.

Chang, A. (2021). How important is nutrition for egg shell quality? The Poulty Site (https://www.thepoultrysite.com/articles/how-important-is-nutrition-for-egg-shell-quality); accessed November 26th 2023.

De Reu, K., et al. (2006). Eggshell factors influencing eggshell penetration and whole egg contamination by different bacteria, including Salmonella enteritidis. *International Journal of Food Microbiology*, 112(3), 253–260. https://doi.org/10.1016/j.ijfoodmicro.2006.04.011.

Dierenfeld, E. S. (2023). Sustainable feeding solutions across the animal kingdom: identifying and implementing. *Journal of Animal Science*, 101(Supplement_3), 116–117. https://doi.org/10.1093/jas/skad281.142.

DSM. (2022). DSM egg quality manual (https://www.dsm.com/content/dam/dsm/anh/en/documents/dsm-egg-quality-manual.pdf); accessed February 13th 2024.

Dunn, I. C., et al. (2021). No evidence that selection for egg production persistency causes loss of bone quality in laying hens. *Genetics Selection Evolution*, 53(1). https://doi.org/10.1186/s12711-021-00603-8.

Ebeid, T. and Al-Homidan, I. (2022). Organic acids and their potential role for modulating the gastrointestinal tract, antioxidative status, immune response, and performance in poultry. *World's Poultry Science Journal*, 78(1), 1–2. https://doi.org/10.1080/00439339.2022.1988803.

European Union. (2018). Regulation (EC) No 589/2008 – Rules on marketing standards for eggs (https://eur-lex.europa.eu/EN/legal-content/summary/rules-on-marketing-standards-for-eggs.html); accessed December 26th 2023.

FAO. (2003). Egg marketing - A guide for the production and sale of eggs. (https://www.fao.org/3/Y4628E/y4628e00.htm#Contents); accessed February 14th 2024.

Farghly, M. F. A., et al. (2019). Intermittent lighting regime as a tool to enhance egg production and eggshell thickness in Rhode Island Red laying hens. *Poultry Science*, 98(6), 2459–2465. https://doi.org/10.3382/ps/pez021.

FarmingUK Team. (2021). Automation in the egg industry: where's it going next? (https://www.farminguk.com/news/automation-in-the-egg-industry-where-s-it-going-next-_57304.html); accessed December 26th 2023.

FDA. (2022). What you need to know about egg safety. (https://www.fda.gov/food/buy-store-serve-safe-food/what-you-need-know-about-egg-safety); accessed December 26th 2023.

FDA. (2023). Egg guidance, regulation, and other information. (https://www.fda.gov/food/guidance-documents-regulatory-information-topic-food-and-dietary-supplements/egg-guidance-regulation-and-other-information); accessed December 26th 2023.

Ghasemi, H. A., et al. (2022). Effect of advanced chelate compounds-based mineral supplement in laying hen diet on the performance, egg quality, yolk mineral content, fatty acid composition, and oxidative status. *Food Chemistry*, 366, 130636. https://doi.org/10.1016/j.foodchem.2021.130636.

Haugh, R. R. (1937), The Haugh unit for measuring egg quality. *U.S. Egg and Poultry Magazine*, 43, 522–555.

He, S., et al. (2022). The relationship between animal welfare and farm profitability in cage and free-range housing systems for laying hens in China. *Animals*, 12(16), 2090. https://doi.org/10.3390/ani12162090.

Hincke, M. T., et al. (1995). Purification and immunochemistry of a soluble matrix protein of the chicken eggshell (ovocleidin 17). *Calcified Tissue International*, 56(6), 578–583. https://doi.org/10.1007/BF00298593.

Hunton, P. (1982). Genetic factors affecting egg shell quality. *World's Poultry Science Journal*, 38(2), 75–84. https://doi.org/10.1079/WPS19820006.

IEC. (2024). Brown to white ratio, interactive statics. (https://marketdata.internationalegg.com/graphs/); accessed February 13th 2024.

Jones, D., et al. (2018). Impact of egg handling and conditions during extended storage on egg quality. *Poultry Science*, 97(2), 716–723. https://doi.org/10.3382/ps/pex351.

Kashimori, A. (2021). *The Illustrated Egg Handbook Second Edition*. Context Publications, Leicestershire, UK.

Khaliduzzaman, A., et al. (2020). A nondestructive eggshell thickness measurement technique using terahertz waves. *Scientific Reports*, 10(1), 1052. https://doi.org/10.1038/s41598-020-57774-5.

Kim, D. H., et al. (2022). Impact of relative humidity on the laying performance, egg quality, and physiological stress responses of laying hens exposed to high ambient temperature. *Journal of Thermal Biology*, 103, 103167. https://doi.org/10.1016/j.jtherbio.2021.103167.

King, N. R. and Robinson, D. S. (1972). The use of the scanning electron microscope for comparing the structure of weak and strong egg shells. *Journal of Microscopy*, 95(3), 437–443. https://doi.org/10.1111/j.1365-2818.1972.tb01046.x.

Kodani, S., et al. (2022). Implications of diet and quality consistence of feed on poultry layers egg quality. *African Journal of Agricultural Research*, 18(8), 617–631. https://doi.org/10.5897/AJAR2022.15948.

Liu, X., et al. (2017). Effects of conjugated linoleic acid on the performance of laying hens, lipid composition of egg yolk, egg flavor, and serum components. *Asian - Australasian Journal of Animal Sciences*, 30(3), 7. https://doi.org/10.5713/ajas.15.1036.

Liu, Z., et al. (2018). Genetic variations for egg quality of chickens at late laying period revealed by genome-wide association study. *Scientific Reports*, 8(1), 10832. https://doi.org/10.1038/s41598-018-29162-7.

Lopez-Bote, C. J., et al. (1998). Effect of free-range feeding on n - 3 fatty acid and α-tocopherol content and oxidative stability of eggs. *Animal Feed Science and Technology*, 72(1–2), 33–40. https://doi.org/10.1016/S0377-8401(97)00180-6.

Mekonnen, M. M. and Hoekstra, A. Y. (2012). A global assessment of the water footprint of farm animal products. *Ecosystems*, 15(3), 401–415. https://doi.org/10.1007/s10021-011-9517-8.

Meng, Y., et al. (2022). UHPLC-Q-Orbitrap-based untargeted lipidomics reveals the variation of yolk lipids during egg storage. *Journal of the Science of Food and Agriculture*, 102(13), 5690–5699. https://doi.org/10.1002/jsfa.11916.

Najimudeen, S. M., et al. (2021). Molecular characterization of 4/91 infectious bronchitis virus leading to studies of pathogenesis and host responses in laying hens. *Pathogens*, 10(5), 624. https://doi.org/10.3390/pathogens10050624.

Nys, Y., et al. (2001). Biochemical and functional characterisation of eggshell matrix proteins in hens. *World's Poultry Science Journal*, 57(4), 410–413. https://doi.org/10.1079/wps20010029.

Okonkwo, J. C., et al. (2021). Effect of storage method and storage duration on chicken egg quality. *International Journal of Environmental and Agriculture Research*, 7(8), 13–16. https://doi.org/10.5281/ZENODO.5335999.

Palmer, P. S. (1967). Wilhelm von Nathusius 1821–1899 on Avian Eggshells C. Tyler. *The Auk*, 84(4), 612–612. https://doi.org/10.2307/4083353.

Purswell, J. and Branton, S. (2015). Effect of air velocity on laying hen performance and egg quality. *Transactions of the ASABE*, 58(3), 813–817. https://doi.org/10.13031/trans.58.10982.

Roberts, J. R. (2004). Factors affecting egg internal quality and egg shell quality in laying hens. *The Journal of Poultry Science*, 41(3), 161–177. https://doi.org/10.2141/jpsa.41.161.

Roland, D. A. and Bryant, M. (1994). Influence of calcium on energy consumption and egg weight of commercial leghorns. *The Journal of Applied Poultry Research*, 3(2), 184–189. https://doi.org/10.1093/japr/3.2.184.

Romanoff, A. L. (1929). Study of the physical properties of the hen's eggshell in relation to the function of shell-secretory glands. *Biological Bulletin*, 56(5), 351–356. https://doi.org/10.2307/1537075.

Rowland, K., et al. (2019). Genetic analysis of production, physiological, and egg quality traits in heat-challenged commercial white egg-laying hens using 600k SNP array data. *Genetics Selection Evolution*, 51(1), 1–9. https://doi.org/10.1186/s12711-019-0474-6.

Ruxton, C. H. S. (2010). The nutritional properties and health benefits of eggs. *Nutrition and Food Science*, 40(3), 263–279. https://doi.org/10.1108/00346651011043961.

Scanes, C. G. (2007). Contribution of poultry to quality of life and economic development in the developing world. *Poultry Science*, 86, 2289–2290.

Scott, M. L., et al. (1971). The calcium requirements of laying hens and effects of dietary oyster shell upon egg shell quality. *Poultry Science*, 50, 1055–1063. https://doi.org/10.3382/ps.0501055.

Sokołowicz, Z., et al. (2022). The effect of storage duration on some quality traits and composition of eggs from different housing systems. *Annals of Animal Science*, 22(1), 459–475.

Surai, P. F. and Sparks, N. H. C. (2001). Designer eggs: from improvement of egg composition to functional food. *Trends in Food Science and Technology*, 12(1), 7–16. https://doi.org/10.1016/S0924-2244(01)00048-6.

Świątkiewicz, S., et al. (2018). Effect of selected feed additives on egg performance and eggshell quality in laying hens fed a diet with standard or decreased calcium content. *Annals of Animal Science*, 18(1), 167–183. https://doi.org/10.1515/aoas-2017-0038.

Tedeschi, L. O. and Mendes, E. D. M. (2023). Precision livestock farming tools for climate-smart feedyard operations. *Journal of Animal Science*, 101(Supplement_3), 326–327. https://doi.org/10.1093/jas/skad281.390.

USDA. (2000). United States standards, grades, and weight classes for shell eggs. (https://www.ams.usda.gov/sites/default/files/media/Shell_Egg_Standard%5B1%5D.pdf); accessed November 26th 2023.

USDA. (n.d.-a). Egg grading manual. (https://www.ams.usda.gov/sites/default/files/media/Egg%20Grading%20Manual.pdf); accessed November 26th 2023.

USDA. (n.d.-b). Shell egg surveillance. (https://www.ams.usda.gov/rules-regulations/eggs); accessed November 26th 2023.

Vlčková, J., et al. (2018). Effect of housing system and age of laying hens on eggshell quality, microbial contamination, and penetration of microorganisms into eggs. *Czech Journal of Animal Science*, 63(2), 51–60. https://doi.org/10.17221/77/2017-CJAS.

Wan, Z., et al. (2023). Promotion of egg production rate and quality using Limosilactobacillus oris BSLO 1801, a potential probiotic screened from feces of laying hens with higher egg productive performance. *Probiotics and Antimicrobial Proteins*, 15, 535–547. https://doi.org/10.1007/s12602-021-09856-7.

Wiaętkiewicz, S., et al. (2015). Dietary factors improving eggshell quality: an updated review with special emphasis on microelements and feed additives. *World's Poultry Science Journal*, 71(1), 83–93. https://doi.org/10.1017/S0043933915000082.
Wilson, P. (2017). Recent advances in avian egg science: a review. *Poultry Science*, 96(10), 3747–3754. https://doi.org/10.3382/ps/pex187.
Wolc, A., et al. (2023). Application of low-pass sequencing to genomic prediction of egg quality in laying hens. Proceedings of 12th World Congress on Genetics Applied to Livestock Production, 1143–1146. https://doi.org/10.3920/978-90-8686-940-4.
Xin, H., et al. (2011). Environmental impacts and sustainability of egg production systems. *Poultry Science*, 90(1), 263–277. https://doi.org/10.3382/ps.2010-00877.
Yimenu, S. M., et al. (2018). Kinetic modeling impacts of relative humidity, storage temperature, and air flow velocity on various indices of hen egg freshness. *Poultry Science*, 97(12), 4384–4391. https://doi.org/10.3382/ps/pey334.
Zang, H., et al. (2011). Effects of different dietary vitamin combinations on the egg quality and vitamin deposition in the whole egg of laying hens. *Brazilian Journal of Poultry Science*, 13(3), 189–196. https://doi.org/10.1590/S1516-635X2011000300005.

Chapter 9

Advances in non-destructive testing of egg quality

Silvia Grassi and Eleonora Loffredi, University of Milan, Italy

1 Introduction

Eggs are extremely important in the human diet as they are one of the cheapest sources of protein, vitamins A and B12, iron, zinc and riboflavin (Drewnowski, 2010); they contain all the essential amino acids and omega-3 fatty acids, and the presence of vitamin D3 and 25-hydroxyvitamin D3 in egg yolks is extremely important.

For these reasons, eggs are not only a staple food, but they represent an important ingredient for the food industry, thanks to their technological properties (i.e. foaming, gelling, and emulsifying) and their nutritional value, which can be relevant for food enrichment (Pal et al., 2017).

The significant increase in egg production over the past years calls for non-destructive and reliable methods to ensure internal quality, healthiness, and eggshell integrity before egg packing and distribution. Indeed, the proper use of eggs, as such or after technological transformation to become an ingredient, is highly dependent on both external (shell strength and thickness influencing shell integrity) and internal (albumen viscosity and yolk shape band firmness) quality parameters (Fig. 1).

http://dx.doi.org/10.19103/AS.2024.0150.13

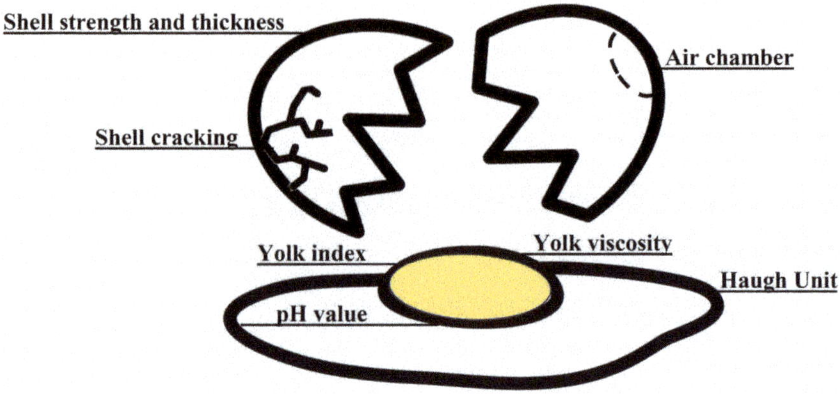

Figure 1 Schematic representation of external and internal quality parameters of eggs.

According to governmental standards, the exterior characteristics evaluation is often made at the primary sorting stage by human graders. The exterior characteristics determining potential acceptability are egg size, shape, presence of contaminants (e.g. feces, blood, and yolk), and shell cracks (Lunadei et al., 2012; Ibrahim et al., 2012); they are generally evaluated by visual inspection. Furthermore, a vernier calliper can measure shell strength and thickness, but only after shelling, thus not allowing their evaluation in primary sorting. However, the prediction of shell strength and thickness, significantly correlated, will enable the reduction of fragile eggs distribution thus leading to the reduction of economic losses and risk of cracking with subsequent bacterial contamination (Sharaf Eddin et al., 2019).

The air chamber, i.e. the space formed between the inner and outer shell membranes, appears at the time of oviposition, presumably because of a reduction in the volume of the egg contents induced by a sharp drop in temperature. During egg storage, the internal water is lost from the pores of the eggshell, increasing the diameter and height of the air chamber, thus its geometry is strictly linked with egg freshness. Indeed, the air chamber height is the only quantitative egg freshness parameter to be evaluated according to the European Union regulation (Commission Regulation No 589/2008).

The freshness of the albumen is usually evaluated by Haugh units (HU) and pH. HU consists of measuring the height of the thick albumen at 1 cm from the yolk with a micrometre mounted on a tripod. An equation then allows the conversion to HU considering the egg weight. The HU value decreases with storage time as a result of the thinning of the egg yolk (Stadelman & Cotterill, 1995). The *pH value* of whole egg, and albumen in particular, is an important freshness marker, having the laid eggs' albumen a pH between 7.6 and 8.5. The pH value of the egg white increases during storage due to the gas exchange with the environmental air through the pores of the shell and the migration of water and minerals between albumen and yolk through the yolk membrane.

The yolk quality is generally evaluated considering the Yolk Index (YI) and yolk viscosity. YI concerns the ratio between the height of the yolk (mm) and its diameter (mm). The staler is the egg, the smaller the YI, as the longer an egg is stored, the lower is the yolk viscosity and the longer the exchange between yolk and albumen of salts/water.

Yolk viscosity concerns the measurement of the distance that the yolk flows when the vitelline membrane is broken. It might be measured with a portable falling-needle viscometer (Kuroki et al., 2017).

The assessment of the mentioned egg quality parameters has been a major challenge in the last century (Stadelman & Cotterill, 1995), leading to the current need for non-destructive testing methods. Indeed, even though destructive analyses for egg quality assessment provide reliable data, they are time-consuming, require trained personnel, and can only be applied in well-equipped laboratories. Finally, they are destructive, thus requiring a representative sampling plan covering the possible variability and the loss of the representative samples during the analysis. These constraints hinder the egg industries, moving researchers toward fast and reliable egg quality assessment methods.

Therefore, a representative discussion on the application of diverse non-destructive testing methods is presented. The considered approaches include spectroscopic methods (Raman, visible and near-infrared (VIS-NIR), dielectric, and Terahertz (THz)), vision systems, acoustic methods, and electronic nose systems.

The chapter aims to provide an overview of the non-destructive testing methods developed to date and the likely future research trends that will lead to non-destructive quality detection in the egg industry.

2 Spectroscopic approaches

Spectroscopic methods are based on the interaction between the light emitted – in the different electromagnetic regions – and the sample, enabling the fingerprint of its chemical and/or physical properties (Fig. 2). The complex features of radiation in the different electromagnetic regions lead

	Very low frequency		Low frequency		Radio waves		Infrared		Visible	Raman
Frequency (Hz)	10	10^2	10^4	10^6	10^8	10^{10}	10^{12}	10^{14}	10^{15}	
Wavelength (nm)			10^4	10^2	10	10^{-2}	10^{-4}	10^{-6}	10^{-7}	

Figure 2 Electromagnetic wave range of the main spectroscopic approaches.

to broad and overlapped signals, which might hinder the direct extraction of information from the raw signal, making the use of chemometric approaches essential. For further details, refer to Grassi et al. (2023a).

The recent success of spectroscopic methods in the food industry for quality control or as an online process analyser has been facilitated by the advances in software algorithms, i.e. the chemometric strategies applied to extract the relevant information from the spectra. Parallelly, the advance in the hardware played an important role, thanks to versatile sampling holders, such as fiber optic probe technology, and to the increasing availability of inexpensive portable devices that can be more easily integrated into the processing lines. Not all the spectroscopic methods currently available reach the same degree of success due to differences in device miniaturisation capability, implementation costs, and modelling performances (i.e. sensitivity, selectivity, and reproducibility).

Among them, a representative selection of their application in the egg field is presented here.

2.1 Raman spectroscopy

Raman spectroscopy relies on the analysis of scattered light, in particular the so-called Raman scattering or inelastic scattering. The scattered light can provide information about substances and their structure by determining the frequency shift, or Raman shift, between the incident beam of light (typically a green laser emitting at 532 nm) and the Raman scattered light. From the Raman shift, the spectrum can be reconstructed, considering that each peak corresponds to a different frequency of light absorbed by the sample, which excites a vibration.

Raman spectroscopy had limited applications in the non-destructive evaluation of egg freshness and quality (Table 1). A remarkable implementation is the one reported by Liu et al. (2020) for the detection of eggshell cuticle modifications correlated to freshness parameters (i.e. HU, YI, albumen pH, air chamber diameter, and height). In the study, 125 Hy-Line Brown eggs were stored under controlled conditions (20°C, 40% RH) for up to 59 days. The Raman spectra (100–3000 cm^{-1}) were collected on the top (pointed area), bottom (blunt area), and equatorial regions. To build Partial Least Squares Regression (PLSR) models, HU, YI, albumen pH, and air chamber height and diameter were analysed by the common destructive methods. The PLSR models were appropriately calibrated by 80% of the samples and validated with a prediction dataset composed of the other 20% of the samples. Good results were obtained for HU, albumen pH, and air chamber dimensions, with a correlation coefficient in prediction (r_p) ranging from 0.81 to 0.89. YI was not satisfactorily predicted (r_p = 0.54), probably due to a weak relationship between the changes in the eggshell surface and the yolk. Considering the different

Table 1 Application of Raman spectroscopy for the evaluation of quality and freshness of shell eggs

Egg parameter	Number of samples (eggs)	Predictive algorithm	Figure of merits of the developed models	Reference
Air chamber height	125 (100, C; 25, P)	PLSR	Air chamber height: r_C = 0.828, r_P = 0.830	Liu et al. (2020)
Air chamber diameter			Air chamber diameter: r_C = 0.903, r_P = 0.915	
Haugh Units			Haugh Units: r_C = 0.944, r_P = 0.925	
Albumen pH			Abumen pH: r_C = 0.945, r_P = 0.935	
Yolk coefficient	26	PLS-DA	Accuracy = 80%	Davari et al. (2024)
Egg shape index				
Yolk index				

C, calibration; P, prediction; PLSR, partial least squares regression; PLS-LDA, partial least squares - linear discriminant analysis; *r*, coefficient of correlation.

distribution of constituents and compartments in the egg, the authors deeply studied the influence of the acquisition area. They assessed an improved performance of PLSR models when considering the spectra collected in the egg-pointed area, probably due to the absence of the air chamber, leading to the contact of the egg content with the eggshell. The authors concluded that increased adaptability of the models is needed to maximise Raman spectroscopy application, thus they envisioned the inclusion of other relevant factors (i.e. breed, hen age, and rearing systems) in future studies. A few other works have been developed for the detection of egg quality and freshness parameters by using Raman spectroscopy (Table 1).

2.2 Visible and near-infrared spectroscopy

Absorptions in the near infrared (NIR) spectral range (14 000 cm^{-1}–4000 cm^{-1}) are associated with the main chemical components of foodstuffs, such as water, proteins, carbohydrates, and fats. Those compounds absorb in the NIR region due to vibration and combination overtones of the fundamental C–H, N–H, O–H, and C=O bonds, with an overlapped signal. Moreover, water strongly characterises NIR spectral information by the O–H first overtone (around 6900 cm^{-1}) and the combination of the asymmetric stretch and bending of the water molecule (around 5300 cm^{-1}) (Workman & Weyer, 2008).

NIR spectroscopy, particularly in combination with the visible range (380–750 nm), has been extensively implemented both at the lab- and production scale to determine the quality of diverse food products. VIS-NIR and NIR spectroscopy have been extensively studied to develop process analytical technology (PAT) approaches for real-time analysis during manufacturing (Grassi & Alamprese, 2018).

In the egg field, VIS-NIR and NIR spectroscopy have been rarely used to directly assess external quality parameters due to the lack of spatial information. On the other hand, VIS-NIR and NIR spectroscopy methods have been explored to measure egg quality and freshness in terms of internal quality, including air chamber diameter and height, HU, albumen and whole egg pH, YI, and apparent yolk viscosity (Lin et al. 2011).

Some relevant implementations will be briefly presented stressing their strong points; the complete list of recent applications is reported in Table 2.

Only Dong et al. (2017b) proposed a PLSR model based on VIS-NIR spectroscopy for *eggshell thickness* prediction with promising results, even if a wider range of variability must be considered to confirm the findings. The work considered 70 white-shelled intact eggs with a thickness range from 0.27 mm to 0.38 mm. Shell thickness was measured by a vernier calliper in the egg equatorial region, i.e. where it has a more homogeneous texture and thickness. The PLSR model, calibrated by the spectral data collected for 52 samples and validated with 18 samples, led to a correlation coefficient (*r*) in prediction of 0.84 and a rootoot mean square error in prediction (RMSEP) of 0.01 mm.

Furthermore, Han et al. (2022) developed a PCA-BP (principal component analysis – back propagation) sorting model to predict egg integrity and color, meaning detecting intact and cracked eggs with different eggshell colors, considering relevant NIR spectral regions (1100–1260 nm, 1648–1698 nm, and 2380–2410 nm). The prediction results showed a sorting accuracy percentage for cracked eggs of 100%, 100%, and 98.75%, for pink, green, and white eggs, respectively.

No other applications have been reported for direct external quality parameter assessment; thus, this remains a relevant topic to be investigated in the future in this spectral range.

Both Dong et al. (2018b) and Aboonajmi et al. (2016) developed methods to predict air chamber characteristics. Aboonajmi et al. (2016) proposed a comprehensive experimental design, considering 300 eggs stored in different conditions (temperatures of 25°C and 5°C, with RH of 40% and 75% respectively, up to 30 days); however, no information about predictive errors was reported, thus making difficult to evaluate the real applicability of the approach.

Concerning albumen quality, VIS-NIR and NIR spectroscopy have been used to construct prediction models for albumen freshness, mainly predicting HU and pH.

The prediction of those parameters has been primely investigated by Kemps et al. (2007), who reported the use of VIS-NIR transmission spectroscopy to assess egg HU in 60 eggs stored at 18°C (55% RH) up to 18 days. The

Table 2 Infrared spectroscopy for the evaluation of quality and freshness of shell eggs

Analytical technology	Egg parameter	Number of samples (eggs)	Predictive algorithm	Figure of merits of the developed models	Reference
VIS-NIR spectroscopy	Eggshell thickness	70 (52, C; 18, P)	PLSR	$r_C = 0.86$, $r_P = 0.84$	Dong et al. (2017b)
	Eggshell integrity	820 (520, C; 270, P)	PCA-BP	Accuracy $_P$ = 98.75–100%	Han et al. (2022)
	Air chamber diameter	90 (68, C; 22, P)	PLSR	$r_C = 0.87$, $r_P = 0.85$	Dong et al. (2018b)
	Air chamber height	300 (no test set)	ANN (RBF)	Air chamber height: $R^2_C = 0.941$, $R^2_{CV} = 0.844$ (5°C) $R^2_C = 0.918$, $R^2_{CV} = 0.835$ (25°C)	Aboonajmi et al. (2016)
	Haugh Units			Haugh Units: $R^2_C = 0.898$, $R^2_{CV} = 0.767$ (5°C) $R^2_C = 0.871$, $R^2_{CV} = 0.745$ (25°C)	
	Haugh Units	60	PLSR	$r_P = 0.89$	Kemps et al. (2007)
	Haugh Units Yolk index Albumen pH	91 (68, C; 23, P)	PLSR	Equatorial region: Haugh Units: $r_C = 0.897$; $r_P = 0.881$ Yolk index: $r_C = 0.903$; $r_P = 0.855$ Albumen pH: $r_C = 0.936$; $r_P = 0.888$	Dong et al. (2018a)
	Albumen pH	80 White Leghorns (53, C; 27, P) 87 Bantam (58, C; 29, P)	PLSR	White leghorns $r_C = 0.918$, $r_P = 0.907$ Bantam $r_C = 0.955$, $r_P = 0.947$	Dong et al. (2019)
	Albumen pH Whole egg pH	178 (133, C; 45, P)	PLSR	Albumen pH: $r_C = 0.943$, $r_P = 0.923$ Whole egg pH: $r_C = 0.776$, $r_P = 0.752$	Dong et al. (2017a)

(*Continued*)

Table 2 (*Continued*)

Analytical technology	Egg parameter	Number of samples (eggs)	Predictive algorithm	Figure of merits of the developed models	Reference
NIR spectroscopy	Apparent yolk viscosity	88 (no test set)	PLSR	R^2_C = -, R^2_{CV} = 0.89	Kuroki et al. (2017)
	Marked day of lay	120 (90, C; 30, P)	PLSR	r_C = 0.83; r_P = 0.89 (28°C) r_C = 0.86; r_P = 0.91 (4°C)	Akowuah et al. (2020)
			LDA	Accuracy $_P$ = 100% (28°C) Accuracy $_P$ = 96% (4 °C)	
	Egg storage time	30 (21, C; 9, P)	ANN	R^2_C = 0.865, R^2_{CV} = 0.873	Coronel-Reyes et al. (2018)
	Egg storage time	700 (490, C; 210, P)	PLSR	R^2_{CV} = 0.67–0.84 SEP = 5.42–796	Cozzolino et al. (2024)

VIS-NIR, visible and near-infrared; NIR, near-infrared; C, calibration; P, prediction; PLSR, partial least squares regression; LDA, linear discriminant analysis; RBF, radial basis function; ANN, artificial neural network; LR, linear regression; *r*, coefficient of correlation; CV, cross-validation; RPD, residual prediction deviation; R^2, coefficient of determination; SEP, standard error of prediction.

developed PLS model reached an r_P of 0.89 with an error (RMSEP) of 8.45 for HU (within a considered range of 21–98 HU).

VIS-NIR spectra-based regression models were built by Dong et al. (2017a) on 178 white-shelled eggs stored for up to 3 weeks at 30°C and 65% RH to predict pH. For albumen pH (ranging from 7.06 to 10.11), the PLSR model got an optimised *r* in prediction of 0.92, with a RMSEP of 0.17, whereas the whole egg pH model yielded a lower prediction accuracy ($r = 0.75$; RMSEP = 0.27).

Finally, it is worth citing that some authors applied NIR spectroscopy to assess marked day of lay (Akowuah et al., 2020) and egg storage time (Coronel-Reyes et al., 2018; Cozzolino et al., 2024). However, it should be borne in mind that the initial quality and storage conditions have a major influence on the changes in the external and internal characteristics of the eggs so these predictions can be very inaccurate and do not necessarily correlate with the actual quality level of the product.

The nature of the shell egg structure together with the high development of instrumentation in the NIR field brought different researchers to evaluate the egg orientation effect during non-destructive measurements using different VIS-NIR instruments (Dong et al., 2018a; Dong et al., 2019). The obtained results highlighted a higher correlation with the freshness parameters for spectra recorded in the equatorial region. The findings are consistent with a possible interference of the air chamber in the blunt end, together with a more homogeneous texture and thickness of the eggshell in the equatorial region. Furthermore, some researchers tried to face eggshell color differences, especially when the visible region is included in the analysis. Even though, most of the work considered just one type of shell color – mainly white-shelled egg – it is advisable to consider both the shell egg types to assess the method's reliability on a larger scale, as done by Dong et al. (2019). For instance, Dong et al. (2019) considered a total of 192 eggs, 96 from White Leghorn hens and 96 from Bantam hens, stored under controlled conditions (30°C, 60% RH) for up to 3 weeks. They developed PLSR models considering separately or jointly the eggs of the two hen breeds. For the Bantam eggs, the best prediction model was obtained with Multivariate Scatter Correction (MSC) pretreatment (coefficient of determination (R^2) of 0.95 and RMSEP of 0.115), and similar results were obtained for White Leghorn eggs (R^2 of 0.91 and RMSEP of 0.123). They were not able to transfer the calibration from one hen breed to the other, whereas they developed a global calibration, considering all the eggs together, and obtaining good performances (R^2 of 0.94 and RMSEP of 0.119).

Even if VIS-NIR and NIR spectroscopies have been widely explored, compared to other spectroscopic approaches, none of the reported works led to enough robust models to be implemented in the food industry, but certainly, they did the groundwork to reach that goal in the future.

2.3 Terahertz spectroscopy

Terahertz (THz) spectroscopy (electromagnetic radiation with submillimeter wavelength) is understudied due to various limitations, such as low accuracy, spatial resolution limitations, penetration limitations, and particle size influence (Møller et al., 2009). However, THz pulses seem successful in the study of crystalline components with minimal moisture content. This is the reason underlying the recent applications of THz spectroscopy for eggshell quality analysis. Khaliduzzaman et al. (2020) proposed a non-destructive approach based on THz reflectance spectroscopy to evaluate eggshell thickness based on the THz interference signals dependency on eggshell thickness. Indeed, it is reasonable that THz waves are reflected from the outer air–eggshell contact as well as the interior eggshell–membrane boundary. On these bases, the frequency-domain oscillation distance was used to measure shell thickness. The developed PLSR models reached an R^2 of 0.93 and a RMSEP of 0.009 mm.

2.4 Dieletric spectroscopy

Dielectric spectroscopy is basically used to study the response of a sample subjected to an applied electric field of fixed or changing frequencies (from 10^3 Hz to 10^{15} Hz). It is based on the interaction of an external electromagnetic field with the dipole of the sample, generally expressed as permittivity. Several factors influence the choice of measurement method, such as dielectric material, frequency, accuracy, costs, and suitability of the method for online application. Few research papers applied dielectric methods for non-destructive measurement of various quality indices of shell eggs (Table 3). The research work by Akbarzadeh and coworkers (Akbarzadeh et al., 2019) is the more comprehensive aiming at developing a dielectric spectroscopy approach (in the microwave frequencies) based on a waveguide and a network analyser instrument for the non-destructive prediction of several egg freshness indices (i.e. air chamber height, thick albumen height, HU, albumen pH, and yolk coefficient (YC)). For this purpose, they stored 244 white-shelled eggs at room temperature (25°C) for up to 24 days. The dielectric reflection and transmission spectra (0.9–1.7 GHz microwave range) were acquired before the destructive evaluation of freshness parameters. Then, regression (PLSR) and classification (Soft Independent Modelling of Class Analogy, SIMCA) models were developed and validated. The return loss reflection spectra gave the best predictive regression models for all quality indices, except for albumen pH, as well as the best classification model considering six different classes of freshness. In conclusion, despite the good results obtained, the authors highlighted the necessity to develop a more economical system, since the network analyser used was expensive to implement the technique for online

Table 3 Dielectric spectroscopy techniques applied to the evaluation of quality and freshness of shell eggs

Egg parameter	Number of samples (eggs)	Predictive algorithm	Figure of merits of the developed models	Reference
Air chamber height	244 (196, C; 48, P)	PLSR	Air chamber height:	Akbarzadeh et al. (2019)
Thick albumen height			$R^2_C = 0.900$, $R^2_P = 0.893$	
Haugh units			Thick albumen height:	
Albumen pH			$R^2_C = 0.839$, $R^2_P = 0.826$	
Yolk coefficient			Haugh units	
			$R^2_C = 0.814$, $R^2_P = 0.804$	
			Albumen pH	
			$R^2_C = 0.787$, $R^2_P = 0.779$	
			Yolk coefficient	
			$R^2_C = 0.883$, $R^2_P = 0.869$	
Freshness	244 (196, C; 48, P)	SIMCA	Accuracy P = 100%	
Air cell height	150 (110, C; 40, P)	ANN	r_P ANN = 0.817	Soltani and Omid (2015)
		SVM	r_P SVM = 0.920	
		DT	r_P DT = 0.906	
Haugh units	287 (124, C; 163, P)	FFNN	Haugh units:	Soltani et al. (2015)
			$R^2_C = 0.998$, $R^2_P = 0.998$	
Yolk/Albumen			Yolk/Albumen:	
			$R^2_C = 0.996$, $R^2_P = 0.998$	
Yolk weight			Yolk weight	
			$R^2_C = 0.994$, $R^2_P = 0.994$	
Yolk index			Yolk Index:	
			$R^2_C = 0.994$, $R^2_P = 0.998$	

C, calibration; P, prediction; PLSR, partial least squares regression; LDA, linear discriminant analysis; ANN, artificial neural network; SVM, support vector machine; DT, decision tree; FFNN, feedforward neural network; SIMCA, soft independent modelling of class analogy; R^2, coefficient of determination; *r*, coefficient of correlation.

applications by decreasing the number of scans so that the data acquisition is faster and to design a special waveguide to be placed on egg conveyor and dedicated for this purpose.

Moving to the range of radio frequency (40 kHz–20 MHz) Soltani and Omid (2015) and Soltani et al. (2015) built a robust model for the non-destructive classification of eggs based on freshness (100% of accuracy in prediction) as well as regression models to predict HU ($R^2_P = 0.998$), the ratio yolk/albumen ($R^2_P = 0.998$), yolk weight ($R^2_P = 0.994$), and YI ($R^2_P = 0.998$). The mean absolute percent errors obtained in prediction were 5.41%, 6.84%, 8.79%, and 4.24% for HU, YI, yolk/albumen, and yolk weight, respectively.

2.5 Ultrasound spectroscopy

Ultrasound technology profits from differences between sound velocities, scattering coefficient, and other characteristics measured in the sample to correlate them to specific quality parameters. Ultrasound region can be split into high-power/low-frequency ultrasound (20–100 kHz), medium-power/medium-frequency ultrasound (100 kHz–1 MHz), and high-frequency/low-power ultrasound (100 kHz–20 MHz). The waves are generated by a pulse-echo sensor, or transmitter, and received by a wave receiver. Ultrasound spectroscopy has been extensively used for solid foreign material detection as the propagation speed increases with the increase of the material density (Laugier & Hayat, 2011). Ultrasound applications for chemicals, structure, and physical study are less extensive but seem promising in the field of egg quality evaluation from the results obtained by Aboonajmi et al. (2010). The work studied the phase velocities of the ultrasonic signal travelling through the shell eggs after storage under various conditions (ambient temperature or refrigerated). The same eggs were also evaluated by destructive methods for a wide range of freshness parameters (HU, air cell height, yolk height, yolk diameter, YI, and thick white height of the eggs). The results demonstrated that it was possible to distinguish fresh and old eggs by the difference in the ultrasonic phase velocity recorded.

Further investigations should be done to support ultrasound spectroscopy reliability as well as its scalability to assess eggs' freshness in a non-destructive way.

3 Machine vision systems

Spectroscopic methods have been widely investigated, and some of them led to relevant results, especially at the research level. No matter the electromagnetic range considered, one relevant disadvantage in their application is the lack of spatial representativeness. Indeed, the acquired signals are punctual, and this is a relevant disadvantage when considering a heterogeneous distribution of components, such as in shell eggs. This disadvantage could be overcome by machine vision technology now being implemented in a variety of fields, including food industries.

Machine vision systems might consider different regions of the electromagnetic spectrum (ultraviolet, visible, and NIR, Raman) profiting from the spectral information of each specific region with the addition of spatial resolution. Generally speaking, a vision system is composed of a lighting system (source of light in the desired spectral range), a sensor combined with a camera

(CCD), an image receiver (frame grabber or digitiser), a computer hardware, and a software (Amigo, 2019). Indeed, after the image acquisition, the software administers image preprocessing, image segmentation, feature extraction, and identification according to the defined purpose (Grassi et al., 2023b).

Machine vision systems based on digital cameras are convenient, highly efficient, accurate, fast, economically affordable, and so reliable for industrial implementation. Thus, they have been exploited to solve external feature assessments such as eggshell cracks on white unwashed eggs (Priyadumkol et al., 2017) or brown shell eggs (Wu et al., 2018). However, the relevant information retrievable from digital imaging is limited to external features. VIS-NIR hyperspectral imaging (HSI) integrates digital computer vision and spectral analysis, thus being able to simultaneously acquire spectral and spatial information from the intact product (Amigo, 2019). Different applications have been the object of scientific research (Table 4). Dai et al. (2020) proposed a method based on NIR-HSI technology to predict egg freshness (expressed in HU) by studying the influence of incident angles of the light source (scattering, reflection, transmission) on the model accuracy. Furthermore, they implemented a variety of classifiers to assess the combination with the highest accuracy. They ended up with the highest accuracy (100%) when scattering hyperspectral was used in combination with metamodels of stacking ensemble learning. Indeed, the model accuracy resulted inversely proportional to the angle of incidence, suggesting that the system can collect more chemical information from a larger proportion of light, i.e. scattering light, than from reflection and transmission modes.

One of the advantages of HSI in transmission in the VIS-NIR range has been highlighted by Yao et al. (2022). Their work presented the application of VIS-NIR HSI for simultaneously predicting egg freshness, scattered yolk, and eggshell cracks. This means that from a single measure, by properly programmed software, the whole concept of freshness might be described.

Moving to the thermal infrared range, it is worth mentioning that thermography has been used to correlate egg freshness features by differences in the temperature emitted by the samples. Freni et al. (2018) developed a methodology for egg freshness assessment based on active thermography through pulsed thermal stimulation in reflection. The thermal images identified the air chamber at the dull pole due to differences in thermal propagation between the air chamber and the liquid part of the egg. However, the sampling has been poorly representative of the possible variability as the same eighteen eggs were analysed day by day along a conservation period of 20 days at 28°C and 30% of RH.

Vision systems offer a high level of flexibility and repeatability, however when considering electromagnetic regions different from the visible range, it is not guaranteed a reasonable cost nor a proper time of analysis for industrial

Table 4 Vision systems applied for the evaluation of quality and freshness of shell eggs

Analytical technology	Egg parameter	Number of samples (eggs)	Predictive algorithm	Figure of merits of the developed models	Reference
CCD digital camera	Geometric indexes	760	-	Accuracy = 90.29%	Alikhanov, J., et al. (2019)
	Cracking	50	BPANN	Accuracy = 94.00%	Priyadumkol et al. (2017)
	Egg exterior characteristics	400	Linear regression	r = 0.93; error = 1.05%	Vasileva et al. (2018)
	Cracking	200	SVM	Accuracy C = 94.00%; Accuracy CV = 93.00%	Wu et al., 2018
VIS/NIR-HSI	Haugh units	188 (141, C; 47, P)	SVM	Accuracy P = 97.87%	Yao et al. (2022)
	Haugh units	350 (200, C; 150, P)	SPA-SEL	Accuracy up to 100%	Dai et al. (2020)
	Haugh units Scattered yolk Air bubbles	645 (200, C; 150, P)	SPA-SVR	Haugh units: R^2_C = 0.89, R^2_P = 0.87 Scattered yolk: Accuracy = 90% Air bubbles: Accuracy = 96.3%	Zhang et al. (2015)
NIR-HSI	Haugh units	91 (58, C; 33, P)	PLSR	R^2_C = 0.91, R^2_P = 0.85	Suktanarak & Teerachaichayut (2017)
Pulsed infrared thermography	Air chamber	18 (no test set)	-	-	Freni et al. (2018)
Thermal camera	Air chamber	234 (174, C; 60, P)	YOLOv4, YOLOv5, EfficientDet	Model Precision YOLO v4 c = 0.99, P = 1; YOLO v5 c = 0.99, P = 1; EfficientDet c = 0.99, p=1 Recall (R): YOLO v4 c = 0.99, P = 0.53; YOLO v5 = 0.99, P = 0.81; EfficientDet c = 0.99, P = 86	Nakaguchi & Ahamed (2022)

NIR-HIS, near-infrared hyperspectral imaging; VIS-NIR-HIS, visible and near-infrared hyperspectral imaging; C, calibration; P, prediction; PLSR, partial least squares regression; BPANN, back-propagation artificial neural network; SVM, support vector machine; SPA-SEL, successive projection algorithm-stacking ensemble learning; SPA-SVR, successive projection algorithm-support vector regression; MLR, multiple linear regression; YOLO, deep learning-based object detection algorithms; *r*, coefficient of correlation; CV, cross-validation; R^2, coefficient of determination.

implementation. Those issues must be faced in the future by simplified systems, i.e. multichannel imaging and fast calculating algorithms.

4 Acoustic systems

Acoustic systems could record sound emission, absorption, or oscillation of phase. They are generally composed by an exciting system, a microphone, a signal conditioning circuit, a signal acquisition, and transmission and an analysis system (Aboonajmi et al., 2015a).

After signal collection, the results are generally processed to extract and select relevant features in the frequency domain to further define patterns characterising the quality classes to be discriminated. The discrimination phase could be performed by a univariate approach, i.e. comparing feature by feature with a reference threshold or by chemometric approaches.

Among the acoustic methods developed, the one by Sun et al. (2013) presented a real industrial application for eggshell integrity evaluation. Indeed, Sun et al. (2013) implemented an acoustic resonance system along a roller conveyor belt with a working speed from 28 cm s^{-1} to 45 cm s^{-1}. The system comprises an exciting system, a microphone (sensitivity: 46 ± 3 dB; response frequency: 80–12 500 Hz), a signal conditioning circuit with a band-pass filter to select the frequency band between Hz 1200 and 6800 Hz, and a digital signal processing connected to a computer. The excitation system was composed of an infrared trigger set, an electromagnetic driver, a power supplier, and a light metallic stick with an optimised impacting intensity of 2.09 N. The robustness of the developed classification model was guaranteed by a calibration (with 200 intact eggs and 200 cracked eggs) followed by an independent test (considering 100 intact eggs and 103 cracked eggs) randomly analysed. After transforming the time domain signal to a frequency signal by using a Fast Fourier Transformation, specific frequency characteristics were selected: mean of the amplitude, value of first peak frequency, index of first peak frequency, and mean and standard magnitude values from the top three peak frequencies. The discrimination was based on a quite simple approach: the five selected frequency characteristics were compared with a critical value one by one, and if three or more features exceeded the critical threshold, the egg was recognised as intact. The prediction results demonstrated that the model was able to correctly discriminate all the intact eggs, whereas the discrimination rate of crack eggs was 96.11%, i.e. four cracked eggs were misclassified as intact. Most importantly, the calculating time was less than 10 ms.

Other relevant research findings support the reliability of acoustic systems, based on sound or vibration signals of stimulated eggs, for shell egg crack detection in a fast and reliable manner (Table 5), thus confirming the feasibility

Table 5 Acoustic spectroscopy applied to the evaluation of quality of shell eggs

Analytical technology	Egg parameter	Number of samples (eggs)	Predictive algorithm	Accuracy of the best models obtained for each algorithm	Reference
Acoustic spectroscopy	Eggshell integrity	705 (prediction not specified)	QDA	Error P = 2.1% FPR = 0.87%	Kertész et al. (2021)
Homemade acoustic system	Eggshell integrity	200 (120 C; 60 P)	LDA	Correct classification: 83.75–93.75%	Lin et al. (2018)
Acoustic resonance system	Eggshell integrity	603 (400 C; 203 P)	Critical value of 5 specific features	Correct classification: 100–96.1%	Sun et al. (2013)
Homemade acoustic system	Eggshell integrity	750 (250 C; 500 P)	Neural networks	Accuracy up to 99.2%	Wang et al. (2016)

P, prediction; FPR, false-positive rate.

of industrial online implementation of acoustic systems for eggshell integrity detection accomplishing with the conveyor line requirements.

5 Electronic nose systems

Electronic nose (e-nose) devices originated in the 1980s by Persaud and Dodd (1982) and were developed in different systems were able to detect simple and complex odours mimicking human olfactory perception (Gardner & Bartlett, 1994). The systems are commonly composed of a sampling system, a detecting system composed of an array of specific or partially specific gas sensors, and a computer for the acquisition and processing of the data. The most used detecting systems consist of metal oxide semiconductor sensors (MOS) or metal oxide semiconductor field-effect transistor sensors (MOSFET) (Grassi et al., 2023b).

An inexpensive e-nose system based on four commercial tin-oxide sensors was first used by Dutta et al. (2003). The work showed good potential for commercial exploitation even though eggs were analysed after the execution of tiny holes in the shells, thus being far from a non-destructive application.

More recently a non-destructive approach was investigated by Li et al. (2017). They placed each egg in a sealed container for 1 h at room temperature before purging the headspace gas into the electronic nose for 70 s. The regression models developed aimed at YI prediction leading to high performances (R^2_P above 0.78 and RMSEP lower than 0.03). The authors associated the correlation between e-nose signals and YI to an increased response of sensor W5S over storage, even though it is not clear how high concentration of nitrogen oxides are correlated to low YI. However, the prediction set was built with eggs bought and stored together with the eggs used to build the model. Thus, it is reasonable that the prediction reaches high performances due to a restricted variability of the analysed samples compared to the eggs possibly present in the market. A real independent test set would be advisable to assess the real robustness of the proposed approach.

A more robust sampling procedure was performed by Yongwei et al. (2019), who considered a total of 200 eggs, coming from two groups stored at 22°C or 4°C up to 24 weeks. The groups themselves originated from five independent sets (40 eggs each). Each sample was inserted in an airtight glass jar for 30 min before collecting the headspace by the e-nose system for 90 s. Subsequently, the eggs were opened and evaluated for HU and YC. The best prediction models were obtained by quadratic polynomial step regression (QPSR), with a percentage error in prediction of 5.87 and 2.89 for HU and YC, respectively.

Even if reliable results have been obtained (Table 6) e-nose systems seem not to be exportable to online process control, as they require a headspace equilibrium step before the analysis.

6 Conclusion

Parallel to the advancement of computer and electronic technology, the application and accuracy of non-destructive techniques are rising, allowing for the development of different solutions for quality evaluation in the egg industry.

Among spectroscopic techniques, VIS-NIR spectroscopy seems to be the most reliable for the evaluation of egg internal quality. The extensive application is probably due to its maturity as an analytical technique together with the accomplishment of industrial needs, such as versatility, miniaturisation, and affordable costs. Raman spectroscopy is also gaining interest, thanks to the progress in instrumentation answering the industrial need of reduced analytical time and costs. THz, dielectric, and ultrasound spectroscopy are the youngest technologies in terms of non-destructive analysis of food quality, having received great attention only in the last decades, but their application for the assessment of internal quality characteristics in shell eggs showed promising results.

Being the spectroscopic techniques able to analyse a restricted and specific area of the shell egg, a relevant advantage, i.e. spatial information, is

Table 6 Electronic nose systems applied to the evaluation of quality and freshness of shell eggs

Egg parameter	Number of samples (eggs)	Predictive algorithm	Accuracy of the best models obtained for each algorithm	Reference
Storage time	63 (35 C; 28 P)	PLSR	R^2_P = 0.9441, RMSEP = 0.887	Yimenu et al. (2017)
Haugh units		PLSR	R^2_P = 0.9511, RMSEP = 1.24	
Odor score		PLSR	R^2_P = 0.9725, RMSEP = 0.0.626	
Overall acceptability		PLSR	R^2_P = 0.9530, RMSEP = 0.629	
Yolk index	140 (112 C; 28 P)	BPANN	R^2_C = 0.8629 RMSEC = 0.227; R^2_P = 0.7863, RMSEP = 0.0286	Li et al. (2017)
		SVM	R^2_C = 0.8641 RMSEC = 0.0123; R^2_P = 0.8339, RMSEP = 0.0275	
Haugh units	200 (120 C; 80 P)	QPSR	R^2_C = 0.93 Error% = 2.55; R^2_P = 0.91, Error% = 5.37	Yongwei et al. (2019)
Yolk coefficient		QPSR	R^2_C = 0.94 Error% = 1.5; R^2_P = 0.93, Error% = 2.89	

C, calibration; P, prediction; PLSR, partial least squares regression; BPANN, back-propagation artificial neural network; SVM, support vector machine; SIMCA, soft independent modelling of class analogy; QPSR, quadratic polynomial step regression; R^2, coefficient of determination; RMSEC, root mean square error in calibration; RMSEP, root mean square error in prediction.

gained when moving to vision systems, profiting from diverse electromagnetic ranges. This principle is the basis for the success of digital vision systems based on VIS-NIR and thermal spectroscopy. The gained spatial resolution allows us to analyse the whole egg at a glance and not just a single spot, thus better predicting external quality parameters, such as eggshell integrity. Even though by acoustic spectroscopy, the most reliable results were obtained for cracked egg discrimination. Indeed, the application of an acoustic resonance system directly on a conveyor belt represents the unique industrial application reported in the literature at present. The online application seems not to be feasible when considering another popular non-destructive system, such as e-nose. Even though its applicability looks promising for non-destructive applications, the need for equilibrium phase steps limits its usage to at-line testing.

7 Future trends in research

Not all the important egg quality parameters have been properly tested by non-destructive approaches. Therefore, much work is still needed to widen possible applications. Solution-oriented devices, providing easy, rapid, and cost-effective measurements are probably the next future of non-destructive methods applications for shell egg quality and freshness evaluation, in view of both portable devices and industrial online equipment. However, adapting the developed methods to an industrial setting may lead to challenges that were not considered in most of the research results revised. Even though some of the presented methods have demonstrated high potential in other food sectors, the barriers to industrial implementation are too high and little investigated. It is envisioned to answer the industrial need for fast, cheap, and robust hardware, together with solution-oriented hardware before considering any other advance in the field. Indeed, the PAT approach had great success in the pharmaceutical industry, but it still looks 'promising' for the whole food sector, which urgently needs to move to the concept of Industry 4.0. The achievement of such a goal requires the collaboration of researchers working in the same field together with an inclination to technological transfer.

8 Where to look for further information

In case the reader is interested in further information about the principle and application in the food field of the mentioned non-destructive techniques, it is possible to look for information about Raman spectroscopy (Wang et al., 2021), VIS-NIR spectroscopy (Williams & Norris, 2001; Workman & Weyer, 2008), teraherz spectroscopy (Khushbu et al., 2022), dieletric spectroscopy (Routray & Orsat, 2018), vision systems (Patel et al., 2012; Amigo, 2019), ultrasound spectroscopy (Laugier & Hayat, 2011), acoustic spectroscopy (Aboonajmi et al., 2015b), and e-nose (Grassi et al., 2023b).

Being interested in PAT applications in the food sector have a look to Grassi and Alamprese (2018) and van den Berg et al. (2013).

Lastly, if interested in knowing more about chemometric methods in the food sector please consider the book by Marini (2013).

9 References

Aboonajmi, M., et al. (2010). An ultrasound based technique for the determination of poultry egg quality. *Research in Agriculture Engineering*, *56*(10), 26–32.

Aboonajmi, M., et al. (2015a). A review on application of acoustic analysis in quality evaluation of agro-food products. *Journal of Food Processing and Preservation*, *39*, 3175–3188.

Aboonajmi, M., et al. (2015b). A review on application of acoustic analysis in quality evaluation of agro-food products. *Journal of Food Processing and Preservation*, *39*, 3175–3188.

Aboonajmi, M., et al. (2016). Quality assessment of poultry egg based on visible–near infrared spectroscopy and radial basis function networks. *International Journal of Food Properties*, *19*(5), 1163–1172. https://doi.org/10.1080/10942912.2015.1075215

Akbarzadeh, N., et al. (2019). Microwave spectroscopy based on the waveguide technique for the nondestructive freshness evaluation of egg. *Food Chemistry*, *277*, 558–565. https://doi.org/10.1016/j.foodchem.2018.10.143

Akowuah, T. O. S., et al. (2020). Rapid and nondestructive determination of egg freshness category and marked date of lay using spectral fingerprint. *Journal of Spectroscopy*, *2020*, 1–11. https://doi.org/10.1155/2020/8838542

Alikhanov, J., et al. (2019). Design and performance of an automatic egg sorting system based on computer vision. *TEM Journal*, *8*(4), 1319–1325. https://doi.org/10.18421/TEM84-31.

Amigo, J. M. (2019). *Hyperspectral imaging*. New York: Elsevier. ISBN: 9780444639776

Commission Regulation (EC) No 589/2008, 163 OJ L (2008). http://data.europa.eu/eli/reg/2008/589/oj/eng

Coronel-Reyes, J., et al. (2018). Determination of egg storage time at room temperature using a low-cost NIR spectrometer and machine learning techniques. *Computers and Electronics in Agriculture*, *145*, 1–10. https://doi.org/10.1016/j.compag.2017.12.030

Cozzolino, D., et al. (2024). Predicting egg storage time with a portable near-infrared instrument: effects of temperature and production system. *Foods*, *13*(2), 212.

Dai, D., et al. (2020). Nondestructive detection for egg freshness based on hyperspectral scattering image combined with ensemble learning. *Sensors*, *20*, 5484.

Davari, M., et al. (2024). Developing a non-destructive method for the detection of egg quality and freshness using micro-Raman spectroscopy. *Applied Food Research*, *4*, 100453. https://doi.org/10.1016/j.afres.2024.100453

Dong, X., et al. (2017a). Comparative study of albumen pH and whole egg pH for the evaluation of egg freshness. *Spectroscopy Letters*, *50*(9), 463–469. https://doi.org/10.1080/00387010.2017.1360357

Dong, X., et al. (2017b). *Nondestructive assessment of eggshell thickness by VIS/NIR spectroscopy*. St. Joseph, MI: ASABE. https://doi.org/10.13031/aim. 201700888. ASABE Paper No 1700888

Dong, X., et al. (2018a). Nondestructive egg freshness assessment from the equatorial and blunt region based on visible near infrared spectroscopy. *Spectroscopy Letters*, *51*(10), 540–546. https://doi.org/10.1080/00387010.2018.1525409

Dong, X., et al. (2018b). Nondestructive egg freshness assessment of air chamber diameter by VIS-NIR spectroscopy. *2018 Detroit, Michigan July 29 - August 1, 2018*. 2018 Detroit, Michigan July 29 – August 1, 2018. https://doi.org/10.13031/aim.201801022

Dong, X., et al. (2019). Maintaining the predictive abilities of egg freshness models on new variety based on VIS-NIR spectroscopy technique. *Computers and Electronics in Agriculture*, *156*, 669–676. https://doi.org/10.1016/j.compag.2018.12.012

Dutta, R., et al. (2003). Non-destructive egg freshness determination: an electronic nose based approach. *Measurement Science and Technology*, 14(2), 190.

Drewnowski, A. (2010). The nutrient rich foods index helps to identify healthy, affordable foods. *The American Journal of Clinical Nutrition*, *91*(4), 1095S–1101S. https://doi.org/10.3945/ajcn.2010.28450D

Freni, F., et al. (2018). Assessment of eggs freshness by means of pulsed infrared thermography. *Proceedings of the 14th quantitative infrared thermography Conference* (pp. 632–639). QIRT COUNCIL. Quantitative Infrared Thermography. https://doi.org/10.21611/qirt.2018.068

Gardner, J. W. & Bartlett, P. N. (1994). A brief history of electronic noses. *Sensors and Actuators B: Chemical*, *18*, 210–211. https://doi.org/10.1016/0925-4005(94)87085-3

Grassi, S. & Alamprese, C. (2018). Advances in NIR spectroscopy applied to process analytical technology in food industries. *Current Opinion in Food Science*, 22, 17–21. https://doi.org/10.1016/j.cofs.2017.12.008

Grassi, S., et al. (2023a) How chemometrics can fight milk adulteration. *Foods*, *12*, 139. https://doi.org/10.3390/foods12010139

Grassi, S., et al. (2023b). E-sensing systems for shelf life evaluation: a review on applications to fresh food of animal origin. *Food Packaging and Shelf Life*, *40*, 101221. https://doi.org/10.1016/j.fpsl.2023.101221

Han, X., et al. (2022). Study on egg sorting model based on visible-near infrared spectroscopy. *Systems Science & Control Engineering*, *10*(1), 733–741.

Ibrahim, R., et al. (2012). Eggs grade classification and dirt inspection using image processing techniques. *Proceedings of the World Congress on Engineering*, Vol. 2.

Kemps, B. J., et al. (2007). Albumen freshness assessment by combining visible near-infrared transmission and low-resolution proton nuclear magnetic resonance spectroscopy. *Poultry Science*, *86*, 752–759.

Kertész, I., et al. (2021). Development of a novel acoustic spectroscopy method for detection of eggshell cracks. *Molecules*, *26*(15), 4693.

Khaliduzzaman, A., et al. (2020). A Nondestructive eggshell thickness measurement technique using terahertz waves. *Science and Reports*, *10*, 1052.

Khushbu, S., et al. (2022). Recent advances in terahertz time-domain spectroscopy and imaging techniques for automation in agriculture and food sector. *Food Analytical Methods*, *15*, 498–526.https://doi.org/10.1007/s12161-021-02132-y

Kuroki, S., et al. (2017). Nondestructive measurement of yolk viscosity in lightly heated chicken shell eggs. *Journal of Food Engineering*, *205*, 18–24. https://doi.org/10.1016/j.jfoodeng.2017.02.030

Laugier, P. & Hayat, G. (2011). *Introduction to the physics of ultrasound*. Bone Quantitative Ultrasound Ed. 468pp. Dordrecht: Springer.

Li, X. C., et al. (2019). Non-destructive testing method of egg quality based on machine vision. *Journal of Food Safety and Quality*, *10*(2), 489–493.

Li, J., et al. (2017). Prediction of egg storage time and yolk index based on electronic nose combined with chemometric methods. *LWT-Food Science and Technology*, *82*, 369–376.

Lin, H., et al. (2011). Freshness measurement of eggs using near infrared (NIR) spectroscopy and multivariate data analysis. *Innovative Food Science Emerging Technology*, *12*(2), 182–186.

Lin, H., et al. (2018). Identification of eggshell crack using multiple vibration sensors and correlative information analysis. *Journal of Food Process Engineering*, *41*(8), e12894.

Liu, Y., et al. (2020). Non-destructive and online egg freshness assessment from the egg shell based on Raman spectroscopy. *Food Control*, *118*, 107426. https://doi.org/10.1016/j.foodcont.2020.107426

Lunadei, L., et al. (2012). Automatic identification of defects on eggshell through a multispectral vision system. *Food and Bioprocess Technology*, *5*(8), 3042–3080. https://doi.org/10.1007/s11947-011-0672-x

Marini, F. (2013). *Chemometrics in food chemistry*. New York: Elsevier. ISBN: 9780444595287

Møller, U., et al. (2009). Terahertz reflection spectroscopy of Debye relaxation in polar liquids. *Journal of the Optical Society of America B: Optical Physics*, *26*, A113.

Nakaguchi, V. M. & Ahamed, T. (2022). Fast and non-destructive quail egg freshness assessment using a thermal camera and deep learning-based air cell detection algorithms for the revalidation of the expiration date of eggs. *Sensors*, *22*(20), 7703. https://doi.org/10.3390/s22207703

Pal, G. K., et al. (2017). Inclusion of poultry based food ingredients in the formulation of noodles and their effects on noodle quality characteristics. *Journal of Food Measurement and Characterization*, *11*(3), 939–947. https://doi.org/10.1007/s11694-017-9467-8

Patel, K. K., et al. (2012). Machine vision system: a tool for quality inspection of food and agricultural products. *Food Science and Technology*, *49*(2), 121–141.

Persaud, K. & Dodd, G. (1982). Analysis of discrimination mechanisms in the mammalian olfactory system using a model nose. *Nature*, *299*(*5881*), 352–355. https://doi.org/10.1038/299352a0

Priyadumkol, J., et al. (2017). Crack detection on unwashed eggs using image processing. *Journal of Food Engineering*, *209*, 76–82. https://doi.org/10.1016/j.jfoodeng.2017.04.015

Routray, W. & Orsat, V. (2018). Recent advances in dielectric properties–measurements and importance. *Current Opinion in Food Science*, *23*, 120–126. https://doi.org/10.1016/j.cofs.2018.10.001

Sharaf Eddin, A., et al. (2019). Egg quality and safety with an overview of edible coating application for egg preservation. *Food Chemistry*, *296*, 29–39. https://doi.org/10.1016/j.foodchem.2019.05.182

Soltani, M. & Omid, M. (2015). Detection of poultry egg freshness by dielectric spectroscopy and machine learning techniques. *LWT - Food Science and Technology*, *62*(2), 1034–1042. https://doi.org/10.1016/j.lwt.2015.02.019

Soltani, M., et al. (2015). Egg quality prediction using dielectric and visual properties based on artificial neural network. *Food Analytical Methods*, *8*(3), 710–717. https://doi.org/10.1007/s12161-014-9948-x

Stadelman, W. J. & Cotterill, O. J. (1995). *Egg science and technology* (4th ed.). Binghamton: The Haworth Press, Inc. https://doi.org/10.1201/9780203758878

Suktanarak, S., & Teerachaichayut, S. (2017). Non-destructive quality assessment of hens' eggs using hyperspectral images. *Journal of Food Engineering*, *215*, 97–103.

Sun, L., et al. (2013). On-line detection of eggshell crack based on acoustic resonance analysis. *Journal of Food Engineering*, *116*(1), 240–245.

van den Berg, F., et al. (2013). Process Analytical Technology in the food industry. *Trends in Food Science and Technology*, *31*, 27–35.

Vasileva, A. V., et al. (2018). Assessing exterior egg quality indicators using machine vision. *British Poultry Science*, *59*(6), 636–645. https://doi.org/10.1080/00071668.2018.1523535

Wang, H., et al. (2016). Acoustic feature extraction and optimization of crack detection for eggshell. *Journal of Food Engineering*, *171*, 240–247. https://doi.org/10.1016/j.jfoodeng.2015.10.030

Wang, K., et al. (2021). Raman spectroscopic techniques for nondestructive analysis of agri-foods: a state-of-the-art review. *Trends in Food Science & Technology*, *118*, 490–504.

Williams, P. & Norris, K. (2001). *Near infrared technology in the agricultural and food industries*. Saint Paul, USA: ACCC.

Workman, J. & Weyer, L. (2008). *Practical guide to interpretative near-infrared spectroscopy*. Boca Raton: CRC Press.

Wu, L., et al. (2018). Detection of crack eggs by image processing and soft-margin support vector machine. *Journal of Computational Methods in Sciences and Engineering*, *18*, 21–31. https://doi.org/10.3233/JCM-170767

Yao, K., et al. (2022). Non-destructive detection of egg qualities based on hyperspectral imaging. *Journal of Food Engineering*, *325*, 111024.

Yimenu, S. M., Kim, J. Y., & Kim, B. S. (2017). Prediction of egg freshness during storage using electronic nose. *Poultry science*, *96*(10), 3733–3746.

Yongwei, W., et al. (2019). Monitoring storage time and quality attribute of egg based on electronic nose. *Analytica Chimica Acta*, *650*, 183–188.

Zhang, W., et al. (2015). Non-destructive internal quality assessment of eggs using a synthesis of hyperspectral imaging and multivariate analysis. *Journal of Food Engineering*, *157*, 41–48. https://doi.org/10.1016/j.jfoodeng.2015.02.013

Chapter 10

Advances in understanding eggshell formation and its composition in table eggs

Joël Gautron, INRAE – UMR BOA – Université de Tours, France; and Alejandro Rodiguez-Navarro, Universidad de Granada, Spain

1 Introduction

The natural role of an egg is to be an enclosed self-sufficient chamber for the development of a chick. This is why the egg has so many components (> 1000) with a wide range of biological properties, allowing life to emerge (Gautron et al., 2019). When an embryo develops, all its constituents are used to create a chick. There will be no opportunity to bring in nutrients or eliminate waste. This is also why egg molecules possess properties in terms of digestibility and balance of these constituents that make an egg a perfectly balanced food rich in proteins and vitamins with high nutritional value and that is widely consumed throughout the world (Réhault-Godbert et al., 2019).

Eggshell is the only non-consumable part of an egg, but its quality is crucial for the marketing of the egg. The properties of the eggshell contribute to the chick's development by providing physical and antimicrobial protection, as well as thermal protection and allowing gas exchange with the atmosphere.

http://dx.doi.org/10.19103/AS.2024.0150.14

It is also a source of calcium for the developing embryo's skeleton. For table eggs, shell quality has a major impact on the economics of this agri-food sector. Downgraded eggs due to dirty, cracked, or broken shells cause substantial economic losses to producers. Its role as a physical barrier that protects the internal part of the egg is essential to avoid bacterial penetration to maintain the internal part free of pathogens and safe for its consumption by humans.

Shell quality is also of paramount importance in the current context of major changes in egg production methods. Indeed, over the past 20 years, there has been a strong consumer demand to take welfare into account in animal production, including table eggs (Gautron et al., 2021b). This is particularly true in Europe and North America but increasingly around the world. As a consequence, the production modes have evolved from cage systems to floor systems and possibly to free range with access to outdoor pasture. The latter alternative systems have to maintain the egg quality and particularly its eggshells properties (Gautron et al., 2021b). Similarly, to limit the number of animals and increase the sustainability of the consumer egg production system, laying hens have been selected to increase their persistence and the number of eggs produced. Breeders have selected strains of laying hens capable of laying sustainability up to 100 weeks of age for a total production of almost 500 eggs per birds (Bain et al., 2016). Consequently, the extension of the production cycle for layers faces the challenge of maintaining a good egg quality in older hens. For all these aspects, shell quality, meaning shell mechanical properties, are most relevant for marketing eggs.

The hen eggshell is the outermost layer of the egg. This mineralized structure plays a protective role against external physical and bacterial threats. It is a composite biomineral highly ordered that give them its strong and remarkable mechanical properties. Its ability to withstand static pressure of over 4 kg on its equator is all the more remarkable given its small thickness of just 0.35–0.4 mm. Its crystalline structure results from the rapid mineral deposition (<20 hours) with a highly ordered structure forming at ambient conditions. It is therefore one of the fastest biomineralization processes in nature. When intact, the shell prevents bacterial ingress but still is permeable to water and gases essential for embryo development.

In this chapter, we will review the recent knowledges on the mechanisms on shell formation. We will report on the structure and mineralogy of avian eggshells, with a focus on mechanisms and molecular factors that are involved in supplying the large quantity of ions necessary for its mineralization. Moreover, we will provide an update on this biomineralization process. Finally, we will describe how this knowledge can be used to maintain and improve shell quality in this rapidly changing production system.

2 The egg development process

The egg is an oocyte surrounded by nutritional reserves and a protective envelope (the shell) where the embryo will develop without maternal contact. The organs responsible for egg synthesis of the laying hen comprises the liver that synthetizes yolk precursors, the ovary that supports steroid synthesis, gametogenesis, folliculogenesis, and the synthesis of the inner layer of the vitelline membrane and the oviduct that sequentially deposits the remaining egg components (outer vitelline membrane, egg white (albumen), eggshell membranes, and the eggshell) around the ovulated yolk. The formation of the egg is characterized by two phases of different duration. The process starts with a long phase beginning from the embryonic life with the oogenesis till the development of mature follicles on the ovary. This folliculogenesis is supported by the liver, which produces the majority of yolk components, as precursors. Once mature, the yolk (mature oocyte) is ovulated to be captured by the oviduct in the infundibulum. This step is the start of a rapid phase of deposition of the other constituents in the oviduct (Nys and Guyot, 2011; Sauveur and De Reviers, 1988). This oviductal phase lasts approximately 24 h and corresponds to the well-known adage 'an egg per day'.

The laying oviduct is characterized by five different regions, namely the infundibulum (deposition of the outer layer of the vitelline membrane and site of fertilization), the magnum (deposition of egg white components), the isthmus (formation of eggshell membranes), the uterus (formation of the calcified eggshell and cuticle), and the vagina (passage of the egg before oviposition, without deposition of supplementary constituents). The yolk is travelling down the oviduct passing through theses functionally distinct regions and acquires the outer vitelline membranes in the infundibulum (15 min), the egg white in the magnum (3.5 h), and the eggshell membrane in the isthmus (1 h). Approximately, 5 h after the ovulation, the forming egg will enter the red isthmus (tubular shell gland) and uterus (shell gland pouch) where the egg will be mineralized in a process lasting about 19 h. Organic mammillary knobs are deposited on the surface of the outer eggshell membranes. These organic cores are the seeding sites in which calcium carbonate mineralization is initiated. When the egg enters the uterus, it acquires its final ovoid shape by hydration of albumen and the shell will be mineralized in the lumen of the uterus. The egg bathes in an acellular milieu (uterine fluid) and about 5–6 g of mineral will be deposited to form the eggshell during a sequential process in which different phases are identified, i.e. the initiation of mineralization (5–10 h post-ovulation, p.o.), rapid growth phase (10–22 h p.o.), and terminal phase (22 h p.o. to oviposition) (Gautron et al., 2021b; Hincke et al., 2012; Nys et al., 1991, 2004). All these phases are detailed below in the chapter.

3 Eggshell structure and composition

All avian shells are mainly composed of calcium carbonate mineral in the form of calcite, one of the crystalline polymorphs of calcium carbonate which is most stable at room temperature. Most bird species share the same structure for their eggshell (columnar layer, calcite mineral composition) though there are variations among species regarding their microstructure (Choi et al., 2023; Rodríguez-Navarro et al., 2024). The most studied species is the chicken eggshell, because of its interest as a foodstuff. We will therefore concentrate on the description of the eggshell of this species in this manuscript.

The avian eggshell is a thin mineral layer (about 350–400 µm thick in chicken) made of calcite crystals permeated with a small amount of organic matter composed of five different layers (Fig. 1):

- Shell membranes are the innermost layers of the shell. There are two of them, made entirely of organic matter. The inner shell membrane, 20 µm thick, is in contact with the albumen. It is from the outer shell membrane (50 µm) that mineralization of the shell begins.
- The mammillary layer (approx. 70 µm) is the innermost part of the calcified layer. Its base is made up of mamillary knobs, which are organic clusters

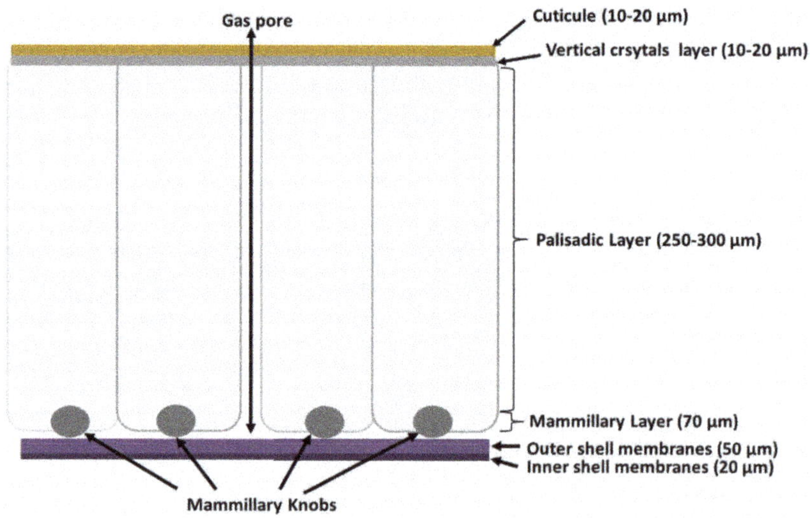

Figure 1 Scheme of a transversal layer of the chicken eggshell structure representing the two shell membranes, the mammillary layer with their organic knobs, the palisade layer made of columnar units, the vertical crystal layer, and the organic cuticle (Kulshreshtha et al., 2022). The cuticle of chicken eggs is a relatively thin layer with variable thickness (0.5–12 mm), with patchy and uneven distribution on the eggshell surface (Ketta and Tůmová, 2016; Rodríguez-Navarro et al., 2013; Samiullah and Roberts, 2014). Gas pores are formed when there is an absence of fusion of adjacent cones.

deposited on the surface of the outer shell membrane and from which mineralization is initiated. The eggshell mineral part is attached to the eggshell membrane's fibers by the tips of the mammillary cones which are distributed pseudo-periodically (Fig. 1). From these sites, the mammillary cones fan out and meet at the base of the palisade layer forming calcite columnar units arranged vertically.

- The palisade layer begins after the spherulitic growth of the mamillary layer cones when adjacent cones fuse and columnar units grow vertically. The palisade layer thus corresponds to a compact layer of minerals associated with an organic framework. The palisade layer comprises most of the thickness of the eggshell mineral part and has the greatest contribution to eggshell mechanical properties. This continuity is broken at the level of the pores that traverse the shell from the inner to the outer side to allow the gaseous exchanges necessary for the embryo's development.
- The palisade layer ends at the vertical crystal layer, which is a thin layer of small and thin calcite crystals arranged vertically.
- Finally, the outer surface of the eggshell is coated by the cuticle, a very thin (a few microns thick) organic layer that plugs the pores to control shell permeability and prevent bacterial ingress through the shell (Bain et al., 2013). Thus, the cuticle, if present, is an effective barrier against bacteria penetration and is of great importance for the food safety of eggs. Gas exchange is made possible by cracks in the dried cuticle.

The ultra and microstructure characteristics of the eggshell can be studied in detail by scanning electron (SEM) and optical microscopy (Fig. 2). In cross-section, the eggshell mineral region is dense with the palisade layer columnar unit arranged vertically. At the base of the palisade layer are the mammillary cones firmly attached to the outer eggshell membrane fibers (Fig. 2b). The cuticle deposited on the eggshell surface is very thin (a few microns thick) and often has an irregular patchy coverage. It has been shown that the amount of the cuticle changes with hen age, normally decreasing at the end of the laying cycle. SEM reveals an eggshell outer surface that is coated by a thin organic layer (cuticle) that shows cracks from drying and that is rich in proteins and polysaccharides as shown by infrared spectroscopy (FTIR) (Fig. 2c, d). Thin sections of eggshell under an optical microscopy demonstrate an eggshell mineral layer formed by columnar calcite single crystal units arranged vertically (palisades), about 70–80 μm wide, extending across the shell thickness (Fig. 2f). Calcite crystals show varying degrees of light extinction under cross-polarized light illumination due to differences in their crystallographic orientation. They also possess significant amounts of occluded organic matter distributed across the eggshell thickness with a greater concentration at the mammillary layer, as can be observed under parallel light illumination (Fig. 2e).

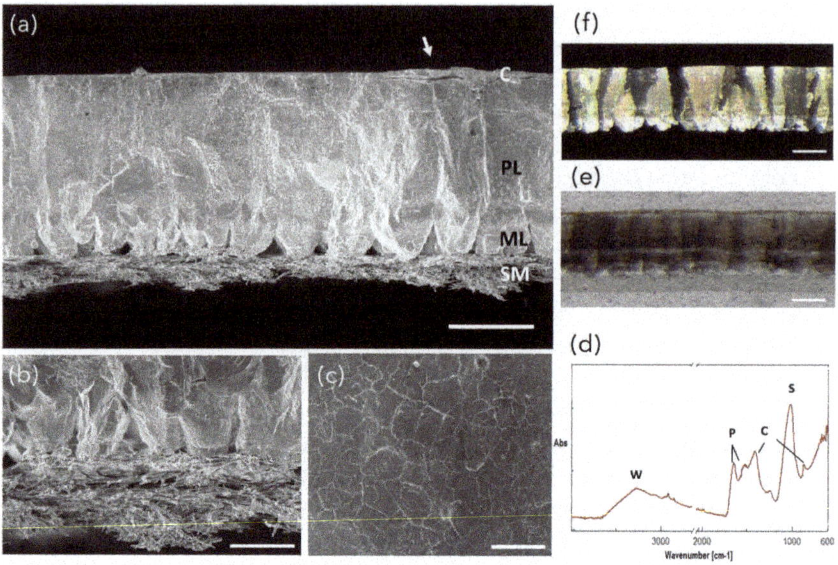

Figure 2 Eggshell ultrastructure and microstructure. (a) SEM images of the cross-section of the eggshell showing the shell membranes (SM), mammillary layer (ML), palisade layer (PL), and the cuticle (c), with a pore plug (arrow); (b) close-up image of the shell membranes and mamillary knobs attached to the outer membrane; (c) image of the outer eggshell surface coated with the cuticle showing cracks from drying; (d) ATR-FTIR spectra of the outer eggshell surface showing infrared absorption peaks from cuticle proteins (P), water (W) and polysaccharides (S), and carbonate (C) from the underlaying mineral; (e) optical microscopy images of an eggshell cross-section viewed under parallel light showing the distribution of the internal organic matter within the shell mineral; (f) the same sample observed under cross-polarized illumination, showing the columnar calcite crystal units of the mineral part. Scale bars are (A, C, E, F) 100 μm. (B) 50 μm.

In *Gallus gallus* eggshell, the mineral part constitutes 95% of the total part of the shell. Calcium carbonate on calcitic form accounts for 98.4% of shell mineral. Other minerals present in small quantities are phosphate mainly present in the outer structure and trace elements such as magnesium, copper, and zinc. There is no storage of mineral precursors in the uterus prior to mineralization. Calcium and bicarbonate ions are therefore taken from the blood and transferred by the uterus into the uterine fluid that constantly bathes the egg during its formation. This fluid contains all the mineral and organic constituents needed to form the eggshell, in the form of precursors. The concentration of ionic calcium in uterine fluid ranges from 5.5 mM to 9.5 mM, depending on the egg's stage of formation. The concentration of bicarbonates varies between 110 mM and 90 mM (Nys et al., 1991). In this fluid, the ionic product of calcium carbonate is much higher than the solubility product of calcium carbonate (permanently exceeded by more than 60 times), allowing calcium carbonate to precipitate (Nys et al., 1991).

The shell also contains 3.5% organic matter (mainly proteins and complex sugars), which is unevenly distributed between the different layers composing the eggshell. In fact, the innermost layers (shell membranes) and the outermost layer (cuticle) are exclusively made up of organic matter. The calcified parts of the shell contain 2–2.5% protein and proteoglycans constituents (Panheleux et al., 2000). The organic matrix of the hen's eggshell plays a fundamental role in the manufacture of the shell and therefore in establishing its mechanical properties. It interacts with the forming calcium carbonate to produce crystalline modifications that explain the shell's remarkable mechanical properties. Finally, it also contains 1.6% water, making it a particularly dehydrated structure. Over the last decades, our knowledge of its production has progressed considerably, thanks to the identification of the constituents of its organic matrix and the demonstration that these influence its crystalline texture and, consequently, the mechanical properties of this biomaterial built on a membrane support. These complex mechanisms involved the supply of minerals necessary for shell formation and the role of amorphous calcium carbonate (ACC) that will be detailed below in the manuscript.

4 Origin and regulation of calcium and carbonates in the shell

4.1 Calcium metabolism, a challenge for laying hen

A laying hen lays around 320 eggs a year, with a shell weight of 5–6 g, giving a cumulative mass of 1600–1920 g of shell per production cycle. This is a considerable quantity and corresponds overall to the weight of the hen. Bearing in mind that the shell is 40% calcium, no less than 640 g of calcium must be transferred from the feed to the uterus and uterine lumen during the egg production cycle to form the shell. This is a considerable metabolic effort for birds. It involves mechanisms and regulation of calcium homeostasis that are highly exacerbated when compared to mammals. This effort is all the more important because there is a desynchronization between calcium requirements and shell mineralization. The hen has access to calcium-rich feed during the day, while the active shell mineralization takes place from 10 h to 12 h p.o. in the middle of the night. As a result, there is insufficient intestinal calcium during the night to meet the requirements for shell formation and, some of the calcium needed for shell formation comes from the hen's long bones, which provide up to 20–40% of the shell's calcium (Nys et al., 2022; Sinclair-Black et al., 2023). In fact, medullary bone resorption and subsequent formation is in synchrony with eggshell formation. Thus, there is an intense osteoclast-mediated resorption of medullary bone followed by intense osteoblastic activity to form new medullary bone before the next cycle of eggshell formation (Nys et al., 2022; Nys and Le Roy, 2018). In order to ensure this function, several organs have a role in the maintenance of calcium homeostasis in the hen, allowing for the mineralization of the egg (Fig. 3).

Regulation of calcium metabolism in laying hens will be detailed elsewhere in this book. In short, calcium comes from food and is absorbed by the gut for transfer to the bloodstream. Part of this calcium from the intestine is transferred to the uterus for mineralization of the shell. The other part is transferred to the bone, and more specifically to the medullary part, which is daily mobilized in birds to supply calcium for shell mineralization. These absorption and transfer mechanisms are dependent on vitamin D, specifically its active metabolite, cholecalciferol or 1,25 $(OH)_2D_3$. Vitamin D is mainly supplied by the hen's diet. It undergoes initial hydroxylation to $25OHD_3$ in the liver. A second hydroxylation to the active vitamin D metabolite (1,25 $(OH)_2D_3$) takes place in the kidney. This hydroxylation is mediated by parathyroid hormone secreted by the parathyroid glands. 1,25 $(OH)_2D_3$ then regulates calcium absorption in the intestine and bones. Its role in uterine calcium absorption is controversial, but the identification of vitamin D response elements in the regulatory zones of genes involved in uterine calcium homeostasis (Nys and Le Roy, 2018), as well as their differential expression depending on the form of vitamin D (Gautron et al., 2022a), and the regulation of genes in response to vitamin D in the shell gland (Sinclair-Black et al., 2024) suggest that the uterus is also sensitive to the active metabolite of vitamin D.

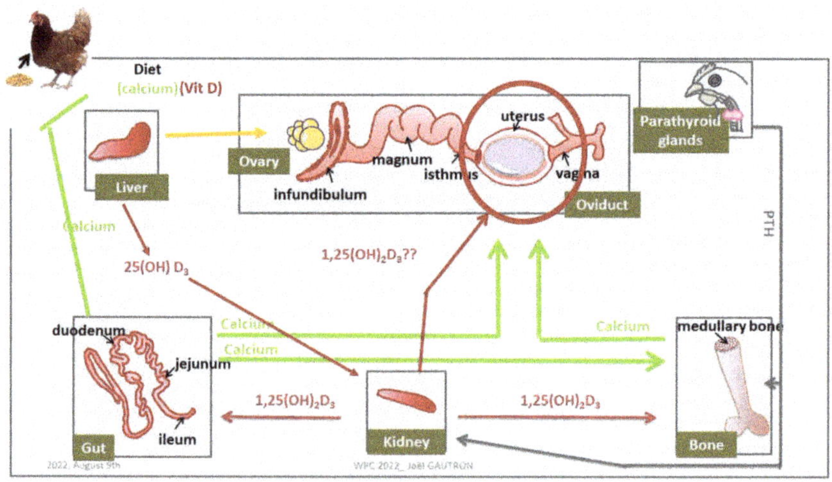

Figure 3 General diagram of calcium exchange mechanisms in laying hen during shell mineralization. This mechanism involves intestinal absorption of both calcium and vitamin D. The latter is hydroxylated to its active metabolite 1,25(OH)2D3 in the liver and kidneys under the control of parathyroid hormone. Calcium is partially stored in the medullary bone before being redistributed to the uterus to allow shell formation during the nocturnal phase.

4.2 Transfer of ions in the mineralizing milieu

The shell is formed in the lumen of the uterus, where the developing egg is immersed in an acellular fluid, the uterine fluid, which contains all the mineral and organic components necessary for shell formation in the form of soluble precursors (Gautron et al., 1997, Nys et al., 1991). Daily shell formation is associated with a massive transfer of calcium and carbonates during the 17–18 h of mineralization. A unique feature of ion transfer in laying hens is its discontinuity. In fact, shell mineralization occurs only over a limited period of time, from 5 h to 22 h p.o. . From 5 h to 10 h p.o., the mineralization rate is approximately 0.1 g/h (Marie et al., 2015), increasing to 0.33 g/h from 10 h to 22 h p.o.. This corresponds to 40 mg and 120 mg of calcium deposited per hour, respectively. This discontinuous calcium transfer occurs against an unfavorable concentration gradient. In fact, the ionic calcium concentration in the blood is 1.2 mM, whereas it ranges from 6 mM to 10 mM in the uterine fluid, depending on the stage of calcification (Nys et al., 1991). Calcium and bicarbonates constituting the eggshell calcite are transferred from blood to the uterine fluid following a three main step process: the entry in the uterine cells, the transfer through the cytosol, and finally the extrusion in the uterine fluid. It involved transcellular, vesicular, and paracellular pathways (Fig. 4).

4.2.1 Transcellular pathway

Active transcellular mineral transport via the transcellular pathway is the most studied because its activity must be high to allow calcium transport against an unfavorable gradient. This active transfer has been reviewed by numerous authors (Bar, 2009; Gautron et al., 2021b; Nys et al., 2022; Nys and Le Roy, 2018). It uses ion channels and exchangers not limited to calcium and bicarbonates as other ions (Na^+, K^+, H^+, Cl^-) are used to maintain a low level of calcium in the uterine cell to keep it viable.

Calcium entry into the uterine cell has been attributed to transient receptor potential cation channels (TRPVs). TRPV6 has been identified by western blot (Yang et al., 2013) and microarray methods (Jonchere et al., 2012). However, more recent microarray studies did not confirm the presence of TRPV6, but suggested other members of this family, in this case TRPV2, 3 or 4 (Brionne et al., 2014; Zhang et al., 2019). Otopetrin 2 (OTOP2) and ATPase secretory pathway Ca^{2+} pumps (ATP2C2) were also identified as potential actors of the Ca^{2+} plasma transfer into uterine cells (Nys and Le Roy, 2018; Sah et al., 2018). Clearly, further studies are needed to identify the channels that allow active calcium entry into the uterine cell. Once inside the cell, intracellular calcium levels must remain low to maintain uterine cell viability. In addition, calcium must be transported for subsequent excretion into the uterine lumen.

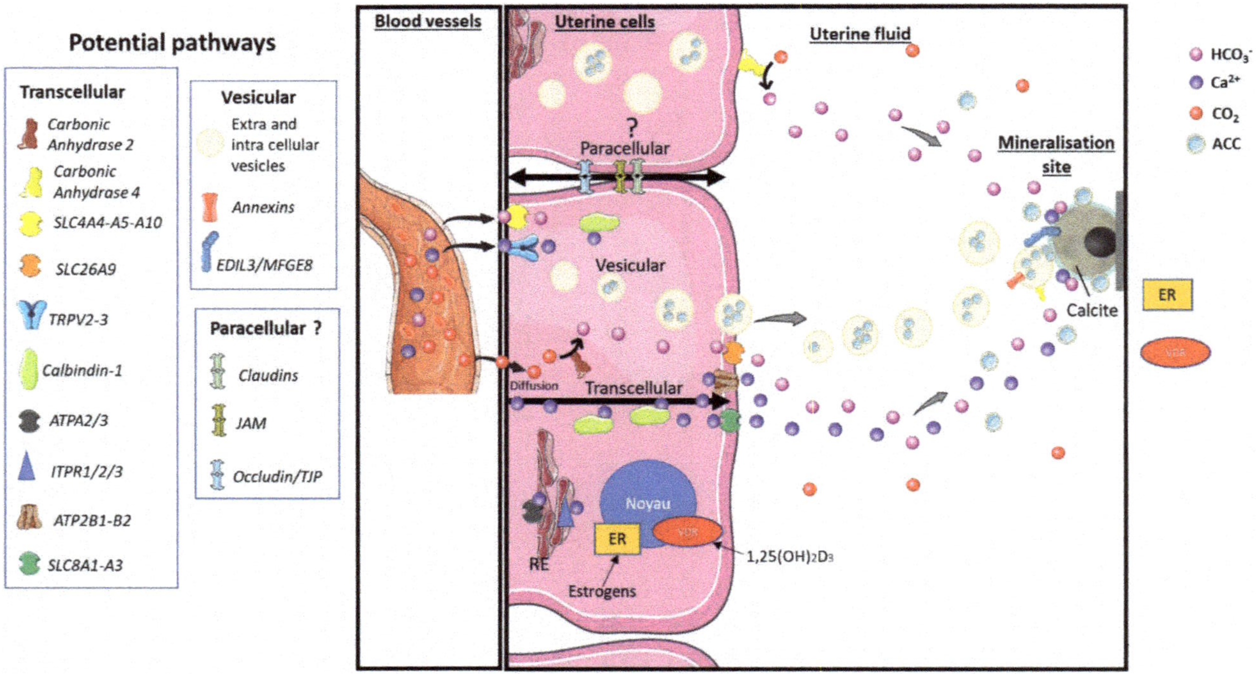

Figure 4 Transfer of ions from blood to the uterine cells to allow shell mineralization in the uterine fluid. The ions exchange involved an active transcellular as well as paracellular and vesicular pathways. Source: Modified and adapted from: (Gautron et al., 2021a; Stapane, 2019).

Calbindin (CaBP28k) has been extensively studied and is strongly stimulated during the active phase of shell calcification to buffer and transport the intracellular calcium (Bar, 2009; Gautron et al., 2021b; Nys et al., 2022; Nys and Le Roy, 2018; Sinclair-Black et al., 2024). Other Ca^{2+} pumps (ATP2A2/3 and ITPR1/2/3) associated with the reticulum endoplasmic could also be involved in the maintenance of a low level of calcium and its transfer within the cells (Gautron et al., 2021b; Nys et al., 2022; Nys and Le Roy, 2018; Sah et al., 2018). Otopetrin gene (*OTOP2*) is also expressed in the uterus to participate to the transepithelial transport of Ca^{2+} (Khan et al., 2019; Sah et al., 2018). A recent study using RNA-Seq identify *EDN2*, *IKZF1*, *PRKCB*, and *UBASH3B* as genes involved at the cellular Ca ion homeostasis and positive regulation of cytosolic Ca ion concentration (Fu et al., 2024). Calcium is extruded from the uterine glandular cell through the membrane by calcium pumps (ATP2B1/2) and Ca^{2+}/Na^{+} exchangers (SLC8A1 and A3) (Brionne et al., 2014; Jonchere et al., 2012; Sah et al., 2018).

The second component of shell calcite, in addition to calcium, is bicarbonate, which must also be present in sufficient amounts in the uterine fluid to allow mineralization of the shell. The main source of bicarbonates is carbon dioxide (CO_2), which diffuses passively through the plasma membrane into the uterine cell (Nys et al., 2022). Type 2 carbonic anhydrase is present in the uterine cell where it can catalyze the conversion of CO_2 to bicarbonates (HCO_3^-). Bicarbonates also actively enter the uterine cell via SLC-type sodium bicarbonate cotransporters (SLCA4-A5 and A-10). Finally, bicarbonates are secreted into the external environment (uterine fluid) via an HCO3$^-$/Cl^- exchanger (SLC26A9) (Brionne et al., 2014; Jonchere et al., 2012; Nys et al., 2022). Also noteworthy is the presence of type 4 carbonic anhydrase, which is membrane-bound and may allow the conversion of dissolved CO_2 into bicarbonate in uterine fluid (Nys et al., 2022). These transfers of calcium and bicarbonate involve exchanges with other ions (Na^+, Cl^-, K+). Cellular homeostasis is maintained by reabsorption of sodium and chloride from the uterine fluid and concomitant secretion of potassium. These mechanisms involve numerous channels and ion exchangers (Brionne et al., 2014; Jonchere et al., 2012).

Eggshells are formed by the following chemical reaction: $Ca^{++} + HCO_3^- \rightarrow CaCO_3 + H^+$. The result of this reaction is the production of protons that acidify the uterine fluid. These protons must be eliminated by active transfer via the uterine gland cells into the plasma. This mechanism is thought to involve two subunits of the V-type H^+ ATPase (ATPB6V1B2 and ATPB6V1C2) (Brionne et al., 2014; Jonchere et al., 2012). Eggshell formation is thus accompanied by metabolic acidosis in the hen, which is partially compensated by hyperventilation of the animal (Nys et al., 2022; Nys and Le Roy, 2018).

4.2.2 Paracellular pathway

While the transcellular pathway is well documented for the hen's glandular uterine cell, this is not the case for the paracellular pathway. Its role remains controversial. The calcium gradient (6–10 mM in uterine fluid vs. 1.2 mM in plasma) does not favor diffusion of this ion between cells. In contrast, calcium flux is favorable in the intestine, where this paracellular transfer is well described. It involves claudins (CLDN), occludins (OCN), junctional adhesion molecules (JAP), and tight junction proteins (TJP) (Gloux et al., 2019). Uterine RNA-Seq studies have also identified Tjp1, Cldn1, Cldn 10 Ocln, and Jam 2 in the uterus (Sah et al., 2018; Yin et al., 2020). Although one study suggested that this gradient could be reversed by an electrical potential difference to promote a paracellular calcium pathway (Bar, 2009), the most likely hypothesis is that this paracellular pathway would contribute to the secretion of water and K, Na, and Cl^- ions, which the concentrations remain stable in uterine fluid at a lower concentration than in plasma (290 vs. 320 mM) (Nys et al., 1999).

4.2.3 Vesicular pathway

More recently, the vesicular transfer of ACC during the formation of the egg shell has also been demonstrated (Stapane, 2019; Stapane et al., 2020). ACC is a metastable polymorph of calcium carbonate first demonstrated in mollusks and echinoderms to provide the ions necessary for rapid shell mineralization (Addadi et al., 2003; Gomez-Morales et al., 2015). ACC is a highly reactive transitional phase for the formation of calcium carbonate biominerals and vesicles stabilize and transport ACC to the mineralization front. In the calcifying chicken eggshell, ACC has been demonstrated with a calcite short-range order favoring the transformation of ACC to calcite which is the shell calcium carbonate polymorph (Rodriguez-Navarro et al., 2015).

Transmission electron microscopy (TEM) was used to reveal intracellular vesicles in the ciliated cells of the surface epithelium of the hen's uterus that accumulate in the apical region to form buds before being secreted into the uterine fluid (Stapane, 2019; Stapane et al., 2020). Physical methods evidenced that these vesicles contain ACC (Stapane, 2019; Stapane et al., 2020). Subsequently, molecular proteomic and immunohistochemical methods were used to investigate the contents of these vesicles and propose a mechanistic model for them. Three annexins (Anxa 1, 2 and 8) were identified as calcium channels to allow the uptake of calcium in the vesicle. Carbonic anhydrase 4 (CA4) was detected in the vesicles and would catalyze the reversible hydration of CO_2 to bicarbonate ions that will form ACC in the vesicles. Using proteomics, other proteins were identified as cargo in the vesicle to serve as stabilizers for ACC. Finally, EDIL-3, a calcium binding protein would be attached in the vesicle membrane and will be a guidance molecule to the mineralization sites (Gautron et al., 2021b; Stapane, 2019; Stapane et al., 2020) (Fig. 4)

5 The organic matrix

5.1 Composition of the matrix protein

The organic matrix plays a fundamental role in the mineralization of the shell, the determination of its ultrastructure and, consequently, the determination of the mechanical properties of this biomaterial and consequently its composition has stimulated lot of researches. Initially, proteins in the organic matrix were studied using classical biochemical approaches based on separative techniques (chromatography, precipitation, electrophoresis), complemented in the 1990s by approaches based on molecular biology techniques (Gautron et al., 2011). These approaches allowed the identification of about 50 proteins in the egg, including only 11 in the shell. A major breakthrough came with the publication of the chicken genome sequence (Int Chicken Genome Sequencing, 2005), which led to the development and expansion of high-throughput techniques (transcriptomics and proteomics) to study egg genes and proteins (Gautron et al., 2019; Gautron et al., 2011). The first major proteomic analysis of egg came in 2006, with the eggshell proteome. Mann et al. identified the acid-soluble eggshell organic matrix and completed this approach with a phosphoproteome of the soluble eggshell proteins (Mann et al., 2006, 2007). Both studies identified 528 different proteins as constituents of the soluble eggshell matrix. The insoluble fraction of the eggshell matrix was also investigated (Miksik et al., 2007, 2010, 2014). Rose-Martel et al. performed a proteomic analysis of the outermost layer of the shell (cuticle), suspected to play a major role in preventing microbial penetration (Rose-Martel et al., 2012). A more recent proteomic survey allowed the identification of novel components (Ahmed et al., 2017; Sun et al., 2013) and the quantification of about 300 eggshell proteins at the key steps of shell calcification (Marie et al., 2015). Finally, a proteomic survey on shell membranes characterized about 300 eggshell membrane proteins (Makkar et al., 2015). All these studies were performed by different research groups using different databases, some of which have since disappeared (International Protein Index-IPI). Published protein data contain thousands of sequences that may be redundant, but a lack of integration prevented a non-redundant view of the unique shell proteome. To overcome this difficulty, all sequences were loaded and aligned using the Blast algorithm to obtain a non-redundant shell proteome consisting of 924 proteins (Gautron et al., 2019).

Bioinformatic analyses have been performed to determine the potential function of these proteins during eggshell biomineralization (Gautron et al.,

2011, 2021b, Marie et al., 2015). These proteins were classified into 2 groups according to their mode of action. The first group includes proteins with a direct role in mineralization. This group is made of proteins directly interacting with the mineral phase to stabilize ACC and/or to modify the morphology of crystals that determine the ultrastructure of avian eggshells. These include proteins for which *in vitro* data on the interaction of the protein with calcium carbonate are available (See below). Bioinformatics investigations have also associated calcium-binding proteins (CaBPs) that interact with calcium to favor nucleation and to specify the morphology of crystals. The second group are proteins involved in the regulation of proteins during mineralization which interact in the uterine fluid with proteins directing mineralization. As the mineralization takes place in an acellular milieu, proteins belonging to this group will inhibit or activate proteins of the mineralization. They are proteins involved in the folding of shell proteins to ensure an appropriate template for calcium and mineral interaction. They are also proteases and protease inhibitors that play a specific and controlled role during the calcification process, either by regulating processing of proteins in their mature forms or by degrading active proteins.

5.2 Role of the matrix protein

The first evidence of the role of organic matrix proteins in the eggshell mineralization is the observation of the progressive change of uterine fluid proteins during the three main stages of eggshell calcification (initiation, growth and terminal phase) (Gautron et al., 1997). This fluid contains the shell matrix proteins on soluble form and several of them present calcium binding affinity which is essential for the interaction with the mineral phase (Gautron et al., 1997). This affinity for calcium has also been demonstrated for the protein fractions of the shell (Abatangelo et al., 1978, Cortivo et al., 1982, Hincke et al., 1992) as well as for the proteoglycans contained in the organic matrix (Arias et al., 1992, 1993, Fernandez et al., 1997). *In vitro* calcite crystallization assays demonstrate that uterine fluid increases the number of crystals produced, decreases their size, accelerates the mineralization kinetics and promotes the exclusive formation of calcite (Dominguez-Vera et al., 2000; Hernandez-Hernandez et al., 2008a). The morphology of calcite crystals is strongly modified by the presence of organic constituents extracted from the shell (Gautron et al., 1996; Hernandez-Hernandez et al., 2008a) and by uterine fluid collected at the three main stages of shell biomineralization (Gautron et al., 1997; Hernandez-Hernandez et al., 2008a). Similar effects have been observed with single proteins isolated from the shell or uterine fluid. Calcite crystals are highly modified when grown in the presence of lysozyme, ovotransferrin, ovalbumin or ovocleidin-17 (Dombre et al., 2017; Gautron et al., 2001; Hincke et al., 2000; Jimenez-Lopez et al., 2003; Reyes-Grajeda et al., 2004). The ability to stabilize unstable phase of calcium carbonate (ACC) was ascribed to ovalbumin. Calcium binds to ovalbumin and this accumulation creates a nucleation center for mineralization. Calcium ions are bound to the protein by complexation through acidic groups leading to protein rearrangement. The calcium ions are the starting point for the subsequent formation of ACC nuclei, which then undergo a series of phase transitions to the stable crystalline polymorphs (Pipich et al., 2008; Schwahn et al., 2004). The role of lysozyme in the stabilization of ACC has been investigated but remains controversial. Lysozyme from eggshells of quail did not induced the precipitation of ACC under in vitro conditions (Lakshminarayanan et al., 2006). However, metastable ACC particles were obtained *in vitro* in the presence of chicken egg white lysozyme. Lysozyme considerably decreased the average diameter of the metastable ACC particles and promoted a network of associated particles, with the protein becoming incorporated into the precipitate. In addition, lysozyme-ACC particles reorganized exclusively into crystalline calcite (Voinescu et al., 2007; Wang et al., 2010). In another study lysozyme was shown to be ineffective in the stabilization of ACC (Wolf et al., 2011) demonstrating that further studies are still necessary to determine its role in this process. Osteopontin is also a favourite candidate in the control of shell mineralization, since it modifies the morphology of calcite crystals and is correlated with eggshell hardness (Athanasiadou et al., 2018; Chien et al., 2008; Hincke et al., 2008).

The involvement of shell matrix proteins on eggshell mechanical properties has been investigated using eggshell of various mechanical properties. These *in vivo* studies shown the variation of the quantity of specific proteins with the eggshell strength (Ahmed et al., 2005; Panheleux et al., 2000; Zhang et al., 2019). Evidence that structural organization of chicken eggshell is under genetic control has reinforced the observed role of shell matrix proteins. Polymorphisms of genes encoding shell matrix proteins were associated to eggshell mechanical traits (Dunn et al., 2009; Romé and Le Roy, 2016). Single nucleotide polymorphisms of ovalbumin, ovocleidin-116 and ovocalyxin-32 explain variations on eggshell thickness, crystal size and orientation as well as shell mechanical properties. Additionally, Quantitative Trait Loci (QTL) were associated to shell strength (Dunn et al., 2009; Jiang et al., 2010a; Romé and Le Roy, 2016; Takahashi et al., 2009, 2010; Tuiskula-Haavisto et al., 2011)

6 Temporal and spatial deposition of the shell

The temporal and spatial deposition of shell is well documented (Gautron et al., 2021b; Hincke et al., 2012; Nys et al., 2004). The first stage is the initiation phase, from 5 h to 10 h p.o. that corresponds to the slow phase of mineralization with a deposition rate of about 0.1 g of shell per hour. During this initial phase, there is a massive deposition of ACC on the shell membrane that progressively transformed in oriented calcite crystals (Rodriguez-Navarro et al., 2015). ACC deposited around these sites dissolves rapidly, providing a continuous supply of ions to form calcite crystals on specific nucleation sites. These units coalesce to form larger crystals in the mammillary layer, and then during the following rapid growth phase (10-22 h p.o.), they form the compact shell palisade layer made of columnar crystals with preferred orientation. The rate of shell deposition is of 0.33 g/h during this phase (Nys et al., 2004). The terminal phase (22 -24 h) is characterized by an arrest of mineralization and the cuticle deposition. Calcite crystals are consequently formed by the aggregation of ACC particles that support the rapid mineralization of the eggshell and there is evidence that this non-crystalline form of calcium carbonate is present throughout the phases of shell formation (Rodriguez-Navarro et al., 2015) (Fig. 5).

In the initial mineralization process, spherulitic growth of calcite crystals radiating from the nucleation sites produces the mammillary cones or the lower part of the mineralized shell (Fig. 2a, e). As crystals continue to grow, adjacent crystals meet and impinge on each other's. Subsequently, calcite crystals oriented with their faster growth directions, in this case the c-axis, perpendicular to the surface, can grow freely, whereas less favourably orientated crystals are buried by neighboring crystals. Thus, calcite crystals compete for the available space, and crystals develop a preferential orientation. This process known as competitive crystal growth results in the development of a columnar ultrastructure (Fig. 6) (Rodriguez-Navarro and García-Ruiz, 2000). Although this process occurs in the eggshell of all avian species, it is most evident in those of ostrich and other palaeognathae birds in which the spacing between mammillary cones is small relative to the total eggshell thickness, resulting in the formation of slender calcite columnar units with a strong fiber texture in which calcite crystals are oriented with their c-axes aligned and perpendicular to the eggshell outer surface (Rodriguez-Navarro et al., 2007). This process is

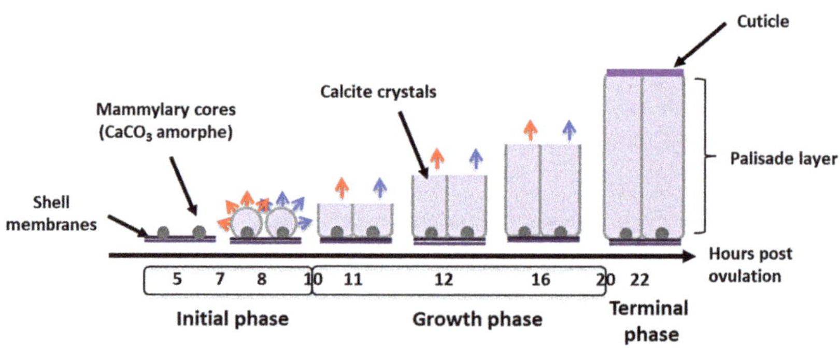

Figure 5 The temporal deposition of chicken eggshell comprises three main phases (initial, growth, and terminal) with involvement of ACC and development of a gradual orientation calcite crystals perpendicular to the surface.

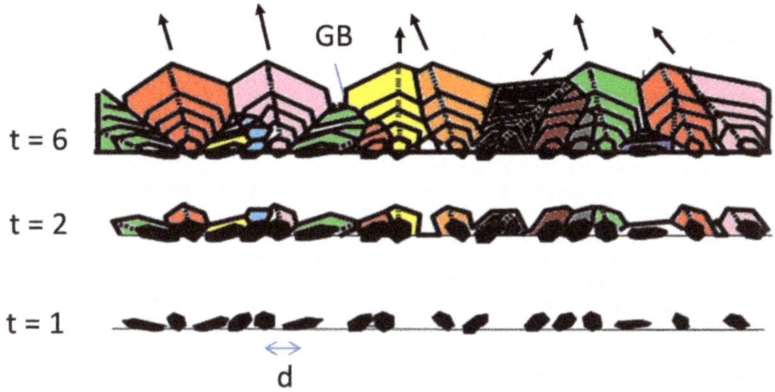

Figure 6 Competitive crystal growth model. Starting from randomly oriented crystals, crystals develop a preferential orientation and a characteristic columnar ultrastructure. GB is grain boundaries; t is the time steps; d is the distance between crystals.

also enhanced by the anisotropic growth of calcite crystals that are elongated along the c-axis as the growth in other directions, i.e. (100) prismatic faces, is inhibited by the absorption of proteins in these faces (Hernandez-Hernandez et al., 2008b). In any case, the characteristic columnar ultrastructure of avian eggshells (palisade region) is due to this mechanics and is common to every eggshell species examined to date. However, there are specific microstructure characteristics (crystal sizes and orientation) that are different for the different bird's species but all has the same mineral composition (calcite). Also, eggshell microstructure is modified by different physiological conditions (i.e. age, diet). For instance, older hens often lay eggs with shells made of abnormally large and highly oriented crystals with a fan-like shape (Rodriguez-Navarro et al., 2002). These changes might be in part responsible for the marked decrease of eggshell mechanical properties at the end of lay.

The evolution of the mineral deposited on the eggshell membrane can be followed by SEM, FTIR, and X-ray diffraction (XRD) (Fig. 7). In eggshell samples collected before 5 h p.o., SEM observation shows the outer eggshell membrane and in specific sites some protuberances which correspond to the mammillary cores, which are organic rich structures made of proteins and sulphated proteoglycans, that have a strong calcium-binding capacity. FTIR shows only absorption bands of proteins and polysaccharides from the eggshell membrane fibers but no carbonate peaks could be detected indicating that no mineral has been deposited yet (Fig. 7c).

At 5 h p.o., in some eggshell samples, it is possible to observe by SEM an incipient mineralization of flat-disk particles coating the eggshell fibers and massive mineral deposits that accumulate around the mammillary knobs. In infrared spectra (Fig. 7c), there is CO_3 v_2 peak at of 873 cm−1, but there is

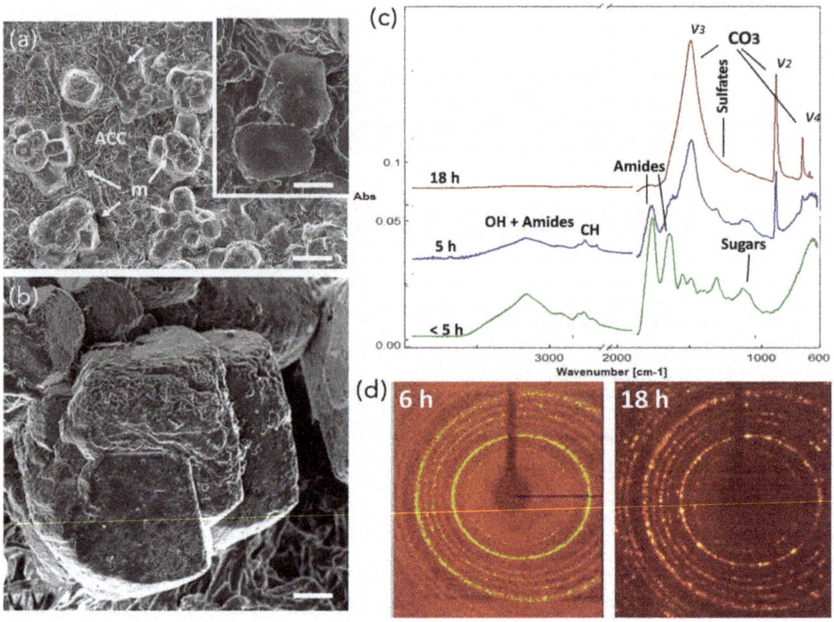

Figure 7 (a) SEM images of eggshell collected at an early stage of mineralization (6 h p.o.) showing calcite crystals aggregates nucleating on the mammillary cores (m) and flat-disk shaped ACC particles deposited on the shell membrane surface (inset); (b) At later stages (7 h p.o.), the ACC particles has been dissolved and the calcite crystals have developed well-defined rhombohedral faces. (c) ATR-FTIR spectra of the eggshell collected at different stages of eggshell mineralization: before mineralization (<5 h p.o.) showing only proteins and sugars peaks from the eggshell membrane at 5 h (p.o.), the outer surface is coated with amorphous calcium carbonate as shown by the presence of carbonate v3 at 1400 cm^{-1} and v2 at 873 cm^{-1} peaks, whereas the v4 peak at 713 cm^{-1} is very weak. At 18 h, all characteristic infrared peaks of crystalline calcium carbonate (calcite) are present. (d) 2D-XRD patterns of eggshells collected at different stages (6 h and 18 h p.o.) of mineralization showing almost continuous rings produced by randomly oriented calcite microcrystals (6 h) and spotty rings produced by larger calcite crystal units (18 h). Scale bars are (a) 20 μm. Inset: 5 μm. (b) 10 μm.

no CO_3 v_4 peak (characteristic of crystalline calcium carbonate). This indicates that the mineral deposited is made of ACC and that this has a proto-calcitic structure. At this time, no calcite peaks could be detected by XRD confirming that the mineral deposited is amorphous. However, after heating the samples at 350°C, calcite peaks appear indicating that the crystallization of ACC into calcite has been induced (Rodriguez-Navarro et al., 2015).

At 6 h p.o., the massive mineral deposits of ACC have disappeared and give way to small rounded calcite microcrystals that does not show well-defined crystal faces. Still there are some flat disk-shaped ACC particles distributed on the eggshell membrane surface (Fig. 7a). At this time, in the infrared spectra, there is a single CO_3 v_4 peak (characteristic of crystalline calcium carbonate) that appears at 710 cm^{-1} (the same as calcite). 2D XRD data show weak and continuous diffraction rings formed by microscopic calcite crystals that are randomly oriented (Fig. 7d).

At 7 h p.o., ACC particles on the eggshell membrane have disappeared and only calcite single crystals with well-defined rhombohedral faces are observed on mammillary cores (Fig. 7b) (Rodriguez-Navarro et al., 2015). Still calcite crystals show a granular nanostructure at their surface supporting the idea that crystals formed by the aggregation of ACC nanoparticles.

At 14 h and later, calcite crystals have fused together and the eggshell is a continuous mineral layer. Still, the mineral surface has a granular nanostructure. At this time, infrared spectra only shows absorption bands from crystalline calcite. 2D XRD data show spotty diffraction rings with strong reflection produced by larger calcite crystal units that are randomly oriented (Fig. 7d).

Calcite is the only calcium carbonate polymorph that is detected in the mature eggshell. ACC and calcite crystals formed by aggregation of nanoparticles, an efficient growth mechanism which can explain the rapid mineralization of the eggshell. Ionic speciation data suggest that the uterine fluid is equilibrated with ACC, throughout the duration of eggshell mineralization process, supporting that this mineral phase is constantly forming at the shell mineralization front (Rodriguez-Navarro et al., 2015). Moreover, calcite crystals formation through dissolution of an intermediate amorphous mineral phase maintains a high supersaturation in the uterine fluid and contribute to a continuous supply of calcium to sustain fast and continuous growth of calcite crystals during the linear phase of eggshell formation (Nys et al., 1991). In infrared spectra (Fig. 6c), even at 5 h when there is only ACC, there is single CO_3 v_2 peak at a constant position of 873 cm^{-1}, which is the same as in calcite (Fig. 7c). This indicates that ACC has a proto-calcitic structure. The calcitic short-range order of ACC facilitates its transformation directly into calcite crystals without the occurrence of any intermediate mineral phase (Rodriguez-Navarro et al., 2015). The proto calcitic structure has been also confirmed from other techniques such as TEM or XRD data. Also, the same transformation process has been observed in eggshells from other bird species (Guinea fowl) (Rodríguez-Navarro et al., 2024). As in other controlled biomineralization processes, the organic matrix plays a key role in the modulation of shell mineralization. It has been shown that specific proteins (i.e. ovocleidin-17 and ovocleidin-116) that became concentrated in specific regions of the eggshell structure (i.e. mammillary cores) are involved in the nucleation and stabilization of calcium carbonate mineral (ACC, calcite) and in the regulation of eggshell mineralization (Gautron et al., 2021b; Marie et al., 2015).

7 Changes in layer production methods and how they affect shell quality

The different ways in which eggs are produced for consumption have evolved considerably since the Second World War in occidental countries and more particularly in Europe. Before the war, production was exclusively backyard and mainly for personal consumption. After the war, agriculture developed and had to cope with a drastic reduction in the number of people working in the primary sector (from 50% of the population to just a few percent today in France and more generally in Europe) (Gautron et al., 2021a, 2022b). There are fewer and fewer farmers, and they have to produce not only for a limited number of people but in larger quantities. Production therefore evolved to meet demand (in terms of quantity) and to control health risks, leading to a model of 'productive egg production' in confinement and cages. In the 1980s and 1990s, this was virtually the only production model. Since the late 1990s, new consumer demands for animal welfare have emerged, leading to market segmentation (Gautron et al., 2021a).

At the European level, public awareness of agricultural production systems in general and animal production in particular, including poultry and eggs, has increased significantly. This consumer demand has led to a strong diversification of farming methods. The current European production models are the result of this social demand. They are regulated by the Directive on the Welfare of Laying Hens (UE, 1999). This regulation is also the result of scientific research to satisfy the five freedoms of animal welfare: absence of hunger, absence of thirst, possibility of movement, absence of fear/distress, and while allowing the expression of natural behaviors. This policy defines caged hens (code 3) and alternative systems. These alternative systems include hens raised in a barn on litter or in aviaries (code 2), hens raised on the ground or aviaries with outdoor access (code 1) and organically produced eggs (hens raised on the ground with outdoor access and fed organically produced feed) (code 0). (As a result of this increase in animal welfare awareness, there has been a significant evolution in production systems in recent years. In 1996, non-cage systems accounted for 8% of the EU laying hen population, 30% in 2009, 46% in 2017, and 51% in 2019 (80% of laying hens were in cages in 2003, 49% in 2019, and 60% today (https://agriculture.ec.europa.eu/document/download/9bdf9842-1eb6-41a2-8845-49738b812b2b_en)) (ITAVI, 2019).

These developments have a significant impact on the quality of eggs, in particular, the quality of the shell. The mode of production of eggs can influence microbiological contamination. Eggs from floor and aviary systems generally have more aerobic bacteria on the egg surface than eggs from conventional systems and enriched cages (Englmaierova et al., 2014; Samiullah et al., 2014), but the differences are small and farmer practices seem to be more important than the production method (Mallet et al., 2010). The egg shell plays an essential role in protecting the egg content from microbes. It is a physical barrier that prevents bacteria from entering the egg, thanks in particular to its surface, which consists of a thin layer of organic matter (the cuticle) that blocks the pores (see section 'Eggshell structure and composition') and the improvement of this systems remains important to maintain shell quality in this changing environment. That is why washing eggs is forbidden in EU to preserve this protective coating intact.

Another important evolution is the increase of the laying period. Currently, laying hens start laying at around 18 weeks of age and peak laying is around 25 weeks of age. The majority of the flocks are depopulated when the laying rate is reducing to unprofitable levels (less than 75% of marketable eggs). This often occurred at around 72 weeks (Bain et al., 2016). However, selection for egg production has successfully reduced the effects of age on the reproductive system. Poultry breeders have extended the laying periods at an advanced age (Bain et al., 2016). Breeders indicated 10 years ago, that they wanted to select strains of laying hens capable of laying sustainability up to 100 weeks of age, for a total production of almost 500 eggs per birds by 2020 (Bain et al., 2016), which has been already achieved (Gautron et al., 2021a). However, any improvement in egg-laying persistence must be achieved with consistent egg quality. Egg weight increases by 70 mg per week between 60 weeks of age and 80 weeks of age (Molnar et al., 2016). While the other egg constituents remain constant, there is a decrease in shell thickness (about 0.23 μm per week) and a decrease in breaking strength with increasing age of the hen.

Also, there are notable changes with hen age in the eggshell ultra- and microstructure characteristics (i.e. decreased mammillary density with poor membrane attachment, increased crystal size and orientation; lesser amount of cuticle) that are the main causes of eggshell quality and integrity to diminish in old hens (Park and Sohn, 2018; Roberts et al., 2013; Rodriguez-Navarro et al., 2002). Now that the industry is aiming to extend the egg-laying period until hens age reaches 100 weeks and could achieve a production of 500 eggs per hen, there is renewed interest in finding solutions to maintain hen laying performance and egg quality for longer periods of time (Bain et al., 2016; Nys, 2017).

8 Future trends: improving eggshell mechanical properties

The marked decrease in eggshell breaking strength at the end of the laying period is not only due to a reduction in eggshell thickness but also to abrupt changes in the eggshell ultra- and microstructure characteristics that was documented in studies of egg-laying in aged hens (Benavides-Reyes et al.,

2021; Park and Sohn, 2018; Roberts et al., 2013; Rodriguez-Navarro et al., 2002). Additionally, there is a decrease in the amount of cuticle with hen age that should have a negative impact in egg safety and egg internal quality. All in all, these results clearly indicate that not only the shell thickness but other characteristics need to be considered to asses eggshell quality. Also, some of these characteristics have an important genetic component and can be used in genetically assisted selection programs aimed at improving eggshell quality and safety of eggs (Dunn et al., 2009; Nys, 2017). It is also important to assess, the amount, composition, and coverage of the cuticle as this organic coating is an effective barrier against bacterial penetration through the shell (Bain et al., 2013). The cuticle also reduces moisture and CO_2 loss through the shell, which slows the natural decline of egg internal quality during storage.

Quantitative genetics has been widely used in the poultry industry, especially for the improvement of the quality of eggs and shells. QTLs (a quantitative trait locus is a region of the genome that may or may not contain genes affecting a quantitative trait) associated with shell morphological parameters have also been identified on chicken chromosomes (Romé and Le Roy, 2016; Takahashi et al., 2009; Tuiskula-Haavisto et al., 2002). 118 QTLs were associated with shell strength (Romé and Le Roy, 2016). Twenty-four of them appear to be related to shell breaking strength and 33 in shell stiffness. The identification of the organic matrix proteins involved in shell formation allows genetic approaches to be strengthened by using functional information from these proteins. Indeed, associations have been found between polymorphisms in genes encoding matrix proteins and phenotypic shell measurements associated with shell quality. The protein osteopontin was associated with shell hardness, ovocleidin-116 with shell elasticity and thickness, and ovocalyxin-32 with mammillary thickness (Dunn et al., 2009; Romé and Le Roy, 2016). Esr1 (Dunn et al., 2009) is a gene thought to affect shell stiffness, while Pth (Jiang et al., 2010b), Lrp8 (Yao et al., 2010), and Ovocalyxin-32 are genes thought to be involved in fracture resistance. Recently, hundreds of genes encoding organic matrix proteins involved in shell formation have been mapped on the genome and their polymorphisms associated with shell quality traits (Hérault et al., 2021). This identification of causal polymorphism is a key step to increase the evaluation accuracy for a genomic selection. It allows to get a link between genes, biological functions, and traits. These strategies using candidate genes coupled with quantitative genetics to improve egg quality is developing fast and are already used by geneticists (Druet et al., 2020). These new strategies are likely to lead to significant improvements in shell quality, taking into account not only the genes involved in shell formation but also new precision phenotypic measures that integrate morphological and ultrastructural shell data, as well as the hen's environment and age to maintain bone integrity in older hens.

Another important factor in improving eggshell quality in changing production environments is the use of precision nutrition. Adequate nutrition (trace metals such as Mn) can significantly improve eggshell quality by modifying the eggshell ultrastructure (increasing mammillary density) (Zhang et al., 2017) as Mn enhances the synthesis of proteoglycans for mammillary knobs formation on the eggshell membrane. In addition, the influence of vitamin D is also an important factor to evaluate to improve the shell. In the past, work on calbindin showed that vitamin D did not regulate this protein (Bar, 2009), which is important for calcium transfer in the uterine fluid, suggesting that the uterus is not sensitive to vitamin D and its active metabolites. Recent work has shown that this protein is regulated by vitamin D in the uterus (Sinclair-Black et al., 2024) and that its expression differs depending on the form of vitamin D (Gautron et al., 2022a).

In conclusion, the eggshell, although not consumed, remains a critical structure in the marketing of eggs. The many advances in our understanding of the regulation of the supply of its constituents, as well as the discoveries made about its mineralogy, ultrastructure, and the role of amorphous forms in calcification, are important steps forward. All this information, combined with new approaches in genetics and nutrition (especially vitamin D), will allow us to maintain shell integrity in the production systems (barn, free range, and extended laying cycle) that have been constantly and rapidly evolving in recent years.

9 Where to look for further information

The following articles provide a good overview on the following topics:

Calcium homeostasis in laying hens:

- Nys, Y., et al. 2022. Chapter 32 - Mechanisms and hormonal regulation of shell formation: supply of ionic and organic precursors, shell mineralization. In: Scanes, C. G. & Dridi, S. (eds.) *Sturkie's Avian Physiology* (7th ed.). San Diego: Academic Press.

The role of organic matrix proteins during shell calcification:

- Gautron, J., et al. 2021b. Avian eggshell biomineralization: an update on its structure, mineralogy and protein tool kit. *BMC Molecular and Cell Biology*, 22, 11.
- Hincke, M. T., et al. 2012. The eggshell: structure, composition and mineralization. *Frontiers in Bioscience*, 17, 1266–1280.
- Nys, Y., et al. 2004. Avian eggshell mineralization: biochemical and functional characterization of matrix proteins. *Comptes Rendus Palevol*, 3, 549–562.

The importance of transient amorphous forms of minerals for biomineralization:

- Addadi, L., et al. 2003. Taking advantage of disorder: amorphous calcium carbonate and its roles in biomineralization. *Advanced Materials*, 15, 959–970.
- Gomez-Morales, J., et al. 2015. Biological crystallization. In: Nishinaga, T. (ed.) *Handbook of Crystal Growth*. Amsterdam: Elsevier.

Future trends in the egg market:

- Gautron, J., et al. 2021. Avian eggshell biomineralization: an update on its structure, mineralogy and protein tool kit. *BMC Molecular and Cell Biology*, 22, 11.

Evaluation of the quality of animal products:

- Prache, S., et al. 2022. Review: Quality of animal-source foods. *Animal*, 16, 100376.

10 References

Abatangelo, G., et al. 1978. Some observations on the calcium-ion binding to the eggshell matrix. *Calcified Tissue International*, 26, 247–252.

Addadi, L., et al. 2003. Taking advantage of disorder: amorphous calcium carbonate and its roles in biomineralization. *Advanced Materials*, 15, 959–970.

Ahmed, A. M. H., et al. 2005. Changes in eggshell mechanical properties, crystallographic texture and in matrix proteins induced by moult in hens. *British Poultry Science*, 46, 268–279.

Ahmed, T. A., et al. 2017. In-depth comparative analysis of the chicken eggshell membrane proteome. *Journal of Proteomics*, 155, 49–62.

Arias, J. L., et al. 1992. Partial biochemical and immunochemical characterization of avian eggshell extracellular matrices. *Archives of Biochemistry and Biophysics*, 298, 293–302.

Arias, J. L., et al. 1993. Biomineralization and eggshells: cell-mediated acellular compartments of mineralized extracellular matrix. *International Review of Cytology*, 145, 217–250.

Athanasiadou, D., et al. 2018. Nanostructure, osteopontin, and mechanical properties of calcitic avian eggshell. *Science Advances*, 4, eaar3219.

Bain, M. M., et al. 2013. Enhancing the egg's natural defence against bacterial penetration by increasing cuticle deposition. *Animal Genetics*, 44, 661–668.

Bain, M. M., et al. 2016. Increasing persistency in lay and stabilising egg quality in longer laying cycles. What are the challenges? *British Poultry Science*, 57, 330–338.

Bar, A. 2009. Calcium transport in strongly calcifying laying birds: Mechanisms and regulation. *Comparative Biochemistry and Physiology a-Molecular & Integrative Physiology*, 152, 447–469.

Benavides-Reyes, C., et al. 2021. Research note: changes in eggshell quality and microstructure related to hen age during a production cycle. *Poultry Science*, 100, 100965.
Brionne, A., et al. 2014. Hen uterine gene expression profiling during eggshell formation reveals putative proteins involved in the supply of minerals or in the shell mineralization process. *BMC Genomics*, 15, 220.
Chien, Y. C., et al. 2008. Ultrastructural matrix-mineral relationships in avian eggshell, and effects of osteopontin on calcite growth in vitro. *Journal of Structural Biology*, 163, 84–99.
Choi, S., et al. 2023. Microstructural and crystallographic evolution of palaeognath (Aves) eggshells. *Elife*, 12, e81092.
Cortivo, R., et al. 1982. Chemical characterization of the hen eggshell matrix - isolation of an alkali-resistant peptide. *Journal of Chromatography*, 237, 127–135.
Dombre, C., et al. 2017. Egg serpins: The chicken and/or the egg dilemma. *Seminars in Cell & Developmental Biology*, 62, 120–132.
Dominguez-Vera, J. M., et al. 2000. The effect of avian uterine fluid on the growth behavior of calcite crystals. *Poultry Science*, 79, 901–907.
Druet, D. P., et al. 2020. Reliability of genomic evaluation for egg quality traits in layers. *BMC Genetics*, 21, Article 17.
Dunn, I. C., et al. 2009. Polymorphisms in eggshell organic matrix genes are associated with eggshell quality measurements in pedigree Rhode Island Red hens. *Animal Genetics*, 40, 110–114.
Englmaierova, M., et al. 2014. Effects of laying hens housing system on laying performance, egg quality characteristics, and egg microbial contamination. *Czech Journal of Animal Science*, 59, 345–352.
Fernandez, M. S., et al. 1997. Eggshells are shaped by a precise spatio-temporal arrangement of sequentially deposited macromolecules. *Matrix Biology*, 16, 13–20.
Fu, Y., et al. 2024. Decreased eggshell strength caused by impairment of uterine calcium transport coincide with higher bone minerals and quality in aged laying hens. *Journal of Animal Science and Biotechnology*, 15, 37.
Gautron, J., et al. 1996. Soluble matrix of hen's eggshell extracts changes in vitro the rate of calcium carbonate precipitation and crystal morphology. *British Poultry Science*, 37, 853–866.
Gautron, J., et al. 2022a. Control of shell calcification with different sources of vitamin D3 supply: regulation of calcium and organic matrix protein transfers in the uterus. In *14èmes Journée de la Recherche Avicole et Palmipèdes à Foie Gras*. Tours: ITAVI.
Gautron, J., et al. 2022b. Review: Production factors affecting the quality of chicken table eggs and egg products in Europe. *Animal*, 16, 100425.
Gautron, J., et al. 2019. Chapter 14 bioactive minor egg components. In: Wu, J. (ed.) *Eggs as Functional Foods and Nutraceuticals for Human Health*. Cambridge: The Royal Society of Chemistry.
Gautron, J., et al. 1997. Precursor matrix proteins in the uterine fluid change with stages of eggshell formation in hens. *Connective Tissue Research*, 36, 195–210.
Gautron, J., et al. 2001. Ovotransferrin is a matrix protein of the hen eggshell membranes and basal calcified layer. *Connective Tissue Research*, 42, 255–267.

Gautron, J., et al. 2011. Use of High-throughput technology to identify new egg components. In: Nys, Y., et al. (eds.) *Improving the Safety and Quality of Eggs and Egg Products.* Cambridge: Woodhead Publishing limited.

Gautron, J., et al. 2021a. Review: what are the challenges facing the table egg industry in the next decades and what can be done to address them? *Animal*, 15, 100282.

Gautron, J., et al. 2021b. Avian eggshell biomineralization: an update on its structure, mineralogy and protein tool kit. *BMC Molecular and Cell Biology*, 22, 11.

Gloux, A., et al. 2019. Candidate genes of the transcellular and paracellular calcium absorption pathways in the small intestine of laying hens. *Poultry Science*, 98, 6005–6018.

Gomez-Morales, J., et al. 2015. Biological crystallization. In: Nishinaga, T. (ed.) *Handbook of Crystal Growth.* Amsterdam: Elsevier.

Hérault, F., et al. 2021. Identification of polymorphisms in Protein Coding Genes which affect eggshell quality traits in layers. *Webinar egg quality - World Poultry Congress.* Online: Woprld Poultry Science Association.

Hernandez-Hernandez, A., et al. 2008a. Identification of some active proteins in the process of hen eggshell formation. *Crystal Growth & Design*, 8, 4330–4339.

Hernandez-Hernandez, A., et al. 2008b. Influence of model globular proteins with different isoelectric points on the precipitation of calcium carbonate. *Crystal Growth & Design*, 8, 1495–1502.

Hincke, M. T., et al. 1992. Soluble protein constituents of the domestic fowl's eggshell. *British Poultry Science*, 33, 505–516.

Hincke, M. T., et al. 2008. Colloidal-gold immunocytochemical localization of osteopontin in avian eggshell gland and eggshell. *Journal of Histochemistry & Cytochemistry*, 56, 467–476.

Hincke, M. T., et al. 2000. Identification and localization of lysozyme as a component of eggshell membranes and eggshell matrix. *Matrix Biology*, 19, 443–453.

Hincke, M. T., et al. 2012. The eggshell: structure, composition and mineralization. *Frontiers in Bioscience*, 17, 1266–1280.

Int Chicken Genome Sequencing, C. 2005. Sequence and comparative analysis of the chicken genome provide unique perspectives on vertebrate evolution (vol 432, pg 695, 2004). *Nature*, 433, 777–777.

Itavi. 2019. *Situation du marché des oeufs et des ovoproduits: édition mars*. Paris: Service économie Itavi.

Jiang, R. S., et al. 2010a. A single nucleotide polymorphism in the parathyroid hormone gene and effects on eggshell quality in chickens. *Poultry Science*, 89, 2101–2105.

Jiang, R. S., et al. 2010b. A single nucleotide polymorphism in the parathyroid hormone gene and effects on eggshell quality in chickens. *Poultry Science*, 89, 2101–2105.

Jimenez-Lopez, C., et al. 2003. Influence of lysozyme on the precipitation of calcium carbonate: a kinetic and morphologic study. *Geochimica Et Cosmochimica Acta*, 67, 1667–1676.

Jonchere, V., et al. 2012. Identification of uterine ion transporters for mineralisation precursors of the avian eggshell. *BMC Physiology*, 12, 10.

Ketta, M. & Tůmová, E. 2016. Eggshell structure, measurements, and quality-affecting factors in laying hens: a review. *Czech Journal of Animal Science*, 61, 299–309.

Khan, S., et al. 2019. Rna-sequencing analysis of shell gland shows differences in gene expression profile at two time-points of eggshell formation in laying chickens. *BMC Genomics*, 20, Article 89.

Kulshreshtha, G., et al. 2022. Properties, genetics and innate immune function of the cuticle in egg-laying species. *Frontiers in Immunology*, 13, 838525.

Lakshminarayanan, R., et al. 2006. Formation of transient amorphous calcium carbonate precursor in quail eggshell mineralization: an in vitro study. *Biomacromolecules*, 7, 3202–3209.

Makkar, S., et al. 2015. Chicken egg shell membrane associated proteins and peptides. *Journal of Agricultural and Food Chemistry*, 63, 9888–9898.

Mallet, S., et al. 2010. Laying hen breeding systems and hygienic status of the eggs. *Inra Productions Animales*, 23, 183–191.

Mann, K., et al. 2006. Proteomic analysis of the acid-soluble organic matrix of the chicken calcified eggshell layer. *Proteomics*, 6, 3801–3810.

Mann, K., et al. 2007. Phosphoproteins of the chicken eggshell calcified layer. *Proteomics*, 7, 106–115.

Marie, P., et al. 2015. Quantitative proteomics provides new insights into chicken eggshell matrix protein functions during the primary events of mineralisation and the active calcification phase. *Journal of Proteomics*, 126, 140–154.

Miksik, I., et al. 2007. Proteins of insoluble matrix of avian (Gallus Gallus) eggshell. *Connective Tissue Research*, 48, 1–8.

Mikšík, I., et al. 2014. Proteomic analysis of chicken eggshell cuticle membrane layer. *Analytical and Bioanalytical Chemistry*, 406, 7633–7640.

Miksik, I., et al. 2010. Determination of insoluble avian eggshell matrix proteins. *Analytical and Bioanalytical Chemistry*, 397, 205–214.

Mikšík, I., et al. 2007. Comparison of Ce-Ms and Lc-Ms analyses of avian eggshell matrix proteins. *Chromatographia*, 67, 89–96.

Molnar, A., et al. 2016. Changes in egg quality traits during the last phase of production: is there potential for an extended laying cycle? *British Poultry Science*, 57, 842–847.

Nys, Y. 2017. Laying hen nutrition: optimising hen performance and health, bone and eggshell quality. *In:* Rober, J. R. (ed.) *Achieving Sustainable Production of Eggs*. Cambridge, Uk: Burleigh Dodds

Nys, Y., et al. 2004. Avian eggshell mineralization: biochemical and functional characterization of matrix proteins. *Comptes Rendus Palevol*, 3, 549–562.

Nys, Y., et al. 2022. Chapter 32 - Mechanisms and hormonal regulation of shell formation: supply of ionic and organic precursors, shell mineralization. In: Scanes, C.G. & Dridi, S. (eds.) *Sturkie's Avian Physiology* (7th ed.). San Diego: Academic Press.

Nys, Y. & Guyot, N. 2011. Egg formation and chemistry. In: Nys, Y., et al. (eds.) *Improving the Safety and Quality of Eggs and Egg Products*. Cambridge: Woodhead publishing.

Nys, Y., et al. 1999. Avian eggshell mineralization. *Poultry and Avian Biology Reviews*, 10, 143–166.

Nys, Y. & Le Roy, N. 2018. Calcium homeostasis and eggshell biomineralization in female chicken. In: Feldman, D. (ed.) *Vitamin D* (4th ed.). Cambridge: Academic press.

Nys, Y., et al. 1991. Whitening of brown-shelled eggs: mineral composition of uterine fluid and rate of protoporphyrin deposition. *Poultry Science*, 70, 1236–1245.

Panheleux, M., et al. 2000. Extraction and quantification by Elisa of eggshell organic matrix proteins (ovocleidin-17, ovalbumin, ovotransferrin) in shell from young and old hens. *Poultry Science*, 79, 580–588.

Park, J. A. & Sohn, S. H. 2018. The influence of hen aging on eggshell ultrastructure and shell mineral components. *Korean Journal for Food Science of Animal Resources*, 38, 1080–1091.

Pipich, V., et al. 2008. Nucleation and growth of Caco3 mediated by the egg-white protein ovalbumin: a time-resolved in situ study using small-angle neutron scattering. *Journal of the American Chemical Society*, 130, 6879–6892.

Réhault-Godbert, S., et al. 2019. The golden egg: nutritional value, bioactivities, and emerging benefits for human health. *Nutrients*, 11, 684.

Reyes-Grajeda, J. P., et al. 2004. Crystal structure of ovocleidin-17, a major protein of the calcified Gallus gallus eggshell: implications in the calcite mineral growth pattern. *Journal of Biological Chemistry*, 279, 40876–40881.

Roberts, J. R., et al. 2013. Egg quality and age of laying hens: implications for product safety. *Animal Production Science*, 53, 1291–1297.

Rodriguez-Navarro, A. & García-Ruiz, J. M. 2000. Model of textural development of layered crystal aggregates. *European Journal of Mineralogy*, 12, 609–614.

Rodriguez-Navarro, A. B., et al. 2002. Influence of the microstructure on the shell strength of eggs laid by hens of different ages. *British Poultry Science*, 43, 395–403.

Rodriguez-Navarro, A. B., et al. 2007. Analysis of avian eggshell microstructure using X-ray area detectors. *European Journal of Mineralogy*, 19, 391–398.

Rodríguez-Navarro, A. B., et al. 2013. Change in the chicken eggshell cuticle with hen age and egg freshness. *Poultry Science*, 92, 3026–3035.

Rodriguez-Navarro, A. B., et al. 2015. Amorphous calcium carbonate controls avian eggshell mineralization: a new paradigm for understanding rapid eggshell calcification. *Journal of Structural Biology*, 190, 291–303.

Rodríguez-Navarro, A. B., et al. 2024. Guinea fowl eggshell structural analysis at different scales reveals how organic matrix induces microstructural shifts that enhance its mechanical properties. *Acta Biomaterialia*, 178, 244–256.

Romé, H. & Le Roy, P. 2016. Chromosomal regions influencing egg production and egg quality traits in hens. *Inra Productions Animales*, 29, 117–127.

Rose-Martel, M., et al. 2012. Proteomic analysis provides new insight into the chicken eggshell cuticle. *Journal of Proteomics*, 75, 2697–2706.

Sah, N., et al. 2018. Rna sequencing-based analysis of the laying hen uterus revealed the novel genes and biological pathways involved in the eggshell biomineralization. *Scientific Reports*, 8, 16853.

Samiullah, S. & Roberts, J. R. 2014. The eggshell cuticle of the laying hen. *World's Poultry Science Journal*, 70, 693–708.

Samiullah, S., et al. 2014. Effect of production system and flock age on egg quality and total bacterial load in commercial laying hens. *Journal of Applied Poultry Research*, 23, 59–70.

Sauveur, B. & De Reviers, M. 1988. *Reproduction des volailles et production d'oeufs*. Paris: Institut National de la Recherche Agronomique (Inra).

Schwahn, D., et al. 2004. Crystallization of the $Caco_3$ mineral in the presence of the protein ovalbumin. *Physica B-Condensed Matter*, 350, E947–E949.

Sinclair-Black, M., et al. 2023. Physiological regulation of calcium and phosphorus utilization in laying hens. *Frontiers in Physiology*, 14, 1112499.

Sinclair-Black, M., et al. 2024. Circadian regulation of calcium and phosphorus homeostasis during the oviposition cycle in laying hens. *Poultry Science*, 103, 103209.

Stapane, L. 2019. *Biominéralisation de la coquille d'œuf de poule: mise en évidence d'un transport vésiculaire du minéral impliquant les protéines Egf-like repeats and discoidinlike domains 3 (Edil3) et Milk fat globule Egf-factor 8 (Mfge8).* PhD. Université de Tours.

Stapane, L., et al. 2020. Avian eggshell formation reveals a new paradigm for vertebrate mineralization via vesicular amorphous calcium carbonate. *Journal of Biological chemistry*, 22, 11.

Sun, C., et al. 2013. Differential label-free quantitative proteomic analysis of avian eggshell matrix and uterine fluid proteins associated with eggshell mechanical property. *Proteomics*, 13, 3523–3536.

Takahashi, H., et al. 2010. Association between ovocalyxin-32 gene haplotypes and eggshell quality traits in an F-2 intercross between two chicken lines divergently selected for eggshell strength. *Animal Genetics*, 41, 541–544.

Takahashi, H., et al. 2009. Mapping of quantitative trait loci affecting eggshell quality on chromosome 9 in an F-2 intercross between two chicken lines divergently selected for eggshell strength. *Animal Genetics*, 40, 779–782.

Tuiskula-Haavisto, M., et al. 2011. Quantitative trait loci affecting eggshell traits in an F-2 population. *Animal Genetics*, 42, 293–299.

Tuiskula-Haavisto, M., et al. 2002. Mapping of quantitative trait loci affecting quality and production traits in egg layers. *Poultry Science*, 81, 919–927.

UE. 1999. Council Directive 1999/74/Ec of 19 July 1999 laying down minimum standards for the protection of laying hens.

Voinescu, A. E., et al. 2007. Mineralization of Caco3 in the presence of egg white lysozyme. *Langmuir*, 23, 12269–12274.

Wang, X. Q., et al. 2010. Influence of ovalbumin on Caco3 precipitation during in vitro biomineralization. *Journal of Physical Chemistry B*, 114, 5301–5308.

Wolf, S. E., et al. 2011. Strong stabilization of amorphous calcium carbonate emulsion by ovalbumin: gaining insight into the mechanism of ,Polymer-induced liquid precursor'processes. *Journal of the American Chemical Society*, 133, 12642–12649.

Yang, J. H., et al. 2013. Expression of Trpv6 and Cabp-D28k in the egg shell gland (uterus) during the oviposition cycle of the laying hen. *British Poultry Science*, 54, 398–406.

Yao, J. F., et al. 2010. Low-density lipoprotein receptor-related protein 8 gene association with egg traits in dwarf chickens. *Poultry Science*, 89, 883–886.

Yin, Z. T., et al. 2020. The transcriptome landscapes of ovary and three oviduct segments during chicken (Gallus gallus) egg formation. *Genomics*, 112, 243–251.

Zhang, J. C., et al. 2019. The differences of gonadal hormones and uterine transcriptome during shell calcification of hens laying hard or weak-shelled eggs. *BMC Genomics*, 20.

Zhang, Y. N., et al. 2017. Dietary manganese supplementation modulated mechanical and ultrastructural changes during eggshell formation in laying hens. *Poultry Science*, 96, 2699–2707.

Chapter 11

Advances in understanding and exploiting the genetics of egg quality

Petek Settar, Hy-Line International, USA; and Jesus Arango, Cobb Genetics, USA

1 Introduction

Today's poultry genetic principles and applications of breeding programs are guided by Mendel's milestone work on inheritance and Darwin's selection principles. Their foundational work was brought to poultry by Robinson (1912). He detailed principles of poultry selection, breeding applications and mating strategies in his work. Around the same time, Raymond Pearl (1911) conducted extensive work on the heritability of egg quality traits such as egg weight, egg shape, albumen height and shell quality. His work helped us to understand the genetic bases of today's egg quality traits, their relationships with other traits and the genetic basis of egg production. J. L. Lush (1945) laid the foundation for modern genetic selection in animals. He applied principles of population genetics, statistics and biometrics to improve traits in livestock. I. M. Lerner (1950) used poultry both as an animal model and as a commercial species to test use of part-year laying records on sibs to replace progeny testing of males and to reduce generation interval and increase selection intensity. His work outlines genetic and population genetic principles and practices. During the mid-twentieth-century, breeding companies pioneered the application of genetics to egg production and egg quality to select breeding lines. Their advancement has led to a significant period of progress in understanding and selecting egg quality traits in poultry programs.

http://dx.doi.org/10.19103/AS.2024.0150.18

In the late 90s, publication of the Chicken Genome (Wallis et al., 2004) and the development of genomic techniques, such as genome-wide association studies (GWAS) and quantitative trait loci (QTL) mapping, allowed researchers to identify specific regions of the chicken genome associated with egg quality traits, providing insights into the genetic bases of many traits (Dekkers, 2004; Dunn, 2015).

Today, understanding of the genetic mechanisms with genomic technologies helps to identify genetic markers associated with egg quality traits, and with that technology geneticist accelerate the breeding process and select individual birds based on their DNA profile for desirable egg quality characteristics on internal and external egg traits.

Additionally, emerging gene editing technologies such as CRISPR-Cas9 offer precise modifications to be made on specific genes associated with egg quality, potentially leading to the development of chicken lines and commercial hybrids with enhanced egg production and quality opportunities.

Today's egg layer industry has been shaped by the implementation of some of the genetic advancements mentioned above, along with market feedback, which has guided egg quality trait definitions aligned to specific customer demands. Long-term genetic changes in laying hens have been achieved through the implementation of truncation selection programs in a pyramidal population structure from pure lines (apex) to commercial layers (base). These selection programs determine which individuals become parents, how many offspring they produce, and how long they remain in the breeding population (Bourdon, 2000). Breeders evaluate large populations across multiple purebred lines by collecting large groups of phenotypic and genotypic data. Following genetic evaluation of pure lines, selected individuals are used to reproduce future generations of pure lines and they reproduce future generations down the breeding pyramid by using crossbreeding between these purebred lines, in the paternal and maternal sides of the final commercial layer. The process from pure lines to grandparents, parents and finally to the final commercial product takes 3–4 years. Even though significant data collection happens at pure line level, the subsequent stages of the multiplication process also contribute data back to the pure line level for selection purposes. This slow and steady progress in pure line breeding programs has resulted in today's highly prolific layers. Achievement of breeding programs has improved both the quality and quantity of the egg components.

In this chapter, we will discuss the definition of egg quality traits from the genetic perspective.

2 Definition of egg quality traits

Egg quality traits which were defined many centuries ago have always focused on producing healthy chicks. The definition has evolved over time toward more specific phenotypes such as egg size, shell quality, albumen height, shell color, internal quality (albumen height, Haugh Units, etc.) and nutritional values of each part of the egg. Earlier research focused on physical characteristics and on quantifying their variations (Robinson, 1912; Romanoff and Romanoff, 1949; Hutt, 1949) while today's work has become more sophisticated including genetic, environmental, health, nutrition and management aspects of egg quality traits (Hunton, 1987; Muir and Aggrey, 2003; Aggrey et al., 2010; Liu et al., 2020). The biggest impact of shifting focus was to use shell quality parameters to extend the laying cycle. Breeding strategies for hens are applied globally. Recently, breeding programs have adopted diversified strategies to tailor products for specific regions while considering industry requirements.

From the industry perspective, there are many definitions of egg quality, depending on the end product. The final product can be defined as hatching egg, table egg, ingredient for baking, pharmaceutical or cosmetics ingredient and its acceptability depends on the characteristics of each market and its preferences. From the table egg perspective, there are markets that prefer or accept eggs of a certain color or a particular size, and these constitute the key quality indicators. The preference for egg size is very marked according to the modality of trade and sale of the final product. Markets that sell eggs by the unit prefer them small (e.g. Brazil, India, Mexico, Peru, etc.); conversely, other markets can have complex size classification systems, and some give a preferential value to larger size eggs (large, extra, jumbo, etc.). Markets also differ greatly in their preference for egg color; generally, the white ones produced by strains of the white *Leghorn-WL* breed and the brown ones are produced by commercial hybrids that combine the *White Plymouth Rock-WPR and Rhode Island Red-RIR* breeds. In Asian markets (Japan and China), it is common to see pink 'tint' eggs (product of various crosses of the three breeds mentioned above). There are also niche markets for eggs of other colors such as those produced by Araucana breed (green and blue), and of other colors, such as intense brown in Iberian breeds or the exotic black eggs in some black feather varieties (Indonesian Ayam Cemani and others). There are markets that place a lot of importance on internal quality, particularly the quality of albumen (*Haugh Units*), especially those where raw egg consumption is common, such as Japan and South Korea. For the breeder side of the industry, quality of eggs is defined with hatchability and chick quality which are directly related to egg size, shell, yolk and albumen quality (Finkler et al., 1998; Onagbesan, 2007).

Most of the egg quality traits are influenced by a combination of genetic factors that affect various aspects of egg formation, structure and composition as well as environmental factors such as management, nutrition and health (Table 1).

3 External quality

The external quality groups together all the characteristics that can be measured without breaking the egg; its weight, color, texture and some shape properties. Listed below are the most important external characteristics, their measurement, determination, and genetic background with some relevant genetic correlations with other economic traits.

3.1 Egg weight

This is one of the most important external quality characteristics because, as mentioned above, in most markets eggs are classified mainly by their size (weight or mass). Egg weight is primarily determined by the genetics of hens, although environmental factors and nutrition also play a role in influencing the weight and composition of the egg, including the proportions of yolk and white (Table 1). Genetic factors impact the size of egg produced by layers. Selection for increased or decreased egg size involves genes controlling follicle development, ovulation rate and nutrient allocation to egg formation.

The EW is taken on individual eggs, sampling several eggs (2–4) at different ages, starting with the first eggs laid at the age of sexual maturity, and then on young, adult and mature birds (20–30, 40–50, 65–70 and >90 weeks of age). The estimates of heritability of egg weight are quite similar in different breeds (strains or lines) and it fluctuates between moderate and high. It tends

Table 1 Selection and environment responses on egg quality traits, created based on breeding practices (+ low impact; ++++ high impact).

Traits	Selection	Environment	Impact factors
Egg Weight	+++	++++	Genetic, age, nutrition, management
Shell color	++++	+	Genetic, health, UV light
Shell quality			Genetic, age, nutrition, health, management,
Elasticity	+	++++	
Breaking strength	+++	++++	
Cracks	++	++++	
Albumin	+++	++++	Genetic, health, storage, ambient temperature
Yolk	+++	++	Genetic, age, health
External/internal defects	+	++++	Health, management, age, ambient temperature

+ low impact; ++++ high impact.

to increase with the age of bird, being lower for the first eggs at sexual maturity (h^2: 0.32–0.45), with intermediate values at mature ages (h^2: 0.42–0.58) and the highest values in older birds (h^2: 0.55–0.61). Repeatability of egg weight tends to be high at all ages (RE: 0.65–0.78). This characteristic is always analyzed using multivariate models, which include egg weight at different ages, as well as other egg quality characters, especially yolk weight and albumen height. Random regression methods using Splines-like functions have also been used to describe weights at different ages as a trajectory (Arango et al., 2009). The genetic parameters obtained using multivariate and covariance function models are quite consistent and similar when estimated at the same ages.

Egg weight measured at different ages has positive genetic correlations, which also tend to increase with the age of the bird; being lower when the bird begins laying (rg: 0.75–0.90) than in adult and mature birds (rg: 0.80–0.96). Genetic correlations with other quality traits include (1) positive and moderate with yolk weight at different ages (rg: 0.41–0.53); (2) positive and low to moderate with albumen quality (rg: 0.20–0.25); and (3) low-to-moderate negative with shell quality (rg: 0 to −0.09). It also tends to exhibit low-to-moderate negative correlations with egg production at different ages (−0.20 to −0.45).

Another important aspect in relation to egg weight is its uniformity and variability during the production cycle. It is well known that the EW tends to increase with the age of the bird, but the market requires and rewards eggs of intermediate weight. In other words, it is advisable to keep the weight of the egg in a band that maximizes its acceptability in the market and avoids extreme weights. The selection program measures the variability of egg weight using different parameters; one of these measurements is the coefficient of variation of EW which measures the percentage variability in the EW of the eggs produced by the same bird in its productive life. The heritability estimates for this trait are moderate (h^2: 0.20–0.30); so it is a characteristic that can be used efficiently in the breeding program to increase egg weight uniformity (egg size stability) in elite populations.

3.2 Shell color

Eggshell pigmentation presents quite a bit of variation in hues, ranging from white to dark brown. Color separation in the layer industry and table egg markets is a result of long-term selection strategies applied by breeding companies to achieve uniform color. As is the case with egg weight, markets tend to segregate by preference for shell color. Those markets that prefer white eggs want them very clear and without spots, although in white leghorn birds, it is common for the eggs to be a little creamy in young layers, and this coloration tends to disappear a few days or weeks after the onset of production. Some markets prefer brown or tinted eggs, preferring them uniformly dark and free of speckles or with speckles. In nature, birds gradually developed the process for synthesizing pigments that were deposited in eggshells, some suggested as a function of protecting eggs from the predators or attracting mates (Soler et al., 2005; Stoddard et al., 2011).

Pigmentation is determined by genetic factors influencing pigment deposition in the shell gland. This pigment is deposited in the cuticle and the outer portion of the calcified eggshell. Premature interference of cuticle formation, release of stress hormones or disease challenges such as bronchitis will result in pale color eggs (Cook, 1986; Hughes, 1986; Mills, 1987). Main pigments such as protoporphyrin IX, biliverdin-IX and zinc chelate are metabolized as a pigment deposited onto the eggshell 90 minutes prior to oviposition. Therefore, such a late addition of pigment to already developed shell structure shows non-uniform distribution throughout the thickness of the eggshell. Additionally, the intensity of the color is given by the concentration of pigment that the bird excretes.

There are different ways to measure shell color. Earlier studies used photoperiodic equipment to measure shell color in selection programs (Hunton, 1962; Gowe, 1965). The most common method uses the 'L-a-b' system, which combines 3 parameters: 'L' (luminosity, refraction of light emitted by the equipment) and two parameters ('a-b') that describe the wavelength of color spectrum of the sample. In many cases, these three values are combined into a color index (CO), which allows the comparison of eggs from any color using the same scale (Fig. 1). In the index, the spectrum ranges from 5 (bleached white) to 110 (dark brown).

Another color aspect of interest for the egg industry is the presence of surface spots, speckles in brown eggs and translucent spots in white eggs. A semi-quantitative classification system (0–5) is used for its measurement, which goes from a completely spot-free egg (0) and increases according to size and distribution (proportion of the surface covered with spots) to a maximum value of '5'. In initial studies of this characteristic, positive genetic correlations were found between color intensity and the incidence of spots or speckles (Arango et al., 2006). Therefore, both characteristics are co-analyzed using multivariate models at different ages. Shell color is the characteristic that shows the highest

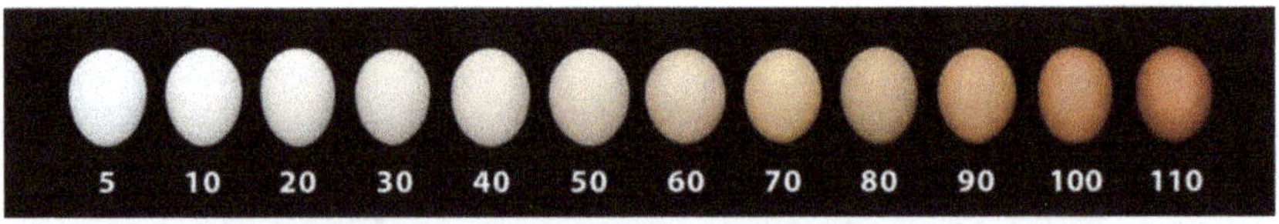

Figure 1 Eggshell color index spectrum, generated by using l-a-b index. Source: Photo courtesy of Hy-Line International, Iowa, USA.

rates of inheritance in laying birds, showing estimates of heritability that range between 0.40 and 0.65 depending on the age of the bird and the genetic line. Heritability is lower in white eggs than in color egg lines. Shell color is also a highly repeatable trait (0.60–0.80). Speckles or surface spots, on the other hand, are less heritable (0.20–0.35) and repeatable (0.35–0.60). Genetic correlations for color measurements at different ages are generally high, ranging between 0.75 and 0.98. As noted above, the genetic correlations between shell color and incidence of surface spots tend to be positive and moderate (0.20–0.40), in other words, care must be taken when trying to remove the incidence of speckles as it can indirectly reduce the intensity of the shell color, which is a valued trait in brown-egg markets.

3.3 Shell quality

The eggshell is the natural physical barrier which allows the egg to resist microbial contamination. It also functions to regulate the transfer of metabolic gases and water. Hence, shell quality is one of the key traits for the poultry breeding industry and is critical for maintaining egg integrity from lay to final customer. The complex structure of the trait is mostly related to shell breaking resistance, plasticity and permeability of the whole structure. The shell not only protects the internal components of the egg from the action of the elements or against microbial contamination, but also allows the transport, sorting and distribution of the product to its final consumer. In the case of embryonated (hatching) eggs from breeder, the shell provides key minerals for the development of the embryo and facilitates gas exchange during incubation. For all the above reasons, shell quality is crucial to produce commercial 'salable eggs' and is one of the most important traits in selection of layer lines, just after egg production, which is without doubt the trait with the highest economic importance.

The influence of genetics is an important factor on variation of shell thickness, strength and texture. Genes that are involved in calcium metabolism, uterine calcium secretion and in the mineralization of the shell process play crucial roles in determining shell quality.

Collagen is present at a very low level in eggshells. The primary protein in shell membrane is not collagen; rather, it is a specific protein that shares some similarities with collagen but is unique to eggs. In addition to the shell mass egg shell matrix proteins that control the mineralization process, the shell texture and mechanical properties are important (Hincke et al., 2012). Recent studies have elucidated some molecular mechanisms involved in shell formation, including the secretion of calcium amorphous carbonate by pinocytosis in the shell gland (Nys et al., 2021).

As illustrated in Table 1, non-genetic (environmental) factors play a major role in determining shell quality, particularly those related to hen's nutrition and management. While both major factors, there are also many impacts on shell quality related to the health of the bird. Many diseases of microbial origin have important effects on the quality of the shell, especially epithelial-trophic viruses such as infectious bronchitis or Newcastle, among others. The management of poultry and the infrastructure of farms and sorting points, as well as storage, transport and distribution of the product also have important effects on the integrity of the shell to the end consumer. Age of the birds also plays a major role, as the shell quality tends to deteriorate as the bird ages in a roughly linear fashion. Egg laying cycles are getting longer each year, which places added emphasis on the persistency of eggshell quality. Solomon (2010) reviewed traditional selection methods to improve shell quality. In general, shell quality measurements can be grouped into destructive vs. non-destructive and static vs. dynamic. Static measurements using nondestructive methods, such as the deformation test, use the displacement of the shell upon applying pressure that simulates point of impact, without cracking the shell. Destructive ones, such as breaking strength (BS), apply blunt force on the surface of the eggshell until it is cracked (Bain, 2005). The BS measurement is highly correlated to shell thickness (Washburn, 1992; Zhang et al., 2005; Blanco et al., 2014). BS can be measured at the equator (BSe) and poles (BSp) of the egg (Arango, 2016). Genetic correlations between BSe and BSp were reported as between 0.51 and 0.68 (Arango et al., 2016). More comprehensive evaluations can be provided by dynamic methods such as the acoustic egg test (AET) developed by Coucke et al. (1999). AET measurement is non-destructive and has proven to be a good predictor of the occurrence of cracks (Bain et al., 2006; Amer Eissa, 2009; Mertens et al., 2006). The acoustic egg test combines shell resonance profile with egg mass to calculate dynamic stiffness (Kdyn), a quantitative indicator of integral shell resistance. This novel method allows direct detection of both cracks and micro-cracks, making it possible for use in selection programs (Arango et al., 2016). The BS traits were moderately correlated with Kdyn reported in 0.23–0.51 range, and negatively correlated (−0.36) with micro cracks. Heritability estimates reported for shell quality traits (Dunn et al., 2005; Icken et al., 2006; Blanco et al., 2014; Arango et al., 2016) were at a moderate level. Also, moderate correlation between those traits confirms that they are measuring different aspects of shell quality; therefore, some selection applications use a shell biocomplex model to improve overall shell quality by simultaneously using BSe, BSp and Kdyn (Arango et al., 2016).

Recent studies focusing on QTL detection for various shell quality traits have been performed, and chromosomes Z, 2, 4, 8, 17, 20 and 24 have been the most frequently reported to impact these traits (Roberts, 2017; Wolc et al., 2020a). Using genomic information in prediction of shell bio-complex breeding values resulted in higher accuracies compared to classic pedigree-based models (Wolc et al., 2020c). GWAS recently reported multiple candidate genes related to eggshell strength in Rhode Island Red chickens of different age (Chen et al., 2024).

3.4 External defects

There are several external defects that affect the quality of the egg. The most important ones are those affecting the integrity of the egg, which include microfractures, breakages and stains, texture, egg shape and deformities. A comprehensive study on the genetic determination of egg defects was presented by Wolc et al. (2012), who reported results of the following defects: bloodstained, broken, calcium deposits, dirty, deformed, too small (pee-wee), shell-less, weak or thin shell, and double-yolk. The latter, despite being an internal defect, is also grouped with external defects as it can be visibly identified by the size and shape of the egg, as well as using an ovoscope. They reported that the highest incidence of defects was observed in young birds at the beginning of laying and their incidence was very low in mature birds. Heritability estimates for these traits were relatively low (about 10%) for all defects except for calcium deposits and double-yolk eggs (about 20%) (Wolc et al., 2012). Defects were analyzed with multivariate models which included age at sexual maturity, body weight of the bird and its egg production. Overall, a negative correlation was consistently found between the incidence of egg defects and egg production. In other words, birds with a poor production profile tend to have a higher proportion of defective eggs. Similarly, a negative correlation was found between double yolk and weights of bird and egg as expected, and to lesser extent with the other defects. Estimates of genetic correlation between defects related to shell quality were positive (0.24–0.73). This suggested that, instead of using individual defects, they could be grouped into a single category (Wolc et al., 2012). In this way, it was possible to avoid problems associated with extreme values (distribution inflated by zeros in the cases of very low frequency defects) and apply selection on a unique trait with a higher heritability. This approach is easier to implement thus achieving greater genetic progress in the control of egg defects than the one that could be achieved by targeting different defects separately. Because of the genetic correlations, prediction models for defects should always include production traits. In addition, egg production should be evaluated and selected for in terms of salable eggs (i.e. excluding defective eggs in production counts).

4 Internal quality

Groups the measurable characteristics once the egg has been broken. Genetic variations impact internal egg characteristics such as yolk size and score, albumen height, air cell size, membrane integrity and presence of internal inclusions (blood spots or meat spots). Genes involved in nutrient metabolism and vascular development contribute to internal egg quality traits.

4.1 Quality of albumen and Haugh Units

One of the most important characteristics of internal egg quality are those related to the albumen. It contains a firm or thick portion and a watery or thin portion. The albumen height (AH, mm) of the thick portion is the most commonly used measurement in genetic selection because it is easy and quick to measure, and it shows a moderate heritability (0.20–0.32) and is highly correlated with the measurement used in the industry, the Haugh Units (HU). Introduced by Raymond Haugh in 1937, HUs are nothing more than the height of albumin adjusted for egg weight, using the following formula:

$= 100 \times \log10\ (7.57 + AH - 1.7 \times {}^{0.37})$, where HU = Haugh Units, AH = albumen height (mm), EW = egg weight (g).

The industry uses HU as an internal quality reference to indicate the freshness conditions of the product, since it shows its maximum value when the eggs are fresh (recently laid) and deteriorates both over time and with the increase of storage temperature. That is, the highest HU values are obtained in fresh eggs or eggs recently stored under refrigeration, and these values decrease as the eggs are stored for a longer duration of time and/or at higher temperatures. In addition, the HU also decreases with the age of the birds, being its maximum value in young birds and observing a linear deterioration with the age of the hen. The comparison between varieties or strains of hens must be observed by comparing the HU values of birds of the same age and subjected to the same storage regimes (time and temperature). Birds that retain the highest HU values maintain the internal appearance of fresh eggs for a longer period of storage time. Some comparisons can be reviewed in the literature, with examples in Curtis et al. (2005), Karcher et al. (2015) and Jones et al. (2018).

Genetic factors influence albumen quality aspects such as albumen viscosity, pH and protein composition, which affect the functional properties of the egg white. Albumen antimicrobial property are relatively stable and heritability is reported low (0.01–0.09) for protein traits (Sellier et al., 2007). Genes controlling protein synthesis, secretion and post-translational modifications contribute to albumen quality.

4.2 Yolk weight and dry matter (solids, SO) content

Egg yolk is a complicated quantitative trait affected by many factors, such as breed and hen age (Suk, 2001; Nangsuay, 2011). Yolk traits are well established and categorized as yolk weight, percent yolk, yolk dry matter and percent yolk concentration (Akbar, 1983; Hartmann, 2001). Through the application of both management and breeding approaches, it is possible to influence yield of yolk weight, proportion and composition. There is strong evidence of the existence of additive genetic variation in several yolk traits, a synergy which would make breeding a viable alternative to the modification of yolk production (Hartmann, 2001). Selection responses and correlated selection responses have been studied extensively for yolk proportion and yolk weight (Miyoshi, 1980; Hartmann et al., 2000; Hartmann, 2001; Icken et al., 2014).

Dry matter content is of great importance for industrial eggs but measuring it in individual eggs is very difficult and expensive because it is time-consuming and requires specialized equipment and materials (Ahn et al., 1997). The individual measurement requires taking the weight of the egg and each of its components, i.e. yolk, albumen and shell with their membranes. Subsequently it requires producing separate homogeneous solutions of the yolk and albumen, taking an aliquot of each and preparing samples followed by drying, using industrial microwaves, then the solids content of each component is calculated by moisture loss, i.e. difference in weight of the wet sample minus the weight of the dry sample. Finally, the solid content of the egg is expressed as a percentage by considering the dry matter content of each component and its proportion of the total weight of the egg. As can be seen, the measurement is not complex but it requires several time-consuming steps, equipment and materials. These processes make it impractical as a routine evaluation trait for the genetic evaluation and improvement program. As such, the individual yolk weight is a quick measurement trait that by itself, or as a proportion of the total weight, is a good marker to select for SO content.

Icken et al. (2014) reported a higher heritability estimate for yolk ratio (h^2: 0.44) than for solids content (h^2: 0.39) and a genetic correlation between them of 0.91, indicating that indirect selection to increase solids content using yolk weight or ratio is more efficient and practical than direct selection base on egg solids content.

There are other measures to assess the internal quality of the egg, such as the strength of the yolk membrane and its elasticity (Jones personal communication) and the yolk index. But they are still in the research and exploration phase and are not widely recognized or used in the industry. Preliminary studies (J. Arango, unpublished data) in three *White Leghorn* strains showed acceptable repeatability for the strength (0.22–0.25) and elasticity (0.19–0.23) of the vitelline membrane, and average phenotypic and genotypic correlations of 0.92 and 0.95, respectively. These values suggest that strength is a better candidate than elasticity as an indicator of internal yolk membrane quality.

Heritability estimates were lower than those observed for albumen height and Haugh units; therefore, the latter measurements are still preferred for the assessment of internal egg quality.

4.3 Internal defects

Internal inclusions (blood and meat spots) are perhaps the most important internal defect traits in layers from the consumer perspective. The end consumer tends to reject the product if they consistently find these inclusions, which also have negative connotations in some cultures and religions. Incidence of these traits is measured during the evaluation of the internal quality of the egg. A scoring scale is generally used that measures inclusions by their presence and size (0 = absence to 5 = large inclusions), some examples are shown in Fig. 2.

In general, blood inclusions are due to the occurrence of micro-hemorrhages during ovulation; therefore, they tend to be visible on the surface of the yolk (Nalbandov, 1944; Shirley, 1965). Meat spots are brown or white spots that can appear in the egg white (albumen). These spots are primarily caused by the shedding of small pieces of the oviduct's epithelium as the egg moves through the magnum section of the oviduct, where the albumen is deposited. To achieve a good differentiation between blood and meat inclusions, the evaluation should be done in fresh eggs (preferably in the first 24 h after laying), since blood stains begin to degrade (digest) or change color due to dilution and it is difficult to differentiate them from meat in stored eggs. The heritability estimates for both types of inclusions are low (10% or

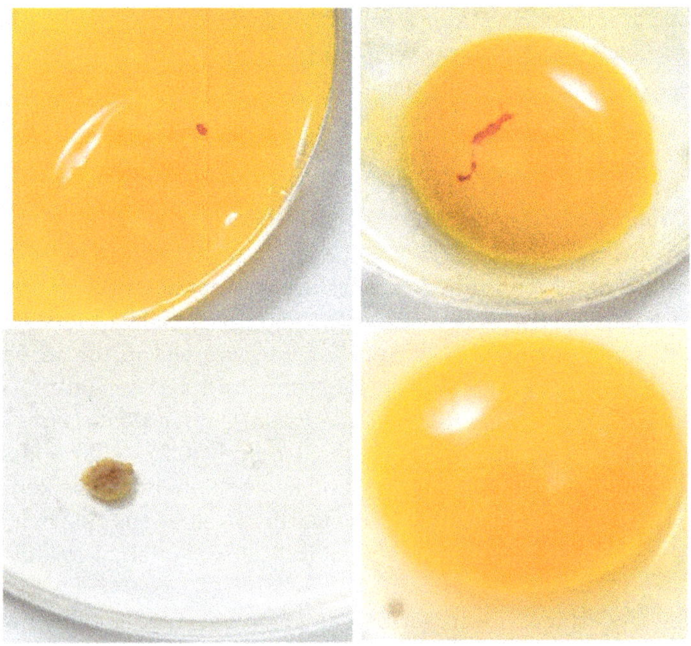

Figure 2 Example of blood inclusions (top photos) and flesh (bottom photos). Source: Photo courtesy of Hy-Line International, Iowa, USA.

less) (Becker, 1973; Wolc et al., 2012; Honkatukia, 2011). Breeding response to selection is very inefficient and requires measurement over long periods of time, multiple generations and in large numbers of eggs. The incidence of inclusion level greatly varies between brown and white shell eggs. Brown shelled eggs may show an incidence as high as 18%, whereas it is only 1% in white egg layers (Campo, 1998; Wolc et al., 2012). The incidence of inclusions is associated with stress factors such as biotic (disease) and thermal stress, sudden changes in ventilation and lighting, metabolic disease or nutritional factors such as vitamin A deficiency or presence of vitamin K antagonists and/or toxic factors like a presence of mycotoxins or anticoagulant compounds such as those found in some rodent control products. Therefore, it is more efficient to control these inclusions through the implementation of good management practices and environmental control than through the genetic improvement program. Management and feeding practices to reduce the incidence of internal inclusion is out of the scope of the chapter herein.

In respect to internal quality, genetic variations impact internal egg characteristics such as yolk consistency, air cell size and presence of blood spots or meat spots. Honkatukia et al. (2011) published the first report on the genome location of a QTL region on chromosome Z that is associated with blood and meat inclusion in chickens.

5 Functional characteristics of the egg

Many functional characteristics of egg components are important for their use in the food, pharmaceutical and cosmetic industries, as well as for culinary applications. Some of these include foaming, crystallization, emulsification and coagulation (Kerth et al., 2005; Jones et al., 2018). These characteristics are more complex to measure and are generally not considered in commercial or breeding company egg quality laboratories. Many of the functional characteristics can be positively affected by selection for physical traits, such as albumen quality (HU) and are beyond the scope of this chapter. Differences in egg functional characteristics have been described between commercial layer varieties (Kerth et al., 2005), indicating that genetic factors do affect them.

6 Conclusion

The genetics of egg quality plays a critical role in determining the commercial success of egg production. Genetic selection programs target many different egg quality traits as detailed in this chapter. These traits are influenced by genetic factors controlling aspects like follicle development, hormone regulation and nutrient metabolism in the hen. Breeding efforts focused on enhancing traits associated with superior egg quality while maintaining or improving other desirable production traits like egg production and feed efficiency. Genomic technologies have revolutionized poultry breeding by enabling the identification of genetic markers associated with egg quality, facilitating more precise selection strategies and accelerating genetic improvement. New technologies can make significant advancements in egg quality traits, ultimately enhancing the profitability and sustainability of egg production systems.

7 References

Ahn, D. U., et al. 1997. Effect of egg size and strain and age of hens on the solids content of chicken eggs. *Poultry Science*, 76, 914–919.

Aggrey, S. E., et al. 2010. Genetics of egg quality traits in chickens. In *Poultry Science Symposium Series* (Vol. 26, pp. 27–35). Poultry Science Association. https://doi.org/10.3382/ps.2009-00293

Akbar, M. K., et al. 1983. Changes in the composition of egg solids during storage. *Poultry Science* 62, 925–933.

Amer Eissa, A. H. 2009. Comparative eggshell stability assessment using three different non-destructive sensing instruments and breakage force strength. *Journal of Food Engineering*, 93(4), 444–452.

Arango, J. 2016. Egg focus. 13. Improving shell quality. *International Poultry Production*, 24(7), 49.

Arango, J., et al. 2006. Relationship between shell color and the incidence of speckles in brown egg lines [Poster]. *Proceedings XII European Poultry Conference*, Verona, Italia. http://www.cabi.org/animalscience/worlds-poultryscience-association-wpsa/wpsa-italy-2006

Arango, J., et al. 2009. A random regression approach to model egg weight trajectory in two layer populations [conferencia]. *Proceedings 6th European Poultry Genetic Symposium*, Poznan, Polonia. http://www.up.poznan.pl/epgs2009/session2/selection/Arango.pdf

Arango, J., et al. 2016. Model comparison to evaluate a shell quality bio-complex in layer hens. *Poultry Science*, 95(11), 2520–2527. http://dx.doi.org/10.3382/ps/pew286

Bain, M. M. 2005. Recent advances in the assessment of eggshell quality and their future application. World's *Poultry Science Journal*, 61, 268–277.

Bain, M. M., et al. 2006. Probability of an egg cracking during packing can be predicted using a simple non-destructive acoustic test. *British Poultry Science*, 47, 462–469.

Becker, W. and Bearse, G. 1973. Selection for high and low percentages of chicken eggs with blood spots. *British Poultry Science*, 14(1), 31–47.

Blanco, A. E., et al. 2014. Genetic parameters of egg quality traits on different pedigree layers with special focus on dynamic stiffness. *Poultry Science*, 93, 2457–2463.

Bourdon, R. M. 2000. *Understanding Animal Breeding* (2nd ed.). Prentice Hall.

Campo, J. L. and Garcia, G. M. 1998. Internal inclusions in brown eggs: relationships with fearfulness and stress. *Poultry Science*, 77(12) 1743–1747.Cavero, D., et al. (2011). Improving hatchability in white egg layer strains through breeding. *Lohmann Info*, 46, 44–54.

Chen, X., et al. 2024. Genetic patterns and genome-wide association analysis of eggshell quality traits of eggtype chicken across an extended laying period. *Poultry Science*, 103, 103458.

Cook, J. K. A. 1986. Pale shelled eggs can be caused by IB virus. *Misset International Poultry*, 2, 38–39.
Coucke, P., et al. 1999. Measuring the mechanical stiffness of an eggshell using resonant frequency analysis. *British Poultry Science*, 40, 227–232.
Curtis, P. A., et al. (2005). Quality and compositional characteristics of layer hens as affected by bird age [conferencia]. *Proceedings XIth European Symposium on the Quality of Eggs and Egg Products. Egg Quality and Composition.* Doorwerth, Países Bajos.
Dekkers, J. C. (2004). Commercial application of marker- and gene-assisted selection in Livestock: strategies and lessons. *Journal of Animal Science*, 82(E-Suppl), E313–E328. https://doi.org/10.2527/2004.8213_supplE313x
Dunn, I. C. and Paton, I. R. 2015. Molecular genetics of egg quality. In I. C. Dunn and I. R. Paton (Eds.), *Improving the Safety and Quality of Eggs and Egg Products* (pp. 97–115). Elsevier.
Dunn, I. C., et al. 2005. Heritability and genetic correlation of measurements derived from acoustic resonance frequency analysis; a novel method of determining eggshell quality in domestic hens. *British Poultry Science*, 46, 280–286.
Finkler, M. S., et al. 1998. Experimental manipulation of egg quality in chickens: influence of albumen and yolk on the size and body composition of near-term embriyos in a precocial bird. *Journal of Comparative Physiology B*, 168, 17–24.
Gowe, R. S., et al. 1965. Effect of inbreeding on laying hens. Poultry Science, 44(1), 264–270.
Hartmann, C. and Wilhelmson, M. 2001. The hen's egg yolk: a source of biologically active substances. *World's Poultry Science*, 57, 3–28.
Hartmann, C., et al. 2000. One generation divergent selection on large and small yolk proportions in White Leghorn line. *British Poultry Science*, 41, 280–286.
Haugh, R. R. 1937. The Haugh unit for measuring egg quality. *U.S. Egg and Poultry Magazine*, 43, 552555.
Hincke, M. T., et al. 2012. The eggshell: structure, composition, and mineralization. *Frontiers in Bioscience*, 17, 1266–1280.
Honkatukia, M., et al. 2011. Mapping of QTL affecting incidence of blood and meat inclusions in egg layers. *BMC Genetics*, 12, 55.
Hughes, B. O., et al. 1986. Categorization and causes of abnormal eggshells: relationship with stress. Brit. *Poultry Science*, 27, 325–337.
Hunton, P. 1962. Genetics of egg shell colour in a Light Sussex flock. *British Poultry Science*, 3, 189193.
Hunton, P. 1987. Laboratory evaluations of egg quality. In R. G. Wells and C. G. Belyavin (Eds.), Egg Quality – Current Problems and Recent Advances (Poultry Science Symposium No. 20, pp. 87–102). Butterworths.
Hutt, F. B. 1949. *Genetics of the Fowl*. McGraw-Hill Book Company Inc.
Hy-line Technical bulletin. 2017. *La Ciencia de la calidad del Huevo.* https://www.hyline.com/Upload/Resources/TU%20EQ%20SPN.pdf
Icken, W., et al. 2006. Dynamic stiffness measurements with the "crack detector": a new method to improve egg shell strength. *Lohmann Information*, 41, 13–19.
Icken, W., et al. 2014. Implications of genetic selection on yolk proportion on the dry matter content of eggs in a White Leghorn population. *British Poultry Science*, 55(3), 291–297. http://dx.doi.org/10.1080/00071668.2014.917272
Jones, D. R., et al. 2018. Hen genetic strain and extended cold storage influence on physical egg quality from cage-free aviary housing system. *Poultry Science*, 97, 2347–2355. http:/dx.doi.org/10.3382/ps/pex052

Karcher, D. M., et al. 2015. Impact of commercial housing systems and nutrient and energy intake on laying hen performance and egg quality parameters. *Poultry Science*, 94, 485–501.

Kerth, L. K., et al. 2005. Impact of bird age on the functionality of eggs from current layer strains [conferencia]. *Proceeding of XIth European Symposium on the Quality of Eggs and Egg Products*. Doorwerth, Países Bajos.

Lerner, I. M. 1950. *Population Genetics and Animal Improvement*. University Press.

Liu, W. and Xu, G. 2020. Advances in genetics and breeding of egg quality traits in poultry. *Animals*, 10(7), 1167. https://doi.org/10.3390/ani10071167

Lush, J. L. 1945. *Animal Breeding Plans* (3rd ed.). Iowa State University Press.

Mertens, K., et al. 2006. Monitoring of eggshell breakage and eggshell strength in different production chains of consumption eggs. *Poultry Science*, 85(9), 1670–1677.

Mills, A. D, Marche, M., and Faure, J. M., 1987. Extraneous eggshell calcification as a measure of stress in poultry. *British Poultry Science*, 28, 177–181.

Miyoshi, S. and Mitsumoto, T. 1980. Selection for high and low yolk-albumen ratio in chicken II. The direct responses to selection on the yolk-albumen ratio for seven generations. *Japanese Poultry Science*, 17, 219–227.

Muir, M. W. and Aggrey, S. E. 2003. *Poultry Genetics, Breeding and Biotechnology*. CABI Publishing.

Nalbandov, A. and Carl, L. 1944. The problem of blood clots and meat spots in chicken eggs. *Poultry Science*, 23, 170–180.

Nangsuay, A., et al. 2011. Yolk absorption and embryo development of small and large eggs originating from young and old breeder hens. *Poultry Science*, 90(11), 2648–2655.

Nys, Y., et al. 2021. Mechanism and hormonal regulation of shell formation: supply of ionic and organic precursor, shell mineralization. In C. G. Scanes and S. Dridi (Eds.), *Sturkie's Avian Physiology* (Chapter 32, pp 813–861). Elsevier Inc.

Onagbesan, O., et al. 2007. Gas exchange during storage and incubation of Avian egg: effects on embryogenesis, hatchability, chick quality and post-hatch growth. *World's Poultry Science Journal*, 63, 557–573.

Pearl, R. 1911. Inheritance of fecundity in the domestic fowl. *The American Naturalist*, 45(534), 321–345.

Roberts, J. 2017. *Achieving Sustainable Production of Eggs* (Volume 2). Burleigh Dodds Science Publishing.

Robinson, J. H. 1912. *Principles and Practice of Poultry Culture*. Ginn and Company.

Romanoff, A. L. and Romanoff, A. J. 1949. *The Avian Egg*. John Wiley & Sons.

Samiullah, S., et al. 2015. Eggshell color in brown-egg laying hens- a review. *Poultry Science*, 94(10), 2566–2575. http://dx.doi.org/10.3382/ps/pev202.

Sellier, N., et al. 2007. Estimations of repeatability and heritability of egg albumen antimicrobial activity and of lysozyme and ovotransferrin concentrations. *British Poultry Science*, 48(5), 559–566.

Shirley, H. V. 1965. An observed blood spot formation. *Poultry Science*, 44, 1139.

Soler, J. J., et al. 2005. Blue and green egg-color intensity is associated with parental effort and mating system in passerines: support for the sexual selection hypothesis. *Evoluation*, 59, 636–644.

Solomon, S. E. 2010. The eggshell: strength, structure and function. *British Poultry Science*, 51(Suppl. 1), 52–59.

Stoddard, M. C., et al. 2011. Imperfect camouflage avian eggs: artifact or adaptation? *Avian Biology Research*, 4, 196–213.

Suk, Y. O. and Park, C. 2001. Effect of breed and age of hens on the yolk to albumen ratio in two different genetic stocks. *Poultry Science*, 80, 855–858.

Tullett, S. G. 1987. Egg shell formation and quality. In R. G. Wells and C. G. Belyavin (Eds.), Egg Quality: Current Problems and Recent Advances (Poultry Science Symposium No. 20, pp. 123–146). Butterworths.

Washburn, K. W. 1992. *Factors Affecting Selection Progress for Shell Quality*. Poultry Breeders of America Roundtable. Accessed March 2016. http://www.poultryscience.org/docs/pba/1952-2003/1992/1992%20Washburn.pdf

Wallis, J. W., et al. 2004. A physical map of the chicken genome. *Nature*, 432(7018), 761–764. https://doi.org/10.1038/nature03030

Wolc, A., et al. 2012. Genetic parameters of egg defects and egg quality in layer chickens. *Poultry Science*, 91, 1292–1298. http://dx.doi.org/10.3382/ps.2011-02130

Wolc, A., et al. 2020a. Genetic parameters of egg quality traits in 2 purebred and 2 commercial layer lines. *Poultry Science*, 99(5), 25042510.Wolc, A., et al. 2020b. Breeding chickens for egg functionality and quality. *Poultry Science*, 99(12), 6663–6670. https://doi.org/10.1016/j.psj.2020.09.002

Wolc, A., et al. 2020c. Accuracy of genomic prediction of Shell quality in White leghorn line. *Poultry Science*, 99, 2833–2840.

Wolc, A., et al. 2010. Estimates of genetic parameters and genetic change for production traits of two selection lines of chickens. *Poultry Science*, 89(3), 428–437. https://doi.org/10.3382/ps.2009-00324

Zhang, L. C., et al. 2005. Heritabilities and genetic and phenotypic correlations of egg quality traits in Brown-egg dwarf layers. *Poultry Science*, 84, 1209–1213.

Chapter 12

Regulation of calcium metabolism in laying hens

Michel J. Duclos, Baraa Ezzo, and Agnès Narcy, INRAE – UMR BOA, France; and Fabien Hervo, Laval University, Canada

1 Introduction

Calcium (Ca) is a crucial element for life (Brini et al., 2013) and is required for a number of cellular and molecular mechanisms. It is also the main constituent of the skeleton, being complexed with phosphorus (P) to form hydroxyapatite. Apart from its mechanical properties, the skeleton also acts as a metabolic reserve of Ca, which can be exchanged with the plasmatic pool through bone remodelling. In laying hens, Ca is of great importance in the eggshell, which requires the deposition of about 2.2 g of Ca in the form of Ca carbonate ($CaCO_3$). As egg production is repeated daily over a long period (50–80 weeks), the hen exports up to 50 times its body Ca. Moreover, as eggshell

http://dx.doi.org/10.19103/AS.2024.0150.20

synthesis occurs mainly during the dark period when the digestive tract is empty, an additional Ca reserve is mobilized. This is devoted to a unique form of bone, the medullary bone which is prone to very rapid bone turnover and develops at the onset of lay. The daily eggshell calcification and the rapid medullary bone turnover represent a huge challenge to Ca homeostasis and its underlying regulations.

2 Eggshell calcification and calcium (Ca) requirements in the laying hen

In all vertebrates, Ca is provided through dietary intake, and its homeostasis is very variable depending on physiological status. For example, an adult woman (Fig. 1A) of about 50 kg ingests 1.3 g/day just for the maintenance of her Ca pool (Fleet, 2017). Absorption is limited to 0.3 g (25%), whereas the serum pool can exchange about 0.5 g/day with the skeleton, with resorption being equal to formation. About 0.2 g will be excreted in urine and 1.1 g in the faeces (non-absorbed fraction ~1.1 g + endogenous loss ~0.2 g). By contrast, a laying hen (Fig. 1B), of 1.5 kg body weight, ingests close to 4 g of Ca, of which 2.2 g will be targeted to the uterus for eggshell formation (Bar, 2009). Laying hens exhibit a high capacity to absorb Ca, thus about 2.6 g, i.e. 65% (Hurwitz et al., 1973), is absorbed, while excretion in droppings remains limited to about 1.6 g, including 0.25 g in urine (Manangi et al., 2018). Due to the mobilization of Ca from the medullary bone, the exchange with the bone reserves is about 0.9 g, i.e. about 40% of the Ca deposited in the eggshell (Bar, 2009).

Laying hens are classically reared on a 16 h light and 8 h night regimen, and the oviposition mainly occurs in the morning. Around 45 min later, the hen expels a new oocyte in the reproductive tract, which reaches the eggshell gland (uterus) at 10 h post ovulation (PO). The calcification process lasts over a 12 h period during which about 0.33 g/$CaCO_3$/h is deposited (Nys and Le Roy, 2018). Thus, the laying hen has to develop strong systems of transfer for massive quantities of Ca across the intestinal and uterine epithelia. The mechanisms of Ca transport across the uterine epithelium and of eggshell biomineralization in the uterine fluid are the object of a specific chapter in this book (see Gautron and Rodriguez-Navarro, 2025). The process is a physiological priority, which requires a renewal of the plasmatic pool of Ca in the range of 12-fold/h (Nys and le Roy, 2018). It is also a huge challenge as it partly occurs during the dark period, when the hen does not feed and the intestine is depleted in ionic Ca.

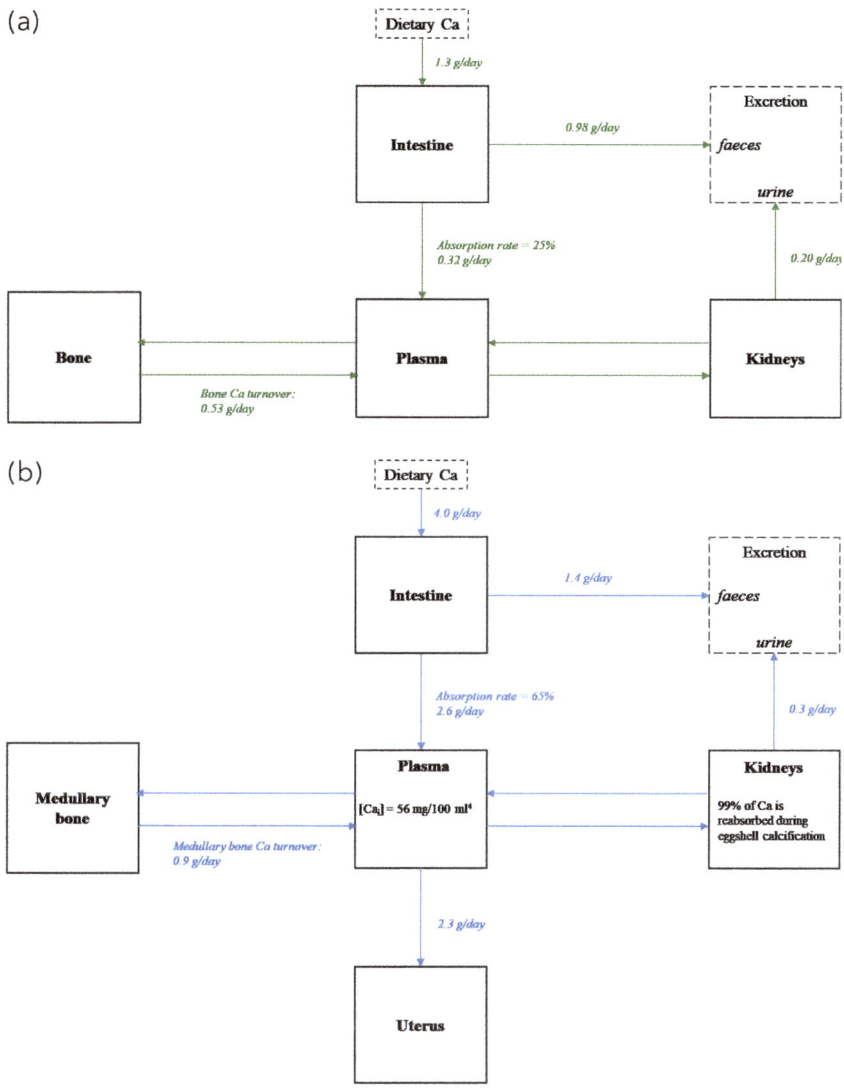

Figure 1 (A) Quantification of daily calcium flows in adult woman; (B) quantification of daily calcium flows in laying hen (during an eggshell forming day). Data issued from Fleet (2017) for Fig. 1A. For Fig. 1B, data are issued from NRC (1994) for Ca intake, from Hurwitz et al. (1973) for intestinal uptake, from Parsons and Combs (1981) for plasmatic concentrations, from Bar (2009) for bone flows, from Manangi et al. (2018) for urinary flows, and from Nys (2018) for Ca requirement for egg formation.

3 Underlying and regulatory mechanisms in the laying hen

3.1 Ingestion

In the diet, Ca is found mainly in the form of $CaCO_3$, provided by limestone. The high Ca requirement of laying hens for the eggshell requires high dietary Ca concentrations of around 35–40 g/kg (Nutrient Requirements of Poultry: Ninth Revised Edition, 1994). While other vertebrates increase Ca intake only in the instance of deficiency, the hen behaves differently. Towards sexual maturity, the hen develops a specific appetite for Ca, increasing its intake simultaneously with the development of her medullary bone (Fleming et al., 1998; Meyer et al., 1970). A similar mechanism occurs subsequently each day when an egg is produced (Nys and Le Roy, 2018).

3.2 Calcium transfer across the small intestine

Once ingested, dietary Ca is solubilized in the stomach (proventriculus and gizzard in birds) prior to its absorption in the small intestine. Only Ca^{2+} ions can cross the epithelium, by two alternative pathways, transcellular or paracellular. Across all species, similar mechanisms are operating although the degree of knowledge can differ (Fleet, 2017).

The transcellular pathway implies transfer through the epithelial cells. It requires a specific active mechanism for entry, a protecting mechanism to maintain intracellular levels of free Ca at levels avoiding cellular damages, in the range of 100 nM (Fleet, 2017). The transcellular pathway mediates an active and saturable transport. The entry of Ca at the apical pole is possible thanks to specific ion channels operating at low Ca concentrations. In mammals, the protype Ca channel is TRPV6 (transient receptor potential vanilloid type 6) (Fig. 2). To avoid excessive intracellular free Ca levels, specific mechanisms are requested to sequester or buffer it. One of them is a specific carrier protein, calbindin (CALB), which releases Ca ions at the basal side through plasma membrane Ca pumps (PMCAs, encoded by *atp2b1, atp2b2,* and *atp2b4* genes), or a Na^+/Ca^{2+} exchanger (NCX1 encoded by the *slc8a1* gene).

This system has been actively studied in birds (Nys and Le Roy, 2018) and the available data are presented in Fig. 3. There is no doubt about the presence and importance of calbindin CALB and PMCA in those species. Nevertheless, the presence and involvement of a TRP channel is more controversial. Some studies reported the expression of a *trpv6* ortholog in chicken intestine (Jonchère et al., 2012; Yang et al., 2011), while others could not detect it (Gloux et al., 2019). Although the expressions of other genes encoding TRP channels (*trpv2, trpc1,* and *trpm7*) were detected in the later study and the increased expression of *trpc1* and *trpm7* with sexual

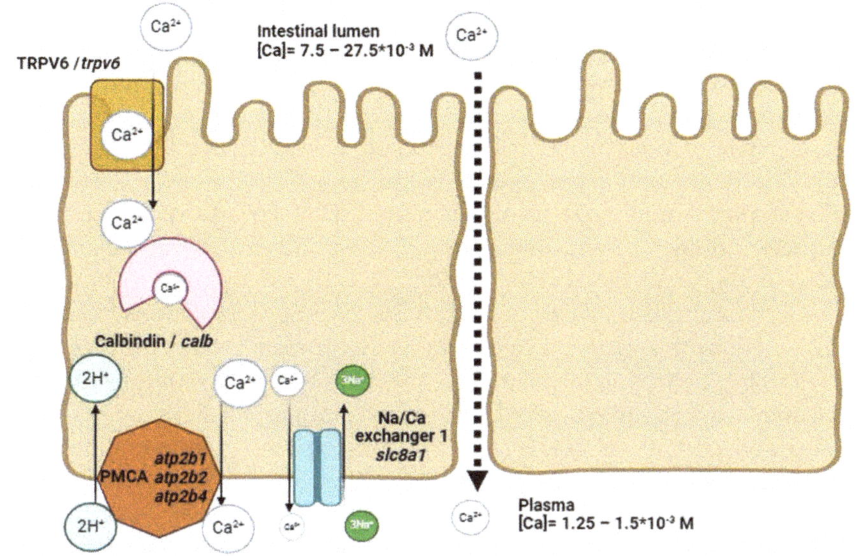

Figure 2 Simplified model listing the molecular actors (proteins/genes) implicated in the active pathways of Ca absorption across the intestinal brush border in mammalian studies. The dotted arrow indicates the paracellular pathway of Ca absorption (passive). The name of the corresponding gene is indicated in italics (protein/gene).

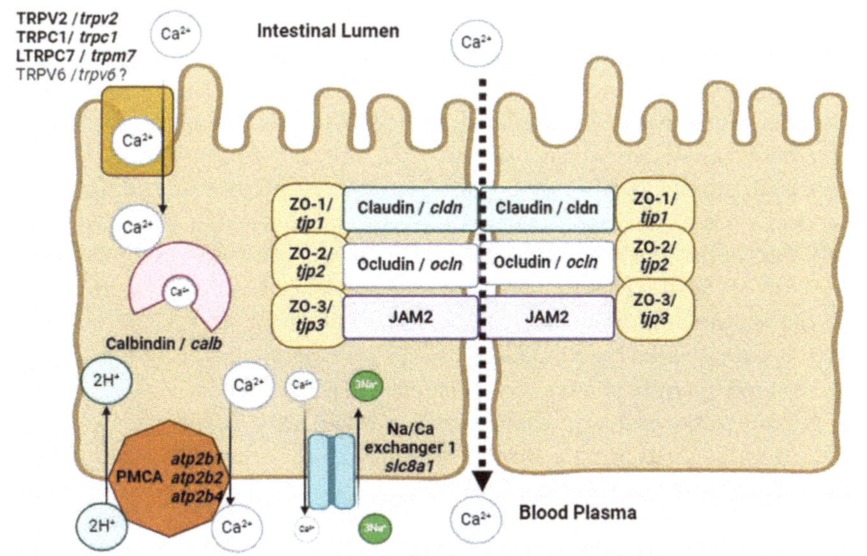

Figure 3 A simplified model summarizing the possible involvement of candidate proteins in Ca transfer across the small intestine's epithelium of the laying hen. The name of the corresponding gene is indicated in italics (protein/*gene*; drawn from data by Gloux et al., 2019).

maturity led the authors to suggest their possible involvement (Gloux et al., 2019), a functional demonstration is still missing. In the laying hen, sexual maturity has been shown to increase the expression of genes involved in Ca absorption (Nys and Le Roy, 2018), including *calb*. The recent observation that its mRNA expression increased with sexual maturity and the onset of lay in the duodenum and in the jejunum (Gloux et al., 2019) was consistent with earlier studies. Additionally, several genes encoding PMCA showed increased expression in laying hens, namely *atp2b1, atp2b2* in the duodenum, and *atp2b4* in the ileum (Gloux et al., 2019).

A non-saturable transport of Ca ions has long been identified in the hen's intestine (Hurwitz and Bar, 1969), but the associated candidate genes and proteins have only been identified recently. The paracellular pathway relies on tight junctions, a complex scaffold of structural proteins, which allows water and solutes transport across epithelia (Fig. 2). Among the different components, the claudins confer the junction its specificity for the different ions. Data from studies in mammals indicate that claudin 2, 10, and 12 (CLDN2, 10, 12) (Fleet, 2017) favour Ca^{2+} transport. Orthologous genes have been identified in the chicken genome and the expression of the corresponding mRNA quantified in the hen's intestine, the available data (Gloux et al., 2019) are summarized in Fig. 3. The coordinated up-regulation of *tjp1-3* and *cldn2* at peak of lay suggest a possible role of this pathway in the increased Ca uptake by the duodenum and jejunum.

These two pathways most likely cooperate for intestinal Ca absorption depending on the gradient of concentrations between the intestinal lumen and the plasma. It is accepted that the transcellular pathway is mainly involved when Ca concentration in the lumen is low, while the paracellular pathway is mainly involved when it is high, such as after a meal (Fleet, 2017). Gene expression studies in the laying hen show that actors of both pathways are present across all three segments of the small intestine (duodenum, jejunum, and ileum) and for most of them vary with the described variations of Ca uptake (Gloux et al., 2019, 2020b). Nevertheless, the functional contribution of the different candidates in the hen remains to be fully demonstrated, an objective which is complicated by the difficulty of genetic manipulations in the hen and the absence of appropriate avian intestinal cell lines.

It cannot be excluded that other pathways are involved. For example, the part of Ca crossing through the cells could be sequestered in the Golgi apparatus or in vesicles (Fleet, 2017).

3.3 The plasmatic pool of calcium

Calcium circulates in the plasma as free Ca^{2+} ions (iCa) or as Ca bound to proteins in the millimolar range. It plays an important role in blood clotting, muscles contraction, and regulation of heart rhythms and nerve functions. Therefore, iCa concentration, about half of total Ca in mammals (Fleet, 2017), needs to be tightly regulated, and any deviation will trigger regulatory mechanisms. The parathyroid gland, which bears specific Ca-sensing receptor (CaSR), reacts to low Ca^{2+} concentrations by secreting parathyroid hormone (PTH) (Chen and Goodman, 2004), which exerts a number of hypercalcaemic effects (see below). On the other hand, calcitonin is secreted by specific cells in the thyroid gland (C cells) when iCa is high, to exert hypocalcaemic effects (Felsenfeld and Levine, 2015).

While iCa values are similar between immature pullets and laying hens, around 1.2 mM, total Ca markedly increases from 2.2 mM to 5.5 mM (Parsons and Combs, 1981). This increase is triggered by sexual maturity, before the onset of lay (Bar et al., 1978). It is unknown how this extra total Ca, which is essentially bound to plasma proteins, can be mobilized, either transferred to the uterus to ensure Ca deposition during eggshell synthesis or released as free Ca^{2+}.

The Ca regulatory mechanism are very well conserved in birds and operative as in mammals for the PTH system. The secretion of calcitonin is devoted to the ultimobranchial gland in birds (Treilhou-Lahille et al., 1984). It has been the object of relatively few studies and there is a belief that this regulation is of minor importance in birds.

4 Bone as a major Ca reservoir for the laying hen

While bone remodelling is a relatively slow process requiring months in mammals, female birds have evolved a specific form of bone so that the process can occur within hours and repeat each day of lay. It is present in the medullary cavity of long bones and therefore called medullary bone. It is synthesized during the first weeks after the onset of lay, while the synthesis of the cortical bone is achieved (Hudson et al., 1993). Its structure and composition differ from that of cortical bone, with a higher mineral content (Dunn et al., 2021) but a lower level of crystallization (Kerschnitzki et al., 2014) and a lower content of collagen. These characteristics likely contribute to its 10-to-15-fold higher rate of bone remodelling compared to cortical bone (Dacke et al., 1993). The consensus is that 30–40% of the Ca incorporated into the eggshell has passed through the bone reservoir since the previous oviposition (Bar, 2009; Nys and Le Roy, 2018).

The mechanisms of bone formation and resorption appear similar in birds and mammals. Osteoblasts ensure bone formation by synthetizing the organic and mineral fractions, while osteoclasts ensure bone resorption by promoting acidification of the osteoclastic niche and subsequently secreting a number of proteolytic enzymes. Osteoclast precursors differentiate into osteoclasts under the effect of receptor activator of nuclear factor (NF)-κB ligand (RANKL), which activates the receptor activator of nuclear factor (NF)-κB (RANK) as a first step towards osteoclastogenesis (Tsai et al., 2023). The difference between laying and non-laying female birds appears to reside in the kinetics rather than the cellular and molecular mechanisms of the process. It is most likely driven by the requirement of Ca for eggshell biosynthesis, which results in hypocalcaemia and thus activates hormonal regulations to return to normocalcaemia. The release of P from bone hydroxyapatite clearly signs the peak of osteolysis occurring during the active phase of eggshell mineralization (Kerschnitzki et al., 2014; Sauveur, 1992), when the surface occupied by osteoclasts increases by nine-fold (van de Velde et al., 1984).

5 Clearance by the kidney

Excess blood Ca is cleared by urinary excretion. In mammals, all ultra-filterable Ca (ionized and complexed Ca, but not protein bound Ca) is filtered through the glomerulus, but 99% of it is subsequently reabsorbed along the nephron (Blaine et al., 2015). Most of it occurs in a passive manner through the paracellular pathway in proximal convoluted tubule, the cortical segment of the loop of Henle, and the distal convoluted tubules. Only a small part (about 5%) is mediated by the transcellular pathway in the connecting tubule. The TRPV5 transporter mediates Ca^{2+} absorption at the apical side, which is then transported through the cytosol by CALB and finally extruded at the basal side through PCMA and Na^{+}/Ca^{2+} exchanger. This last step is regulated by PTH and calcitriol to keep Ca^{2+} in case of hypocalcaemia.

This process in much less documented in birds. The anatomy of the nephron is different from mammals (Nishimura, 2008; Wideman, 1987): the cortical and medullary parts of the kidney are not clearly separated. Avian kidneys have long nephrons with Henle's loop and reptilian type nephrons, i.e. short nephrons without Henle's loop. The proportion between the two types of nephrons varies from species to species, but on average, reptilian nephrons account for 85% (Braun and Dantzler, 1972). The exact function of these different parts is not known. Nevertheless, Wideman (1987) reviewed avian renal Ca metabolism and noted maximal reabsorption at the end of the eggshell calcification phase. Recently, it was observed that *trpv5*, *calb*, and *atp2b1* expression could be detected in the kidney without dissection of the organ. At peak of lay, *calb* expression was high at 0–1 h PO and that of *trpv5* at 18–19 h PO (Gloux et al., 2020a), when Ca reabsorption increased (Wideman, 1987).

6 Regulatory factors

6.1 Calcium sensing and physiological response

As mentioned above, calcaemia is the main signal driving a number of short-term regulatory mechanisms. The parathyroid glands harbour CaSR. These receptors are G protein-coupled receptors, which inhibit *pth* gene expression and PTH secretion when plasma iCa concentration is high. On the contrary, the decrease of plasma iCa levels removes this inhibitory effect, thus resulting in PTH liberation and number of studies indicate that this occurs in birds as in mammals (Nys and Le Roy, 2018). In the laying hen (Fig. 4), the decline of iCa during the active phase of eggshell calcification induces increased PTH levels. The hormone will act directly on the kidney, where it stimulates the conversion of 25-OH-vitamin D (25-OH-D, also designated as hydrocholecalciferol or calcidiol) into 1,25-OH_2-vitamin D (1,25-OH_2-D, also designated as dihydrocholecalciferol or calcitriol). PTH also acts on the bone, where it stimulates osteolysis and results in hyperphosphataemia during the active phase of eggshell calcification.

6.2 The central role of vitamin D

Vitamin D is a central regulator of Ca metabolism and can be synthesized from steroids or administered in the hens' diet, as vitamin D_3. The active form results from two successive hydroxylation of vitamin D occurring sequentially in the liver (25 hydroxylation catalysed by the 25 hydroxylase enzyme and leading to 25-OH-D_3) and in the kidney (1 alpha hydroxylation catalysed by 1- alpha hydroxylase) leading to the active form 1,25-OH_2-D_3 (Fleet, 2017). The first enzyme is the product of the *cyp27a1* gene, which is expressed in the liver (Gloux et al., 2020b). The second one is the product of the *cyp27b1* gene in mammals, for which no chicken ortholog has been identified so far (Sinclair-Black et al., 2023). Nevertheless, the activity of the enzyme has been measured in kidney homogenates (Fraser and Kodicek, 1970; Shanmugasundaram and Selvaraj, 2012) and some claimed to measure the expression of the corresponding gene (Shanmugasundaram and Selvaraj, 2012), but the sequence corresponds to that of the *cyp27c1* ortholog encoding the cytochrome P450 27C1 isoform, involved in steroid hormone biosynthesis.

The hormone 1,25-OH_2-D_3 binds to specific nuclear vitamin D receptors (VDR), which dimerize with retinoid X receptors (RXR) to exert a number of effects on Ca metabolism (Fleet, 2017). It stimulates Ca absorption by the intestine, Ca release by the bone, and Ca reabsorption by the kidney tubule. A number of classical references list its effect in birds and particularly in the laying hen (Bar, 2009; Nys and Le Roy, 2018). Due to its requirement for Ca metabolism, vitamin D_3 is routinely supplemented to layers' diets, in the native,

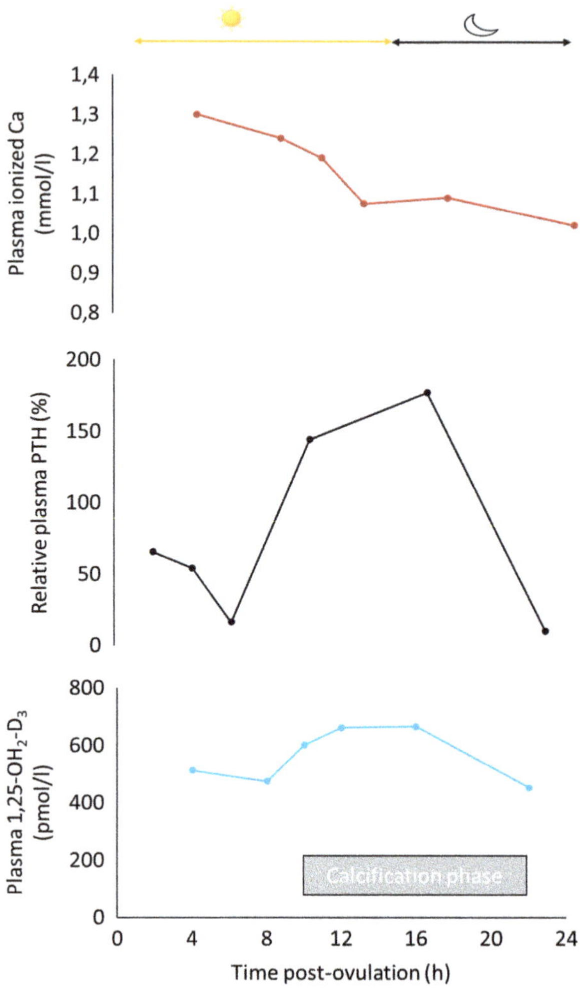

Figure 4 Blood parameters during the oviposition cycle of laying hens. Drawn from data by Velde et al. (1984) for PTH and by Nys et al. (1992) for iCa and 1,25-OH_2-D_3

mono or di-hydroxylated form, with positive effects on egg production and quality, as well as on bone strength (Świątkiewicz et al., 2017).

6.3 A link between calcium and phosphorus metabolisms and their regulation

Generalities about Ca and P metabolisms can be found in Shaker and Deftos (2000). Ca and P are intimately linked, as they share similar absorption pathways, interact with each other throughout the digestive tract and are stored in the same organs. In the diet, a large proportion of P is in the organic form of phytic phosphorus (Humer et al., 2015), while the remaining is non-phytic (NPP) or inorganic P (Pi), mainly in the form of Ca phosphate. In the digestive tract, organic P forms can complex with Ca, depending on pH conditions and on their relative concentrations (Selle et al., 2012). Additionally, Pi and iCa can form de novo Ca–phosphate complexes depending on pH values. The formation of any of these complexes can affect the transepithelial transfer, which can only occur in the soluble form of Ca and phosphate ions (respectively, Ca^{2+}, $H_2PO_4^-$, or HPO_4^{2-}).

Moreover, P and Ca are both involved in bone structure and metabolism. The hard external part of bone (cortical bone) consists largely of proteins, such as collagen, and a mineral substance called hydroxyapatite ($Ca_{10}(PO_4)_6(OH)_2$), which is largely responsible for its strength and density. Bone is constantly the object of a remodelling process, implying successive phases of osteogenesis and osteolysis liberating Ca and P in the ratio of its contribution to hydroxyapatite (~1.67).

The regulatory mechanisms controlling homeostasis of both minerals are also linked. The hormone PTH, while stimulating osteolysis in response to hypocalcaemia, favours the liberation of both minerals, the urinary elimination of P, and the reabsorption of urinary Ca. Through its effect on 1-alpha hydroxylation, PTH also increases the concentration of circulating active vitamin D, which in turn stimulates intestinal absorption of both minerals.

A secreted Fibroblast Growth factor, FGF23, is another actor in this interaction (Erben, 2018; Martin et al., 2012). This peptide is secreted by the osteocytes, in response to hyperphosphataemia (Fig. 5). Only its intact form (iFGF23, 251 amino acids) is active and can be quickly inactivated by cleavage under the effect of a FURIN, which liberates an inactive N terminal peptide and an inhibitory C terminal peptide (Ho and Bergwitz, 2021). The main action of iFGF23 is exerted on the kidney, where it stimulates urinary excretion of P, by inhibition of P reabsorption in the nephron. This effect is directly mediated by inhibition of the Na/P cotransporter NTP2a, a transporter which otherwise reabsorbs most filtrated P. It also acts by inhibiting the activity of the enzyme 1-alpha hydroxylase and stimulating the activity of 24-hydroxylase, thus reducing the available concentrations of active vitamin D. As a consequence of decreased 1,25-OH_2-D, the intestinal absorption of Ca and P is blunted. Some authors provided data suggesting that FGF23 could also directly act on the intestine, the bone. and the parathyroid gland (Ho and Bergwitz, 2021). The canonical cellular response to FGF23 relies on KLOTHO-FGFR1 double heterodimers (KL-FGFR1c in mouse and human) creating a high-affinity binding site for the FGF23, and thus resulting in activation of the ERK signalling pathway. Non-canonical responses can occur at elevated FGF23 concentrations through interactions with other FGFR (notably 3 or 4), in the absence of KL. The canonical

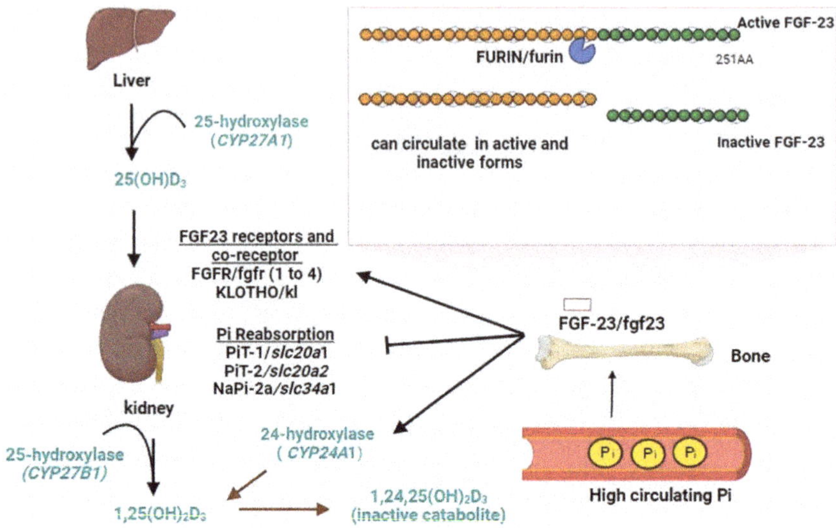

Figure 5 General knowledge on FGF3, its receptors, and main effectors. From data obtained in mammalian species and reviewed by Ho and Bergwitz, 2021.

pathway mediates the effects of FGF23 on the kidney (Ho and Bergwitz, 2021). This hormonal system has deserved a lot of attention in human and in animal models, while much less is known in birds.

7 FGF23: a new research topic in poultry calcium and phosphorus metabolism

The pioneer works of Professor Cook's group on this subject provided indirect evidence of FGF23 existence and involvement in the regulation of P metabolism in chickens (Bobeck et al., 2012) and hens (Ren et al., 2017) through immunization against this peptide. Hadley et al (2016) detected *fgf23* mRNA expression in the cortical and the medullary bone of the laying hen (Hadley et al., 2016). In the medullary bone, it was up-regulated during the active phase of eggshell synthesis (Gloux et al., 2020a). The mRNA expression of *fgf23* and the *kl* gene, encoding its coreceptor KL, was quantified in different tissues of the laying hen (Wang et al., 2018). It was observed that *fgf23* was expressed at a higher level in the liver than in different bones of adult laying hens, while the expression profile was opposite in chicken embryo. In the hen, the level of bone *fgf23* expression followed dietary P supply and phosphataemia. By contrast, liver expression did not vary with dietary P. The *kl* gene was expressed at its highest level in the kidney, and also detected in the bone, but not regulated by dietary P intake.

Numerous publications report measures of FGF23 by commercially available chicken FGF23 immunoassays, as the low similarity between the chicken and mammalian FGF23 peptide sequences prohibits the use of assays validated for human or mouse samples. Nevertheless, a full validation applied to available chicken FGF23 immunoassays should follow the recommendations published for FGF23 assay in human medicine (Heijboer and Cavalier, 2023). The aim would be to accurately distinguish and quantify intact rather than total (intact and cleaved forms) FGF23. Nevertheless, the available immunoreactive FGF23 values were listed in a recent study (Poorhemati et al., 2023) and submitted to metanalysis yielding values in the range of 250 pg/ml for both broilers and layers (Poorhemati et al., 2023). Further observations originating from these studies are discussed below.

A study considering the effect of dietary P supply in broiler chickens (Horvat-Gordon et al., 2019) observed no relationship between circulating immunoreactive FGF23 levels and P utilization. By contrast, the comparison of hens fed diets with regular or decreased level of inorganic P suggested that dietary intake of P could drive bone *fgf23* mRNA expression and circulating immunoreactive FGF23 (Ren et al., 2020a). In other studies, low phosphataemia resulting from dietary supplementation of laying hens with nicotinamide (Ren et al., 2020b) or vitamin D3 deprivation (Yan et al., 2022) were also associated with decreased levels of circulating immunoreactive FGF23 but without modification of *fgf23* gene expression. In one study, conducted during the oviposition cycle of laying hens, the highest circulating concentrations of serum immunoreactive FGF23 were recorded around 9 h PO, at the beginning of the calcification phase, and declined thereafter (Ren et al., 2019), while minimal values were recorded around oviposition. These data suggest a positive link between phosphataemia, circulating immunoreactive FGF23 and urinary excretion of P, at least in the laying hen, if not in the immature broiler.

The available studies suggest that transcriptional regulation of *fgf23* in the bone is important to account for the effects of variable dietary supply of P and possibly to the variation of phosphataemia during the oviposition cycle. By contrast, *fgf23* gene expression in the liver does not appear to be regulated under these conditions. Its high level appears as a species specificity of the chicken, while its expression in humans and laboratory mice is minimal under physiological conditions. In humans, increased hepatic expression has been observed in rare cases of tumour-induced osteomalacia (Dadoniene et al., 2015).

The available literature has also considered the receptors and effectors of FGF23 action in the kidney and other tissues. Several genes encoding FGF receptors (*fgfr1,2,3* and *4*) and the coreceptor KL (*kl*) are present in the avian genome and their expression can be detected in several tissues of the domestic hen, including bone, kidney (Gloux et al., 2020a, 2020b; Ren et al., 2020a,

2020b), and small intestine (Ren et al., 2020a, 2020b). The Na:P exchanger NPT2a, encoded by the *slc34a1* gene, is also expressed in the kidney (mRNA and protein), where it could mediate the biological response to FGF23 (Gloux et al., 2020a, 2020b; Ren et al., 2020a, 2020b). Its variation in physiological conditions where P excretion and immunoreactive FGF23 levels (Ren et al., 2020a, 2020b) vary, with the limits discussed above, suggest that it could be physiologically important in the hen as in other species. As mentioned above, *cyp27b1*, another target of FGF23 has not yet been identified in the chicken genome (Sinclair-Black et al., 2023).

8 Case study: alteration of Ca homeostasis by diet or age in laying hens

Classical studies on Ca metabolism involving the use of Ca isotopes allowed for the identification of physiological factors altering Ca absorption and its post-absorptive fate. Some established a relationship with the expression of molecular actors of this process. It was thus possible to show that intestinal absorption of Ca^{2+} was more intense in the duodenum than in the jejunum and ileum and that it was accompanied by variation of the calbindin protein, thus suggesting a major involvement of the transcellular pathway (Hurwitz and Bar, 1966; Morrissey and Wasserman, 1971). It was also observed that, at a given age, Ca absorption by the small intestine increased during eggshell calcification (Hurwitz and Bar, 1969), while it decreased with age (Al-Batshan et al., 1994). The use of radioactive isotopes is now restricted, so that these classical observations will not be repeated but considered as demonstrated. Researchers now consider that variations in the level or activity of genes or proteins allow to formulate hypotheses on their role. The availability of high throughput RT-qPCR makes this approach very powerful, given the possibility of rapidly assessing a high number of genes and samples. In two successive studies, this approach was applied to candidate genes involved in Ca homeostasis across a wide range of tissues. The factors of variation were the form of dietary Ca at peak of lay (Gloux et al., 2020b), or the age of the hens (23 vs 90 weeks) for the same form of Ca supply (Gloux et al., 2020a).

9 Comparing optimal and suboptimal diets affecting Ca homeostasis in laying hens

Supplying $CaCO_3$ in the form of fine particles (i.e. <0.6 mm) results in rapid solubilization of Ca but only transient release, as they are not retained in the gizzard for long (Sauveur, 1992; Zhang and Coon, 1997): the retention time was recently estimated around 18 min (Hervo et al., 2024). By contrast, feeding a mixture of fine and coarse particles of $CaCO_3$ results in a delayed release of soluble Ca, over a much longer period of time: coarse limestone particles retention time in the gizzard was around 256 min (Hervo et al., 2024). Thus, coarse limestone particles are believed to prolong dietary Ca availability during the night period and therefore decrease bone mobilization to ensure the eggshell deposition. Moreover, the increase in phosphataemia during the active phase of shell calcification (Fig. 6) is greater in hens fed $CaCO_3$ as fine particles only, compared to hen given a separate access to coarse particles (Mongin and Sauveur, 1979). Nowadays, the hens are routinely fed a mixture of fine and coarse particles of $CaCO_3$ considered as optimal rather than fine particles of $CaCO_3$ considered as suboptimal. The comparison of the two diets therefore provides an excellent model for studying the mechanisms of bone remodelling and its regulation.

The expression of specific candidate genes marks the relative contribution of osteogenesis and osteolysis in this process and it shows that both components are similarly increased when $CaCO_3$ is fed as fine rather than as large particles. The similar expression of RANK between the two groups of hens indicated that osteoclasts were active at 18–19 h PO, as early as 4 h after light extinction, regardless of the diet (Gloux et al., 2020b).

Although the hens fed fine particles of $CaCO_3$ show higher 1,25-OH_2-D3, the expression of several intestinal Ca^{2+} transporters is reduced (Gloux

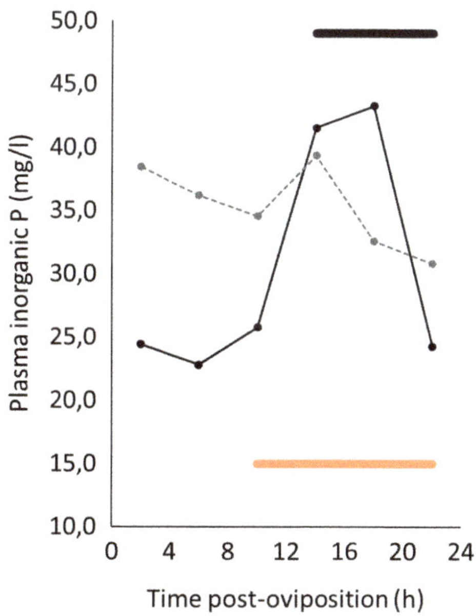

Figure 6 Plasma inorganic P during the oviposition cycle of laying hens fed $CaCO_3$ as fine particles (solid line) only or with separate access to coarse particles (broken line) of $CaCO_3$. Drawn from data by Mongin and Sauveur (1979).

et al., 2020b). Modulation of renal Ca^{2+} reabsorption was also suggested by an increase in ATP2B1 expression in hens fed fine particles of $CaCO_3$ (Gloux et al., 2020b). The results are summarized in Fig. 7.

10 Comparing hens at the peak versus the end of lay

In a further study conducted in our group (Gloux et al., 2020a), hens at the peak of lay were compared to hens at the end (23 vs 90 weeks of age), all fed the same diet, providing Ca as a mix of fine and coarse particles (30:70). The challenge to Ca homeostasis during the active phase of eggshell synthesis was more important in older hens, since it induced earlier resorption of medullary bone, as attested by the higher expression of RANK at 9–10 h PO. In addition, bone resorption was likely more active in older hens, as the expression of carbonic anhydrase 2, which is essential for maintaining an acidic environment in the resorption plate, was higher. Meanwhile, in aged hens, increased resorption was not compensated by increased bone formation, as COL1A1 mRNA was under-expressed. The circulating levels of active vitamin D3 were lower in older hens than in hens at the peak of lay, which may explain reduced expression of several Ca^{2+} transporters. This could be a consequence of the increased expression of FGF23 and its effect on the increased renal expression of CYP24A1 (24-hydroxylase). In fact, the induction of FGF23 expression in the medullary bone occurred earlier during the ovulatory cycle in older hens. Our hypothesis is that the repetition of phases of osteolysis, over a prolonged period, could result in repeated peaks of elevated plasma P although not detectable with a limited number of repeated samplings. This could trigger FGF23 over-expression and exacerbate a disorder in the P and Ca homeostasis. A decrease of TRPV5 expression was observed in the kidney of older hens, possibly resulting from the lower concentrations of active vitamin D3. The reduced expression of the renal P transporters SLC20A1 and SLC34A1, likely in response to FGF23, would reduce renal reabsorption of Pi in order to maintain normal phosphataemia in older hens.

As mentioned earlier, the liver plays a central role in the supply of 25(OH) D3. In older hens, although 25(OH)D3 did not vary, the hepatic expression of CYP27A1 (25-hydroxylase), ESR1, and ESR2 decreased compared to young hens consistent with previous work in rats showing that 25-hydroxylase activity was dependent on the oestrogen status (Saarem and Pedersen, 1987).

Therefore, while facing a major challenge to their calcium homeostasis, peak laying hens fed fine particles of $CaCO_3$ manage to maintain their eggshell strength and meet the need for Ca^{2+} export to the uterus. This is permitted by a strong bone remodelling activity. This is not the case for older hens, which show a state of calcium imbalance associated with low Ca retention and high bone resorption. Further studies unravelling the underlying mechanisms should contribute to design new nutritional strategies preventing this decline.

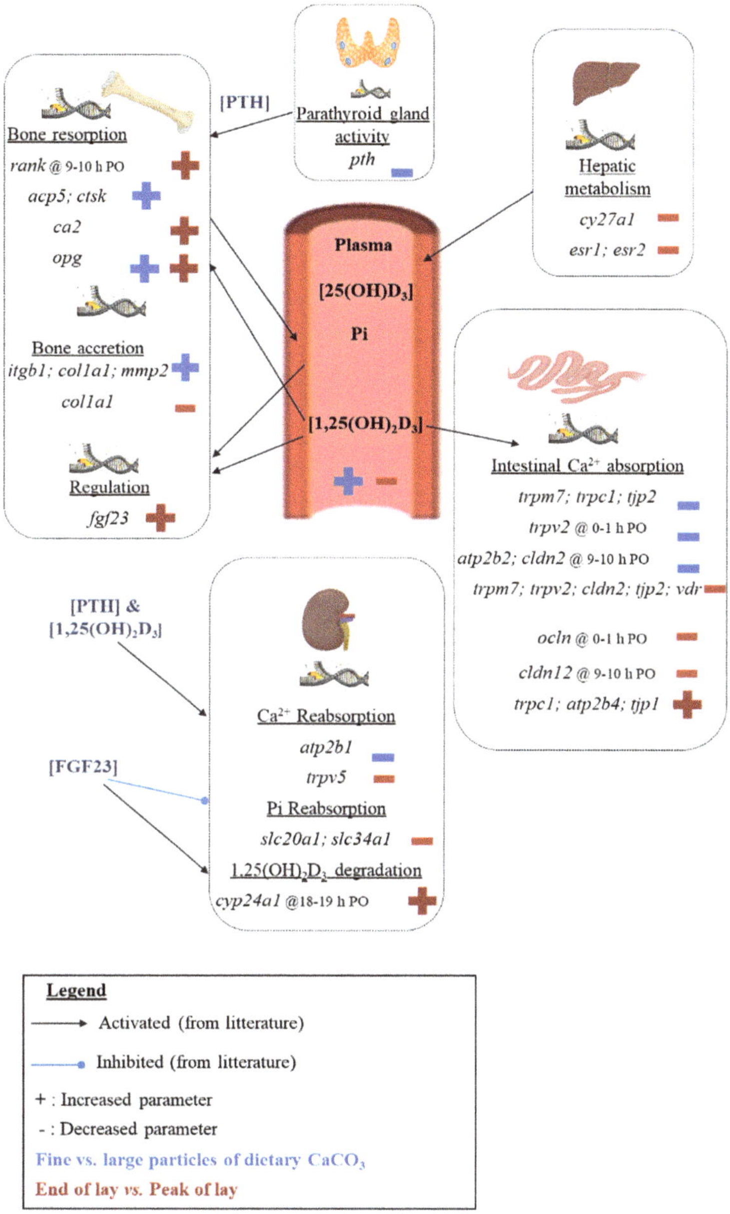

Figure 7 Alterations of candidate gene expression in link with Ca/P homeostasis in laying hens submitted to a nutritional stressor at peak of lay or to ageing. Drawn from data by Gloux et al. (2020a, 2020b).

11 Conclusion and future trends

The available bibliography and our recent studies show that the general mechanisms involved in the control of Ca homeostasis are conserved in *Gallus* compared to the most studied mammalian models. Female birds, with the necessity to face a huge demand of Ca export for eggshell synthesis, have developed several adaptations. The major and most specific one is probably the development of the medullary bone, having a strong impact on the hen's Ca and P metabolism.

As evidenced from the literature, some genes described in mammals have not yet found their ortholog in *Gallus* and would thus deserve further studies. For example, this is the case for TRPV6 and for CYP27B1. This could be due to technical problems, if the genes are present within regions of micro chromosomes with high guanine-cytosine content, or alternatively the chicken could mobilize alternative mechanisms for similar functions. For genes with an already identified ortholog in the chicken, further knowledge and functional studies remain in some case necessary to decipher their exact mode of action. As an example, whereas the orthologous chicken FGF23 gene has been identified and its expression shown to be regulated by circulating P levels, a full characterization of the peptide, with the development of specific and quantitative assays of total and intact chicken FGF23, is necessary to conclude that the peptide is secreted by the bone and acts distantly on the kidney, as it does in human and murine models. Additionally, the significance and consequences of FGF23 expression in the chicken liver remains a puzzle. This fundamental knowledge is a prerequisite to use it as a signature of alterations of P/Ca homeostasis.

A major applied objective would be the development of new feeding strategies allowing for a sustained delivery of soluble Ca to the intestine during the entire period of eggshell mineralization, and thus limiting the dependence upon osteolysis and the loss of a valuable resource, phosphate.

The identified mechanisms, with the development of non-invasive sampling methods and appropriate analytical procedures for quantification of the underlying actors, could serve as biomarkers of the success of these new strategies. They could also suggest targets for the conception of new additives.

Such developments would limit the impairments of Ca/P metabolism observed after the peak of lay and thus improve the sustainability of the egg production chain.

12 Where to look for further information

The following articles provide a good overview on the following topics:

Calcium homeostasis in laying hens

- Bar, A. (2009). Calcium transport in strongly calcifying laying birds: mechanisms and regulation. *Comparative Biochemistry and Physiology Part A: Molecular & Integrative Physiology* 152(4): 447–469.
- Nys, Y. et al. (2022). Chapter 32 - Mechanisms and hormonal regulation of shell formation: supply of ionic and organic precursors, shell mineralization. In: Scanes, C. G. & Dridi, S. (eds.) *Sturkie's Avian Physiology* (7th ed.). San Diego: Academic Press.

FGF23 physiology

- Ho, B.B. & Bergwitz, C. (2021). FGF23 signalling and physiology. *Journal of Molecular Endocrinology* 66(2): R23–R32. https://doi.org/10.1530/JME-20-0178. PMID: 33338030; PMCID: PMC8782161.
- Martin, A. et al. (2012). Regulation and function of the FGF23/klotho endocrine pathways. *Physiological Reviews* 92: 131–155.

13 References

Al-Batshan, H.A., et al. (1994) Duodenal calcium uptake, femur ash, and eggshell quality decline with age and increase following molt. *Poultry Science* 73(10): 1590–1596.

Bar, A. (2009) Calcium transport in strongly calcifying laying birds: mechanisms and regulation. *Comparative Biochemistry and Physiology. Part A, Molecular & Integrative Physiology* 152(4): 447–469.

Bar, A., et al. (1978) Involvement of cholecalciferol metabolism in birds in the adaptation of calcium absorption to the needs during reproduction. *Comparative Biochemistry and Physiology Part B: Comparative Biochemistry* 59(3): 245–249.

Blaine, J., et al. (2015) Renal control of calcium, phosphate, and magnesium homeostasis. *Clinical journal of the American Society of Nephrology: CJASN* 10(7): 1257–1272.

Bobeck, E.A., et al. (2012) Maternally-derived antibody to fibroblast growth factor-23 reduced dietary phosphate requirements in growing chicks. *Biochemical and Biophysical Research Communications* 420(3): 666–670.

Braun, E.J. and Dantzler, W.H. (1972) Function of mammalian-type and reptilian-type nephrons in kidney of desert quail. *The American Journal of Physiology* 222(3): 617–629.

Brini, M., et al. (2013) Calcium in Health and Disease. In: Sigel, A., et al. (eds) *Interrelations between Essential Metal Ions and Human Diseases*. Metal Ions in Life Sciences. Dordrecht: Springer Netherlands, pp. 81–137. Available at: https://link.springer.com/10.1007/978-94-007-7500-8_4 (accessed 4 June 2024).

Chen, R.A. and Goodman, W.G. (2004) Role of the calcium-sensing receptor in parathyroid gland physiology. *American Journal of Physiology. Renal Physiology* 286(6): F1005–F1011.

Dacke, C.G., et al. (1993) Medullary Bone and Avian Calcium Regulation. *Journal of Experimental Biology* 184(1): 63–88.

Dadoniene, J., et al. (2015) Tumour-induced osteomalacia: a literature review and a case report. *World Journal of Surgical Oncology* 14(1): 4.

Dunn, I.C., et al. (2021) No evidence that selection for egg production persistency causes loss of bone quality in laying hens. *Genetics, Selection, Evolution: GSE* 53(1): 11.

Erben, R.G. (2018) Physiological actions of fibroblast growth factor-23. *Frontiers in Endocrinology* 9: 267.

Felsenfeld, A.J. and Levine, B.S. (2015) Calcitonin, the forgotten hormone: does it deserve to be forgotten? *Clinical Kidney Journal* 8(2): 180–187.

Fleet, J.C. (2017) The role of vitamin D in the endocrinology controlling calcium homeostasis. *Molecular and Cellular Endocrinology* 453: 36–45.

Fleming, R.H., et al. (1998) Bone structure and strength at different ages in laying hens and effects of dietary particulate limestone, vitamin K and ascorbic acid. *British Poultry Science* 39(3): 434–440.

Fraser, D.R. and Kodicek, E. (1970) Unique biosynthesis by kidney of a biological active vitamin D metabolite. *Nature* 228(5273): 764–766.

Gautron, J. and Rodriguez-Navarro, A. (2025). Advances in understanding eggshell formation and its composition in table eggs. Chapter taken from: Alamprese, C., Nys, Y., Mine, Y. and Guyonnet, V, (ed.), Advances in improving the safety and quality of eggs, Burleigh Dodds Science Publishing, Cambridge, UK (ISBN: 978 1 80146 875 6; www.bdspublishing.com)

Gloux, A., et al. (2019) Candidate genes of the transcellular and paracellular calcium absorption pathways in the small intestine of laying hens. *Poultry Science* 98(11): 6005–6018.

Gloux, A., et al. (2020a) Increased expression of fibroblast growth factor 23 is the signature of a deteriorated Ca/P balance in ageing laying hens. *Scientific Reports* 10(1): 21124.

Gloux, A., et al. (2020b) Possible roles of parathyroid hormone, 1.25(OH)2D3, and fibroblast growth factor 23 on genes controlling calcium metabolism across different tissues of the laying hen. *Domestic Animal Endocrinology* 72: 106407.

Hadley, J.A., et al. (2016) Bone sialoprotein keratan sulfate proteoglycan (BSP-KSPG) and FGF-23 are important physiological components of medullary bone. *Comparative Biochemistry and Physiology Part A: Molecular & Integrative Physiology* 194: 1–7.

Heijboer, A.C. and Cavalier, E. (2023) The measurement and interpretation of fibroblast growth factor 23 (FGF23) concentrations. *Calcified Tissue International* 112(2): 258–270.

Hervo, F., et al. (2024) Effect of limestone particle size and microbial phytase on phosphorus and calcium digestion kinetics along the gastrointestinal tract in laying hens. *British Poultry Science* 65(5): 605–614.

Ho, B.B. and Bergwitz, C. (2021) FGF23 signalling and physiology. *Journal of Molecular Endocrinology* 66(2): R23–R32.

Horvat-Gordon, M., et al. (2019) Lack of a relationship between plasma fibroblast growth factor-23 and phosphate utilization in young chicks. *Poultry Science* 98(4): 1762–1765.

Hudson, H.A., et al. (1993) Histomorphometric bone properties of sexually immature and mature White Leghorn hens with evaluation of fluorochrome injection on egg production traits. *Poultry Science* 72(8): 1537–1547.

Humer, E., et al. (2015) Phytate in pig and poultry nutrition. *Journal of Animal Physiology and Animal Nutrition* 99(4): 605–625.

Hurwitz, S. and Bar, A. (1966) Rate of passage of Calcium-45 and Yttrium-91 along the intestine, and calcium absorption in the laying fowl. *The Journal of Nutrition* 89(3): 311–316.

Hurwitz, S. and Bar, A. (1969) Intestinal calcium absorption in the laying fowl and its importance in calcium homeostasis. *The American Journal of Clinical Nutrition* 22(4): 391–395.

Hurwitz, S., et al. (1973) Regulation of calcium absorption by fowl intestine. *The American Journal of Physiology* 225(1): 150–154.

Jonchère, V., et al. (2012) Identification of uterine ion transporters for mineralisation precursors of the avian eggshell. *BMC Physiology* 12(1): 10.

Kerschnitzki, M., et al. (2014) Rapid alterations of avian medullary bone material during the daily egg-laying cycle. *Bone* 69: 109–117.

Manangi, M.K., et al. (2018) Calcium particle size effects on plasma, excreta, and urinary Ca and P changes in broiler breeder hens. *Poultry Science* 97(8): 2798–2806.

Martin, A., et al. (2012) Regulation and function of the FGF23/klotho endocrine pathways. *Physiological Reviews* 92(1): 131–155.

Meyer, G.B., et al. (1970) Decreased feed consumption and increased calcium intake associated with Pullet's First Egg1,2. *Poultry Science* 49(5): 1164–1169.

Mongin, P. and Sauveur, B. (1979) Plasma inorganic phosphorus concentration during egg-shell formation: effect of the physical form of the dietary calcium. *British Poultry Science*. Taylor & Francis Group. Epub ahead of print 1 July 1979. https://doi.org/10.1080/00071667908416599

Morrissey, R. and Wasserman, R. (1971) Calcium absorption and calcium-binding protein in chicks on differing calcium and phosphorus intakes. *American Journal of Physiology-Legacy Content* 220(5): 1509–1515.

National Research Council. (1994) Nutrient requirements of poultry: ninth revised edition, 1994. Washington, DC: The National Academies Press. https://doi.org/10.17226/2114

Nishimura, H. (2008) Urine concentration and avian aquaporin water channels. *Pflugers Archiv: European Journal of Physiology* 456(4): 755–768.

Nys, Y. and Le Roy, N. (2018) Calcium Homeostasis and Eggshell Biomineralization in Female Chicken. In: *Vitamin D*. Elsevier, pp. 361–382. Available at: https://linkinghub.elsevier.com/retrieve/pii/B9780128099650000227 (accessed 4 June 2024).

Nys, Y., et al. (1992) Plasma 1,25 dihydroxycholecalciferol and its free index are potentiated by ovulation dependent factors and shell formation induced hypocalcemia in the laying hens. *Domestic Animal Endocrinology* 9(1): 37–47.

Parsons, A.H. and Combs, G.F. (1981) Blood ionized calcium cycles in the chicken. *Poultry Science* 60(7): 1520–1524.

Poorhemati, H., et al. (2023) FGF23 level in poultry chicken, a systematic review and meta-analysis. *Frontiers in Physiology* 14: 1279204.

Ren, Z., et al. (2017) Antibody to fibroblast growth factor 23-peptide reduces excreta phosphorus of laying hens. *Poultry Science* 96(1): 127–134.

Ren, Z., et al. (2019) Dynamics of serum phosphorus, calcium, and hormones during egg laying cycle in Hy-Line Brown laying hens. *Poultry Science* 98(5): 2193–2200.

Ren, Z., et al. (2020a) Phosphorus restriction changes the expression of fibroblast growth factor 23 and its receptors in laying hens. *Frontiers in Physiology* 11: 85.

Ren, Z.Z., et al. (2020b) Supplemental nicotinamide dose-dependently regulates body phosphorus excretion via altering type II sodium-phosphate co-transporter expressions in laying hens. *The Journal of Nutrition* 150(8): 2070–2076.

Saarem, K. and Pedersen, J.I. (1987) Sex differences in the hydroxylation of cholecalciferol and of 5 β -cholestane-3 α, 7 α, 12 α-triol in rat liver. *Biochemical Journal* 247(1): 73–78.

Sauveur, B. (1992) Adaptation des apports alimentaires aux variations journalières des besoins en calcium et phosphore de la poule. *INRAE Productions Animales* 5(1): 19–28.

Selle, P.H., et al. (2012) Protein–phytate interactions in pig and poultry nutrition: a reappraisal. *Nutrition Research Reviews* 25(1): 1–17.

Shaker, J.L. and Deftos, L. (2000) Calcium and Phosphate Homeostasis. In: Feingold, K.R., et al. (eds) *Endotext*. South Dartmouth, MA: MDText.com, Inc. Available at: http://www.ncbi.nlm.nih.gov/books/NBK279023/ (accessed 5 June 2024).

Shanmugasundaram, R. and Selvaraj, R.K. (2012) Vitamin D-1α-hydroxylase and vitamin D-24-hydroxylase mRNA studies in chickens. *Poultry Science* 91(8): 1819–1824.

Sinclair-Black, M., et al. (2023) Physiological regulation of calcium and phosphorus utilization in laying hens. *Frontiers in Physiology* 14: 1112499.

Świątkiewicz, S., et al. (2017) Efficacy of dietary vitamin D and its metabolites in poultry - review and implications of the recent studies. *World's Poultry Science Journal* 73(1): 57–68.

Treilhou-Lahille, F., et al. (1984) Ultimobranchial gland of the domestic fowl. Two types of secretory cells involved in calcitonin metabolism. *Cell and Tissue Research* 235(2): 439–448.

Tsai, J., et al. (2023) Origin of osteoclasts: osteoclast precursor cells. *Journal of Bone Metabolism* 30(2): 127–140.

van de Velde, J.P., et al. (1984) Changes in activity of chicken medullary bone cell populations in relation to the egg-laying cycle. *Metabolic Bone Disease & Related Research* 5(4): 191–193.

Velde, J.-P.V.D., et al. (1984) Parathyroid hormone responses to calcium stress during eggshell calcification*. *Endocrinology* 115(5): 1901–1904.

Wang, R.M., et al. (2018) Fibroblast growth factor 23 mRNA expression profile in chickens and its response to dietary phosphorus. *Poultry Science* 97(7): 2258–2266.

Wideman, R.F. (1987) Renal regulation of avian calcium and phosphorus metabolism. *The Journal of Nutrition* 117(4): 808–815.

Yan, J., et al. (2022) Dietary vitamin D3 deprivation suppresses fibroblast growth factor 23 signals by reducing serum phosphorus levels in laying hens. *Animal Nutrition (Zhongguo Xu Mu Shou Yi Xue Hui)* 9: 23–30.

Yang, J.H., et al. (2011) Localisation and expression of TRPV6 in all intestinal segments and kidney of laying hens. *British Poultry Science* 52(4): 507–516.

Zhang, B. and Coon, C.N. (1997) The relationship of calcium intake, source, size, solubility in vitro and in vivo, and gizzard limestone retention in laying hens. *Poultry Science* 76(12): 1702–1706.

Chapter 13

Biofortification of eggs with micronutrients

Tom Hill, Newcastle University, UK

1 Introduction

1.1 Chapter overview

This chapter highlights the potential of biofortifying eggs as a significant dietary source of micronutrients. It brings to the fore the widespread concern of low dietary intakes of a range of micronutrients, including micronutrient deficiencies, emphasizing the need to enhance dietary intakes in many populations. Specifically, the chapter summarizes recent research on biofortifying eggs with key vitamins and trace minerals of relevance to public health nutrition. It includes a commercial case study of vitamin D biofortification by a leading free-range egg brand in the UK, involving collaborative academia-industry-funded research. The chapter also covers biofortification with trace minerals like iron, zinc, selenium, and iodine, which are known to be deficient in many populations.

While this chapter focuses solely on micronutrients, biofortification of eggs with other nutrients, in particular omega-3 fatty acids, has historically been a very active area of research. The further reading section of the chapter (Fraeye et al., 2012) provides an excellent review of this topic through hen feed supplementation with flaxseed, fish oil, or microalgae. The book also covers other chapters that complement the biofortification material, such as the nutritional composition of egg yolk and albumin, and the health impacts associated with egg intake.

http://dx.doi.org/10.19103/AS.2024.0150.21

1.2 Eggs as a source of micronutrients: The global context

In developed countries, eggs often serve as a complementary source of micronutrients within a varied diet. In developing countries, eggs can be a primary source of micronutrients due to limited dietary diversity. Table eggs play a significant role in providing essential micronutrients globally, with substantial contributions to vitamin A, vitamin D, vitamin B_{12}, folate, choline, and selenium intake. Eggs provide around 10% of the daily recommended intake of vitamin A per egg (about 75 μg of retinol), which is particularly vital in areas with a high prevalence of vitamin A deficiency, such as Sub-Saharan Africa and South Asia.

In Northern Europe and North America, where sunlight exposure can be limited, eggs offer approximately 10% of the recommended daily intake of vitamin D (1.1–1.5 μg), making them essential for maintaining adequate vitamin D intake. For vitamin B_{12}, eggs are a significant source globally, especially in countries with large vegetarian populations like India, providing about 0.6 μg per egg, which is roughly 25% of the daily requirement. Eggs contribute about 6% of the daily requirement of folate (24 μg per egg), which is crucial during pregnancy and in regions like South Asia and Sub-Saharan Africa, where clinical deficiencies are common.

In the United States and Europe, eggs are one of the richest sources of choline, providing approximately 150 mg per egg, or about 27% of the daily adequate intake for women and 22% for men, which is important for brain health. Finally, eggs supply around 22% of the daily selenium requirement (15.4 μg per egg), particularly in regions with selenium-deficient soils, such as parts of China and Europe. Analyses of nutrient intake from the American NHANES Food Survey revealed that children who consume eggs have notably higher levels of choline, as well as vitamins A, D, and pantothenic acid, to those who do not consume eggs (Papanikolaou and Fulgoni, 2021).

1.3 The public health concern of poor dietary micronutrient intakes in the UK

Micronutrients, including vitamins and minerals, are essential for numerous physiological functions and maintaining health. Required in small amounts, they play diverse roles, including enzyme function, immune response, and cellular repair. Micronutrient deficiency exists on a sliding scale, from clinical to subclinical and suboptimal. Clinical deficiency is characterized by overt symptoms and diseases. Subclinical deficiency shows no obvious symptoms but can impair bodily functions. Suboptimal deficiency, though not causing immediate illness, may result in impaired nutrient function over time.

Low dietary intake of micronutrients remains a significant concern across European populations. Recent findings from the 2019 British National Diet and Nutrition Survey for adults highlight deficiencies in vitamin D, with only 28 to 35% of the recommended nutrient intake (RNI) met across all age groups, and 13% of older adults exhibiting inadequate biochemical vitamin D status. Additionally, there are prevalent deficiencies in vitamin A, riboflavin, and folate in significant proportions of adults. Specifically, among older adults in Britain, further data reveal that 95% of participants have vitamin D intakes below recommended levels, and over 80% have inadequate dietary selenium intake. Suboptimal concentrations of key nutritional biomarkers in plasma, notably vitamin D, vitamin B12, and selenium, parallel these low dietary intakes. Addressing these deficiencies is crucial to improving overall public health and reducing the risk of associated health complications.

1.4 The case for food (bio)fortification

There is an ongoing need to develop effective and coordinated approaches to improve micronutrient intake in many countries. Mandatory fortification of staple food items in some countries, including margarines (with vitamins A and D), flour (with folic acid and other B vitamins, calcium, and iron), and dairy products (with vitamin D) has been instrumental in contributing to micronutrient intakes in some countries over the last few decades. For instance, the UK mandated the fortification of white and brown (but not wholemeal) flour with calcium, iron, thiamine, and niacin in 1941, resulting in many breads and baked products serving as a significant dietary source of these nutrients for the entire population, including older adults. Manufacturers offer infrequent and patchy voluntary food fortification with nutrients, such as with breakfast cereals, in accordance with regional laws.

Evidence among UK consumers highlights generally positive views towards fortification, although this is less mandatory as opposed to voluntary fortification. In addition to fortification, which involves directly adding the desired nutrient during food processing, there's also the method of biofortification for animal products. This approach enriches food items by incorporating the desired nutrient(s) into animal feed. Eggs, in particular, have attracted considerable attention for biofortification. Clark et al. (2019) noted that consumers often perceive this process as more natural compared to traditional fortification methods.

It allows for clean labels, which consumers prefer, unlike traditional fortification methods, as pointed out by Asioli et al. in 2017. The concept of biofortification of foods with micronutrients is gaining increasing attention internationally (Avnee et al., 2023).

1.5 Nutritional approaches to egg biofortification

Hen diets are formulated to meet the nutritional needs of poultry, promoting health, growth, and egg production. These diets typically include grains, protein sources, vitamins, minerals, and additives. When the goal is egg biofortification, specific feed additives are incorporated to improve the egg's nutrient profile. Grains such as corn, wheat, and barley provide essential carbohydrates for energy. Protein sources, primarily soybean meal, supply the necessary amino acids. Essential vitamins and minerals are added for metabolic functions, including immunity and bone health. To enhance omega-3 content in eggs, feeds are supplemented with flaxseed, fish oil, and algae. Flaxseed is ground to improve digestibility, while fish oil and algae are encapsulated for stability and palatability. Vitamin D_3 (cholecalciferol) is added to eggs to increase its concentration, aiding calcium absorption. Vitamins E and A are also supplemented in microencapsulated forms to enhance bioavailability. Selenium is biofortified in eggs via organic selenium (selenomethionine), which is more effectively absorbed than inorganic forms. Zinc and iodine are included as chelated minerals for improved bioavailability. Natural antioxidants like lutein and zeaxanthin, from marigold petals, are added to enhance egg antioxidant content and are provided in dried, powdered forms for uniform mixing and stability.

2 Vitamin biofortification

This section will summarize the evidence for vitamin biofortification in eggs. The focus will be on vitamins where evidence supports biofortification and where there is nutritional rationale to improve dietary intakes in consumers. The section covers vitamins A, D, E, and K and folate. Each vitamin will be discussed separately, starting with a general overview of its main nutritional aspects, including its functions, the nutritional requirements for chickens and humans, and the consequences of deficiency or excess. The evidence on biofortification for each vitamin will be explored and summarized.

2.1 Vitamin A

2.1.1 Nutritional aspects

Vitamin A encompasses compounds with retinol's biological activity, measured in retinol equivalents (REs). This standardization assumes efficient absorption of provitamin A carotenoids, but recent research shows lower and variable absorption rates (Tang, 2010). Retinol equivalents account for the conversion of carotenoids to retinol, with 1 μg of retinol equivalent to 12 μg of dietary beta-carotene or 24 μg of other provitamin A carotenoids (i.e. α-Carotene, γ-Carotene, and β-Cryptoxanthin). Vitamin A plays vital roles in vision, immunity, reproduction, and growth, while carotenoids may offer antioxidant benefits and reduce the risks of various diseases in humans. Deficiency in humans leads to night blindness, xerophthalmia, impaired immunity, abnormal growth, and other serious health issues. Vitamin A deficiency in hens leads to stunted growth, reduced egg production, vision problems, weakened immunity, respiratory issues, and dry, thickened skin and mucous membranes prone to infection.

The National Research Council's 1994 recommendation of 3000 IU/kg vitamin A may be outdated; new studies suggest 8000–10,000 IU/kg improves production and immune response, with a maximum tolerance of 20,000 IU/kg to avoid toxicity. In humans, adequacy criteria are based on maintaining liver reserves to prevent deficiency for about four months. The UK dietary reference values for vitamin A are 700 and 600 micrograms (μg) of Retinol Equivalent (RE) per day for adult men and women, respectively. Excessive intake of preformed vitamin A can cause hepatotoxicity and teratogenicity, leading to a tolerable upper intake level (UL) of 3000 μg RE/day for adults to avoid adverse effects. According to the latest UK data on the nutritional composition of regular (unfortified) hen eggs, the total vitamin A content (RE) of raw whole eggs is 126 μg per 100g.

2.1.2 Vitamin A biofortification

Schiavone and Barroeta (2011) have extensively reviewed numerous studies on retinoid and carotenoid supplementation in laying hens. Research overall indicates that as supplemental vitamin A increases, its deposition in the egg and liver also rises, but the extent varies. Naber (1993) classified vitamins by their transfer efficiency from hen diet to egg, placing vitamin A in the "very high" category with a 60–80% transfer rate. For example, increasing dietary retinyl acetate from 4000 to 16,000 IU/kg raised egg vitamin A concentration from 10.3 to 24.0 IU/g (Squires and Naber, 1993). Similarly, Mendonça et al. (2002) reported a 50% increase in egg retinol after supplementation with 25,000 IU/kg dietary retinyl acetate. Jiang et al. (1994) observed a linear increase in egg yolk beta-carotene and a slight increase in retinol with higher dietary beta-carotene, suggesting conversion to retinol. However, high vitamin A levels can negatively interact with other fat-soluble vitamins, such as vitamin E, reducing their absorption and deposition in the yolk (Jiang et al., 1994).

Recent studies have also explored the impact of dietary additives like stevia (*Stevia rebaudiana*) on egg quality and carotenoid content (Pirgozliev et al., 2022). In an experiment involving 90 Hy-Line Brown hens housed in enriched cages, dried stevia leaves were supplemented at 0%, 1%, and 2% levels to assess their effects. The study revealed that yolks from hens fed 2% stevia had significantly higher carotenoid content (5.16 μg/g) compared to those in the 1% (4.23 μg/g) and 0% (2.96 μg/g) groups, with no differences in any egg quality characteristics between the groups (Pirgozliev et al., 2022). This suggests that stevia supplementation could effectively enrich eggs with carotenoids, enhancing their nutritional value and potentially antioxidant properties.

2.2 Vitamin D

2.2.1 Nutritional aspects

Vitamin D, initially recognized for its role in preventing rickets, comprises cholecalciferol (vitamin D_3) from animals and ergocalciferol (vitamin D_2) from fungi and yeasts. Both forms can be synthesized from precursors under sunlight and are effective at raising vitamin D status in humans. However, it is recognised that vitamin D_3 is more bioavailable than vitamin D_2. Vitamin D activity accounts for the conversion of 7-dehydrocholesterol in the skin to cholecalciferol (vitamin D_3) upon exposure to UVB radiation, with 1 IU of vitamin D equivalent to 0.025 micrograms (μg) of cholecalciferol. Vitamin D's primary function is regulating calcium homeostasis through its active hormone, calcitriol. It also influences the absorption of phosphorus and modulates various organ systems. Deficiency, indicated by low concentrations of the liver metabolite 25-hydroxyvitamin D [25-D] levels, increases the risk of rickets, osteomalacia, immunosuppression, and other health issues (SACN, 2016).

Determining dietary requirements is complex due to sunlight exposure's contribution. Vitamin D deficiency in poultry leads to poor bone mineralization, reduced egg production, and weakened immunity. The National Research Council's 1994 recommendation of 300 IU/kg vitamin D_3 for poultry is set as a minimum to avoid bone diseases, but is insufficient to maintain optimal hen health, immunity, and improved production in modern layers. The feed's EFSA upper limit for vitamin D_3 content in layer feed is 3200 IU/kg. Use of 25-D [up to 75μg/kg] is approved for use in poultry feed (as full or partial replacement for vitamin D_3) and is superior to vitamin D_3 for poultry health (EFSA, 2005).

In the UK, the Scientific Advisory Committee on Nutrition (SACN) recommends that most individuals aim for an intake of 10 μg of vitamin D daily. Excessive vitamin D intake from supplements can lead to hypercalcemia and soft tissue calcification, but sunlight exposure regulates its production, with excess converted to inactive metabolites. The EFSA Tolerable Upper Intake Level (UL) of 100 μg/day (4,000 IU/day) is set for adults to avoid adverse effects. Naturally containing vitamin D, eggs typically provide up to 10% of dietary vitamin D intake in the British diet. The Department of Health has indicated that whole, raw eggs contain 3.15 μg of vitamin D per 100 g of egg (Department of

Health 2013). However, additional research has demonstrated the significant potential to enrich eggs with vitamin D, thereby increasing their overall vitamin D content (refer to Section 2.2.2 and the case study in Section 3).

2.2.2 Vitamin D biofortification

Several feeding trials have demonstrated that manipulating vitamin D_3 and/or 25-D in hen diets leads to a considerable increase in egg vitamin D content (Mattila et al., 1999, 2003, 2004, 2011; Browning and Cowieson 2014). According to these studies, consuming fortified feed within approximately 6–12 weeks results in the saturation of egg yolks with vitamin D, potentially leading to a 78% increase in the total vitamin D content of eggs (Mattila et al., 2003; Browning and Cowieson, 2014). However, some of the feeding methods used in these studies surpassed the current upper safety limits for vitamin D set by the European Food Safety Authority (EFSA) for laying hen feed (3200IU/kg diet) (Mattila et al., 2003, 2004; Browning and Cowieson 2014; European Commission 2017), raising questions about the commercial feasibility and applicability of the results. Therefore, considering current EFSA guidelines, it would be prudent to investigate the commercial viability of enriching eggs with vitamin D.

Supplementing adults with calcidiol (25-D) instead of vitamin D_3 raises serum 25-D roughly five-fold, a potency ratio (5:1) that the UK and Ireland already apply when calculating total vitamin D in foods (Cashman et al., 2012). EFSA (2023), however, proposes a different equivalence for intakes ≤ 10 μg/day–0.4 μg calcidiol equals 1 μg vitamin D_3/D_2 (1 μg vitamin-D equivalent = 40 IU); a factor that may reshape future food composition standards. However, in terms of animal nutrition, there is currently no distinction between D_3 and 25-D, as both forms are considered equally effective for hens. Employing bio-fortification principles to increase 25-D content in eggs presents a promising approach to enhancing vitamin D availability in eggs and, consequently, for consumers. Addressing the question of physiological impact in humans, Hayes et al. (2016) demonstrated that consuming eggs enriched with either vitamin D_3 or 25-D helped mitigate the seasonal decline in vitamin D levels among Irish adults during the winter. Moreover, evidence suggests that incorporating 25-D into poultry diets may offer greater benefits for bone health compared to vitamin D_3 (Mattila et al., 2004), highlighting potential advantages for animal welfare that could positively influence consumer acceptance of the practice. Collectively, these findings provide a compelling rationale for exploring the commercial enrichment of eggs with vitamin D to augment dietary intake among UK consumers (see case study in Section 3).

2.3 Vitamin E

2.3.1 Nutritional aspects

Vitamin E encompasses compounds with the biological activity of α-tocopherol (α-toc), measured in α-tocopherol equivalents (α-TE). This standardization assumes efficient absorption of various tocopherols and tocotrienols. Alpha-tocopherol equivalents account for the different activities of tocopherol isomers, with 1 mg of α-tocopherol equivalent to 1.5 mg of γ-tocopherol. Vitamin E plays vital roles in protecting cell membranes from oxidative damage, supporting immune function, and preventing chronic diseases. Trugilho et al. (2024) nicely summarize an excellent, up-to-date review of the potential mechanisms of action of vitamin E and its potential benefits to chronic diseases. Deficiency in humans leads to neuromuscular issues, impaired immune responses, and hemolytic anemia. Vitamin E deficiency in hens leads to encephalomalacia, exudative diathesis, muscular dystrophy, reduced egg production, and compromised immunity.

The National Research Council's 1994 recommendation for layers is 10 mg/kg feed. In humans, adequacy criteria are based on maintaining plasma concentrations of alpha-tocopherol to prevent deficiency. In the UK, dietary reference values for vitamin E are 4 mg/day for adult men and 3 mg/day for adult women, although American recommendations are higher at 15 mg/day for adult males and females. Excessive intake of vitamin E can interfere with vitamin K-dependent processes and cause hemorrhage, leading to a European Tolerable Upper Intake Level (UL) of 300 mg/day for adults to avoid adverse effects. According to the latest UK data on the nutritional composition of regular (unfortified) hen eggs, the total vitamin E content of raw whole eggs is 1.29 mg per 100 g.

2.3.2 Biofortification with different forms of tocopherols

Numerous studies on vitamin E supplementation in laying hens show a linear relationship with total vitamin E in eggs, but with diminishing transfer efficiency as dietary α-toc levels increase (for review see Schiavone and Barroeta, 2011). Efficiency rates range from 41.8% at 50 mg/kg of feed to 26% at 200 mg/kg of feed. The maximum α-toc content in eggs is typically achieved within 2-3 weeks of supplementation, remaining stable for another two weeks before declining by about 50% over the following 4weeks, emphasizing the need for continuous supplementation. Higher unsaturated fat levels in the diet reduce α-toc deposition in eggs due to increased oxidative stress. The interaction between vitamins A and E also has an impact on their absorption and egg deposition. Additionally, vitamin E content remains stable during egg storage but declines significantly during processing methods like boiling, scrambling, and spray-drying, as well as under high-temperature storage conditions. These findings underscore the complexity of enhancing egg vitamin E content and the influence of dietary and processing factors.

2.4 Vitamin K

2.4.1 Nutritional aspects

Vitamin K encompasses compounds like phylloquinone (K_1), menaquinones (K_2), and menadione (K_3), crucial for protein carboxylation, especially in blood clotting and bone metabolism. Severe deficiency in humans leads to impaired blood clotting, resulting in bleeding and hemorrhage. Newborns, particularly breastfed ones, are vulnerable to hemorrhage due to limited placental transfer and low breast milk vitamin K content. While severe deficiency is rare in adults, subclinical deficiency may affect extrahepatic tissues, notably bone. Vitamin K deficiency in hens leads to poor blood clotting, internal bleeding, reduced egg production, and compromised bone health.

Requirements of vitamin K for layers set by the NRC give a minimum requirement of 0.5 mg/kg. In poultry diets, menadione nicotinamide bisulfite (MNB) and menadione sodium bisulfite (MSB) are two commonly used forms of menadione. In the UK, dietary reference values for vitamin K are 1 μg/kg body weight per day, translating to about 75 μg/day for adult men and 65 μg/day for adult women, while American recommendations are higher at 120 μg/day for adult males and 90 μg/day for adult females. Dietary vitamin K intakes in the US are lower than recommendations and have been associated with cardiovascular disease, low bone mineral density, and fracture risk (Shea and Booth, 2016). Excessive vitamin K intake is rare, but it can interfere with anticoagulant medications, leading to a European Tolerable Upper Intake Level (UL) that is not established due to low toxicity risk. According to the latest UK data on the nutritional composition of regular (unfortified) hen eggs, the total vitamin K content (phylloquinone and menaquinone) of raw whole eggs is between 7 and 13 μg per 100 g (Department of Health, 2013).

2.4.2 Vitamin K biofortification

There are a few studies that have explored the menaquinone (MK) biofortification of eggs, each yielding specific results. Suzuki and Okamoto (1997) demonstrated that supplementing hen diets with 10–100 mg/kg of vitamin K1 (phylloquinone) resulted in eggs containing 104–1908 mg of vitamin K1 and 67–192 mg of menaquinone K_2 per 100 g yolk. Supplementing hens with 10–1000 mg/kg of vitamin K3 (menadione) fortified the resulting eggs with menaquinone K2 at levels of 115–240 mg per 100 g yolk. Park et al. (2005)

tested the effect of including vitamin K3 (as menadione sodium bisulfite) at levels of 0.5 (control), 1.5, 5, 7.5, 10, and 12.5 mg/kg on egg vitamin K content in a 30-day hen feeding trial. The vitamin K content in the eggs plateaued after 20 days. Between days 20 and 30, the total vitamin K content of the eggs increased 3.3- to 4.8-fold, with the 7.5 mg/kg feed producing the highest increase (53 µg/100 g whole egg) (Park et al., 2005).

Recently, O'Sullivan et al. (2020) conducted a 12-week feeding trial in Hyline chickens, randomized into four groups and given diets containing vitamin K_3 (as menadione nicotinamide bisulfite) at levels of 12.9, 23.7, and 45.7 mg/kg feed and a control. Vitamin K1, menaquinone (MK)-4, MK-7, and MK-9 were measured in raw whole eggs. MK-4 was the most prevalent form of vitamin K (91–98%) in all eggs. Increasing the vitamin K_3 content in hen feed significantly ($p < 0.001$) increased the MK-4 content in eggs (mean range: 46–51 µg/100 g, approximately 42–56% of the US Adequate Intake values).

2.5 Water soluble vitamins

2.5.1 Nutritional aspects

B vitamins are essential for energy metabolism, nerve function, DNA synthesis, and amino acid metabolism. Severe deficiencies can lead to various conditions: thiamin deficiency causes beriberi and Wernicke's encephalopathy; riboflavin deficiency results in mouth lesions; niacin deficiency causes pellagra; vitamin B_6 deficiency leads to convulsions and nerve damage; vitamin B_{12} deficiency results in pernicious anaemia; biotin and pantothenic acid deficiencies are rare but can cause neuromotor defects and gastric disturbances.

Folate deficiency causes megaloblastic anaemia, prevents neural tube defects (NTDs) and is crucial for women from pre-conception through early pregnancy. Many countries have mandated folic acid fortification of flour to prevent NTDs during pregnancy. Emerging evidence also suggests folate protects against cardiovascular disease, stroke, cognitive impairment, and osteoporosis, highlighting its importance for the general population.

Folate and folic acid are both forms of folate, found naturally in foods like leafy greens and liver, and folic acid in supplements, differing in their chemical structure and bioavailability. Public health authorities are concerned about overexposure to synthetic folic acid but consider natural folate sources, like eggs, safe. Unlike synthetic folic acid, natural folate in eggs poses no health risks, and hens can efficiently convert supplemented folic acid into natural folates.

Eggs are devoid of vitamin C but are an excellent dietary source of a number of B vitamins, in particular vitamin B_{12}, biotin, and choline.

Table 1 Water-soluble vitamin content of unfortified whole British eggs (per 100 g) and percentage contribution to UK reference intakes

Vitamin	Amount per 100g whole egg (raw)	% of UK recommended intake for adults
Thiamin	0.08 mg	8%
Riboflavin	0.50 mg	38%
Niacin	3.45 mg	20%
Vitamin B6	0.13 mg	9%
Vitamin B12	2.7 µg	180%
Folate	47 µg	24%
Pantothenic Acid	1.35 mg	27%*
Biotin	19.5 µg	49%*
Vitamin C	0 mg	0%
Choline	285 mg	71%*

*EFSA Adequate Intake values, as there is no UK reference intake.

Table 1 summarizes the amount of water-soluble vitamins found in unfortified whole British eggs (per 100 g) and the respective contribution to daily recommended intakes.

Eggs are by far the richest source of choline in the diet. Despite the significant contribution of egg choline, many populations in Europe, America, Canada, and Australia fail to meet the choline intake recommendations. The EFSA has recognized several health benefits associated with choline, including its essential role in lipid metabolism, liver function, and the reduction of homocysteine levels, which are important for cardiovascular health.

2.5.2 Folate biofortification

The biofortification of eggs with B vitamins primarily focuses on folate, with limited evidence on other B vitamins such as vitamin B_{12}. Enriched eggs primarily contain natural folates, offering a safe way to boost folate intake and potentially protecting against diseases without the risks associated with synthetic folic acid. Studies show that hens can efficiently convert dietary folic acid into natural folates, enhancing egg folate content.

House et al. (2002) observed that egg folate content increases linearly with dietary folic acid supplementation of hen diets at levels from 0 to 128 mg of folic acid/kg feed of feed until it plateaus around 30–50 mg/egg. It has been suggested that the optimal folate enrichment levels for hens' diets were determined as 16 mg/kg feed, yielding 144.6 mg/100 g egg folate (Hoey et al., 2009). One study evaluated the stability of natural folates in biofortified eggs during storage and cooking (Altic et al., 2016). The study's findings revealed that enriched eggs had a significantly higher folate content than unenriched eggs, and that storage for up to 27 days, along with cooking methods such as boiling, poaching, frying, and scrambling, did not significantly reduce the folate levels in biofortified eggs (Altic et al. 2016). Therefore, it can be concluded that folate-enriched eggs retain their nutritional benefits, supporting their potential as functional foods to improve dietary folate intake Altic et al. (2016).

3 Case study: commercialising vitamin D-biofortified eggs in the UK

For countries at higher latitudes, like the UK, with limited UV exposure, particularly in winter, obtaining vitamin D from food is crucial for maintaining population's vitamin D status (SACN, 2016). Moderate to good dietary sources include fish, oil-rich fish products, eggs, and meat, along with fortified cereals and dairy (Buttriss and Lanham-New, 2020). However, the UK National Diet and Nutrition Survey shows that adults only achieve vitamin D intakes of 3–4 μg/day, equivalent to 30–40% of the daily recommended intakes. Therefore, urgent food-based strategies are required to fulfill daily recommendations (Buttriss et al., 2022).

In 2015, a project funded by Innovate UK's Agri-Tech catalyst, involving Newcastle University, DSM Nutritional Products, and Noble Foods, aimed to assess the feasibility of commercial-scale vitamin D enrichment of eggs (Clark et al., 2021). The study enriched hens' diets with vitamin D following EFSA safety regulations, testing different formulations to evaluate the extent of vitamin D enrichment in eggs on a large scale [10 flocks of colony (Lohmann) and free-range (Hy-Line) birds]. The intervention lasted six weeks, with hens receiving diets of either vitamin D_3 alone (3000 IU/kg), a combination of D_3 and 25-D (1500 IU of D_3 + 37.5 μg of 25-D/kg), or 25-D alone (75 μg/kg). Eggs were analyzed for D_3 and 25-D content at weeks 0, 3, and 6 using LC-MS/MS. To account for the higher potency of 25-D versus vitamin D_3, total vitamin D content was calculated by multiplying 25-D values by five and adding them to D_3 values.

By week 6, eggs from hens fed only 25-D showed a 43% increase in total vitamin D content compared to baseline, while no significant changes were noted in eggs from hens fed only vitamin D_3 or a combination of vitamin D_3 and 25-D. The vitamin D content in eggs increased steadily, without adversely affecting hen mortality. In addition, there was no impact of housing system (colony vs. free-range) on the vitamin D content of the eggs. These findings led to the re-launch of Noble Foods' "happy egg.co" brand, which now provides 94% of the daily vitamin D requirement per two large eggs according to European food labelling standards. By 2021, the brand was valued at £99.6

million, purchased by 29.5% of UK households, and offered at no extra cost compared to pre-reformulation.

Building on the initial project's success, a second project, "Sunshine Eggs 2," commenced in 2018 with funding from Innovate UK's Innovations in Health and Life Sciences call. This project, involving the same partners plus SME egg business Bryants Court Farm, aimed to maximize the commercial potential of vitamin D-enriched eggs. The project's objectives included evaluating the effects of storage and cooking on vitamin D content, assessing the physiological effects of consuming enriched eggs, determining consumer acceptability, and identifying market opportunities.

Specific research objectives included:

1 Evaluating how storage and cooking affect the vitamin D content of enriched eggs.
2 Investigating the impact of consuming vitamin D enriched eggs on individuals' vitamin D status through an 8-week trial in the North of England.
3 Exploring consumer attitudes toward vitamin D enriched eggs.
4 Identifying market opportunities for these enriched eggs.

The research examined the effects of egg laying timing, storage conditions (ambient vs. refrigeration), and common cooking methods (boiling, frying, scrambling, poaching, and microwaving) on the concentration of vitamin D metabolites in eggs enriched with 25-D_3 (Clark et al., 2023). Samples were freeze-dried and analysed for vitamin D_3 and 25-D_3 using HPLC-MS(/MS). Results indicated that both storage and cooking methods influenced vitamin D metabolite retention. Specifically, 25-D_3 retained 72% to 111% of its content with concentrations of 0.67 to 0.96 μg/100 g of whole egg, while vitamin D_3 retained 50% to 152% with concentrations of 0.11 to 0.61 μg/100 g of whole egg (Clark et al., 2023). Overall vitamin D activity in enriched eggs ranged from 3.45 to 5.43 μg/100 g, a 22% to 132% increase compared to non-enriched eggs. The study suggested that 25-D_3 remains stable in eggs after storage and cooking, indicating its potential as a valuable dietary source of vitamin D (Clark et al., 2023).

The Sunshine Eggs project exemplifies a successful collaboration between academia and industry to drive the commercialization of vitamin D-enriched eggs in the UK. This effort aims to enhance the understanding of vitamin D-enriched eggs' broader science, including processing effects on vitamin D content and physiological effects upon consumption. Additionally, it seeks to address consumer attitudes and overcome barriers to maximize the potential of this novel food product, ultimately helping to close the gap between population's dietary vitamin D intake and requirements.

4 Trace mineral biofortification

This section will summarize the evidence for the dietary mineral biofortification of eggs. The focus will be on trace minerals where evidence supports biofortification and where there is nutritional rationale to improve dietary intakes in consumers. Four trace minerals will be discussed (iron, zinc, selenium, and iodine) separately, providing a general overview of the main nutritional aspects, including functions, nutritional requirements, consequences of deficiency, and excess. The evidence on biofortification for each trace mineral will be explored and summarized.

4.1 Iron

4.1.1 Nutritional aspects

Iron is essential for biological systems, primarily functioning as a component of proteins involved in oxygen transport, energy metabolism, and enzyme activity. Deficiency can result in anaemia, impaired development, and weakened immunity, particularly affecting women of childbearing age, infants, and vegetarians. Bioavailability is influenced by physiological status and dietary factors like vitamin C and phytates. Although iron toxicity from dietary sources is rare, excessive supplementation can cause gastrointestinal discomfort and, in severe cases, organ damage. Animal studies show varying toxic effects depending on the form of iron. EFSA suggests that there is insufficient data to set a Tolerable Upper Level for iron due to the complexities of assessing iron intake and its relationship with adverse effects. Instead, they recently established a safe iron intake level of 40 mg/day for adults, including pregnant and lactating women. The World Health Organization reports that anaemia affects approximately 40% of children aged 6–59 months, 37% of pregnant women, and 30% of women aged 15–49 worldwide. In 2019 alone, anaemia resulted in the loss of 50 million years of healthy life due to disability. The primary contributors to this burden were dietary iron deficiency, thalassemia, sickle cell trait, and malaria.

The UK recommended intake of iron is 8.7 mg for men (19 to 50 years) and 14.8 mg for women (19 to 50 years). The iron requirement for laying hens is 38–50 mg/kg (National Research Council, 1994). Hen diets typically supplement iron in both organic and inorganic forms, either in the ferrous (Fe2+) or ferric (Fe3+) valence states. Iron is commonly supplemented to hen diets in forms such as ferrous sulfate, ferric chloride, ferric citrate, ferric ammonium citrate, and iron amino acid complexes (Klasing, 2000). UK data indicates raw whole eggs contain 1.72 mg of iron per 100g (Department of Health, 2013).

4.1.2 Iron biofortification

Extensive hen feeding trials of iron supplementation have demonstrated low to moderate efficacy in enhancing egg iron content (for review see Schiavone and Barroeta, 2011). Iron-biofortified eggs typically yield 10–20% higher iron content than regular eggs. However, varying outcomes in iron-enriched eggs across different studies emphasize the importance of considering the source of iron supplementation, dietary interactions with other minerals, the impact of bird age and strain, and individual differences in egg iron concentration. The iron and manganese content in hen eggs shows significant variation among individual birds, unlike the concentration of other minerals (Revell and Hughes, 2005). Developing a breeding program that targets 'high-iron' birds, alongside tailored dietary strategies, could be a significant research focus. This includes creating techniques to identify birds with high iron content for specialized markets (Revell and Hughes, 2005).

Recent research also highlights the potential for mineral-amino acid chelates in improving laying hens' production performance and egg compositional quality. Studies show that iron glycine chelate (Fe-Gly) is more effective than ferrous sulfate (FeSO4) in improving iron biofortification due to its higher bioavailability and better absorption (Xie et al., 2019). In an 8-week feeding trial, hens supplemented with 60 mg Fe/kg as $FeSO_4$ achieved an iron concentration of 55.97 mg/kg and 7.57 mg/kg in egg yolk and albumin, respectively, while hens fed 60 mg Fe/kg as Fe-Gly reached iron concentrations of 63.81mg/kg and 8.16mg/kg in egg yolk and albumin, respectively (Xie et al., 2019).

4.2 Zinc

4.2.1 Nutritional aspects

Zinc is vital for over 300 enzymes involved in protein, fat, and carbohydrate metabolism, as well as structural functions. Deficiency, though rare, leads to growth retardation, delayed puberty, and diverse effects in humans. Factors affecting zinc absorption include phytate, iron, copper, and certain amino acids. Bioavailability varies depending on dietary sources, with animal products being the primary contributors. Average dietary intakes in the EU meet recommended levels, but vegetarians may require a higher intake. Zinc toxicity can result in gastrointestinal disturbances and interfere with copper and iron metabolism.

The UK recommended daily amount of zinc is 9.5 mg for men (19 to 64 years) and 7 mg for women (19 to 64 years). Recent evidence suggests that moderately high zinc intakes have no adverse effects on health, with the European Upper Level set at 25 mg/d for adults, including pregnant and lactating women. The zinc requirement for laying hens is 29–44 mg/kg diet (National Research Council, 1994). Hen diets typically supplement zinc in two forms: inorganic (zinc oxide, zinc carbonate, and zinc sulphate) or organic (complexed with amino acids) (Klasing, 2000). UK data indicates raw whole eggs contain 1.12 mg of zinc per 100g (Department of Health, 2013).

4.2.2 Zinc biofortification

Zinc biofortification efforts have used both inorganic (e.g. zinc sulphate, zinc oxide) and organic forms (e.g. amino acid complexes). Research indicates that higher dietary zinc levels lead to increased zinc deposition in egg yolks through mechanisms likely involving vitellogenin, a zinc-binding protein crucial for yolk formation (Richards, 1997). Stahl et al. (1988) demonstrated significant increases in egg zinc content with high and very high zinc diets (up to 1762–1861 mg Zn/g), showing up to 90% more zinc compared to controls. Kim and Patterson (2005) further explored the dose-response relationship between very high dietary zinc levels and egg zinc content, with variations observed between inorganic and organic zinc forms. In the same study, hens fed diets supplemented with 2000 and 3000 mg/kg zinc experienced reductions in body weight, feed consumption, egg production, egg weight, and shell thickness.

Interestingly, Michalak et al. (2011) showed that zinc-enriched macroalgae supplementation in laying hens led to a 171% increase in zinc content in egg albumen. In summary, further research is necessary to optimize supplementation strategies without compromising poultry health or egg quality, even though zinc-biofortified eggs could potentially contribute to dietary zinc intake, especially in regions with zinc deficiency concerns.

4.3 Selenium

4.3.1 Nutritional aspects

Selenium (Se) is vital for thyroid, neural, immune, and gastrointestinal functions. Soil selenium levels affect food content, with low levels in Denmark, Finland, parts of China, and Russia and high levels in the USA, Canada, and Colombia. Se exists in inorganic and organic forms, with the latter essential for selenoproteins like glutathione peroxidase (GPx) and iodothyronine deiodinase. Deficiency can cause Keshan and Kashin-Beck diseases, affecting the heart and bones, respectively, and influencing iodine metabolism and immunity. Selenium-rich foods include Brazil nuts, fish, meat, and eggs. The UK recommended daily amount of selenium is 75 µg for men (19 to 64 years) and 60 µg for women (19 to 64 years).

In the UK, the average selenium intake for adults is 40 µg/day, which is significantly below recommended intakes (SACN, 2013). Toxicity, or selenosis, leads to symptoms such as hair and nail loss, with an upper level of 300–400 µg/

day. Global selenium deficiency remains a concern, particularly in low-selenium soil regions, with marginal intake levels during adolescence and pregnancy posing risks (SACN, 2013). The National Research Council recommends dietary selenium levels of 0.05–0.08 mg/kg, noting selenium's toxicity with a narrow margin between deficiency and excess (10 mg/kg). Chickens efficiently absorb selenite, selenate, and organic selenides with minimal homeostatic regulation. UK data indicates raw whole eggs contain 23 μg of selenium per 100g (Department of Health, 2013).

4.3.2 Selenium biofortification

For over 20 years, selenium biofortification of eggs has been the subject of intense scientific investigation and commercial application. Extensive hen feeding trials using various forms and doses of selenium (either alone or in combination with other nutrients) have demonstrated good efficacy in enhancing egg selenium content (for review see Schiavone and Barroeta, 2011).

Initially, enriching eggs with selenium was challenging due to the relatively low efficiency of transferring inorganic selenium (selenite or selenate) to the egg. However, replacing sodium selenite with selenium-yeast in hen diets substantially increases the selenium deposition in eggs (Fisinin et al., 2009). Selenium enrichment, using organic selenium at concentrations between 0.3 and 0.5 mg/kg in feed, significantly enhances the content of selenium in eggs to between 20 and 40 μg of selenium per egg (Fisinin et al., 2009). This amount can supply up to 70% of the recommended human selenium intake, making these eggs a viable approach to improving selenium in the diet. In a recent study of 576 healthy 43-week-old Hy-Line Brown laying hens (Li et al., 2024), the aim was to compare the effects of sodium selenite (SS), yeast selenium (SY), and methionine selenium (SM) at selenium levels of 0.15 mg/kg and 0.30 mg/kg on egg selenium content. Selenium deposition in the whole egg after 8 weeks of 0.30 mg/kg selenium treatment was 0.56, 0.65, and 0.64 mg selenium/kg of whole egg for SS (inorganic), SY (organic), and SM (organic), respectively. These results further support the superiority of organic over inorganic sources of selenium for selenium egg enrichment (Li et al., 2024).

To ascertain the human bioefficacy of selenium-biofortified eggs, a Ukrainian egg intervention trial found that consuming two selenium-enriched eggs daily for 8 weeks significantly increased plasma selenium levels in volunteers (Surai et al., 2004). Treatment group eggs had 28–32 μg selenium per egg, compared to 7–9 μg selenium per egg in the control group. Plasma selenium increased from low baseline levels (0.055–0.081 μg/mlś) to 0.09–0.14 μg/mL, indicating improved selenium status (Surai et al., 2004). A recent animal study examined the effects of selenium-enriched egg white protein (Se-EWP) on inflammation (Chen et al., 2022). The results showed that serum cytokine levels (IL-6, IL-2, and TNF-α) were significantly reduced in the Se-EWP group compared to the EWP group, indicating that Se-EWP could mitigate cyclophosphamide-induced immunosuppression (Chen et al., 2022). These findings suggest Se-EWP may be a promising functional food with immunomodulatory properties.

Russia and Ukraine have traditionally led innovations in bringing selenium-biofortified eggs to market (Fisinin et al., 2009), and many countries now offer selenium-enriched eggs to consumers. In the UK, there are a number of selenium-enriched egg brands, including Enriched Eggs and Super Eggs which typically contain ~30 µg of selenium per 100g whole egg.

4.4 Iodine

4.4.1 Nutritional aspects

Iodine is essential for thyroid hormone synthesis, regulating metabolic rate, cellular metabolism, and connective tissue integrity. Deficiency can cause hypothyroidism symptoms like lethargy, weight gain, and poor concentration, while prenatal deficiency increases the risk of miscarriage and congenital abnormalities. Iodine deficiency remains a major public health concern globally, with the latest World Health Organization estimate estimating that 1.88 billion people, including 241 million school-age children, have insufficient dietary iodine intake.

In the UK and Western Europe, the prevalence of severe iodine deficiency has significantly decreased over the past century due to improved general nutrition and targeted interventions to boost iodine intake. In the UK, there is now a public health concern that iodine intake levels may once again be marginally adequate or insufficient, particularly during adolescence, reproduction, gestation, and development (SACN, 2014). Iodine-rich foods include marine fish, seaweed, dairy, and iodized salt. Excess iodine can elevate thyroid-stimulating hormone levels, potentially leading to hypothyroidism.

The most up-to-date recommended iodine intake for adults (over 12 years) established by the World Health Organization is 150 µg/day, which increases to 250 µg/day during pregnancy and lactation. Upper levels set by EFSA are 600 µg/day for adults based on marginal TSH changes observed at high intakes. Poultry require 0.3–0.4 mg/kg of iodine in their diet, but the presence of goitrogenic feeds can more than double the iodine requirement. According to the latest UK data on the nutritional composition of regular (unfortified) hen eggs, the iodine content of raw whole eggs is 50 µg per 100g (Department of Health, 2013).

4.4.2 Iodine biofortification

Several older studies have shown that iodine content in whole eggs, egg yolks, and egg albumen increases with increased iodine supplementation (for review see Schiavone and Barroeta, 2011). Opaliński et al. (2012) aimed to assess the impact of iodine yeast (I-yeast) on laying hens' performance, egg traits, and egg iodine content. Sixty Hy-Line Brown hens (25 weeks) were randomly divided into 3 groups (4 replicates) for a 12-week trial. Diets included: control (1 mg I/kg from Ca(IO3)2·H2 O), E1 (1 mg I/kg from I-yeast), and E2 (2 mg I/kg from I-yeast). I-yeast did not have a significant effect on body weight gain. E2 showed the highest egg and albumen weights and eggshell iodine content, nearly three times higher than the control, suggesting I-yeast as an effective method to enrich eggs with iodine (Opaliński et al., 2012).

A pilot population intervention program in Thailand aimed to eliminate iodine deficiency through iodine biofortification of eggs in a rural community setting (Charoensiriwatana et al., 2010). A community hen egg farm in northeastern Thailand was selected, where regular feed was replaced with iodine-enriched feed (4 mg iodine/kg) using potassium iodide (KI). Hens consumed 120–130 g of this feed daily. Eggs produced contained 94 μg and 98 μg of iodine per egg (55–60 g and 60–65 g, respectively). Volunteers consuming these eggs showed an increase in median urine iodine levels from 7 μg/dl to 21 μg/dl in Ban Kew and from 7 μg/dl to 14 μg/dl in Ban Nong Nok Kean, thus showing the bioefficacy of high iodine eggs in reducing iodine deficiency in these regions (Charoensiriwatana et al., 2010).

A study from Jabalpur in Central India aimed to assess iodine supplementation levels in 135 White Leghorn hens, who were fed diets enriched with calcium iodate (5–20 mg/kg; 0.45–13 ppm I_2) for 10 weeks (Sumaiya et al., 2016). Higher iodine levels (up to 13.0 ppm I_2) significantly increased egg iodine content to 120 μg/100g. However, minimum and a significantly lower feeding cost (rupees per dozen or per kg eggs) was observed in hens receiving 6.50 ppm I_2; i.e. biofortified eggs containing 60 μg/100g I_2 and widespread biofortified eggs with iodine should aim to replicate this level of hen feeding for economic reasons (Sumaiya et al., 2016).

5 Conclusion and future trends

Biofortification of table eggs has significant potential for improving micronutrient intakes in human populations. The successful implementation of vitamin D- biofortified eggs, exemplified by the UK case study, underscores the practical benefits and feasibility of this approach in improving vitamin D intake in consumers. Moreover, the enrichment of eggs with other micronutrients, in particular vitamin A compounds and key trace minerals such as selenium, demonstrates additional nutritional benefits, which may lead to improved population health and disease prevention.

Future research should delve deeper into optimizing biofortification techniques, focusing on enhancing the bioavailability and stability of added nutrients. Long-term studies are needed to assess the health impacts on both hens and consumers, ensuring that the biofortified eggs are safe, effective, and beneficial over prolonged periods. Refining the balance of nutrient doses is essential to avoid adverse effects and ensure the welfare of the hens. Innovations in hen feed that focus on sustainability and health, utilizing alternative proteins like insect meal, probiotics for gut health, organic ingredients, and other functional additives, will positively shape biofortification strategies in the future.

Future research should also explore consumer attitudes, preferences, and willingness to pay for biofortified eggs. Effective communication of nutrition and health benefits, as well as marketing strategies, is required to educate the public on the benefits of biofortified eggs and encourage widespread acceptance and adoption. Collaborative efforts between researchers, industry, trade bodies for the egg sector, and policymakers are vital to advancing the field of egg biofortification. Establishing supportive regulatory frameworks will facilitate innovation while ensuring safety and efficacy. Fostering such collaborations is going to be essential to reducing global micronutrient deficiencies and contributing to sustainable food systems.

6 Where to look for further information

The following article provides a good overview of older studies on egg biofortification with vitamins and trace minerals:

- Schiavone, A. and Barroeta, A. C. (2011), Egg enrichment with vitamins and trace minerals. In Y. Nys, M. Bain, & F. Van Immerseel (Eds.), *Improving the Safety and Quality of Eggs and Egg Products* (pp. 289–320). Woodhead Publishing Series in Food Science, Technology and Nutrition. Woodhead Publishing, Delhi.

The following recent article summarizes the bioavailability and environmental impacts of various sources of selenium in animal feeds:

- Ferrari, L., et al. (2023), Advances in selenium supplementation: from selenium-enriched yeast to potential selenium-enriched insects, and selenium nanoparticles. *Anim. Nutr.*, 14, 193–203.

The following recent article provides a comprehensive review on topics related to carotenoid enrichment of eggs:

- Dansou, D. M., et al. (2023), Carotenoid enrichment in eggs: from a biochemistry perspective. *Anim. Nutr.*, 14, 315–333.

The following article provides a comprehensive review on Omega-3 enriched eggs:

- Fraeye, I., et al. (2012), Dietary enrichment of eggs with omega-3 fatty acids: a review. *Food Res. Int.*, 48(2), 961–969.

Online data on the micronutrient content of eggs can be found here:

- USA: FoodData Central (usda.gov).
- United Kingdom: Department of Health. (2012), *Nutrient Analysis of Eggs*. Available at: https://www.gov.uk/government/publications/nutrient-analysis-of-egg.

Further details linked to the case study on vitamin D-biofortified eggs can be found here:

- Clark, A., et al. (2021), Egg enrichment with vitamin D: the sunshine eggs projects. *Nutr. Bull.*, 46, 332–338.
- Clark, A., et al. (2023), The influence of storage and cooking on the vitamin D content of 25-hydroxyvitamin D_3-enriched eggs. *Foods*, 12, 2522.

Details on the UK National Diet and Nutrition Survey can be found here:

- https://www.gov.uk/government/collections/national-diet-and-nutrition-survey.

Nutritional value of eggs from various avian species:

- Réhault-Godbert, S., et al. (2019), The golden egg: nutritional value, bioactivities, and emerging benefits for human health. *Nutrients*, 11(3), 684.

7 References

Altic, L., et al. (2016), Validation of folate-enriched eggs as a functional food for improving folate intake in consumers, *Nutrients*, 8(12), 777.

Asioli, D., et al. (2017), Making sense of the "clean label" trends: a review of consumer food choice behavior and discussion of industry implications, *Food. Res. Int.*, 99, 58–71.

Avnee, S., et al. (2023), Biofortification: an approach to eradicate micronutrient deficiency, *Front. Nutr., 10*, 1233070.

Browning, L. C. and Cowieson, A. J. (2014), Vitamin D fortification of eggs for human health, *J. Sci. Food. Agric.*, 94, 1389–1396.

Buttriss, J. L., et al. (2022), Proceedings of the second Rank Prize Funds Forum on vitamin D: implementation strategies for increasing vitamin D intake in the UK: current controversies and future perspectives, *Br. J. Nutr.,* 127(10), 1567–1587.

Buttriss, J. L. and Lanham-New, S. A. (2020), Is a vitamin D fortification strategy needed? *Nutr. Bull.*, 45, 115–122.

Cashman, K. D., et al. (2012), Relative effectiveness of oral 25-hydroxyvitamin D_3 and vitamin D_3 in raising wintertime serum 25-hydroxyvitamin D in older adults. *Am. J. Clin. Nutr.,* 95(6), 1350–1356.

Charoensiriwatana, W., et al. (2010), Consuming iodine enriched eggs to solve the iodine deficiency endemic for remote areas in Thailand, *Nutr. J.*, 9, 68.

Chen, M., et al. (2022), Identification and immunomodulatory effect on immunosuppressed mice of selenium-enriched peptides of egg white, *J. Agric. Food. Chem.*, 70, 12663–12671.

Clark, A., et al. (2021), Egg enrichment with vitamin D: the sunshine eggs projects, *Nutr. Bull.,* 46, 332–338.

Clark, A., et al. (2023), The influence of storage and cooking on the vitamin D content of 25-hydroxyvitamin D_3-enriched eggs, *Foods.*, 12, 2522.

Clark, B., et al. (2019), Consumer perception of vitamin D and fortified foods, *Br. Food. J.*, 121, 2205– 2218.

Department of Health. (2013), *Nutrient analysis of eggs: analytical report*, Retrieved from https://assets.publishing.service.gov.uk/media/5a74cd56ed915d502d6cb209/Nutrient_analysis_of_eggs_Analytical_Report.pdf

European Commission. (2017), Concerning the authorisation of cholecalciferol as a feed additive for all animal species, *Official. J. Eur. Union.*, 60, 19–22.

European Food Safety Authority (EFSA). (2005), Opinion of the Scientific Panel on additives and products or substances used in animal feed (FEEDAP) on the evaluation of safety and efficacy of "Hy•D" (calcifediol), based on 25-hydroxylcholecalciferol/25-hydroxy-pre-cholecalciferol, as feed additive in accordance with Council Directive 70/524/EEC. *EFSA. J.,* 3(7), 218.

European Food Safety Authority (EFSA) Panel on Nutrition, Novel Foods and Food Allergens (NDA) (2023), Scientific opinion on the tolerable upper intake level for vitamin D, including the derivation of a conversion factor for calcidiol monohydrate. *EFSA J.*, 21(8), e08145.Thank you

Fisinin, V. I., et al. (2009), Producing selenium-enriched eggs and meat to improve the selenium status of the general population, *Crit. Rev. Biotechnol.*, 29(1), 18–28.

Fraeye, I., et al. (2012), Dietary enrichment of eggs with omega-3 fatty acids: A review, *Food. Res. Int., 48*(2), 961–969.

Hayes, A., et al. (2016), Vitamin D- enhanced eggs are protective of wintertime serum 25-hydroxyvitamin D in a randomized controlled trial of adults, *Am. J. Clin. Nutr.*, 104, 629–637.

Hoey, L., et al. (2009), Laying hens can convert high doses of folic acid added to the feed into natural folates in eggs providing a novel source of food folate, *Br. J. Nutr.,* 101(2), 206–212.

House, J. D., et al. (2002), The enrichment of eggs with folic acid through supplementation of the laying hen diet, *Poult. Sci., 81*(9), 1332–1337.

Jiang, Y., et al. (1994), Alpha-tocopherol, beta-carotene, and retinol enrichment of chicken eggs, *Poult. Sci.*, 73, 1137–1143.

Kim, W. K. and Patterson, P. H. (2005), Effects of dietary zinc supplementation on hen performance, ammonia volatilization, and nitrogen retention in manure, *J. Environ. Sci. Health*. Part B., 40, 675–686.

Klasing, K. C. (2000), *Comparative Avian Nutrition*, CABI Publishing, New York, NY.

Li, Y., et al. (2024), Effects of different selenium sources and levels on the physiological state, selenoprotein expression, and production and preservation of selenium-enriched eggs in laying hens. *Poult. Sci.*, 103(2), 103347.

Mattila, P., et al. (1999), Cholecalciferol and 25-hydroxycholecalciferol content of chicken egg yolk as affected by the cholecalciferol content of feed, *J. Agric. Food. Chem.*, 47, 4089–4092.

Mattila, P., et al. (2003), Effect of cholecalciferol-enriched hen feed on egg quality, *J. Agric. Food. Chem.*, 51, 283–287.

Mattila, P., et al. (2004), Effect of vitamin D_2- and D_3-enriched diets on egg vitamin D content, production, and bird condition during an entire production period, *Poult. Sci.*, 83, 433–440.

Mattila, P. H., et al. (2011), Effect of different vitamin D supplementations in poultry feed on vitamin D content of eggs and chicken meat, *J. Agric. Food. Chem.*, 59, 8298–8303.

Mendonça, C. X., et al. (2002), Effect of dietary vitamin A on egg yolk retinol and tocopherol levels, *J. App. Poult. Res.*, 11, 373–378.

Michalak, I., et al. (2011), Effect of macroalgae enriched with microelements on egg quality parameters and mineral content of eggs, eggshell, blood, feathers and droppings, *J. Animal. Physiol. Animal. Nutr.*, 95, 374–387.

Naber, E. C. (1993), Modifying vitamin composition of eggs: a review, *J. Appl. Poult. Res.*, 2, 385–393.

National Research Council. (1994), *Nutrient Requirements for Poultry*, 9th ed., National Academy Press, Washington, DC.

Opaliński, S., et al. (2012), Effect of iodine-enriched yeast supplementation of diet on performance of laying hens, egg traits, and egg iodine content, *Poult. Sci.*, 91(7), 1627–1632.

O'Sullivan, S. M., et al. (2020), Biofortification of chicken eggs with vitamin K-nutritional and quality improvements, *Foods*, 9(11), 1619.

Papanikolaou, Y. and Fulgoni, V. L. (2021), Increasing egg consumption at breakfast is associated with increased usual nutrient intakes: a modeling analysis using NHANES and the USDA child and adult care food program school breakfast guidelines, *Nutrients.*, 13(4), 1379.

Park, S. W., et al. (2005), Enrichment of vitamin D_3, K and iron in eggs of laying hens, *Asian. Aust. J. Animal. Sci.*, 18, 226–229.

Pirgozliev, V. R., et al. (2022), Stevia (Stevia rebaudiana) improves carotenoid content in eggs when fed to laying hens, *Foods.*, 11(10), 1418.

Revell, D. K. and Hughes, R. J. (2005), Between-bird variation in the concentration of iron and other minerals in egg yolk across a laying cycle, *Proc. Austral. Poult. Sci. Symposium., 17*, 287–290.

Richards, M. P. (1997), Trace mineral metabolism in the avian embryo, *Poult. Sci.*, 76, 152–164.

Schiavone, A. and Barroeta, A. C. (2011), Egg enrichment with vitamins and trace minerals. In Y. Nys, M. Bain, & F. Van Immerseel (Eds.), *Improving the Safety and Quality of Eggs and Egg Products* (pp. 289–320). Woodhead Publishing Series in Food Science, Technology and Nutrition. Woodhead Publishing, Delhi.

Scientific Advisory Committee on Nutrition (SACN). (2013), *Selenium and Health*. The Stationery Office, London. Available from: https://www.gov.uk/government/publications/sacn-statement-on-selenium-and-health-2013

Scientific Advisory Committee on Nutrition (SACN). (2014), *Iodine and Health*. The Stationery Office, London. Available from: https://www.gov.uk/government/publications/sacn-statement-on-iodine-and-health-2014

Scientific Advisory Committee on Nutrition (SACN). (2016), *Vitamin D and Health*. The Stationery Office, London. Available from: https://www.gov.uk/government/publications/sacn-vitamin-d-and-health-report

Shea, M. K. and Booth, S. L. (2016), Concepts and Controversies in Evaluating Vitamin K Status in Population-Based Studies, *Nutrients.,* 8(1), 8.

Squires, M. W. and Naber, E. C. (1993), Vitamin profiles of eggs as indicators of nutritional status in the laying hen: vitamin A study, *Poult. Sci.*, 72, 154–164.

Stahl, J. L., et al. (1988), Zinc, iron, and copper contents of eggs from hens fed varying levels of zinc, *J. Food. Comp. Anal.*, 1, 309–315.

Sumaiya, S., et al. (2016), Effect of dietary iodine on production of iodine enriched eggs, *Vet. World.*, 9(6), 554–558.

Surai, P. F., et al. (2004), Consumption of selenium-enriched eggs improves selenium status in human volunteers, *Proc. XXII. World. Poultry. Congress.*, Turkey.

Suzuki, Y. and Okamoto, M. (1997), Production of hen's eggs rich in vitamin K, *Nutr. Res.*, 17, 1607–1615.

Tang, G. (2010), Bioconversion of dietary provitamin A carotenoids to vitamin A in humans, *Am. J. Clin. Nutr.,* 91(suppl),1468S–1473S.

Trugilho, L., et al. (2024), Vitamin E and conflicting understandings in non-communicable diseases: Is it worth supplementing? *Clin. Nutr. ESPEN.*, 59, 343–354.

Xie, C., et al. (2019), Effect of iron glycine chelate supplementation on egg quality and egg iron enrichment in laying hens, *Poultr. Sci.*, 98, 7101–7109.

Chapter 14

One-Welfare approach to assessing egg quality

Angela Trocino, University of Padova, Italy; Francesca Leone and Valentina Ferrante, University of Milan, Italy

1 Introduction

All over the world, 7.9 billion laying hens, most of which are kept in cages, contribute to commercial egg production. In the EU in 2022, 39.7% of hens were farmed in enriched (also called furnished) cages, whereas the remaining 60.3% of hens was kept in alternative and cage-free housing systems, i.e. barns (including floor with single tier and floor with multi-tiers aviaries in conventional farms) (37.8%) and free range (15.5%) and organic systems (7.0%) as extensive productions (European Commission, 2024; Council Directive 1999/74/EC; Commission Regulation 889/2008; Kollenda et al., 2020). Enriched cages, intended to provide additional space and resources for satisfying behavioral needs of hens, have a relatively low risk for diseases and parasitism under controlled conditions, besides high economic efficiency (Lay et al., 2011; Mench et al., 2011; Campbell et al., 2019). Nevertheless, the European Resolution P9_TA(2021)0295, answering the European Citizens' Initiative 'End the Cage Age,' asks for banning any cage system for farmed animals in the EU. Thus, the European Parliament has called for a phasing out of cages by 2027, which

http://dx.doi.org/10.19103/AS.2024.0150.22

is consistent with the EU Green Deal and Farm-to-Fork Strategy, promoting a green transition toward more sustainable agriculture practices and animal-friendly production systems, especially organic ones.

Experience in the field and literature have identified weaknesses besides strength points of cage-free systems for egg production and hen welfare (Hartcher & Jones, 2017; Gautron et al., 2021), which require further insights for the identification of the best on-farm solutions. This is further stressed in the perspective of incoming new regulations on additional issues intended to further improve animal welfare, e.g. ban of beak trimming, provision of natural light, use of litter, forage, and perches.

Since the respect for animals requires a cultural context that motivates the human–animal interactions to protect and care for all animals, a comprehensive evaluation of the green transition in production systems for laying hens requires a One-Welfare approach (Pinillos et al., 2016; Bourque, 2017) as a holistic and ethical approach to sustainability. The One-Welfare approach recognizes interconnections between different pillars (*Animals, Economics, Society, and Environment*) to take care of all relationships between animal welfare, human well-being, and environment with a high level of responsibility of final and intermediate customers. Consistently with the One-Welfare approach, one of the main challenge is nowadays to define frameworks for comparing production and husbandry systems, which should incorporate social perceptions of animal welfare, besides a broad range of key elements, such as environmental impacts, resource management practice, and sustainability measures (Paulović et al.,2024). These frameworks can be used for enforcement and labeling schemes, fostering communication between producers and consumers and highlighting important welfare issues.

When specifically referring to the challenges associated to the transition of the farming system for laying hens, as for *Economics*, the approach should consider that on-farm production results (i.e. farm efficiency) can be affected by differences in the management and equipment associated to the different farming systems (Holt et al., 2011; Matthews & Sumner, 2015; Villanueva et al., 2017). As for *Society*, farmer labor on one side (Brannan & Anderson, 2021) and extrinsic (ethical aspects) and intrinsic (safety, nutritional, and sensory properties) quality of eggs, as perceived and requested by consumers, are affected by feeding and outdoor access, besides management of hens (Gazzotti et al., 2021). Nevertheless, whether the consumers' claim for sustainability corresponds or not to an effective willingness to pay higher prices has still to be proven for eggs obtained in alternative systems in all countries (Rondoni et al., 2020; De Marchi et al., 2024; Estévez-Moreno et al., 2025). As for *Environment*, the impact of alternative cage-free systems, including free range and organic productions, has not yet been fully assessed, whereas conventional poultry production is recognized for its low impact, thanks to the high efficiency in converting feed into food (Xin et al., 2011; Leinonen et al., 2012; Pelletier, 2017; Abín et al., 2018; Costantini et al., 2020).

Thus, based on a One-Welfare approach, the present chapter will address the impacts and research challenges focusing on the *Animal* pillar of this transition, as welfare concerns for animals remain in all farming systems (EFSA AHAW Panel, 2023). Thus, the main welfare concerns and corresponding risk factors and mitigation strategies, together with welfare improvements in all housing systems, will be hereby addressed in view of the impacts that they can have on the intrinsic quality of eggs.

2 Welfare and egg quality issues across housing systems

Several concerns for the welfare of laying hens exist both under conventional systems (enriched cages, floor systems with single tiers and floor systems with multi-tiers in Europe) and in alternative free-range and organic systems. In its latest Scientific opinion about the welfare of laying hens on farm (EFSA AHAW Panel, 2023), EFSA has identified eight highly relevant welfare consequences for laying hens on farm which are listed in Table 1 together with the animal-based measures (ABMs) that can be used to assess their occurrence.

As for the welfare consequences that are shared by most housing systems, bone lesions (intended as keel fractures), skin disorders (infections, inflammation of the skin or sunburn), and soft tissue lesions and integument damages are highly relevant ones. Then, the housing system associated with almost all welfare consequences (but the predation risk) is that in enriched cages.

Indeed, in laying hens, bone health (including keel) is one of the major welfare issues on farm, where several factors related to each other affect the development of the bone apparatus. Out of bone lesions, keel fractures have been proven to produce negative states, such as pain, distress, and/or discomfort (Nasr et al., 2012; Rokavec & Zupan Šemrov, 2020; Edgar et al., 2023). Different hazards for this welfare consequence have been identified related to animals (e.g. genetics, age, and persistently high egg production and large egg size), housing and equipment (e.g. cage vs. free cage systems, perches, ramps), and management (e.g. training of pullets for navigation ability and feeding) (Whitehead & Fleming, 2000; Rufener & Makagon, 2020; Toscano et al., 2020; EFSA AHAW Panel, 2023), even if outcomes of the different experimental studies and field reports are not always consistent.

With reference to the effect of the housing system on bone health, in enriched cages, poor bone quality is the cause usually identified for keel bone deviations and fractures, where the restriction of movement (included among welfare consequences in this system) is considered to favor osteoporosis and

Table 1 Highly relevant welfare consequences (and corresponding animal based measures, ABMs) in laying hens under different housing systems (X = indicates in which housing system the welfare consequence is identified as highly relevant)

Highly relevant welfare consequence (related ABMs)	Enriched cages	Floor systems with single tier	Floor system with multitier	Mobile housing*
Bone lesions, keel fractures (keel bone fracture)	X	X	X	X
Group stress (aggressive interactions, fear responses, injurious pecking, physiological stress indicators, piling behavior, plumage damage)	X	X	X	
Inability to perform comfort behaviors (lack of dust bathing, preening, wing flapping, and wing and leg stretching)	X			
Inability to perform exploratory or foraging behaviors (injurious pecking, plumage damage; walking, scratching, and pecking)	X			
Predation stress (dead birds or survivor birds with signs of predation or remains of birds; excess missing birds; fear response, physiological stress indicators; reluctance to use free range				X
Restriction of movement (reduced bone quality; locomotor behaviors, pushing and jostling behaviors, plumage damage)	X			
Skin disorders (restlessness)	X	X	X	X
Soft tissue lesions and integument damages (beak shape and length, bruises, plumage damage, wounds)	X	X	X	X

*Mobile housing: always combined with access to a free range.
Source: Modified from EFSA AHAW Panel (2023).

bone fragility with advancing age (Fleming et al., 2006; Lay et al., 2011; Toscano et al., 2020; EFSA AHAW Panel, 2023). Conversely, movement possibility is expected to affect skeletal development (Fleming et al., 2006; Regmi et al., 2015) as increased exercise can favor bone growth in pullets (Biewener & Bertram, 1994) and laying hens (Isaksson et al., 2009). However, the occurrence of keel lesions is high in all farming systems. In cage-free systems, collisions that can occur among animals or between animals and the environment, due to a poor ability of laying hens to move, are considered co-responsible for keel lesions (EFSA AHAW Panel, 2023). In this context, the occurrence of keel bone lesions can change with management and structures, both during the pullet and laying phase, where also the presence and type of perches can play a negative role (Hester et al., 2013; Stratmann et al., 2015; Wilkins et al., 2011). In free-range systems, the high occurrence of keel bone lesions has been attributed to the lower feed (and thus mineral) intake compared to hens kept in cages (Pulcini et al., 2023). Finally, genotype has also been found to play a role on the occurrence of keel lesions, where both production level (Käppeli et al., 2011; Candelotto et al., 2017; Pulcini et al., 2023) and movement ability change with genotypes (Ali et al., 2020; Ciarelli et al., 2023) and an interaction between housing systems and genotype can be expected.

Notably, the presence of keel bone fracture has been found to negatively affect productivity and egg quality. As for production, in a small-scale study, Nasr et al. (2013) found that oviposition rate was lower and feed intake was higher in brown hens with keel fractures compared to healthy ones. They also measured a significant positive association between keel fracture severity and water consumption, and a significant negative association between keel fracture severity and egg weight and keel bone strength. A negative relationship between the severity of keel bone fractures and egg laying performance with increasing age was also measured by Rufener et al. (2019) in two selected genotypes of laying hens, whereas egg quality traits (as for egg mass, shell breaking strength, and shell width) were not affected by the presence or severity of keel bone fractures. Differently, Wei et al. (2020) reported that keel fracture in laying hens reduced production performance, feed intake, and eggshell quality, besides welfare. Then, Candelotto et al. (2017) highlighted some relationships between egg quality and bone strength in four distinct crossbred and one pure line, where an increased susceptibility to keel bone fractures was associated with thinner eggshells and reduced egg breaking strength regardless of the genotype.

With reference to the other two welfare consequences highly relevant in all housing systems, skin disorders are related to infections, inflammation of the skin or sunburn, whereas soft tissue lesions and integument damages can depend on management practices (e.g. beak trimming), interaction with housing environment (e.g. feeder chains, sharp objects, sharp edges and abrasive surface, defects of the equipment) and among animals depending also on group size (e.g. unwanted sexual behaviors) (EFSA AHAW Panel, 2023). In cage-free systems, freedom of movement may cause injuries and fractures due to collisions among animals and between animals and equipment, depending also on hens'navigation ability (Stratmann et al., 2015; Ali et al., 2016; Campbell et al., 2016; Armstrong et al., 2020; Ciarelli et al., 2023). The impacts of the occurrence of these two welfare consequences on egg production are rather indirect as they are associated with negative affective status, i.e. pain, discomfort, and/or distress, which are expected to be associated with a reduction in feed intake and likely in oviposition rate when associated with high severity and duration.

The remaining welfare consequences listed in Table 1 are behavioral restrictions with 'inability to perform comfort behaviors, inability to perform exploratory or foraging behaviors, and restriction of movement' specific for enriched cages and with 'group stress' shared by all housing systems but the mobile ones.

In fact, in enriched cages, as for comfort behaviors, dust bathing is limited because of the dimensions of the area and the non-continuous availability of suitable materials; exploration or foraging cannot take place because of the absence of substrates; movement is still restricted and flight prevented, despite the presence of perches and nests. Then, in all systems, group size and stocking densities may restrict or challenge some behaviors to a different extent depending on the genotype: hens may experience different stress and competition for resources, which can end with aggression, feather pecking and cannibalism, fearful status, piling and smothering, animal losses or suffering, injuries, lesions, and illness (Ali et al., 2020; Armstrong et al., 2020; Wei et al., 2020).

Based on the ABMs listed in Table 1, aggressive interactions and related injuries, which can result in cannibalism and high mortality, are the main consequence of all the listed behavioral restrictions and remain the main issues to be faced in cage-free systems to guarantee a suitable welfare to farmed laying hens.

Indeed, gentle feather pecking (GFP) and severe feather pecking (SFP) are different forms (Rodenburg et al., 2013) and have very different consequences on hen welfare. Gentle feather pecking to the feathers tips is a widespread behavior observed in pullets and laying hens (Lambton et al., 2010; Gilani et al., 2013; Rodenburg et al., 2013); it is preferentially directed towards unknown conspecifics as a form of social exploration (Riedstra & Groothuis, 2002); it may have a stereotyped, repetitive, and invariant form. In both cases, the resulting feather damages are often minor, making GFP a concern primarily for its unclear relationship with more severe forms of feather pecking. In younger birds, correlations between GFP and SFP have been reported (Nicol et al., 2013) and some studies have hypothesized that GFP might precede or develop into SFP in older hens. However, most of the studies conducted in commercial farms have not demonstrated a clear link between these two behaviors (Newberry et al., 2007; Rodenburg et al., 2013).

On the other hand, SFP involves pecking or removing feathers of a conspecific; it has more impact on animal welfare and productivity compared to GFP as it is painful; victims often react negatively, moving away. SFP can lead to substantial feather loss, skin damage, major susceptibility to infections (Drake et al., 2010), reduced egg production, increased feed requirements, and higher mortality rates (Lambton et al., 2010; Rodenburg et al., 2013). Cloacal pecking is directed toward the tissues around the cloaca and it especially occurs at the time of egg laying. When tissues are ingested, the term cannibalism is used (Rodenburg et al., 2013). Overall, SFP, cloacal pecking, and cannibalism are all harmful pecking behaviors, which share many risk factors, where episodes of cannibalism or cloacal pecking can also occur without being preceded by SFP (Green et al., 2000; Lambton et al., 2015).

The probability of developing SFP significantly increases when egg production begins (Lambton et al., 2010), since it is strongly correlated with the increasing demand for nutrients and foraging. In fact, according to EFSA (EFSA AHAW Panel, 2023), the most important behavioral restrictions which can cause injurious pecking are the 'inability to perform exploratory and foraging behavior' and the 'inability to perform comfort behavior' and mitigation strategies should be adjusted accordingly.

Nevertheless, the exact mechanisms leading to the development of injurious pecking and cannibalism in laying hens are not yet fully understood, as several factors can play a role. Low dietary provisions of protein and fibre have been identified as for feeding, where other management practices can contribute, such as nest lighting (which can increase the attractiveness of the cloacal region while hens are laying), the use of bell drinkers, early deposition, and a high number of diet changes during the laying period (Lambton et al., 2015), besides the provision of pelleted feed and specific perch configurations (Green et al., 2000).

Due to this complexity, the current mitigation strategy still involves beak trimming, which consists of amputating part of the beak using infrared or thermal cautery. This practice can cause acute and chronic pain due to the innervated tissue, causing a loss of normal function and ability to sense materials (Pötzsch et al., 2001; Janczak & Riber, 2015). According to EFSA (EFSA AHAW Panel, 2023), alternative approaches to beak trimming for reducing the sharpness of the beak should be implemented.

As for relationships between injuries pecking and egg quality, while a reduction in egg production can be expected on the individual hen, associated to a higher feed intake to compensate for thermoregulation in case of severe feather damages, at a farm level, high mortality due to cannibalism produces a direct economic damage. On the other hand, no major effects have been reported as for egg quality, where an increase of mislaid eggs can be hypothesized in cage-free systems where aggression occurs in collective nests as well as an increase of eggs with dirty shells due to blood because of wounds at the cloaca. Indeed, there is a genetic component related to the behavior of laying hens in the nest which can favor negative behaviors (such as pecking and piling associated with gregarious nesting) and can impact as a consequence on egg quality, due to differences in the pattern and preferences for nest use besides behavior in the nest (Giersberg et al., 2019; Bécot et al., 2023).

3 Improvements and challenges in cage-free systems: hen welfare versus egg quality

Cage-free systems provide laying hens with different possibilities for moving and navigating which difficulties can increase from floor systems with single tiers to floor systems with multi-tiers and systems with outdoor access (from covered verandas to outdoor free-range space). In fact, hens must navigate easily within the housing systems to reach feed and water for their health and body development; they must distribute homogeneously in all levels of any multi-tier system for a full and safe space use; and they must quickly identify the different areas of the system for the expression of all their specie-specific behaviors and for functional use of the different areas to prevent floor egg laying. Differences in navigation ability and space use in a cage-free system can depend on genotype, aviary design, the management of pullets at housing and, last but not least, the early experience of pullets (Purdum et al., 2020; Pufall et al., 2021; Du et al., 2022) where several studies are available on animal distribution and space use of laying hens at housing in the new shed (Colson et al., 2008; Zheng et al., 2019; Ciarelli et al., 2023) and during the laying period (Ali et al., 2016, 2020; MacLachlan et al., 2020; Purdum et al., 2020).

As for movement restriction, generally speaking, while in cage-free systems, this issue is solved, other welfare consequences, as discussed above, can occur together with floor egg laying. This latter finally results in negative impacts on egg safety (as for contamination) (De Reu et al., 2005; Jones et al., 2015), egg losses (as for the higher number of dirty and broken eggs), farmer labor (as for egg collections from the floor), and has thus a negative impact on farm economy (Oliveira et al., 2016; Villanueva et al., 2017; Hemsworth, 2021; Michel et al., 2022). The development of floor egg laying is multifactorial, that is why results from single surveys or single experimental studies are not always consistent. For example, based on data collected on commercial Australian brown-egg-laying flocks, only the use of different brown strains of hens was associated with an incidence of floor eggs higher than 2%, whereas no association was recorded as for the type of cage-free system, indoor housing infrastructure, presence of perches and platforms, lighting color, and expression of social behavior.

On the other hand, with a comprehensive approach which updated the early review of Appleby (1984) and combined information from egg producers in Australia, Campbell (2023) developed a list of recommendations validated on the base of literature and field farmer experience to control floor eggs in cage-free systems which are reported in Table 2.

Then, in cage-free systems, the outdoor access (either a veranda or an outdoor area) is undoubtedly associated with a positive effect on welfare because of the possibility of foraging behavior (Campbell et al., 2020), but it brings about further challenges besides those already discussed above

Table 2 Conclusions and applications for management measures to prevent floor egg laying

Conclusions and applications
The development of floor egg laying is multifactorial and can be affected by individual preferences as well as the hen strain. Thus, specific strains of hens may be best suited to particular housing system designs.
The occurrence of the behavior can vary across flocks in the same housing system which means a good understanding of bird behavior is critical for optimal management of the undesirable behavior.
The rearing period is critical for training pullets to travel upwards. Perches, elevated tiers, or rearing like-to-like will improve pullets' physical and behavioral development.
The pullets need time to learn the laying housing system before they come into lay, transfer before 16 week may aid in this.
The period following transfer is critical for ensuring the birds are learning to move up off the floor and into the system to access the nest box.
Labor investment when the birds first start laying to keep them up off the floor, which prevent floor egg laying from becoming established in the flock.
The nest box needs to be the only comfortable, quiet, enclosed place for the hen to lay her eggs.
Nest box use can be affected by individual hen preferences as well as social factors in the flock.
Placement of feeders and drinkers should not make nest access difficult nor encourage more hens onto the litter, particularly during the morning egg-laying period. Feed times should not coincide directly with oviposition times.
New technologies may assist with detecting and managing egg floor laying, but good labor investment early on by knowledgeable staff can result in negligible floor egg percentages.
Preventing floor egg laying from developing is critical, as once the behavior is established in a flock, there is very little (or nothing) that can be done to change this.

Source: Adapted from: Campbell (2023).

when referring to movement possibilities in cage-free systems. In details, the openings from indoor toward outdoor should be correctly designed; outdoor area should be managed in order to increase its attractiveness compared to indoor and guarantee health, where it could be simply a grass-covered area or a more complex environment including trees and other equipment; stocking density and weather conditions can affect the use and the maintenance of the area; and birds of different genotypes have a different aptitude to use and explore the outdoor area (Gilani et al., 2014; Pettersson et al., 2016; Campbell et al., 2017; EFSA AHAW Panel, 2023; Fiorilla et al., 2024). As for egg quality, all these issues make the control of floor eggs even more challenging compared with indoor cage-free systems, which call for specific farmer knowledge and experience due to the variability or results between and within the same flocks (Campbell, 2023).

Additionally, the use of the outdoor area and the corresponding grass intake, in terms of quantity and quality, have a direct effect on egg quality traits. In fact, while the nutritional composition of eggs, in terms of protein and fat contents, is rather stable (Gautron et al., 2022) because of the physiological role of this product, the fat quality, in terms of fatty acid composition, and the content in other bioactive compounds such as carotenoids, vitamins, and minerals can be improved by grass intake and quality and exposure to sunlight in the outdoor area.

Firstly, pasture intake has been identified as a strategy to increase the value of eggs as a functional food for human consumption where the content of n-3 fatty acids and several other bioactive compounds (vitamins and pigments) can be manipulated (Hammershøj & Johansen, 2016; Cartoni Mancinelli et al., 2022). The direct relationships in poultry, as in other monogastrics, between the dietary fat quality and the fat quality of final products (both meat and eggs) and the associated physiological mechanisms are well-known (Cartoni Mancinelli et al., 2022), where grass composition and season of foraging can affect the contents of the different bioactive compounds (Table 3) (Mugnai et al., 2014; Popova et al., 2020).

Indeed, the increase in egg carotenoids associated with grass intake has also direct consequences on the yolk color. Then, the increased vitamin contents in eggs of laying hens reared outdoor is associated on one side with the grass contents and, as for vitamin D, with animal exposure to sun light (Schutkowski et al., 2013; Kühn et al., 2014) which can have positive effects on the eggshell characteristics, besides bone development. In fact, Sirri et al. (2023) confirmed that the amounts of vitamin D3 and calcidiol (25-OH-D_3) in the egg yolk of laying hens reared outdoor increased by 2–4 times compared to that of hens kept in cages, even if with some differences depended on the genotype, where eggshell breaking strength was also generally higher in eggs produced outdoor.

4 Welfare and egg quality in different housing systems: results from commercial farms

Based on a recent study under the Italian conditions (De Marchi et al., 2024), food choices by consumers in the case of eggs are especially related to sustainability attributes, where rearing system (and thus expected animal welfare on farm) and local production are more valued than biodiversity protection. On the other hand, other authors have highlighted that the average willingness to pay elicited for sustainability attributes is higher compared to the other traits (Tully & Winer, 2014). In other words, animal welfare is an egg quality indicator regardless of the differences in egg quality according to the production systems.

Table 3 Egg quality in laying hens kept in cage systems (control), organic systems, and organic-plus system at different seasons

	Control				Organic				Organic plus			
	W	Sp	Su	A	W	Sp	Su	A	W	Sp	Su	A
Carotenoids (mg/kg yolk)	7.02^{a}	7.61^{a}	6.85^{a}	7.38^{a}	7.85^{a}	8.73^{a}	5.95^{a}	7.81^{a}	$15.10^{a,b}$	$17.88^{b,c}$	$11.07^{a,b}$	19.86^{c}
Flavonoids (g/kg yolk)	0.15^{a}	0.21^{a}	0.09^{a}	0.11^{a}	0.14^{a}	0.20^{a}	0.10^{a}	0.12^{a}	0.52^{b}	0.56b	$0.23^{a,b}$	0.47^{b}
Fatty acids content (g/kg yolk)												
SFA	85.7^{b}	82.1^{b}	85.0^{b}	85.0^{b}	$68.1^{a,b}$	60.1^{a}	63.5^{a}	62.4^{a}	$73.6^{a,b}$	$68.1^{a,b}$	$75.5^{a,b}$	$75.0^{a,b}$
MUFA	106^{a}	112^{a}	108^{a}	108^{a}	112^{b}	115^{b}	115^{b}	115^{b}	118^{b}	125b	1167^{b}	118^{b}
PUFA	38.1b	$34.1^{a,b}$	$36.1^{a,b}$	$35.1^{a,b}$	32.9^{a}	37.5^{b}	$35.0^{a,b}$	$35.5^{a,b}$	28.1^{a}	33.3^{a}	29.9^{a}	30.0^{a}
C18:2n-6	27.5b	26.1^{b}	25.8^{b}	26.9^{b}	24.1^{b}	27.1^{b}	25.6^{b}	27.0^{b}	20.7^{a}	18.8^{a}	19.0^{a}	19.2^{a}
C20:4n-6	4,14b	5.88^{b}	5.14^{b}	5.15^{b}	4.47^{b}	5.04^{b}	4.74^{b}	4.52^{b}	0.78^{a}	1.10^{a}	0.70^{a}	0.88^{a}
C18:3n-3	1.25a	1.30^{a}	$1.84^{a,b}$	1.74^{a}	1.50^{a}	$1.74^{a,b}$	1.20^{a}	1.41a	2.48^{b}	3.34^{c}	$2.92^{b,c}$	$2.89^{b,c}$
C20:5n-3	0.11	0.09	0.14	0.12	0.28	0.29	0.25	0.23	0.25	0.31	0.23	0.28
C22:5n-3	0.08	0.09	0.08	0.07	0.12	0.15	0.13	0.14	0.15	0.15	0.13	0.12
C22:6n-3	0.95^{a}	0.87^{a}	1.05^{a}	0.64^{a}	1.10^{a}	1.11^{a}	1.08^{a}	1.09^{a}	1.66^{a}	5.01^{c}	2.70^{b}	3.45^{b}

Organic: the hens were kept in four (one per season) adjacent covered, straw-bedded houses (6 hens m^{-2}) with access to four separate pens (one per season) with natural pasture; each pens was divided into three sub-pens (4 m^2 per hen). Organic-plus: the hens were kept in four identical (one per season) adjacent covered, straw-bedded houses (6 hens m^{-2}) with access to four separate pens (one per season) with natural pasture; each pen was divided into three sub-pens (10 m^2 per hen; 12 replications).

W: winter; Sp: spring; Su: summer; A: autumn.

SFA, saturated fatty acid; MUFA, monounsaturated fatty acids; PUFA, polyunsaturated fatty acids.

Different superscript letters (a,b,c) indicate significant differences between means ($P < 0.05$).

Source: Modified from: Mugnai et al. (2014).

As just reviewed above, studies conducted under controlled conditions have highlighted that the main changes that we can expect in eggs produced in welfare-friendly cage-free systems compared to caged systems are related to the occurrence of floor eggs (and thus to egg safety and farm efficiency) (in all cage-free systems) and eggs enrichment with bioactive compounds (in systems where outdoor access is previewed). Indeed, Samiullah et al. (2017) have found that when pullets of the same origin were farmed in different commercial housing and production systems (cage, barn, and free range), differences in eggshell and egg internal quality were relatively a few. In fact, under practical and commercial conditions, the quality of the eggs from hens kept in different housing system is the result of several management factors, from genotype (Rizzi, 2020; Castillo et al., 2024) to feeding, for which literature data about what consumers buy can provide different results.

In market eggs produced in Italy from hens housed in different systems (alternative, enriched cage, litter floor, and organic), significant differences were measured for physical traits (i.e. whole egg weight, albumen weight and yolk weigh) as well as fatty acid profile, where the contents in polyunsatured fatty acids and linoleic acid, and n6/n3 ratio were the discriminant traits among systems (Marelli et al., 2021). On the other hand, under the condition of Israeli farms, no difference in internal quality of eggs was measured depending on the housing system (cage vs. cage-free) (Wiseman et al., 2024). As for bioactive compounds, in UK table eggs from commercial systems, vitamin D3, B2, or B9 contents resulted similar regardless from the housing systems, whereas nutritionally desirable FA, carotenoid, and vitamins A and E were increased from eggs produced in caged systems to those coming from free-range, organic and extensive organic systems (Chatzidimitriou et al., 2024).

Then, available data collected on commercial eggs also confirm a higher resistance of egg shells in eggs produced from brown hens kept in cage-free and free-range systems compared to eggs from other systems (barren and enriched cages), where in the case of free-range eggs both egg yolks and shells were darker (Alig et al., 2023). Indeed, eggs from free-range systems have been found to contain a higher quantity of macrominerals (P, Mg, and Na) and a lower quantity of microminerals (Cu, Fe, K, S, and Mn) compared to eggs from cage systems (Dao et al., 2020), where a high individual variability has been reported.

Last, but not least, some concerns have been raised about egg contamination by undesired chemicals that have been released in the natural environment by human activities. As an example, Gazzotti et al. (2021) measured a relatively uniform contamination from per- and poly-fluoroalkyl substances in Italian eggs from backyard chickens and a significantly higher contamination in these eggs compared to those from commercial laying hens, where the consumption of eggs from backyard chickens was found to contribute significantly to Per Fluorinated Alkylated Substances dietary intake. Higher contamination levels of polychlorinated dibenzo-p-dioxins, polychlorinated dibenzofurans, dioxin-like and non-dioxin-like polychlorinated biphenyls were also measured in eggs collected from hens farmed under free-range conditions compared to those from hens kept in indoor systems (Lambiase et al., 2022). Thus, while foraging represents a great improvement for hen welfare and an enrichment for bioactive compounds, it can be challenging when animals are raised in areas characterized by soil contamination.

5 Future research priorities

Based on current knowledge and field experience, hen welfare on farm is the results of the combination of multiple factors which pertains to the animal (e.g. genotype, age at start of laying, and production level), to the housing system (e.g. cage vs. cage free, access to outdoor area, provisions of different equipments, and nests), and to the management (e.g. lighting conditions, feeding, and enrichments) and, last but not least, farmer's knowledge and ability.

The main challenge in the different phases, from pullets to laying hens, is to combine in a proper way the hen abilities and aptitudes (as for behavioral pattern, exploration activity, navigation ability, nesting behavior) with the housing environment to guarantee the best welfare by maximizing the expression of specie-specific behaviors and avoiding negative affective status which could increase animal stress and produce unwanted and dangerous behaviors (aggression, pecking, piling, smothering) besides negative productive results (reduction of oviposition rate, increase of feed intake, floor eggs).

In this context, on one side, research is expected to focus on genetic selection for reducing unwanted behaviors, such as injurious pecking, even if a low heritability and a high environmental role have been identified (Fernyhough et al., 2020). Out of the different behaviors, selection based on nesting behaviors should be encouraged as a genetic influence on traits related to nest preference and laying duration has been identified, which could have effects on piling and pecking behaviors besides floor eggs (Bécot et al., 2023). Lastly, research in the evaluation of different genotypes, as pure breeds and/or crosses as for behavior and egg production and quality, is also expected as the transition toward alternative housing systems which use outdoor areas will likely require more resilient genotypes which could adapt to different conditions, which can be even more challenging under the actual climate change scenario (Sirri et al., 2023). The results of these evaluations could contribute to preserve biodiversity in poultry and egg production (Pulcini et al., 2023; Menchetti et al., 2024).

On the other side, to improve welfare and productive results of laying hens under cage-free systems, research is expected to focus on the pullet phase, looking at relationships between early experience and later space use and navigation ability as well as pullet training. In the phase of laying, all strategies to reduce aggression among animals and undesired and harmful piling and smothering of hens will be the focus. Precision livestock farming could offer a contribution to this regards on farm, if efficient monitoring and predictive systems are developed.

Finally, it is worth mentioning that the approach with respect to animal welfare is shifting: providing animals with opportunities for rewarding experiences and situations in which they feel satisfied is recognized as a key for their welfare, beyond the alleviation of any suffering. Positive feelings could outweigh negative ones in order to achieve good overall welfare, even if how this could be achieved is still up for debate (Rault et al., 2023). Nevertheless, several factors (e.g. genetics, pre-birth and early experiences, etc.) account for individual differences among animals (LIFT, 2024), where a more comprehensive approach to animal welfare will take into account both physical and emotional aspects (Turner, 2019; Paulović et al., 2024). While positive welfare in poultry and, specifically in laying hens (Papageorgiou et al., 2023; Paul et al., 2023), is under study, it will be challenging in the future to establish and evaluate relationships between positive affective states and egg production (both for quantity and quality), where the same aspects have been already investigated as for negative affective status, as discussed in previous chapters.

6 Conclusion

The contribution of research to increased sustainability of egg production should address the tools and terms of a safe transition from caged to cage-free housing systems, where impacts on animals, society, and environment need to be evaluated. Key issues have already been raised as for the welfare consequences that impact animals in all housing systems which have also indirect consequences on egg production and quality and, definitively, farm efficiency. The development of systems based on outdoor access cannot be done without taking into account soil consumption, health issues as for illness such as aviary influence, besides the positive and negative impacts that grass intake can have on the nutritional values of eggs and on their safety for human consumption. In this panorama, the development of precision livestock farming tools could represent an advancement both under indoor and (even if with more difficulties) outdoor conditions for monitoring animal activity, health, and welfare and for adopting any corrective measure in a short time when needed, where exposure to climate condition further challenge farming conditions. Under this scenario, further research in the best animal genotype for the different housing environment and management conditions is encouraged.

7 Where to look for further information

- The European Food Safety Authority (EFSA) has recently and extensively reviewed the welfare of laying hens on farm including difefrent production systems.EFSA AHAW Panel,et al.(2023). Scientific Opinion on the welfare of laying hens on farm. *EFSA Journal*, 21(2), 7789, 188 pp. https://doi.org/10.2903/j.efsa.2023.7789.
- Other valuable scientific information can be retreat from the Proceedings of the main Poultry conferences organized by the World's *Poultry Science* Association (WPSA) (https://www.wpsa.com/) and listed below
- European Symposium on the Quality of Eggs and Egg Products.
- World's Poultry Congress.
- European Symposium on Poultry Welfare.

There are a number of current research projects designed to give insights in relationships between hens' welfare and egg quality under different production systems, including:

- https://ppilow.eu/.
- https://omelette.nweurope.eu/.
- https://www.eupahw.eu/.

Centers of expertize include:

- Dr. Dana L. M. Campbell, Commonwealth Scientific and Industrial Research Organisation (CSIRO) Canberra, Australia.
- Prof. Cesare Castellini, Alessandro Dal Bosco, Department of Agricultural, Food and Environmental Science, University of Perugia, Italy.
- Dr. Michael Toscano, Group Leader of Poultry Housing and Welfare at University of Bern, Switzerland.
- Prof. Federico Sirri, Department of Agricultural and Food Sciences (DISTAL), University of Bologna, Ozzano Emilia, Italy.
- Dr. Inma Estévez, Ikerbasque Research Professor, NEIKER, Department of Animal Production, IKERBASQUE. Basque Foundation for Science, Bilbao, Spain.

8 References

Abín, R., et al. (2018). Environmental assessment of intensive egg production: a Spanish case study. *Journal of Cleaner Production*, 179, 160–168. https://doi.org/10.1016/j.jclepro.2018.01.067

Ali, A.B.A., et al. (2016). Influence of genetic strain and access to litter on spatial distribution of 4 strains of laying hens in an aviary system. *Poultry Science*, 95(11), 2489–2502. https://doi.org/10.3382/ps/pew236

Ali, A.B.A., et al. (2020). A risk assessment of health, production, and resource occupancy for 4 laying hen strains across the lay cycle in a commercial-style aviary system. *Poultry Science*, 99(10), 4672–4684. https://doi.org/10.1016/j.psj.2020.05.057

Alig, B.N., et al. (2023). Evaluation of physical egg quality parameters of commercial brown laying hens housed in five production systems. *Animals*, 13, 716. https://doi.org/10.3390/ani13040716

Appleby, M.C. (1984). Factors affecting floor laying by domestic laying hens: a review. *Worlds Poultry Science Journal*, 40, 241–249. doi.org/10.1079/WPS19840019

Armstrong, E.A., et al. (2020). Keel bone fractures induce a depressive-like state in laying hens. *Scientific Reports*, 10(1), 3007. https://doi.org/10.1038/s41598-020-59940-1

Bécot, L., et al. (2023). Nest preference and laying duration traits to select against floor eggs in laying hens. *Genetics Selection Evolution*, 55, 8. https://doi.org/10.1186/s12711-023-00780-8

Biewener, A.A. & Bertram, J.E. (1994). Structural response of growing bone to exercise and disuse. *Journal of Applied Physiology*, 76(2), 946–955. https://doi.org/10.1152/jappl.1994.76.2.946

Bourque, T. (2017). One welfare. *The Canadian Veterinary Journal*, 58(3), 217–218.

Brannan, K.E. & Anderson, K.E. (2021). Examination of the impact of range, cage-free, modified systems, and conventional cage environments on the labor inputs committed to bird care for three brown egg layer strains. *Journal of Applied Poultry Research*, 30(1), 100118. https://doi.org/10.1016/j.japr.2020.100118

Campbell, D.L.M., et al. (2016). Laying hen movement in a commercial aviary: enclosure to floor and back again. *Poultry Science*, 95(1), 176–187. https://doi.org/10.3382/ps/pev186

Campbell, D.L.M., et al. (2017). Outdoor stocking density in free-range laying hens: effects on behaviour and welfare. *Animal*, 11(6), 1036–1045. https://doi.org/10.1017/S1751731116002342

Campbell, D.L.M., et al. (2019). A review of environmental enrichment for laying hens during rearing in relation to their behavioral and physiological development. *Poultry Science*, 98(1), 9–28. https://doi.org/10.3382/ps/pey319

Campbell, D.L.M., et al. (2020). Free-range egg production: its implications for hen welfare. *Animal Production Science*, 61(10), 848–855. https://doi.org/10.1071/AN19576

Campbell, D.L.M. (2023). Floor egg laying: can management investment prevent it? *Journal of Applied Poultry Research*, 32(4), 100371. https://doi.org/10.1016/j.japr.2023.100371

Candelotto, L., et al. (2017). Susceptibility to keel bone fractures in laying hens and the role of genetic variation. *Poultry Science*, 96(10), 3517–3528. https://doi.org/10.3382/ps/pex146

Cartoni Mancinelli, A., et al. (2022). Poultry meat and eggs as an alternative source of n-3 long-chain polyunsaturated fatty acids for human nutrition. *Nutrients*, 2022(14), 1969. https://doi.org/10.3390/nu14091969

Castillo, A., et al. (2024). Physical and chemical characteristics of eggs from eight Italian chicken breeds. *Italian Journal of Animal Science*, 23(1), 342–347. https://doi.org/10.1080/1828051X.2024.2314149

Chatzidimitriou, E., et al. (2024). Variation in nutritional quality in UK retail eggs. *Food Chemistry*, 454, 139783. https://doi.org/10.1016/j.foodchem.2024.139783

Ciarelli, C., et al. (2023). Space use and navigation ability of hens at housing in the aviary for the laying phase: Effect of enrichment with additional perches and genotype. *Poultry Science*, 102(11), 102962. https://doi.org/10.1016/j.psj.2023.102962

Colson, S., et al. (2008). Influence of rearing conditions of pullets on space use and performance of hens placed in aviaries at the beginning of the laying period. *Applied Animal Behaviour Science*, 111(3), 286–300. https://doi.org/10.1016/j.applanim.2007.06.012

Commission Regulation (EC) No 889/2008 of 5 September 2008 laying down detailed rules for the implementation of Council Regulation (EC) No 834/2007 on organic production and labelling of organic products with regard to organic production, labelling and control. OJ L 250, 18.9.2008, pp. 1–84. Available: https://eur-lex.europa.eu/legal-content/EN/TXT/?uri=celex%3A32008R0889. Accessed: August 28, 2024.

Costantini, M., et al. (2020). Environmental sustainability assessment of poultry productions through life cycle approaches: a critical review. *Trends in Food Science & Technology*, 110, 201–212. https://doi.org/10.1016/j.tifs.2021.01.086

Council Directive 1999/74/EC of 19 July 1999 laying down minimum standards for the protection of laying hens. OJ L 203, 3.8.1999, pp. 53–57. Available: https://eur-lex.europa.eu/legal-content/EN/TXT/?uri=CELEX%3A31999L0074

Dao, H.T., et al. (2020). A comparison of eggshell mineral composition between cage and free-range eggs via inductively coupled plasma-optical emission spectrometry. *Animal Production Science*, 60(17), 2060–2067. https://doi.org/10.1071/AN19705

De Marchi, E., et al. (2024). What is the role of environmental sustainability knowledge in food choices? A case study on egg consumers in Italy. *Journal of Cleaner Production*, 441, 141038. https://doi.org/10.1016/j.jclepro.2024.141038

De Reu, K., et al. (2005). Bacterial eggshell contamination in conventional cages, furnished cages and aviary housing systems for laying hens. *British Poultry Science*, 46(2), 149–155. https://doi.org/10.1080/00071660500065359

Drake, K.A., et al. (2010). Influence of rearing and lay risk factors on propensity for feather damage in laying hens. *British Poultry Science*, 51(6), 725–733. https://doi.org/10.1080/00071668.2010.528751

Du, X., et al. (2022). Effects of key farm management practices on pullets welfare: a review. *Animals,* 12, 729.

Edgar, J.L., et al. (2023). Fear, anxiety, and production in laying hens with healed keel bone fractures. *Poultry Science*, 102(4), 102514. https://doi.org/10.1016/j.psj.2023.102514

EFSA AHAW Panel, et al. (2023). Scientific opinion on the welfare of laying hens on farm. *EFSA Journal*, 21(2), 7789. https://doi.org/10.2903/j.efsa.2023.7789

Estévez-Moreno, L.X., et al. (2025). Do Mexican consumers really care about hen welfare? Understanding their attitudes, constraints and willingness to pay for cage-free eggs.

Food Quality and Preference, 122, 105292. https://doi.org/10.1016/j.foodqual.2024.105292

European Commission. (2024). Egg market situation. Available: https://agriculture.ec.europa.eu/document/download/9bdf9842-1eb6-41a2-8845-49738b812b2b_en?filename=eggs-dashboard_en.pdf Accessed August 28, 2024.

European Parliament. (2021). P9_TA(2021)0295 European Citizens' Initiative "End the cage age" European Parliament resolution of 10 June 2021 on the European Citizens' Initiative 'End the cage age' (2021/2633(RSP)). Available: https://www.europarl.europa.eu/doceo/document/TA-9-2021-0295_EN.pdf

Fernyhough, M., et al. (2020). The ethics of laying hen genetics. *Journal of Agricultural and Environmental Ethics*, 33, 15–36. https://doi.org/10.1007/s10806-019-09810-2

Fiorilla, E., et al. (2024). Effects of housing systems on behaviour and welfare of autochthonous laying hens and a commercial hybrid. *Applied Animal Behaviour Science*, 274, 106247. https://doi.org/10.1016/j.applanim.2024.106247

Fleming, R.H., et al. (2006). Relationships between genetic, environmental and nutritional factors influencing osteoporosis in laying hens. *British Poultry Science*, 47(6), 742–755. https://doi.org/10.1080/00071660601077949

Gautron, J., et al. (2021). Review: what are the challenges facing the table egg industry in the next decades and what can be done to address them? *Animal*, 15(Supplement 1), 100282. https://doi.org/10.1016/j.animal.2021.100282

Gautron, J., et al. (2022). Review: production factors affecting the quality of chicken table eggs and egg products in Europe. *Animal*, 16(Supplement 1), 100425. https://doi.org/10.1016/j.animal.2021.100425

Gazzotti, T., et al. (2021). Perfluoroalkyl contaminants in eggs from backyard chickens reared in Italy. *Food Chemistry*, 362, 130178. https://doi.org/10.1016/j.foodchem.2021.130178

Giersberg, M.F., et al. (2019). Pecking and piling: the behaviour of conventional layer hybrids and dual-purpose hens in the nest. *Applied Animal Behaviour Science*, 214, 50–56. https://doi.org/10.1016/j.applanim.2019.02.016

Gilani, A.M., et al. (2013). The effect of rearing environment on feather pecking in young and adult laying hens. *Applied Animal Behaviour Science*, 148(1), 54–63. https://doi.org/10.1016/j.applanim.2013.07.014

Gilani, A.M., et al. (2014). Factors affecting ranging behaviour in young and adult laying hens. *British Poultry Science*, 55(2), 127–135. https://doi.org/10.1080/00071668.2014.889279

Green, L.E., et al. (2000). Cross-sectional study of the prevalence of feather pecking in laying hens in alternative systems and its associations with management and disease. *Veterinary Record*, 147(9), 233–238. https://doi.org/10.1136/vr.147.9.233

Hammershøj, M. & Johansen, N.F. (2016). Review: the effect of grass and herbs in organic egg production on egg fatty acid composition, egg yolk colour and sensory properties. *Livestock Science*, 194, 37–43. http://dx.doi.org/10.1016/j.livsci.2016.11.001

Hartcher, K.M. & Jones, B. (2017). The welfare of layer hens in cage and cage-free housing systems. *World's Poultry Science Journal*, 73(4), 767–782. https://doi.org/10.1017/S0043933917000812

Hemsworth, P.H. (2021). Cage production and laying hen welfare. *Animal Production Science*, 61, 821–836.

Hester, P.Y., et al. (2013). The effect of perch availability during pullet rearing and egg laying on musculoskeletal health of caged White Leghorn hens. *Poultry Science*, 92(8), 1972–1980. https://doi.org/10.3382/ps.2013-03008

Holt, P.S., et al. (2011). The impact of different housing systems on egg safety and quality. *Poultry Science*, 90(1), 251–262. https://doi.org/10.3382/ps.2010-00794

Isaksson, H., et al. (2009). Physical exercise improves properties of bone and its collagen network in growing and maturing mice. *Calcified Tissue International*, 85(3), 247–256. https://doi.org/10.1007/s00223-009-9273-3

Janczak, A.M. & Riber, A.B. (2015). Review of rearing-related factors affecting the welfare of laying hens. *Poultry Science*, 94(7), 1454–1469. https://doi.org/10.3382/ps/pev123

Jones, D.R., et al. (2015). Microbiological impact of three commercial laying hen housing systems. *Poultry Science*, 94(3), 544–551. https://doi.org/10.3382/ps/peu010

Käppeli, S., et al. (2011). Effects of housing, perches, genetics, and 25-hydroxycholecalciferol on keel bone deformities in laying hens. *Poultry Science*, 90(8), 1637–1644. https://doi.org/10.3382/ps.2011-01379

Kollenda, E., et al. (2020). *Transitioning towards cage-free farming in the EU: assessment of environmental and socio-economic impacts of increased animal welfare standards*. Policy report by the Institute for European Environmental Policy, Brussels & London, pp. 1–65.

Kühn, J., et al. (2014). Free-range farming: a natural alternative to produce vitamin D-enriched eggs. *Nutrition*, 30(4), 481–484. https://doi.org/10.1016/j.nut.2013.10.002

Lambiase, S., et al. (2022). Bioaccumulation of PCDD/Fs and PCBs in free-range hens: congener fingerprints and biotransfer factors. *Chemosphere*, 309, 136602. https://doi.org/10.1016/j.chemosphere.2022.136602

Lambton, S.L., et al. (2010). The risk factors affecting the development of gentle and severe feather pecking in loose housed laying hens. *Applied Animal Behaviour Science*, 123(1), 32–42. https://doi.org/10.1016/j.applanim.2009.12.010

Lambton, S.L., et al. (2015). The risk factors affecting the development of vent pecking and cannibalism in free-range and organic laying hens. *Animal Welfare*, 24(1), 101–111. https://doi.org/10.7120/09627286.24.1.101

Lay, D.C., et al. (2011). Hen welfare in different housing systems. *Poultry Science*, 90(1), 278–294. https://doi.org/10.3382/ps.2010-00962

Leinonen, I., et al. (2012). Predicting the environmental impacts of chicken systems in the United Kingdom through a life cycle assessment: egg production systems. *Poultry Science*, 91(1), 26–40. https://doi.org/10.3382/ps.2011-01635

LIFT. (2024). Positive Animal Welfare - LIFT - COST ACTION CA21124 [WWW Document]. Available: https://liftanimalwelfare.eu/about/positive-animal-welfare/ Accessed August, 2024.

MacLachlan, S.S., et al. (2020). Influence of later exposure to perches and nests on flock level distribution of hens in an aviary system during lay. *Poultry Science*, 99(1), 30–38. https://doi.org/10.3382/ps/pez524

Marelli, S.P., et al. (2021). Egg production systems, open space allowance and their effects on physical parameters and fatty acid profile in commercial eggs. *Animals*, 11(2), 1–10. https://doi.org/10.3390/ani11020265

Matthews, W.A. & Sumner, D.A. (2015). Effects of housing system on the costs of commercial egg production. *Poultry Science*, 94(3), 552–557. https://doi.org/10.3382/ps/peu011

Mench, J.A., et al. (2011). Sustainability of egg production in the United States: the policy and market context. *Poultry Science*, 90(1), 229–240. https://doi.org/10.3382/ps.2010-00844

Menchetti, L., et al. (2024). Effect of genotype and nutritional and environmental challenges on growth curve dynamics of broiler chickens. *Poultry Science*, 103(10), 104095. https://doi.org/10.1016/j.psj.2024.104095

Michel, V., et al. (2022). The relationships between damaging behaviours and health in laying hens. *Animals*, 12, 986. https://doi.org/10.3390/ani12080986

Mugnai, C., et al. (2014). The effects of husbandry system on the grass intake and egg nutritive characteristics of laying hens. *Journal of the Science of Food and Agriculture*, 94(3), 459–467. https://doi.org/10.1002/jsfa.6269

Nasr, M.A.F., et al. (2012). Do hens with keel bone fractures experience pain? *PLoS One*, 7(8), e42420. https://doi.org/10.1371/journal.pone.0042420

Nasr, M.A.F., et al. (2013). The effect of keel fractures on egg production, feed and water consumption in individual laying hens. *British Poultry Science*, 54(2), 165–170. https://doi.org/10.1080/00071668.2013.767437

Newberry, R.C., et al. (2007). Behaviour when young as a predictor of severe feather pecking in adult laying hens: the redirected foraging hypothesis revisited. *Applied Animal Behaviour Science*, 107(3–4), 102801. https://doi.org/10.1016/j.applanim.2006.10.010

Nicol, C.J., et al. (2013). The prevention and control of feather pecking: application to commercial systems. *World's Poultry Science Journal*, 69(4), 775–788. https://doi.org/10.1017/S0043933913000809

Oliveira, J., et al. (2016). *Nesting behavior and egg production pattern of laying hens in enriched colony housing*. ASABE Annual International Meeting Orlando, FL, USA.

Papageorgiou, M., et al. (2023). Positive welfare indicators and their association with sustainable management systems in poultry. *Sustainability*, 15, 10890. https://doi.org/10.3390/su151410890

Paul, E.S., et al. (2023). Affective trajectories: are hens influenced by positive and negative changes in their living conditions? *Applied Animal Behaviour Science*, 261, 105883. https://doi.org/10.1016/j.applanim.2023.105883

Paulović, T., et al. (2024). Development of a roadmap for action for the project More Welfare: towards new risk assessment methodologies and harmonised animal welfare data in the EU. *EFSA Supporting Publications*, 21(1), 8566E. https://doi.org/10.2903/sp.efsa.2024.EN-8566

Pelletier, N. (2017). Life cycle assessment of Canadian egg products, with differentiation by hen housing system type. *Journal of Cleaner Production*, 152, 167–180. https://doi.org/10.1016/j.jclepro.2017.03.050

Pettersson, I.C., et al. (2016). Factors affecting ranging behaviour in commercial free-range hens. *World's Poultry Science Journal*, 72(1), 137–150. https://doi.org/10.1017/S0043933915002664

Pinillos, R.G., et al. (2016). One welfare – a platform for improving human and animal welfare. *Veterinary Record*, 179(16), 412–413. https://doi.org/10.1136/vr.i5470

Popova, T., et al. (2020). Quality of eggs from layers reared under alternative and conventional system. *Brazilian Journal of Poultry Science*, 22(1), eRBCA-2019-1172. https://doi.org/10.1590/1806-9061-2019-1172

Pötzsch, C.J., et al. (2001). A cross-sectional study of the prevalence of vent pecking in laying hens in alternative systems and its associations with feather pecking, management and disease. *Applied Animal Behaviour Science*, 74(4), 259–272. https://doi.org/10.1016/S0168-1591(01)00167-8

Pufall, A., et al. (2021). Effects of rearing aviary style and genetic strain on the locomotion and musculoskeletal characteristics of layer pullets. *Animals*, 11, 3. https://doi.org/10.3390/ani11030634

Pulcini, D., et al. (2023). Shape and fractures of carina sterni in chicken genotypes with different egg deposition rates reared indoor or free-range. *Scientific Reports*, 13(1), 22495. https://doi.org/10.1038/s41598-023-49909-1

Purdum, S., et al. (2020). The effects of 2 genetic lines on spatial distribution and use and preference of perch and nest area in an aviary system. *Poultry Science*, 99(7), 3328–3333. https://doi.org/10.1016/j.psj.2020.03.056

Rault, J.L., et al. (2023). Editorial: positive welfare: from concept to implementation. *Frontiers Animal Science*, 4, 1289659. https://doi.org/10.3389/fanim.2023.1289659

Regmi, P., et al. (2015). Effect of rearing environment on bone growth of pullets. *Poultry Science*, 94(3), 502–511. https://doi.org/10.3382/ps/peu041

Riedstra, B. & Groothuis, T.G.G. (2002). Early feather pecking as a form of social exploration: the effect of group stability on feather pecking and tonic immobility in domestic chicks. *Applied Animal Behaviour Science*, 77(2), 127–138. https://doi.org/10.1016/S0168-1591(02)00031-X

Rizzi, C. (2020). Yield performance, laying behaviour traits and egg quality of purebred and hybrid hens reared under outdoor conditions. *Animals*, 10, 584. https://doi.org/10.3390/ani10040584

Rodenburg, T.B., et al. (2013). The prevention and control of feather pecking in laying hens: identifying the underlying principles. *World's Poultry Science Journal*, 69(2), 361–374. https://doi.org/10.1017/S0043933913000354

Rokavec, N. & Zupan Šemrov, M. (2020). Psychological and physiological stress in hens with bone damage. *Frontiers in Veterinary Science*, 7, 589274. https://doi.org/10.3389/fvets.2020.589274

Rondoni, A., et al. (2020). Consumer behaviour, perceptions, and preferences towards eggs: a review of the literature and discussion of industry implications. *Trends in Food Science & Technology*, 106, 391–401. https://doi.org/10.1016/j.tifs.2020.10.038

Rufener, C., et al. (2019). Keel bone fractures affect egg laying performance but not egg quality in laying hens housed in a commercial aviary system. *Poultry Science*, 98(4), 1589–1600. https://doi.org/10.3382/ps/pey544

Rufener, C. & Makagon, M.M. (2020). Keel bone fractures in laying hens: a systematic review of prevalence across age, housing systems, and strains. *Journal of Animal Science*, 98(Supplement_1), S36–S51. https://doi.org/10.1093/jas/skaa145

Samiullah, S., et al. (2017). Effect of production system and flock age on eggshell and egg internal quality measurements. *Poultry Science*, 96, 246–258. http://dx.doi.org/10.3382/ps/pew289

Schutkowski, A., et al. (2013). UVB Exposure of farm animals: study on a food-based strategy to bridge the gap between current vitamin D intakes and dietary targets. *PLoS One*, 8(7), e69418. https://doi.org/10.1371/journal.pone.0069418

Sirri, F., et al. (2023). *Yolk vitamin D and eggshell quality traits as affected by hen genotype and housing system*. XIX European Symposium on the Quality of Eggs and Egg Products - XXV European Symposium on the Quality of Poultry Meat, Krakóv, Polland.

Stratmann, A., et al. (2015). Modification of aviary design reduces incidence of falls, collisions and keel bone damage in laying hens. *Applied Animal Behaviour Science*, 165, 112–123. https://doi.org/10.1016/j.applanim.2015.01.012

Toscano, M.J., et al. (2020). Explanations for keel bone fractures in laying hens: are there explanations in addition to elevated egg production? *Poultry Science*, 99(9), 4183–4194. https://doi.org/10.1016/j.psj.2020.05.035

Tully, S.M. & Winer, R.S. (2014). The role of the beneficiary in willingness to pay for socially responsible products: a meta-analysis. *Journal of Retail*, 90(2), 255–274. https://doi.org/10.1016/j.jretai.2014.03.004

Turner, P.V. (2019). Moving beyond the absence of pain and distress: focusing on positive animal welfare. *ILAR Journal*, 60, 366–372. https://doi.org/10.1093/ilar/ilaa017

Villanueva, S., et al. (2017). Nest use and patterns of egg laying and damage by 4 strains of laying hens in an aviary system1. *Poultry Science*, 96(9), 3011–3020. https://doi.org/10.3382/ps/pex104

Wei, H., et al. (2020). Keel fracture changed the behavior and reduced the welfare, production performance, and egg quality in laying hens housed individually in furnished cages. *Poultry Science*, 99(7), 3334–3342. https://doi.org/10.1016/j.psj.2020.04.001

Whitehead, C.C. & Fleming, R.H. (2000). Osteoporosis in cage layers. *Poultry Science*, 79(7), 1033–1041. https://doi.org/10.1093/ps/79.7.1033

Wilkins, L.J., et al. (2011). Influence of housing system and design on bone strength and keel bone fractures in laying hens. *Veterinary Record*, 169(16), 414–414. https://doi.org/10.1136/vr.d4831

Wiseman, A., et al. (2024). Effects of housing systems, flock variables, and storage conditions on internal egg quality in Israel. *The Journal of Applied Poultry Research*, 33, 100456. https://doi.org/10.1016/j.japr.2024.100456

Xin, H., et al. (2011). Environmental impacts and sustainability of egg production systems. *Poultry Science*, 90(1), 263–277. https://doi.org/10.3382/ps.2010-00877

Zheng, H., et al. (2019). Modification of perchery system: preference for ramps rather than ladders during early adaptation period for cage-reared pullets. *International Journal of Agriculture and Biology*, 12, 34–42. https://doi.org/10.25165/j.ijabe.20191204.4030

Part 4

Improving the sustainability of egg production

Chapter 15

Assessing and reducing the environmental impact of egg production systems

Vincent Guyonnet and Audrey E.M. Guyonnet, FFI Consulting, Canada

1 Introduction

The interest of consumers in sustainably sourced products is rapidly gaining momentum within food supply chains. While sustainability encompasses three broad areas (environmental, economic, and social), limited understanding of these pillars usually restricts a consumer focus to the environmental impact of products. Additionally, an emphasis on topical issues such as greenhouse gas (GHG) emissions, or carbon footprint, often results in the underrepresentation of other major factors, both positive and negative, in sustainability reporting.

http://dx.doi.org/10.19103/AS.2024.0150.23

Environmental impact assessment is particularly complex in larger systems such as animal-based food production, and the role of animal-sourced proteins in healthy diets is debated by governments and civil society organizations due to questions of environmental sustainability (Beal et al., 2023). A broader understanding of environmental impact factors may lead the general public to reconsider their perception of sustainability achievements within animal-based food production systems such as the egg sector.

After introducing life cycle assessment (LCA) methodology used to evaluate the impact of food production on the environment, this chapter will address the effects of egg production on key impact categories: climate change, land use change (LUC), acidifying and eutrophying emissions and water footprint (WF). A series of mitigation strategies will also be presented, including some of the trade-offs required to address all three sustainability pillars. Finally, the growing interest of both consumers and financial markets has increased the need for egg producers to make commitments to reducing their impact on the environment and this trend will stimulate further research and innovation.

2 Life cycle assessment: methodology and impact categories

With the challenge of feeding our growing global population, requiring more land, crops, and improved efficiencies in animal production, it is evident that robust methodologies are needed to evaluate the environmental impact of food production and the environmental consequences of consumer food choices.

2.1 Life cycle assessment methodology

LCA methodology has been adopted to define the environmental footprint (EF) of any product or service. Also known as the 'From cradle to grave' analysis, LCA evaluates and assesses the EF of a product or service during all stages of its existence. The potential impact on the environment is evaluated by quantifying the use of resources or 'inputs' (e.g. energy, raw materials, water) and the environmental emissions or 'outputs' to air, water, and soil, at each stage during the life of a product. The International Organization for Standardization (ISO) has issued a procedure for implementing LCA, outlining four key phases (ISO, 2006):

1 definition of the goal and scope;
2 inventory analysis;
3 impact assessment; and
4 interpretation.

However, definitions of scope, including system boundaries, and the depth and the breadth of LCA can differ considerably between studies. This has resulted in debate within the poultry and livestock sectors on the methodology used as well as on possible mitigation strategies.

The Livestock Environmental Assessment and Performance Partnership (LEAP, 2024), a multi-stakeholder initiative dedicated to improving the economic, social and environmental sustainability of livestock supply chains, has been established under the leadership of the Food and Agriculture Organization of the United Nations (FAO). A series of guidelines has brought some standardization in evaluating the EF of egg production, covering areas such as production (FAO, 2016a), feeds (FAO, 2016b) and feed additives (FAO, 2020), nutrient flow (FAO, 2018), and water use (FAO, 2019). As egg producers, consumers and various other stakeholders are increasingly in need of information on the ecological performance and sustainability of the egg supply chain, these standardized methodologies are essential to identifying the most promising mitigation strategies to reduce the EF of egg production.

2.2 Impact categories and scope of emissions

The EF of a product is assessed through different impact categories, usually related to the use of a resource (e.g. water, land) or to the emission of substances with damaging effects on water or soil quality or human health (e.g. greenhouse gases, harmful chemicals). For each impact category, an indicator must be selected along with a method or model used for quantification (Table 1). The results for each indicator provide information on the environmental issues associated with inputs and outputs of the product system (ISO, 2006; European Union, 2013).

More specifically for GHG emissions, the Greenhouse Gas Protocol Initiative has introduced the concept of different 'scopes' to delineate indirect and direct emissions and improve transparency in reporting (GHG Protocol, 2004):

- Scope 1 includes direct GHG emissions, meaning occurring from sources under the ownership or control of a company (e.g. vehicles, heating and ventilation system, manure);
- Scope 2 includes indirect GHG emissions generated by the purchased energy used by a company; and
- Scope 3 includes all other indirect GHG emissions within the value chain (i.e. upstream or downstream activities) not owned or controlled by a company.

Table 1 Impact categories and relevant impact category indicators used for environmental footprint evaluation

Impact category	Definition	Impact category indicator
Climate change	Relative effects of greenhouse gas emissions in terms of climate change over fixed period compared to same mass of carbon dioxide	kg CO_2 equivalent
Ozone depletion	Emissions causing the destruction of the stratospheric ozone layer	kg CFC-11 equivalent
Ecotoxicity for aquatic and fresh water	Variety of toxic mechanisms caused by the release of substances with a direct effect on the health of the ecosystem	CTUe (comparative toxic unit for ecosystems)
Human toxicity – cancer effects	Adverse health effects caused by the intake of toxic substances insofar as they are related to cancer	CTUh (comparative toxic unit for humans)
Human toxicity – non-cancer effects	Adverse health effects caused by the intake of toxic substances insofar as they are related to non-cancer effects	CTUh (comparative toxic unit for humans)
Particulate matter/ respiratory inorganics	Adverse health effects on human health caused by emissions of particulate matter (PM) and its precursors (NO_x, SO_x, NH_3)	kg $PM_{2.5}$ equivalent
Ionizing radiation	Adverse effect on human health	kg U^{235} equivalent
Photochemical ozone formation	Formation of ozone at the ground level damaging vegetation, human respiratory tract and man-made materials	kg non-methane volatile organic compounds
Acidifcation	Emissions of NO_x, NH_3 and SO_x lead to the release of H+ contributing to the acidification of soils and water	kg mol H+
Eutrophication – terrestrial	Enrichment of the terrestrial ecosystem with nutritional elements, due to the emission of nitrogen-containing compounds	mol N-equivalent
Eutrophication – aquatic	Enrichment of aquatic ecosystems with nutritional elements, due to the emission of nitrogen or phosphorous-containing compounds	kg PO_4-eq (fresh water)kg N-eq (marine water)
Resource depletion – water	Relative use of water based on regionalized water scarcity factors	m^3 water use
Resource depletion – mineral, fossil	Use of natural resources, either renewable or non-renewable, biotic or abiotic	kg Sb (antimony) equivalent
Land use – transformation	Use (occupation) and conversion (transformation) of land area by activities such as agriculture, roads, housing, mining etc.	kg C (deficit) orkg CO_2 equivalent

Source: Adapted from: European Union (2013).

3 Environmental footprint of shell eggs: climate change

While the effects of human actions on the planet may be monitored through 14 impact categories (sometimes up to 19 by subdividing acidification and eutrophication into separate categories for soil, fresh water and marine water), most of the published LCA studies of egg production focused only on 3–5 of these impact categories (McClelland et al., 2018; Costantini et al., 2021):

- climate change;
- land use change; and
- emissions causing acidification and eutrophication.

This section focuses on climate change impacts.

Climate change is probably the best recognized impact category by consumers, mainly given its relation to emissions of GHGs. GHGs include naturally occurring atmospheric gases such as carbon dioxide (CO_2), methane (CH_4), nitrous oxide (N_2O) and ozone (O_3) that are necessary to maintain a habitable atmosphere on Earth. The term 'greenhouse' implies these substances trap heat in the Earth's atmosphere, similar in function to the glass walls of a greenhouse. However, anthropogenic activities have caused an upregulation in GHG emissions, thus resulting in substantial differences inthe composition of Earth's atmosphere and its ability to trap heat.

As a by-product of fossil fuel production, CO_2 is the most ubiquitously produced GHG, but GHGs do not have a similar capacity for influencing climate: for instance, methane and nitrous oxide have respectively a global warming potential 21 and 300 times that of carbon dioxide. The Intergovernmental Panel on Climate Change (or IPCC) defined the Global Warming Potential (GWP) as an indicator reflecting the relative effect of these various GHG in terms of climate change, considered over a fixed time period (usually 100 years), when compared to the effects of the same mass of carbon dioxide (IPCC, 2006). For this reason, CO_2 eq. is used as the standard unit to compare the effects of emissions of different GHGs. The total amount of GHG emissions associated with a product throughout its life cycle is generally expressed in kilograms or tonnes of CO_2 equivalent (eq.) per unit of product, thus resulting in the term 'carbon footprint'.

In 2011, the FAO estimated that the global livestock sector contributed about 14.5% of total anthropogenic GHG emissions, at par with the emissions of the transportation sector (14%), but less than electricity and heat production (25%) and the global industrial sector (21%). Total GHG emissions from the livestock sector were estimated at 7.1 Gigatons of CO_2 eq. per year, with the production of eggs accounting for only 2.8% of these emissions or 200 million

CO_2 eq. per year (Gerber et al., 2013). Within the egg value chain (Gerber et al., 2013):

- 69.4% of these emissions were related to the production and processing of feeds (oxidation of carbon in crops and soil, fossil fuels used to manufacture fertilizers and to produce and transport poultry feeds);
- 20% were related to manure management; and
- 10.6% encompassed all post-farm activities (packing of shell eggs, processing of egg products and distribution).

An LCA study of egg production conducted in the UK showed 80.8% of GHG emissions were related to the egg production phase, while the pullet stage and the maintenance of breeding stocks accounted for 17.5% and 1.7%, respectively (Leinonen et al., 2012). Therefore, mitigation strategies must clearly focus on reducing GHG emissions during the egg production stage.

Numerous LCA studies have been conducted in different countries using national production statistics (Dekker et al., 2011; Wiedemann and McGahan, 2011; Leinonen et al., 2012; Leinonen et al., 2014; Pelletier et al., 2014; Pelletier, 2017; Turner et al. 2022a) or data from individual farms (Mollenhorst et al., 2006; Ghasempour and Ahmadi, 2016; Mainali et al., 2017; Abín et al., 2018; Costantini et al., 2020; Estrada-González et al., 2020; González-Velandia et al., 2020; Guillaume et al., 2022) to calculate GHG emissions associated with egg production (Table 2).

When laying hens were housed in conventional cage systems (11 studies), the production of 1 kg of eggs accounted for GHG emissions ranging from 1.30 kg CO_2 eq. to 5.58 kg CO_2 eq. In enriched cage systems (5 studies), GHG emissions ranged between 2.31 kg CO_2 eq. and 3.40 kg CO_2 eq. per kg of eggs while, in cage-free barn systems, single-tier or multi-tier (6 studies), the GHG emissions varied from 2.40 kg CO_2 eq. to 4.30 kg CO_2 eq. per kg of eggs. When laying hens were given access to the outdoors, GHG emissions varied from 1.60 kg CO_2 eq. to 4.40 kg CO_2 eq. per kg of eggs in free-range systems (7 studies) and from 1.30 kg CO_2 eq. to 3.46 kg CO_2 eq. per kg of eggs in organic production systems (seven studies).

Since the conventional cage system is still dominant at the global level (Guyonnet, 2024), the impact of switching to other housing systems on GHG emissions is presented in Table 3. While a transition from conventional to enriched cages seemed to have a minimal reduction effect on GHG emissions (three studies out of three), these emissions were increased when compared to a barn system (four studies out of five) or free-range system (four studies out of six). When comparing conventional cages to organic systems, two studies from Canada reported lower GHG emissions while two studies in the UK and the Netherlands reported the opposite. More recent studies comparing enriched

Table 2 Greenhouse gas emissions (expressed in kg CO_2 equivalent/kg of eggs) depending on the housing system used for the production of eggs

Country	Conventional cages	Enriched cages	Barn/aviary	Free-range	Organic	Reference
Australia	1.3	–	–	1.6	–	Wiedemann and McGahan, 2011
Bangladesh	2.89	–	–	–	–	Mainali et al., 2017
Canada	2.44	2.31	2.40	2.40	1.37	Pelletier, 2017
Canada	2.36	2.33	2.38	2.91	1.30	Turner et al., 2022a
Colombia	–	–	–	–	2.77	González-Velandia et al., 2020
Czech Rep.	–	2.46	3.45	3.41	3.46	Guillaume et al., 2022
Iran	4.07	–	–	–	–	Ghasempour and Ahmadi, 2016
Italy	–	–	–	–	1.56	Costantini et al., 2020
Mexico	5.58	–	–	–	–	Estrada-González et al., 2020
The Netherlands	3.9	–	4.3	4.4	–	Mollenhorst et al., 2006
The Netherlands	2.23	–	2.67	2.75	2.54	Dekker et al., 2011
Spain	–	3.40	–	–	–	Abín et al., 2018
UK	2.92	–	3.45	3.38	3.42	Leinonen et al., 2012
UK	2.92	2.83	–	–	–	Leinonen et al., 2014
USA	2.08	–	–	–	–	Pelletier et al., 2014

Table 3 Percent variation in greenhouse gas emissions (expressed in kg CO_2 equivalent/kg of eggs) between cage systems (conventional or enriched) and other systems for the production of eggs

Housing system	Enriched cages	Barn	Free range	Organic
Conventional cages	−5.3% to −1.3%	−1.6% to 19.7%	−1.6% to 23.3%	−44.9% to 17.1%
Enriched cages	–	2.1% to 40.2%	3.9% to 38.6%	−44.2% to 40.7%

cages to cage-free systems showed increased GHG emissions with barn or free-range systems (three studies out of three), while again the comparison with organic systems showed conflicting results: two studies in Canada reported significantly lower GHG emissions, while a study in the Czech Republic identified the opposite. It should be noted that birds raised under organic systems in Canada will typically stay indoor for 4–5 months due to inclement winter weather, which means they are raised as barn/aviary hens for a third of the year.

The differences in results between studies are mainly due to variations in system boundaries used for the LCA studies, the origin of feed ingredients (potentially contributing to a greater effect of LUC in the calculations) and the breadth of each dataset.

In most studies, the breakdown of GHG emissions by scope is rarely directly reported. However, the Spanish LCA study evaluating the GHG emissions in the enriched cage system showed that 88% of the GHG emissions were Scope 3 (2.99 kg CO_2 eq. per kg of eggs) while Scope 1 and Scope 2 combined accounted for 12% of the total (0.41 kg CO_2 eq. per kg of eggs) (Abín et al., 2018). At a national level, combining data from all production systems in Canada, it was estimated that 75.2% of the GHG emissions were Scope 3 (1.764 kg CO_2 eq. per kg of eggs) while Scope 1 and Scope 2 accounted for 22.4% (0.527 kg CO_2 eq./kg of eggs) and 2.4% (0.056 kg CO_2 eq. per kg of eggs), respectively (Grassauer et al., 2023a). Given that less than 25% of GHG emissions are directly under the control of producers, the commitments to zero emissions by egg producers or national producer associations will require them to consider mitigation strategies in areas such as cropping, choice of feed ingredients, feed production and energy generation.

Although GHG emissions from egg production are lower than those of most other animal-sourced foods, they are typically higher than those of plant-based foods, based on weight (Gerber et al., 2013). However, it is important to relate the level of GHG emissions of foods to the nutritional value delivered by their consumption, namely in terms of protein and calorie intake. For instance, the GHG emissions of 1 kg of eggs is much higher than those for the same amount of rice or fresh produce, but when expressed on the basis of protein content, the emissions for eggs (24.4 kg of CO_2 eq. per kg of proteins) are comparable to those for rice (21.2 kg CO_2 eq. per kg protein) and much lower than those for fresh produce (37.2 kg CO_2 eq. per kg protein) (Clark and Tilman, 2017).

4 Environmental footprint of shell eggs: land use change

LUC is a complex issue with numerous factors, including the conversion of forest areas for the cultivation of crops or livestock production, logging activities or urban encroachment on rural land. These combined activities affect the ability of soils and vegetation to store carbon. Overall, this impact category relates to the way land areas are used (occupation) and potentially converted (transformation) for these different activities:

- land occupation takes into consideration the area involved and the duration of its occupation (changes in quality multiplied by area and duration); and
- land transformation evaluates the extent of changes (changes in quality multiplied by area).

By definition, GHG emissions linked to LUC are allocated to the crops grown on that land for the next 20 years following the change (European Union, 2013).

In 2013, the FAO estimated that livestock feed contributed to ~70% of GHG emissions from the global egg sector. Direct emissions from feeds, accounting for about 57%, occurred mainly during the production of crops, via the application of fertilizers, the use of fuel for farm equipment and for the transportation of raw ingredients and finished feeds. The remaining 13% were related to LUC (Gerber et al., 2013).

Regions currently most affected in terms of crop expansion are Latin America (soybean), Asia and Africa (maize/corn). Many feed crops originating from these regions have higher LUC emissions than those originating from Europe or North America, where deforestation occurred more than 100 years ago (Guyonnet and Guyonnet, 2024). While the expansion of soybean production in Brazil through deforestation is likely one of the most visible examples of LUC, accounting for it in layer feeds is not an easy task. A recent study in Brazil showed that the emission intensity of soybeans varies by more than 200-fold depending on their geographic origin within the country (Escobar et al., 2020). LUC therefore represents a complex impact category that can be both difficult to measure and to compare based on geographic and socioeconomic factors.

LCA studies will typically report Land Use (Land Occupation) data or the amount of land occupied for the production of eggs. The data, combining pre-farm activities such as the production of crops and the area required to raise

Table 4 Land use (occupation) for the production of eggs under different systems (expressed in m^2 per kg of eggs)

Country	Conventional cages	Barn/ aviary	Free range	Organic	Reference
Canada	7.54	7.89	7.97	4.57	Pelletier, 2017
Mexico	5.2	–	–	–	Estrada-González et al., 2020
The Netherlands	4.5	4.8	5.1–5.7	–	Mollenhorst et al., 2006
The Netherlands	3.26	3.75	4.08	6.75	Dekker et al., 2011
UK	4.0	4.2	5.1	16.9	Leinonen et al., 2012

and keep birds, ranged from 3.26 m^2 to 7.54 m^2 for 1 kg of eggs produced in a conventional cage system and as high as 16.9 m^2 to produce organic eggs (Table 4). Land use was consistently higher in barn (+3.4% to +32.2%) and free-range systems (+5.7% to +27.5%) than in conventional cage systems. In one study using a different reporting unit (and therefore not included in Table 4), land use was 32.2% and 21.2% higher in barn and free-range systems, respectively, than in the enriched cage system (Guillaume et al., 2022). Comparisons between cage systems and organic production yield conflicting results with the study in Canada showing a 39.4% reduction in land use compared to conventional cages while studies in the Netherlands and the UK reported an increase of 107.1% and 322.5%, respectively. When compared to an enriched cage system, land use was increased by 39.4% for the production of organic eggs (Guillaume et al., 2022).

5 Environmental footprint of shell eggs: emissions causing acidification and eutrophication

Similar to GHG emissions, acidifying emissions are naturally occurring but have been recently accelerated by human activities. Various gases including sulfur oxides (SO x), nitrogen oxides (NO x) and ammonia have the ability to donate hydrogen ions (H+ ions), thus their acidifying potential. The release of H+ ions contributes to the acidification of soils and water, especially when their buffer capacity is low. Acidification also depends on the processes involved between emission and deposit and on the various processes in the soil and water.

Although their impact indicator is measured as kg mol H^+, acidifying emissions are combined and reported as acidification potential, expressed in g SO_2 eq. per kg eggs produced. In conventional cage systems, LCA studies show that acidifying emissions related to the production of eggs range from 23.0 g SO_2 eq. to 78.4 g SO_2 eq. per kg of eggs (Table 5). The LCA study from the UK identified that most acidifying emissions were related to the egg production

Table 5 Acidifying and eutrophying emissions from egg production in a conventional cage system

Country	Acidifying emissions (g SO_2^- eq./kg eggs)	Eutrophying emissions (g PO_4 eq./kg eggs)	Reference
Canada	78.4	24.4	Pelletier, 2017
Canada	76.7	26.0	Turner et al., 2022a
Iran	43.9	5.4	Ghasempour and Ahmadi, 2016
The Netherlands	32	25	Mollenhorst et al., 2006
The Netherlands	23.0	–	Dekker et al., 2011
UK	53.1	18.5	Leinonen et al., 2012
UK	55.5	19.02	Leinonen et al., 2014
USA	70	20	Pelletier et al., 2014

stage (over 80%) and broken down into three main sources: manure (47%), housing and land (35%) and feeds (15%) (Leinonen et al., 2012).

Acididying emissions are typically comparable between conventional and enriched cage systems, and slightly lower than emissions from barn/aviary and free-range systems (Table 6). Two Canadian studies demonstrated that the organic production of eggs had the lowest level of acidifying emissions of all production systems while the LCA from the UK, with manure contributing 59% of the total of acidifying emissions, outlined the opposite (Leinonen et al., 2012; Pelletier, 2017; Turner et al., 2022a).

Table 6 Comparison of acidifying and eutrophying emissions from different egg production systems

Reference	Conventional cages	Enriched cages	Barn/aviary	Free-range	Organic
Acidifying emissions (g SO_2^- eq./kg eggs)					
Pelletier, 2017	78.44	82.50	80.63	69.78	46.60
Turner et al., 2022a	76.74	78.56	80.20 - 81.45	84.66	45.73
Mollenhorst et al., 2006	32.0	–	57.0	42.0–65.0	–
Leinonen et al., 2012	53.14	–	59.43	64.13	91.63
Leinonen et al., 2014	55.50	55.02	–	–	–
Dekker et al., 2011	23.0	–	40.0–63.5	41.8–65.0	–
Eutrophying emissions (g PO_4 eq./kg eggs)					
Pelletier, 2017	24.43	26.82	25.56	23.75	14.75
Turner et al., 2022a	26.02	27.71	27.46 – 27.89	31.34	14.73
Leinonen et al., 2012	18.47	–	20.32	22.03	37.61
Leinonen et al., 2014	19.02	18.96	–	–	–

In addition to their role as potentially acidifying substances, ammonia, oxides of nitrogen and sulfur oxide compounds also play a role in the formation of secondary fine particles that influence air quality (see section 'Environmental footprint of shell eggs: other impact categories').

Eutrophying emissions, evaluated for both fresh and marine water ecosystems and the terrestrial ecosystem, are linked to excessive amounts of mineral and nutrient (e.g. nitrate and phosphate) in waterways and the environment. The deposit of minerals and nutrients is often caused by agricultural fertilizer (natural or synthetic) runoff leaching into the natural ecosystems. When these nutrients build up in water systems, they facilitate the growth of algae which in turn depletes the water of oxygen. With less available oxygen, water ecosystems can be altered to the point of becoming uninhabitable to plant and fish species.

Although the impact category indicators for eutrophying emissions are reported in mol N-equivalent (terrestrial), kg N eq. (marine water) and kg PO_4 eq. (fresh water), most LCA studies of egg production expressed these emissions in g PO_4 eq. per kg of eggs produced. In conventional cage systems, these values ranged from 5.4 g PO_4 eq. to 26.0 g PO_4 eq. (Table 5). Most of the eutrophying emissions are linked to the egg production stage (> 80%), of which feed and manure make up 42.2% and 39.1%, respectively (Leinonen et al., 2012). There are few differences in eutrophying emissions among the conventional and enriched cages, barn and free-range production systems, while conflicting results between studies in Canada and the UK were reported for organic production (Table 6).

6 Environmental footprint of shell eggs: water footprint

Animal production has a significant impact on the freshwater resources of the planet and accounts for nearly a third of all water used by the agricultural sector. The water footprint (WF) is estimated by considering the use of three water components:

- Blue water (surface and ground water);
- Green water (water from rainfall); and
- Gray water (amount of fresh water required to dilute pollutants below an acceptable level).

The WF of egg production is calculated by adding water used in feed crop production and other miscellaneous inputs (indirect needs) to the water used during production, such as drinking water for the birds and water required to clean the houses and equipment as well as water used for the washing of eggs where allowed (direct needs). The WF of egg production was broken down into

79.4% Green water, 7.5% Blue water and 13.1% Gray water (Mekonnen and Hoekstra, 2012).

Based on reviewing data from China, India, the Netherlands and the USA, the WF of egg production in cage systems was estimated at 2.872 liters of water per kg of eggs, similar to chicken meat production (2.873 liters of water per kg meat), higher than that of milk production (1.207 liters of water per liter of milk) and much lower than that of intensive beef (10.244 liters per kg meat) and pork (5.225 liters per kg) productions. The WF of egg production compares favorably to some plant-based foods, with egg production requiring about 30% less water than pulse production.

Due to the complexity of WF calculations, only a few LCA studies have reported the WF of egg production as an impact category (Table 7). For conventional cage systems, the WF ranged from 2.19 liters (Australia) to 4.26 liters (USA) per kg of eggs. Two Canadian studies showed no significance difference in WF between conventional and enriched cage systems. Cage-free production systems (barn, aviary, free-range, and organic) typically have higher WF values, with the highest reported in organic egg production. While the higher WF in cage-free systems is mainly related to higher feed consumption during production, variations between countries are usually attributable to differences in management and feed ingredients used.

7 Environmental footprint of shell eggs: other impact categories

Of all the other impact categories, particulate matter/respiratory inorganics ($PM_{2.5}$) have been the most studied, although not in LCA studies. The evaluation of ambient respirable dust in commercial layer houses in France showed that, while dust was comparable between conventional cages (0.11 mg/m^3) and enriched cages (0.15 mg/m^3), these levels were multiplied by 2.8 times and 9.0 times in single-tier and multi-tier barn systems, respectively (Le Bouquin et al., 2013).

Table 7 Water footprint from different egg production systems (liter/kg eggs)

Reference	Conventional cages	Enriched cages	Barn/ aviary	Free-range	Organic
Pelletier, 2017	3.12	2.63	4.05	2.82	3.35
Turner et al., 2022a	3.41	3.45	3.37 – 3.79	3.53	3.97
Guillaume et al., 2022	–	5.04	5.54	4.90	8.09
Pelletier et al., 2014	4.26	–	–	–	–
Wiedemann and McGahan, 2011	2.19	–	–	3.17	–
Mekonnen and Hoestra, 2012	2.8	–	–	3.86	–

Concerned by the consequences of a switch from cages to cage-free systems, the Coalition for Sustainable Egg Supply in the US sponsored some studies to evaluate the $PM_{2.5}$ and overall air quality in conventional cage, enriched cage and aviary systems. In a study comparing air quality between these three housing systems during two consecutive production cycles, the level of $PM_{2.5}$ was overall significantly lower in conventional cages (0.035 mg/m^3) and enriched cages (0.056 mg/m^3) than in the aviary system (0.410 mg/m^3). When analyzed in relation to the temperature outside the houses (and consequently the intensity of air ventilation), the level of $PM_{2.5}$ were consistently higher at lower temperatures: 0.047 mg/m^3 (+34.3 %) in conventional cages, 0.073 mg/m^3 (+30.3 %) in enriched cages and 0.76 mg/m^3 (+85.8 %) in the aviary system (Zhao et al., 2015). Another study evaluating the concentration of airborne bacteria showed that, although levels were above the recommended level for humans in all three housing systems, a much higher concentration was observed in the cage-free system. The concentration and emission rates of airborne total bacteria were positively correlated with PM_{10} and found to be higher in the afternoon when increased flock activity caused disruption to floor bedding (Zhao et al., 2016).

LCA studies reporting any other impact categories (e.g. ozone depletion, human toxicity (cancer effects and non-cancer effects), ecotoxicity aquatic fresh water and marine, resource depletion) identified the production of feeds as the main contributing factor (Ghasempour and Ahmadi, 2016; Abín et al., 2018; Costantini et al., 2020; Guillaume et al., 2022; Turner et al., 2022a). In the LCA of the Canadian egg sector, the organic production system had the lowest effect on all these impact categories when compared to all other production systems, except for the depletion of abiotic resources (e.g. minerals, fossil fuels, and other non-renewable materials) which had the highest value (Turner et al., 2022a). Conventional and enriched cage systems were shown to have less effect than barn (single-tier free-run) and aviary (multi-tier free-run) systems, while the production of eggs in a free-range system had the highest effect on these environmental impact categories.

8 Mitigation strategies: crop production, feed formulation and feed conversion

Two LCA studies of the US and Canadian egg sectors demonstrated the tremendous improvements made over a 50-year period (Pelletier et al., 2014; Pelletier, 2018). The analysis of the three main types of emissions (GHG, acidifying and eutrophying emissions) showed that both the US and Canadian egg sectors were able to reduce emissions by at least 60% while also increasing production by 30% and 50%, respectively. These improvements were attributed to several factors:

- better genetics;
- better nutrition (more precise diet formulation);
- improved feed ingredient quality); and
- better health of laying hens.

Improvements in flock management with the use of effective vaccines and improved biosecurity measures contributed to reductions in flock mortality (57% in the US and 75% in Canada). These improvements also translated into water use reductions exceeding 30%. All these improvements have allowed to produce more eggs with less birds, less feeds and less land use.

Regardless of these great past achievements, the global egg sector must continue to identify and adopt new mitigation strategies under pressure from civil society, large corporations (food service and processing customers), ESG (Environmental, Social, and Governance) investors and, to a lesser extent, consumers demanding further reductions in EF to achieve net-zero targets (Poore and Nemecek, 2018; Grassauer et al., 2023a). This and the following sections look at different mitigation strategies:

- crop production, feed formulation and feed conversion;
- bird management and health;
- manure management; and
- housing systems.

This section starts with crop production, feed formulation and feed conversion.

The intensity of GHG emissions of feedstuffs is linked to the type of crop produced, with wheat contributing less GHG emissions than corn, and corn itself less than corn distiller's dried grains with solubles (DDGS) or soybean. These emissions vary also with crop management practices, and the extent of the adoption of mitigation strategies such as conservation tillage, no tillage, the 4Rs approach toward land fertilization and amendment (right source at the right rate, at the right time and in the right place) and the addition of biochar, nitrification and urease inhibitors to soils, irrigation and cover crops (Ershadi et al., 2020; Young et al., 2021; Grassauer et al., 2023a).

While the production of crops, the processing of feed ingredients and feed production are major contributors to the EF of egg production, most egg producers have no control over these upstream processes (Scope 3) and must rely on long-term changes made by the agriculture sector. However, producers can directly influence the type of feed formulation used and the nature and quality of feed ingredients added, decisions having a two-fold impact on the environment:

- ingredients with a higher impact on the environment can be avoided; and
- ingredients improving feed conversion should be preferred.

These strategies would allow more eggs to be produced from the same amount of feed, thus reducing the overall emissions per kilogram of eggs produced. For instance, the switch from imported soybean to domestic-grown pea in rations fed to laying hens in Spain would reduce LUC by 69%, fresh water ecotoxicity by 28%, terrestrial ecotoxicity by 30% and GHG emissions by 4%. The switch from palm oil, also imported from recently de-forestated regions, to domestic cotton seed oil would reduce LUC by 22%, fresh water ecotoxicity by 32%, terrestrial ecotoxicity by 54% and GHG emissions by 22% (Abín et al., 2018). Organic feeds are produced using crops grown with a limited input of fertilizers and exclude all animal by-products (e.g. fish meal, the incorporation of synthetic amino acids and certain sources of vitamins), and thereby have a lower impact on the environment (Turner et al., 2022a). However, if not used in laying hen rations, the relatively high environmental impact of animal by-products would simply be allocated to other animal productions or sectors, resulting in no overall benefit to the mitigation of climate change.

Innovative formulation methodologies that incorporate the environmental impacts of feed ingredients into the least-cost formulation of poultry feeds have been suggested. However, there are still many issues to resolve. In many cases, the conversion of forested areas to cropland affects LUC and thus the overall environmental impact of feedstuffs. For instance, it was shown that the emissions intensity of soybeans could vary by more than 200-fold depending on their geographic origin within Brazil (Escobar et al., 2020). In addition, the weighing method to reconcile the economic and environmental goals of these new approaches to feed formulation will have to be defined and agreed upon by all stakeholders (Pelletier et al., 2018).

Laying hens typically digest and utilize about 80–90% of their grain-based diet. Feed enzymes, such as proteases, phytase and non-starch polysaccharides degrading enzymes, are therefore important tools to increase the nutritional value of feed ingredients, reduce feed costs and improve egg production. The use of proteases not only improves the conversion of feeds into eggs but will also reduce the protein level in layer rations, directly reducing the amount of nitrogen excreted by the birds (Leinonen and Kyrizakis, 2016). Beyond this traditional use of enzymes, enzymes also have the functional role of reducing antinutritional and toxic compounds and promoting favorable gut microbiota in the digestive system. For instance, proteases act on allergenic proteins in rations, reducing oxidative stress and gut inflammation, and promote a heathy gut for optimum nutrient absorption and conversion (Bedford and Apajalahti, 2021).

9 Mitigation strategies: bird management and health

Feed conversion (kg of feeds/kg of eggs) can also be improved through better health and welfare practices, lower mortality rates, better bone strength and improved genetics, all contributing to a reduction in various impact categories for each kilogram of eggs produced.

For instance, major reductions in the environmental impact of US and Canadian egg production have been linked to very significant decreases in average flock mortality, from an average of 11.7–3.5% in the US (1960–2010) and from 12.9 % to 3.2% in Canada (1962–2012) (Pelletier et al., 2014; Pelletier, 2018). A meta-analysis combining data from 2823 free-range flocks of hens located in the UK, the Netherlands and Sweden evaluated the effects of cumulative mortality at 72 weeks of age on the EF of egg production. While cumulative mortality was as high as 69.3%, the average cumulative mortality was 10% across all flocks. Four impact categories (GHG, acidifying emissions, eutrophying emissions and land use) were assessed for cumulative mortalities ranging from 0% to 25%. A reduction of mortality from 20% to 5% translated to a significant reduction across all impact categories: GHG emissions (24.4%), acidifying emissions (17.9%), eutrophying emissions (17.9%) and land use (22.3%). Even a more modest reduction of cumulative mortality from 10% to 5% translated into significant emission reductions (GHG emissions by 6.7%; acidifying emissions by 6.1%; eutrophying emissions by 6.0%; and land use by 6.4%). This meta-analysis showed these four impact categories could experience a 1–2% reduction for each percent reduction in mortality (Weeks et al., 2016).

The adoption of best practices, often referred to by the FAO as 'bridging the gap,' is a very important mean to significantly reduce the EF of egg production. A survey conducted with 38 countries revealed the variability in the number of eggs produced between countries, despite similar production cycle length. For instance, among four countries with a 60-week production cycle, there was a 28% difference in egg yield between the lowest and highest performing countries. Ideally, the variability in productivity should be as low as possible and for similar length of production cycles, egg producers in each country and across countries should aim to obtain as many eggs per hen housed as the best performing country or producer (Guyonnet and Guyonnet, 2024). A recent study in Canada showed that the adoption of current best practices in terms of nutrition and management of laying hens could have a significant effect on the EF of egg production, reducing GHG emissions by 43%, acidifying emissions by 40 %, eutrophying emissions by 44 %, land use by 42% and water use by 32% (Pelletier, 2017).

Improvement in the genetic lines of laying hens allows for much longer production cycles, with excellent productivity up to 100 weeks of age. With the pullet stage (0–19 weeks of age) accounting for 17–20% of the EF of egg production, an increase by 27 weeks of the period of production (from 73 to 100 weeks of age) will reduce by 32.5% the contribution of the pullet stage to the EF of each kilogram of egg produced (Grassauer et al., 2023a). Challenges in extending the production period include maintaining consistent internal egg quality, shell strength, as well as bone quality and nutrition of older hens (Gautron et al., 2021).

10 Mitigation strategies: manure management

While egg farmers may have limited influence on emissions generated during the growing, harvesting and processing of crops used for layer rations, the emissions produced by manure, the second largest source of emissions from the egg value chain (~20% of total GHG emissions), remain under their direct control. Optimum manure management aims to reduce the fermentation processes by anaerobic bacteria in manure before its application onto fields as fertilizer.

Poultry rations have a direct impact on manure chemistry, which influences the GHGs emitted during manure storage and application to soils. A decrease in nutrient digestibility in layer feeds increases their availability as fermentable organic matters in manure, thus increasing GHG emissions. Mitigation strategies can focus on various areas: manure composition, manure management inside layer houses, manure storage conditions and treatment, and manure application methods (Hristov et al., 2013; Ershadi et al., 2020; Ershadi et al., 2021).

Depending on the composition of the feed, the manure excreted by laying hens may vary in terms of pH, humidity level, the presence of undigested organic substances and its nitrogen content. Manure is not only a source of nitrogen and phosphorus but can also be rich in other minerals including zinc, boron and copper. The inclusion in poultry rations of ingredients or supplements that enhance gut absorption or promote a healthy digestive system will allow egg producers not only to improve feed conversion and profitability on their farm but also to reduce the amount of organic matter in manure that is favored by microorganisms.

Adequate management of manure inside layer houses will lead to significant reductions in ammonia volatilization and indirectly in nitrous oxide emissions, with mechanical ventilation (manure drying) and prompt removal of manure from barns via automated manure belts as simple and effective solutions. In cage-free production systems, the selection of bedding material will also have an impact on emissions, with wood shavings and maize silage reducing ammonia emissions more significantly than other organic bedding substrates. Ammonia scrubbers and biofilters in layer houses will further contribute to the removal of ammonia from the environment (Ershadi et al., 2020; Ershadi et al., 2021).

Proper storage of manure should limit the volatilization of ammonia and methane. Manure is typically stored in one of two forms:

- dry manure; and
- wet manure.

Dry manure storage is most common and involves gathering manure in piles left to dry for several months, ideally under a shed to protect from precipitation. In modern housing systems, manure is partially dried before it is removed from the house, usually 2 or 3 times per week. Older housing systems still use a deep pit under the cages, with manure accumulating throughout the production cycle and removed only at the end of the flock (Meda et al., 2011).

Wet manure storage involves mixing waste with gray water and storing it in outdoor tanks or lagoons, making it more convenient for farmers to apply it as a fertilizer onto fields. In uncovered lagoons, the anaerobic fermentation of the manure leads to the volatilization of methane and ammonia while the application of a cover (cover materials like wood chips or vegetable oil, plastic or other impermeable synthetic materials) will reduce or even eliminate ammonia volatilization but may increase nitrous oxide emissions (Ershadi et al., 2020). Regardless of the housing system used to produce eggs, liquid storage of manure produces more GHG emissions, potentially accounting for more than 50% of the GHG emissions linked to egg production (Guyonnet and Guyonnet, 2024).

Although dry manure storage generates less GHG emission than lagoon storage, some techniques can minimize emissions from liquid manure storage. For instance, it is recommended to empty storage tanks completely, as the presence of 'aged' manure left inside the tank will accelerate the spread of bacteria and the fermentation process and emissions in the next batch of stored liquid manure. Studies have shown a reduction up to 55% in GHG emissions when slurry tanks were completely emptied (Wood et al., 2014).

Various treatments to manure during storage will also contribute to the reduction of emissions. The addition of biochar reduces the emissions of methane and nitrous oxide and absorbs ammonia. The acid treatment of manure also reduces methane emissions, but its effect on nitrous oxide emissions remains uncertain (Ershadi et al., 2020; Grassauer et al., 2023a).

Anaerobic digestion of manure converts organic matter into biogas (a mixture of CH_4 and CO_2) and digestate, the residual manure lower in dry matter and nutrients. The production of biogas reduces methane emissions that typically occur during storage, and the produced digesta, used as fertilizer, has a lower degradable carbon content, thus reducing its potential for nitrification and direct nitrous oxide emissions during application onto soils. Biogas may be used to generate heat and electricity for farm applications, or sold to the electric grid, reducing both scope 1 and scope 2 GHG emissions from egg production (Grassauer et al., 2023a). A study in Bangladesh showed the significant two-fold benefits of using a biodigester to handle manure from laying hens, with a 64.7% reduction in GHG emissions from the generation of biogas and an additional 16.6% reduction from the use of the digesta as fertilizer (Mainali et al., 2017).

Finally, the timing and method of manure application onto fields also impact emissions produced by manure management, especially the amount of nitrous oxide emitted, both through nitrification and denitrification. Likewise, improper techniques or timing of applications (e.g. frozen ground) could have significant eutrophying effects. Applying manure through deep or shallow injections (liquid manure) or incorporating manure during tillage are two practices that significantly reduce ammonia and nitrous oxide emissions in comparison to surface spreading (Ershadi et al., 2020). Optimal manure application relies on the four 'Rs' – the right rate of application, the right time, the right application method and the right product. Manure should be analyzed prior to application, and compounds such as nitrification inhibitors may be added. Spring should be the preferred period for spreading manure, as it coincides with the increased nutrient demands of plants, and results in lower nitrogen losses and emissions than Fall/Autumn spreading. Applying manure on cloudy days should also be preferred to sunny days in order to limit ammonia volatilization (Chadwick et al., 2011).

Overall, reducing the nitrogen concentration in manure, preventing the formation of anaerobic conditions and reducing the presence of degradable carbon in manure are effective strategies to reduce GHG emissions during manure application. Recent studies evaluating the feasibility of implementing these different strategies (technically and financially) showed that the combination of mechanical ventilation with scrubbers, manure belts or scrapers in barns, covered storage of manure with the addition of biochar and the 4R approach to field application are the best options to reduce Scope 1 emissions and, to a lesser extent, Scope 2 emissions on egg farms (Ershadi et al., 2020; Ershadi et al., 2021; Grassauer et al., 2023a).

11 Mitigation strategies: housing systems

As presented in the section addressing climate change, different housing systems have various effects on GHG emissions per kilograms of eggs produced, with a higher feed consumption and lower feed conversion typically reported with cage-free systems, translating into higher GHG per kilograms of eggs produced. The analysis of national production data from the UK showed that the daily feed consumed by laying hens was 8% higher in the barn system and 13% higher in the free-range system when compared to the cage system. Furthermore, egg production was 4.8% lower in the barn system and 7% lower in the free-range system (Leinonen et al., 2012). These results not only affected the cost of egg production in cage-free systems but also increased GHG emissions and other impact categories (Tables 2–7). However, the growing demand for cage-free production systems in Europe and North America is related to animal welfare and ethical concerns and does not consider the environmental impacts or any potential trade-offs in this decision.

A recent study analyzed the potential environmental consequences of changes in egg production systems in Canada, with the combination of a Data Envelopment Analysis (DEA) considering efficiencies in egg production and a LCA study using national production data. From the various scenarios tested, a switch from still predominantly conventional cage production system to a national supply of eggs divided equally among enriched cage, aviary and organic production systems (i.e. 1/3 each) would yield a 20% decrease in GHG, acidifying and eutrophying emissions. A switch to 100% enriched cage production system would account for more reduction in emissions (~10%) than switching to 100% aviaries (i.e. indoor cage-free) or to 50% enriched cage and 50% aviary production systems (Turner et al., 2022b). This research clearly identifies the need to highlight trade-offs between environmental impact and welfare benefits of different production systems. It is evident that large retailers and food service operations committing to the sole supply of eggs from cage-free systems in the US did not use a holistic approach in their decision-making process.

Unable to provide convincing data to select housing systems with the lowest impacts on the environment, recent research has focused on improving energy efficiencies on egg farms (Scope 2 emissions) as well as on construction materials and equipment used in layer barns (Grassauer et al., 2023a). Net zero energy (NZE) buildings are energy efficient structures with NZE consumption, where the total amount of energy used by the building is equal to the amount of renewable energy it can generate on an annual basis. These buildings rely on three approaches: structural considerations to reduce energy use, adoption of energy efficient systems and on-site generation of renewable energy.

In most countries, layer houses are built with timber, plywood, tin and stone wool insulation material. The environmental impact of these materials is called their embodied energy and is the sum of all the energy required to produce them. While adding extra insulation, new renewable energy systems and more energy-efficient equipment in NZE barns will indeed have a small additional effect on the overall environmental impact of the building itself, this increase will easily be offset by the reduction in direct energy consumed for heating and ventilation during egg production. The LCA approach comparing conventional and NZE layer houses showed that the direct cumulative energy input was reduced by 36% per tonne of eggs in the NZE layer house. Depending on the environmental impact of energy production from fossil fuel, gas, coal or hydroelectricity, a switch to NZE layer houses will significant reduce CO_2 emissions and also other impact categories (Li et al., 2021). It is likely that NZE technology will become more common in the future to further reduce the EF of the egg sector.

12 Conclusion and future trends

With consumers, ESG investors and governments showing interest in reducing the EF of food choices, the importance of sustainability will grow exponentially over the coming years. LCA studies are complex estimations of emissions and other impact categories, relying on the accuracy and relevance of a wide range of raw data from many parts of the world and on several assumptions and allocations made by researchers (Costantini et al., 2021; Arulnathan et al., 2022; Fan et al., 2022). The monitoring and management of multiple impact categories is extremely complicated and LCA studies, usually evaluating the EF of egg production at a national level, do not represent the situation on individual farms. For instance, based on the LCA study conducted in Canada, the average mortality of flocks was 3.18% but ranged from 0.90% to 7.30%, representing a 10% difference in GHG emissions between the best and worst performing farms. Similar variation was also seen for feed conversion: 2.00 was reported as the national average, but the values ranged from 1.64 to 2.82 (Pelletier, 2017). This high variability in production efficiencies demonstrates the impossibility of using national values to track and report the EF of egg production on individual farms.

This has led to the development of environmental impact assessment tools by private enterprises (e.g. Carbonbox™ (https://www.carbonbox.app/), Eggbase™ (https://eggbase.co.uk/), or Sustell™ (https://www.sustell.com/)). Not only do these new tools allow egg producers to measure the EF in their own production facilities, but they also offer built-in simulation programs that allow egg producers to identify and test the best mitigation strategies for their farms. Using a similar on-farm monitoring tool, 10 large-scale egg producers in the UK were able to reduce their average GHG emissions by nearly 25% over a 3-year period (Vetter et al., 2018). The selection of the appropriate commercial monitoring tool is essential to accurately evaluate the EF of egg production. There are several key characteristics to consider:

1. the model must follow international rules and guidelines (IPCC and LEAP) discussed previously;
2. default values used by the model must be relevant to local conditions; and
3. the tool must be connected to relevant international databases to ensure consistent, relevant and accurate calculations.

Using these tools, egg producers will be able to report accurately the EF of egg production and take full credit for their investments (e.g. NZE houses, renewable energy, changes in feed ingredients and additives) to reduce the EF of egg production on their farms.

The allocation made during LCA studies (eggs, meat from hens, feed ingredients and animal by-products) are also being questioned and will stimulate more research on improving the relevance of these studies (van Hal et al., 2019; Green et al., 2020; Arulnathan et al., 2022). Of particular significance could be a switch from an economic to a food-based allocation, meaning that animal by-products of no nutritional value to humans (e.g. fish meal, feather meal, bone meal), would be assigned no environmental impact other than that of their own processing (no allocation of upstream burden). This new approach to LCA would consider any competition between food (human nutrition) and feed (animal nutrition) for land use and resources and would reduce, under certain scenarios, the overall EF of egg production (van Hal, 2019).

With feeds representing the biggest contributor to the EF of egg production, mainly through Scope 3 emissions, future research will focus on new cropping methods (e.g. geospatial technologies for precision farming and use of inputs), new feed ingredients (e.g. insect proteins) and natural additives controlling gut flora (e.g. bacteriophages). Feed formulation will also see some significant changes, with the addition of an environmental component to the least-cost formulation approach to allow selection of feed ingredients by monetary cost and cost to the planet. However, more research will be needed to integrate these two components that often conflict (Heidari et al., 2021).

Manure and waste management are the biggest contributor to emissions directly under the control of egg producers (Scope 1). As such, they will receive more focus to identify emission reduction technologies and valorization opportunities (Kanami et al., 2020; Grassauer et al., 2023b). The challenge will be to ensure that any waste valorization does not have a detrimental effect on the planet. For instance, while the conversion of eggshells into limestone or lime would reduce the amount of shell waste sent to landfills, the energy required for the high thermal treatment (~800°C) of shells will in fact increase GHG emissions (among other impact categories) in comparison to using oyster shells (Lee et al., 2020). It will therefore be necessary to take a holistic approach and identify all potential trade-offs in any valorization initiative.

Finally, with the intense scrutiny of governments, ESG investors and civil society on the animal production sector, there will be a need to standardize and harmonize further the models used for reporting progress. This will limit the risk of 'greenwashing', or of passing the burden from one sector to another, with no real benefit to the sustainability of animal production (Toscano et al., 2022; Liu et al., 2023).

13 Where to look for more information

The vast amount of resources available from the LEAP (https://www.fao.org/partnerships/leap/partners/en/) should be consulted by any practitioner wishing to design a new LCA study on egg production. The growing interest in sustainability has also seen a rise of publications in journals often new to poultry scientists: *Journal of Cleaner Production*; *Journal of Environmental Management*; *Resources, Conservation & Recycling Advances*; *Sustainability*. With the European Union having made the commitment to become climate-neutral by 2050, a lot of information is available on the European Green Deal website (https://commission.europa.eu/strategy-and-policy/priorities-2019-2024/european-green-deal_en), with facts, figures and information related to the environmental sustainability of agriculture (https://agriculture.ec.europa.eu/sustainability/environmental-sustainability_en). The Farm-to-Fork strategy (https://food.ec.europa.eu/horizontal-topics/farm-fork-strategy_en) is at the core of the EU Green Deal, with the goal to make food systems fair, healthy and environmentally friendly. The website from the United Nations Environmental Program (UNEP) presents an overview of current climate actions around the world (https://www.unep.org/).

For information about ESG reporting, the website from the United Nations Development Program (UNDP) is an excellent start point (https://www.undp.org/future-development/signals-spotlight/rethinking-governance-esg) to learn about the need for greater transparency and to follow the progress made on the development of international standards for ESG reporting. Finally, the FAIRR initiative (https://www.fairr.org/about) uses a network of investors to evaluate the ESG risks and opportunities in the global food sector. They feature an assessment of animal protein-producing companies, including a few large companies involved in egg production, e.g. Cal-Maine Foods Inc. from the USA, Charoen Pokphand Foods PCL from Thailand, JAPFA from Indonesia and Venky's from India.

14 References

Abín, R., et al. (2018). Environmental assessment of intensive egg production: a Spanish case study. *Journal of Cleaner Production*, 179, 160–168. https://doi.org/10.1016/j.jclepro.2018.01.067

Arulnathan, V., et al. (2022). Internal causality in agri-food life cycle assessments: solving allocation problems based on feed energy utilization in egg production. *Journal of Environmental Management*, 309, 114673. https://doi.org/10.1016/j.jenvman.2022.114673

Beal, T., et al. (2023). Friend or Foe? The role of animal-source foods in healthy and environmentally sustainable diets. *The Journal of Nutrition*, 153(2), 409–425. https://doi.org/10.1016/j.tjnut.2022.10.016

Bedford, M.R. and Apajalahti, J.H. (2021). The role of feed enzymes in maintaining poultry intestinal health. *Journal of the Science of Food and Agriculture*, 102, 1759–1770. https://doi.org/10.1002/jsfa.11670

Chadwick, D., et al. (2011). Manure management: implications for greenhouse gas emissions. *Animal Feed Science and Technology*, 66–167, 514–531. https://doi.org/10.1016/j.anifeedsci.2011.04.036

Clark, M. and Tilman, D. (2017). Comparative analysis of environmental impacts of agricultural production systems, agricultural input efficiency, and food choice. *Environmental Research Letters*, 12, 064016. https://doi.org/10.1088/1748-9326/aa6cd5

Costantini, M., et al. (2020). Investigating on the environmental sustainability of animal products: the case of organic eggs. *Journal of Cleaner Production*, 274, 123046. https://doi.org/10.1016/j.jclepro.2020.123046

Costantini, M., et al. (2021). Environmental sustainability assessment of poultry productions through life cycle approaches: a critical review. *Trends in Food Science & Technology*, 110, 201–212. https://doi.org/10.1016/j.tifs.2021.01.086

Dekker, S.E.M., et al. (2011). Ecological and economic evaluation of Dutch egg production systems. *Livestock Science*, 139(1–2), 109–121. https://doi.org/10.1016/j.livsci.2011.03.011

Ershadi, S.Z., et al. (2020). Improving nitrogen use efficiency in crop-livestock systems: a review of mitigation technologies and management strategies, and their potential applicability for egg supply chains. *Journal of Cleaner Production*, 265, 121671. https://doi.org/10.1016/j.jclepro.2020.121671

Ershadi, S.Z., et al. (2021). Comparative life cycle assessment of technologies and strategies to improve nitrogen use efficiency in egg supply chains. *Resources, Conservation and Recycling*, 166, 105275. https://doi.org/10.1016/j.resconrec.2020.105275

Escobar, N., et al. (2020). Spatially-explicit footprints of agricultural commodities: mapping carbon emissions embodied in Brazil's soy exports. *Global Environmental Change*, 62(May 2020), 102067. https://doi.org/10.1016/j.gloenvcha.2020.102067

Estrada-González, I., et al. (2020). Decreasing the environmental impact in an egg-producing farm through the application of LCA and lean tools. *Applied Science*, 10, 1352–1366. https://doi.org/10.3390/app10041352

European Union (2013). 2013/179/EU: commission recommendation of 9 April 2013 on the use of common methods to measure and communicate the life cycle environmental performance of products and organisations. Retrieved May 12, 2022, from https://eur-lex.europa.eu/legal-content/EN/TXT/?uri=CELEX%3A32013H0179

Fan, J., et al. (2022). Life cycle assessment on agricultural production: a mini review on methodology, application, and challenges. *International Journal of Environmental Research and Public Health*, 19, 9817. https://doi.org/10.3390/ijerph19169817

FAO (2016a). *Greenhouse gas emissions and fossil energy use from poultry supply chains: guidelines for assessment*. Livestock Environmental Assessment and Performance Partnership. FAO, Rome, Italy. Retrieved May 12, 2022, from https://www.fao.org/3/i6421e/i6421e.pdf

FAO (2016b). *Environmental performance of animal feeds supply chains: Guidelines for assessment*. Livestock Environmental Assessment and Performance Partnership. FAO, Rome, Italy. Retrieved May 12, 2022, from https://www.fao.org/3/i6433e/i6433e.pdf

FAO (2018). *Nutrient flows and associated environmental impacts in livestock supply chains. Guidelines for assessment version 1*. Livestock Environmental Assessment and Performance Partnership. FAO, Rome, Italy. Retrieved March 19, 2024, from https://www.fao.org/3/CA1328EN/ca1328en.pdf

FAO (2019). *Water use in livestock production systems and supply chains. Guidelines for assessment. Version 1*. Livestock Environmental Assessment and Performance Partnership (FAO LEAP). Rome. Retrieved March 19, 2024, from https://www.fao.org/3/ca5685en/ca5685en.pdf

FAO (2020). *Environmental performance of feed additives in livestock supply chains – Guidelines for assessment – Version 1*. Livestock Environmental Assessment and Performance Partnership (FAO LEAP), Rome. https://doi.org/10.4060/ca9744en

Gautron, J., et al. (2021). Review: What are the challenges facing the table egg industry in the next decades and what can be done to address them? *Animal*, 15(Supplement 1), 100282. https://doi.org/10.1016/j.animal.2021.100282

Gerber, P.J., et al. (2013). *Tackling climate change through livestock – A global assessment of emissions and mitigation opportunities*. FAO, Rome. https://www.fao.org/3/i3437e/i3437e.pdf

Ghasempour, A. and Ahmadi, E. (2016). Assessment of environment impacts of egg production chain using life cycle assessment. *Journal of Environmental Management*, 183, 980–987. http://dx.doi.org/10.1016/j.jenvman.2016.09.054

GHG Protocol (2004). A corporate accounting and reporting standard. Revised edition. Retrieved March 19, 2024 from, from https://ghgprotocol.org/sites/default/files/standards/ghg-protocol-revised.pdf

González-Velandia, K.D., et al. (2020). Evaluación de impactos ambientales en la cadena de producción de huevos agroecológicos con un enfoque de ciclo de vida. *Revista de ciencias ambientales*, 54(2), 165–179. https://doi.org/10.15359/rca.54-2.9

Grassauer, F., et al. (2023a). Towards a net-zero greenhouse gas emission egg industry: a review of relevant mitigation technologies and strategies, current emission reduction potential, and future research needs. *Renewable and Sustainable Energy Reviews*, 181, 113322. https://doi.org/10.1016/j.rser.2023.113322

Grassauer, F., et al. (2023b). Manure valorization using black soldier fly larvae: a review of current systems, production characteristics, utilized feed substrates, and bioconversion and nitrogen conversion efficiencies. *Sustainability*, 15, 12177. https://doi.org/10.3390/su151612177

Green, A., et al. (2020). Assessing nutritional, health, and environmental sustainability dimensions of agri-food production. *Global Food Security*, 26, 100406. https://doi.org/10.1016/j.gfs.2020.100406

Guillaume, A., et al. (2022). Environmental impacts of egg production from a life cycle perspective. *Agriculture*, 12(3), 355. https://doi.org/10.3390/agriculture12030355

Guyonnet, V. (2024). World egg production and marketing trends. In Y. Mine, V. Guyonnet, H. Hatta, F. Nau, and N. Qiu (Eds.), *Handbook of egg science and technology* (1st ed., pp. 9–26). CRC Press, Boca Raton.

Guyonnet, V. and Guyonnet, A.E.M. (2024). The sustainability of the egg sector – Opportunities and Mitigation strategies. In Y. Mine, V. Guyonnet, H. Hatta, F. Nau, and N. Qiu (Eds.), *Handbook of egg science and technology* (1st ed., pp. 747–764). CRC Press, Boca Raton.

Heidari, M.D., et al. (2021). Proposing a framework for sustainable feed formulation for laying hens: a systematic review of recent developments and future directions. *Journal of Cleaner Production*, 288, 125585. https://doi.org/10.1016/j.jclepro.2020.125585

Hristov, A.N., et al. (2013). Mitigation of greenhouse gas emissions in livestock production – A review of technical options for non-CO2 emissions. In P.J. Gerber, B. Henderson and H.P.S. Makkar (Eds.), *FAO animal production and health paper No. 177*. FAO, Rome, Italy. https://www.fao.org/3/i3288e/i3288e00.htm

IPCC (2006). 2006 IPCC guidelines for national greenhouse gas inventories. In H.S. Eggleston, et al. (Eds.), *Institute for Global Environmental Strategies (IGES)*, Japan. https://www.ipcc-nggip.iges.or.jp/public/2006gl/index.html

ISO (2006). ISO 14040:2006 Environmental management - Life cycle assessment - Principles and framework. Retrieved May 12, 2022, from https://www.iso.org/obp/ui/#iso:std:iso:14040:ed-2:v1:en

Kanami, F., et al. (2020). Waste valorization technology options for the egg and broiler industries: a review and recommendations. *Journal of Cleaner Production*, 262, 121129. https://doi.org/10.1016/j.jclepro.2020.121129

LEAP (2024, March 19). Livestock Environmental Assessment and Performance partnership. https://www.fao.org/partnerships/leap/en/

Le Bouquin, S., et al. (2013). Aerial dust concentration in cage-housed, floor-housed, and aviary facilities for laying hens. *Poultry Science*, 92(11), 2827–2833. https://doi.org/10.3382/ps.2013-03032

Lee, M., et al. (2020). Reusing shell waste as a soil conditioner alternative? A comparative study of eggshell and oyster shell using a life cycle assessment approach. *Journal of Cleaner Production*, 265, 121845. https://doi.org/10.1016/j.jclepro.2020.121845

Leinonen, I. and I. Kyriazakis, I. (2016). How can we improve the environmental sustainability of poultry production? *Proceedings of the Nutrition Society*, 75(3), 265–273 https://doi.org/10.1017/S0029665116000094

Leinonen, I., et al. (2012). Predicting the environmental impacts of chicken systems in the United Kingdom through a life cycle assessment: egg production systems. *Poultry Science*, 91(1), 26–40. https://doi.org/10.3382/ps.2011-01635

Leinonen, I., et al. (2014). The effects of welfare-enhancing system changes on the environmental impacts of broiler and egg production. *Poultry Science*, 93(2), 256–266. https://doi.org/10.3382/ps.2013-03252

Li, Y., et al. (2021). Net zero energy barns for industrial egg production: an effective sustainable intensification strategy? *Journal of Cleaner Production*, 316, 128014. https://doi.org/10.1016/j.jclepro.2021.128014

Liu, Y.S., et al. (2023). Carbon emissions, carbon disclosure and organizational performance. *International Review of Financial Analysis*, 90, 102846. https://doi.org/10.1016/j.irfa.2023.102846

Mainali, B., et al. (2017). Greenhouse gas mitigation using poultry litter management techniques in Bangladesh. *Energy*, 127, 155–166. https://doi.org/10.1016/j.energy.2017.03.103

McClelland, S.C., et al. (2018). Type and number of environmental impact categories used in livestock life cycle assessment: a systematic review. *Livestock Science*, 209, 39–45. https://doi.org/10.1016/j.livsci.2018.01.008

Meda, B., et al. (2011). Influence of rearing conditions and manure management practices on ammonia and greenhouse gas emissions from poultry houses. *World's Poultry Science Journal*, 67(3), 441–455. https://doi.org/10.1017/S0043933911000493

Mekonnen, M.M. and Hoekstra, A.Y. (2012). A global assessment of the water footprint of farm animal products. *Ecosystems*, 15, 401–415. https://doi.org/10.1007/s10021-011-9517-8

Mollenhorst, H., et al. (2006). On-farm quantification of sustainability indicators: an application to egg production systems. *British Poultry Science Journal*, 47(4), 405–417 https://doi.org/10.1080/00071660600829282

Pelletier, N., et al. (2014). Comparison of the environmental footprint of the egg industry in the United States in 1960 and 2010. *Poultry Science*, 93(2), 241–255. https://doi.org/10.3382/ps.2013-03390

Pelletier, N. (2017). Life cycle assessment of Canadian egg products, with differentiation by hen housing system type. *Journal of Cleaner Production*, 152, 167–180 https://doi.org/10.1016/j.jclepro.2017.03.050

Pelletier, N. (2018). Changes in the life cycle environmental footprint of egg production in Canada from 1962 to 2012. *Journal of Cleaner Production*, 176, 1144–1153. https://doi.org/10.1016/j.jclepro.2017.11.212

Pelletier, N., et al. (2018). Sustainability in the Canadian industry – Learning from the past, navigating the present, planning for the future. *Sustainability*, 10(10), 3524–3548. https://doi.org/10.3390/su10103524

Poore, J. and Nemecek, T. (2018). Reducing food's environmental impacts through produces and consumers. *Science*, 360(6393), 987–992. https://doi.org/10.1126/science.aaq0216

Toscano, A., et al. (2022). Sustainability practices and greenwashing risk in the Italian poultry sector: a grounded theory study. *Sustainability*, 14(21), 14088. https://doi.org/10.3390/su142114088

Turner, I., et al. (2022a). Life cycle assessment of contemporary Canadian egg production systems during the transition from conventional cage to alternative housing systems: Update and analysis of trends and conditions. *Resources, Conservation and Recycling*, 176, 105907. https://doi.org/10.1016/j.resconrec.2021.105907

Turner, I., et al. (2022b). Environmental impact mitigation potential of increased resource use efficiency in industrial egg production systems. *Journal of Cleaner Production*, 354, 131743. https://doi.org/10.1016/j.jclepro.2022.131743

van Hal, O., et al. (2019). Accounting for feed-food competition in environmental impact assessment: towards a resource efficient food-system. *Journal of Cleaner Production*, 240, 118241. https://doi.org/10.1016/j.jclepro.2019.118241

Vetter, S.H., et al. (2018). The potential to reduce GHG emissions in egg production using a GHG calculator: a cool farm tool case study. *Journal of Cleaner Production*, 202, 1068–1076. https://doi.org/10.1016/j.jclepro.2018.08.199

Weeks, C.A., et al. (2016). Implications for welfare, productivity and sustainability of the variation in reported levels of mortality for laying hen flocks kept in different housing systems: a meta-analysis of ten studies. *PLoS One*, 11(1), e0146394. https://doi.org/10.1371/journal.pone.0146394

Wiedemann, S.G. and McGahan, E.J. (2011). *Environmental assessment of an egg production supply chain using life cycle assessment*. Final project report. Australian Egg Corporation Limited. https://www.freeranger.com.au/uploads/7/4/2/0/7420102/aecl_carbon_footprint.pdf

Wood, J.D., et al. (2014). Gas emissions from liquid dairy manure: complete versus partial storage emptying. *Nutrient Cycling in Agroecosystems*, 99, 95–105. https://doi.org/10.1007/s10705-014-9620-2

Young, M.D, et al. (2021). Impacts of agronomic measures on crop, soil, and environmental indicators: a review and synthesis of meta-analysis. *Agriculture, Ecosystems & Environment*, 319, 107551. https://doi.org/10.1016/j.agee.2021.107551

Zhao, Y., et al. (2015). Environmental assessment of three egg production systems – Part I: monitoring system and indoor air quality. *Poultry Science*, 94(3), 518–533. https://doi.org/10.3382/ps/peu076

Zhao, Y., et al. (2016). Environmental assessment of three egg production systems – Part III: airborne bacteria concentrations and emissions. *Poultry Science*, 95(7), 1473–1481. https://doi.org/10.3382/ps/pew053

Chapter 16

Improving sustainability of egg production by extending the laying cycle for hens

Teun G.H. van de Braak, Hendrix Genetics Layers, The Netherlands

1 Introduction

Extending the laying cycle of commercially kept laying hens has been the key driver of the breeding programs and breeding goals in laying hen breeding for decades. The extended laying cycles allow the egg producers to generate more eggs per laying hen that they house in their layer barns. These extended cycles will result in less replacement of flocks over the course of time, thereby saving costs on flock replacements and the cleaning of the layer barns.

According to Bain et al. (2016), over 12 000 oocytes are present in the ovary of the female chick on the day of hatch, of which only a small percentage of these will ever reach maturity. Genetic selection has resulted in a growing percentage of these oocytes maturing and eventually developing into a good quality egg during the hen's lifetime. Back in 2010, the first breeding company claimed that by the year 2020, the 'long-life' layer would be capable of producing 500 eggs in a single production cycle of 100 weeks (van Sambeek, 2010). To clarify, a single production cycle means that no forced molting is introduced during the laying cycle, and a production cycle of 100 weeks does include the rearing

http://dx.doi.org/10.19103/AS.2024.0150.24

period, which is in today's cage-free production systems the equivalent of 20 weeks. This claim became reality, in the year 2017, the first to be known flock of commercially kept laying hens reached 503.7 eggs per hen housed in even less than 100 weeks (Dekalb-Poultry website). As of today, a growing number of egg producers have been able to achieve the claim made by van Sambeek back in 2010, with numbers outperforming the 500 eggs per hen housed.

2 Extending the laying cycle: implications for more sustainable production

2.1 Understanding sustainability

Since its introduction, back in 1987 by the UN World Commission on Environment and Development (WCED), the definition of 'sustainability' has been under debate, as it can be interpreted in various ways. Back in 1987, it was stated that development is only 'sustainable' if it 'meets the needs of the present without compromising the ability of future generations to meet their own needs' (WCED 1987). As it has been 36 years since its first introduction, the interpretation of the definition has changed over time and generations, highlighting the dynamic nature of sustainable development. This dynamism underscores that sustainable development is not a fixed concept but evolves in response to societal, environmental, and economic shifts. Broom et al. (2013) came up with the following definition: 'A system or procedure is sustainable if it is acceptable now and if it effects will be acceptable in future, in particular in relation to resource availability, consequences on functioning and morality of action.' Keeling et al. (2019) stated that sustainable development aims to balance different needs toward achieving dignity, peace, and prosperity for people, against an awareness of the environmental, social, and economic limitations we face as a society. As of the differences in cultural, geographical, and socio-economics, there is no one solution for sustainable development of egg production that fits all, fortunately, extending laying cycles fits with many of them. Moulting of laying hens is a method still commonly used in many parts around the globe to extend the laying cycle, and thereby utilize the genetic potential of the commercial laying hen. Traditionally, molting has been one strategy employed to rejuvenate hens' reproductive systems and improve egg quality. However, concerns about animal welfare have led to the prohibition of traditional fasting-based molting methods in some countries and restrictions on non-feed withdrawal methods in others (Widowski and Rentsch, 2022). Non-feed withdrawal methods, which involve nutrient restriction through limitations on feed intake or reducing nutrient density, have emerged as alternatives. Yet, they too are mired in controversy. While they aim to induce molting without the extreme weight loss associated with fasting methods, concerns persist regarding their impact on hen welfare and health. Research indicates that both fasting and non-fasting molts can have deleterious effects on bone mineralization, raising further ethical and practical considerations. In addition to welfare concerns, the effectiveness of non-feed withdrawal methods is challenged by the innate drive of hens to consume feed, making it difficult to achieve the desired physiological changes without significant stress. Moreover, despite efforts to minimize mortality rates during molting, concerns remain about hen welfare and the overall sustainability of these practices. Thus, while molting offers a potential solution to the challenges of extending laying periods, the ethical and practical complexities surrounding its implementation underscore the need for further research and innovative approaches to address the competing demands of productivity, welfare, and sustainability in modern egg production.

2.2 Societal impact on sustainability

From a societal point of view, extended laying cycles address various societal themes, of which animal welfare is among the most prominent themes. Extended laying cycles can contribute in a positive way to the overall image of the egg sector in terms of animal welfare. The genetic potential of laying hens in order to maintain egg production can only be achieved when all requirements of the laying hen itself are met. Bain et al. (2016) concluded that in order to reach the goal of a long-life laying hen, a multifactorial approach including genetics, nutrition, and design of housing systems needs to be adopted to ensure that possible negative effects on correlated traits such as bone health are avoided.

Extending the laying cycle doesn't solely focus on the productivity of laying hens; traits related to animal health and animal welfare are included in the overall approach of extended laying cycles. Star et al. (2008) introduced the concept of adding robustness traits to the breeding goal; robustness traits include individual traits of an animal, which are relevant for health and welfare. They defined robustness as 'a robust animal is an animal that has the potential to keep functioning and take short periods to recover under varying environmental conditions'. Laying hens remain productive only when there is harmony between the laying hens and their environment, and when they are free of diseases and other physical disorders. Laying hens that are capable of maintaining the production of eggs with good eggshell quality need to build up enough medullary bone early in life, as hen skeletal integrity and eggshell quality deteriorate with age (Alfonso-Carrillo et al., 2021). As the formation of medullary bone starts in the weeks prior to the onset of lay, Alfonso-Carrillo et al. (2021) indicated that optimizing the pre-lay diet and delaying the age at first egg can reduce osteoporosis and bone fractures in laying hens, allowing the hens to sustain high egg production with good eggshell quality.

2.3 Environmental impact on sustainability

Extending laying cycles have a direct impact on the environment and varies between housing systems. The study by Van Asselt et al. (2015) showed that enriched cages scored more than double on the environmental dimension compared to cage-free housing systems (barn, free-range, organic). When looking at the social dimension, the free-range housing system scored more than double when compared to enriched cages, barn and organic. As every flock of laying hens is unique, it requires a tailor-made approach to fully understand the environmental impact. Life Cycle Assessment (LCA) can take these differences into account and an LCA assesses the impact of all the relevant production processes during the life cycle of the production of eggs, as shown by Mollenhorst et al. (2006). The 2022 LCA update by Turner et al. on Canadian egg production systems showed that the transition away from cage-housing may be a net negative for the environmental sustainability of egg production, but it is likely that as farmers continue to transition away from cage housing and gain experience working in cage-free housing systems, the resource-use efficiency in cage-free systems would increase, further decreasing the environmental impacts of egg production. Extending the laying cycles, whilst maintaining productivity and keeping mortality under control, will help to decrease the environmental impact of egg production. In the case of extended laying cycles, the impact of the rearing period plays an important role. Extending the laying cycle from 80 weeks to 90 weeks of depletion age will result in fewer cycles. Already in 6 cycles of 90 weeks, an egg producer can save 1 full cycle, which means a reduction of 14% during the rearing period. Bain et al. (2016) estimated that an increase of just 25 eggs per hen housed, which is the equivalent of just 4 weeks of egg production, could potentially reduce the UK flock by 2.5 million birds (including parent stock) per annum.

2.4 Economical impact on sustainability

The majority of literature concerning the extension of laying cycles primarily examines its biological feasibility and assesses its implications for both the environment and the welfare of hens; very little research is available that specifically assesses the economic impacts of extended laying cycles. Traore and Doyon (2023) analyzed the economic impact of extended egg laying cycles in Canada, their results demonstrated that extending the laying cycle (i.e. the period that the laying hens are actually producing eggs, excluding the rearing period) beyond 52 weeks will result in a positive net change in total profit for Canadian egg producers, meaning that the positive financial changes associated with a longer laying cycle outweigh the negative. The authors demonstrated that in the case of aviary/free-run farms, the optimal economic duration of the laying period is 71 weeks, which is by far longer than the current standard practice of 52 weeks in Canada.

Despite the limited scientific literature available, egg producers, in practice, are keeping their laying hen flocks for longer extended periods compared to before. This suggests an economic incentive for extended laying cycles. In volatile egg markets, especially when egg prices are high, there is a clear motivation to extend laying cycles due to the positive financial impact on egg producers. This trend has been evident over the past 12 months in both Europe and the United States.

In Europe, several factors contribute to the overall increase in egg prices (Mulder, 2022). Firstly, the cost of production has risen, driven by higher feed and energy prices, which have been offset by the increased egg prices. Secondly, the lower number of day-old female chicks born in Western Europe, influenced by legislation prohibiting the culling of day-old layer male chicks in major European egg-producing countries such as Germany and France, has had an impact (MEG Marktbalanz, 2023). The German legislation, in particular, has affected the dynamics of the Dutch egg market, as the majority of eggs produced in the Netherlands are exported to Germany, subjecting Dutch egg producers to similar requirements for egg production and animal welfare. The third factor is related to the presence of the Highly Pathogenic Avian Influenza (HPAI) Virus, causing the death of more than 2.2 million laying hens by the virus in the Netherlands alone in 2022 (WUR, 2021/2022/2023). The confluence of these events has resulted in a tight supply of available eggs in 2023, and this tightness is expected to persist in 2024 (MEG, 2024).

3 Extending the laying cycle: the laying hen breeder's point of view

For more than one century, breeding laying hens has been taking place on a professional scale (Tixier-Boichard et al., 2012). But in these hundred years, large consolidation steps have taken place, that has resulted in three main players being active in todays' egg markets: Hendrix Genetics (HQ the Netherlands), EW-group (HQ Germany) and Beijing Huadu Yukou Poultry Industry Co., Ltd. (HQ China), of which the latter is predominantly active in the Chinese market. With so few players, it is important that breeding decisions and breeding goal directions are carefully being made, as the impact on the global egg-value chain doesn't go unnoticed.

Genetic improvement has been driven for decades on economic and environmental relevant traits, such as peak of lay, egg quality, feed conversion, and mortality. As the biological limit of an egg a day at the peak of lay has become a reality, extending the laying cycles allowed the breeding companies to deliver continued benefits to egg producers that fit with these economic and environmental drivers (Fernyhough et al., 2020).

Extending the laying cycle via genetic selection is not something of the recent decade; Anderson and Havenstein (2012) visualized the results of commercial laying hen performance over a 50-year period under controlled conditions at the North Carolina Layer Performance and Management Testing Station. Their results clearly show the evolution of commercial layer breeds that were under genetic selection. In these five decades of selection, both white and brown laying hen breeds evolved significantly, as they became lighter in bodyweight, earlier sexually mature, had a higher hen day production, produced more eggs per hen housed, consumed less feed per hen per day, improved their feed conversion ratio, and reduced in overall mortality.

As Bain et al. (2016) already highlighted and visualized, extending the laying cycle has the potential to significantly reduce the number of multiplier hens and hence the overall shape of the breeding pyramid. This has become reality in mature and saturated egg markets, as less laying hens are needed in order to fulfill the steady demand for eggs, thereby also reducing the number of parent stock. But, as the human population continues to grow, as well as global per capita egg consumption in various regions, there is still growth predicted for the global egg sector. Windhorst (2022) mentioned that the OECD-FAO predicts an increase in egg production of 8.0 million tons between 2020 and 2030, of which more than half (56.4%) will be contributed by Asian countries. This predicted growth offers more than enough reasons for breeding companies to keep on investing in selecting for the improvement in sustainability of egg production, thereby addressing all three aspects of sustainability, societal, environmental, and economical and not solely on one of them.

Today's breeding programs can be best described as well-balanced, as a growing number of stakeholder interests are taken into account when determining the breeding goal for the future selections (Preisinger, 2018). The days when selection was done on just a few economic traits have been left far behind; today's selection indexes often contain more than 40 traits, that all have their own weight in a balanced way. With a growing number of traits, less genetic progress per trait can be expected. However, the introduction of genomic selection (Genomic Evaluation) about one decade ago in the breeding program of laying hens has proven to be successful, as it increases the accuracy of predictions at a very young age, thereby increasing the selection intensity and decreasing the generation interval (Druet et al., 2020; Sitzentsock et al., 2013). Due to its effectiveness at a young age, it allows breeding companies to expand population sizes during the brooding period, introducing an extra step of selection at a relatively low cost since the price of genomic evaluation has come down in recent years. Genomic selection is more effective in the selection of males compared to females. In laying hen breeding, it takes much more time to get own performance records on males, as it takes time to record the data being on the offspring of the males in order to get phenotypic data on their commercial traits, as males are not able to produce eggs themselves but do contribute close to 50% to the next generation.

4 Extending the laying cycle: balancing economic, ethical, and environmental aspects of sustainability

As Preisinger stated in 2017, laying hen breeding companies need to look beyond the current requirements and anticipate at least 5 years ahead to changing needs and opportunities. Therefore, close communication between the breeding companies and the various stakeholders of the egg production chain is necessary to ensure that genetic development is aligned with stakeholder expectations and market requirements. For laying hen breeders, it is important to also balance the three aspects of sustainability, described previous in this chapter: economical, societal, and environmental all deserve the right attention in order to safeguard a sustainable direction of the breeding goal.

Stakeholder expectations evolve over time: a clear example of the changing public debate can be seen in the European Union (Blokhuis et al., 2007; Sandilands and Hocking, 2012; Rahmani et al., 2019) and Northern America (Mench et al., 2011; Thompson et al., 2011). In Europe and North America, there has been a strong consumer demand to take welfare into account in animal production (Gautron et al., 2021). Table egg consumption patterns have seen a major shift from cage-to-non-cage production systems because of societal pressures. This shift likely signals the future direction for the layer sector in developed countries, with the acceleration of the conversion of cage to barn and aviary systems with access to wintergardens or the outdoors. Next to many advantages as the hens are much better able to express their natural behaviors, this shift can come with unintended consequences for the welfare of the hens from bone fracture to increased disease exposure (Lay et al., 2011, Widowski et al., 2016) and for the environment as cage-free housing systems have a bigger negative impact on the environment (Xin et al., 2011; Oliveira et al., 2018), all requiring solutions.

A direction to improve the sustainability of the production of eggs, whilst improving the hens' welfare, has been breeding for the extended laying cycle. As stated earlier, laying hens are born with more than 10 000 oocytes, indicating that there is a potential to extend the laying cycle even further.

To extend the laying period, the focus of today's breeding program is on the hens' health and well-being. Despite the intensive genetic selection for egg production, the decrease in genetic variation in the closed pure lines is not critical (Preisinger and Flock, 2000). Heritabilities for the majority of the classical breeding traits (egg production, egg quality, egg weight) are still moderate to high (Preisinger and Flock, 2000; Underwood et al., 2021). The trait that has seen the largest decline in genetic variation is peak production, as this trait has nearly approached its biological limit of one egg a day, with commercial flocks of laying hens peaking nowadays at 98% of lay and even higher and with female pure lines peaking at 99% or more. As a result of a century of genetic selection, the trait of peak production is almost fixed in today's laying hens, which has resulted in a heritability close to zero for this trait. As selecting for better egg laying persistency, egg quality, egg weight, or body weight has proven itself, the challenge and the focus of today's breeding programs is on the behavioral traits. Unfortunately, heritabilities of behavioral traits are low, and at the same time, phenotypes of behavioral traits are often difficult (Peeters, 2015), expensive to record, and data recording is not 100% accurate (Icken et al., 2012). Icken et al. (2012) showed that heritabilities of the same traits differ between housing systems, adding an extra complexity as more traits need to be added to the selection index. New insights in traits like mortality as a result of cannibalistic behavior have resulted in taking the social interactions among group members into account (Ellen et al., 2008). The research of Ellen (2008) showed that heritable variation in survival days is substantially larger when considering both direct and associative (the effect on an individual on the survival of its group members) effects than suggested by the conventional direct effects model. Allowing for new methods is to reduce one of the major welfare problems associated with egg production in all types of commercial housing systems (Albentosa et al., 2003).

The introduction of cage-free housing systems for individually tagged and pedigreed laying hens has been developed during the last decade (Icken et al., 2012). Rufener et al. (2019) developed a method that allowed for assessing individual laying hen behavior in large and complex aviary systems, which are known for containing a considerable degree of complexity and structure. These developments will contribute to having better tools available to record phenotypes that play an important role in cage-free housing systems in an accurate way. One of these phenotypes is related to keel bone fractures; the Farm Animal Welfare Committee stated already back in 2013 that the high prevalence of keel bone fractures represents one of the greatest welfare problems facing the egg production industry (FAWC, 2013). Wilkins et al. (2011) showed that the type and design of housing systems have a major influence on the prevalence of keel bone fractures, with systems containing multi-level perches having the highest incidence of damage and the highest severity scores. The work by Toscano et al. (2020) suggested that next to elevated and sustained egg production that there are four broad areas that are likely to be the major contributing factors toward fractures: the age at first egg, the late ossification of the keel, predisposing bone diseases, and inactivity leading to poor bone health. Thøfner et al. (2021) identified hen size, age at onset of lay, and daily egg weight at onset of lay to be major risk factors for the development of keel bone fractures in modern laying. From the results of Anderson and Havenstein (2012), it can be seen that pullet bodyweight went down during these 50 years of random sample testing. The pullet body weight of the brown laying hen breeds went down by more than 600 grams (2.11 in 1958 versus 1.39 in 2009), and for the white laying hen breeds by more than 400 grams (1.61 in 1958 versus 1.16 in 2009). From their results, it can be seen that the body weights stabilized from 1999 onwards, i.e. not declining further. From the same study, it can be seen that the age at 50% production also went down, more than 26 days for the brown egg layers, versus more than 33 days for the white egg layers, which indicates that the age at first egg went down. As very little attention has been given to keel bone fractures by both the academic and commercial world, it is difficult to see the evolution of keel bone fractures over time. Dunn et al. (2021) tested the hypothesis that egg production is associated with quality, breaking strength and density of bone. The low genetic correlation that they found between post-peak production with bone quality suggests that selection for increased persistency of egg production may not adversely affect bone quality. They identified the onset of puberty and the mineralization of the medullary bone as important factors associated with the quality of the skeleton later during egg production. As can be concluded from all the research done so far is that (keel)bone health is affected by various factors, next to genetics, also environmental and management factors can positively impact the overall quality of the laying hen skeletal (Dunn et al., 2021).

While skeletal health remains a major welfare concern in laying hens, there are also economic incentives associated with extended laying cycles (Hanlon et al., 2022). Eggshell quality remains one of the most important drivers for egg producers, or egg packing stations, to keep on extending the productive life of their laying hen flocks. Bain et al. (2016) showed that the eggshell is vital for the integrity and protection of the internal components of table eggs. As laying hens age, the eggshell breaking strength decreases and the average egg size increases (Benavides-Reyes et al., 2021; Zita et al., 2009; Roberts, 2004; Nys, 1986), resulting in a higher percentage of eggs containing cracks. Next to applying nutritional strategies like split feeding (Molnár, 2018), breeding companies are selecting to improve the overall egg shell breaking strength as well as delaying the decline in egg shell breaking strength as laying hens get older. One of the strategies that has been implemented by breeding companies in extending the laying cycle is by flattening the egg size curve (Nys, 2017). Flattening the egg size curve will result in higher uniformity of egg weight (Tůmová et al., 2017). The breeders' philosophy behind this theory is that laying hens that can better control their egg weight development as they age can also better control their eggshell quality. Nys (2017) showed that the decrease in eggshell quality that is generally observed in older hens is partly due to a reduction of calcium absorption by the intestine and to the increase in egg size as laying hens age.

Not only will egg shell quality undergo changes as laying hens tend to get older, but also internal egg quality is changing as laying hens age. It is well known that age is among the many factors that affect albumen height and Haugh Units (Roberts, 2004). Induced molting programs have demonstrated to improve the albumen quality and therefore the Haugh Unit (Tona et al., 2002). But also genetic selection has contributed greatly to improving the Haugh Unit of todays laying hen strains. Breeding companies have put extra selection pressure on the strains commonly used for the Asian egg markets like Japan, Korea, and Taiwan. These aforementioned egg markets value high levels for Haugh Units, thereby making persistency for Haugh Unit levels an economic driver for the breeding companies. The heritability of Haugh Units is moderate, in the range of 0.25–0.35, depending on the strain (Hendrix Genetics, 2024). These moderate heritabilities for Haugh Unit will allow the breeding companies to further improve the persistency of Haugh Unit quality as laying hens get older, thereby contributing to the sustainability of the global egg sector.

Improved internal and external egg quality will contribute to ensure the safety and integrity of table eggs, as the prolonged exposure of laying hens to environmental stressors and their natural aging process can compromise the integrity of the eggshell, membranes, and albumen viscosity, thereby jeopardizing egg quality and safety. A recent study by Jabalera (2022) has shed light on the potential vulnerability of antimicrobial defenses in aged hens, further exacerbating concerns surrounding egg hygiene and microbial contamination. Despite these emerging findings, a lack of scientific literature exists in this crucial area, underscoring the need for further exploration and understanding the impact of extended production cycles. Moreover, the current growth of the market for egg products poses additional requirements for laying hen breeds and egg producers, as consumers increasingly turn to convenient and versatile egg products, such as liquid eggs and pre-packaged egg whites (Mine and Guyonnet, 2024). This shift in consumer preferences could result in a reevaluation of the traditional breeding traits to ensure the sustainability and efficiency of egg production systems.

As mentioned earlier in this chapter, the decline in bodyweight in commercial pullets has been stopped around the millennium (Anderson and Havenstein, 2012). The importance of absolute bodyweight (Barboza and Hume, 2006), as well as the body composition during the rearing period (Kwakkel et al., 2017), should be optimized for extended laying cycles (van Eck, 2023). In today's management guidelines, breeding companies put emphasis on the importance of the rearing period and on reaching the target body weights

(Hendrix Genetics, 2021; Lohmann, 2021). Van Eck et al. (2023) discussed in their literature review the interaction between the main metabolic processes and the influence of diet and hormones on sustained egg production. They concluded that nutrition plays a pivotal role in supporting the growth and ovulation of follicles in the ovary. Proper functioning of organs, particularly the liver, is essential for supplying the necessary lipids for follicle development. When laying hens age, the functioning of the liver can be reduced due to a higher chance of liver steatosis and the occurrence of fatty liver disease, thereby negatively impacting egg production (Julian, 2005; Bain et al., 2016).

From the results of Anderson and Havenstein (2012), it can be seen that daily feed consumption dropped by 10 grams a day for brown egg layers, and 14 grams a day for white egg layers. As a result of the lower feed consumption and the improved egg production, feed conversion ratio (shown as gram of egg produced per gram of feed consumed) improved significantly: brown egg layers improved more than 50% in feed conversion efficiency (0.326 in 1958 versus 0.492 in 2009), and white egg layers improved more than 45% in feed conversion efficiency (0.345 in 1958 versus 0.501 in 2009) (Anderson and Havenstein, 2012). The world egg production reached 93 million tons of eggs in 2021 (FAO, 2022), imagine the extra feed that was needed if genetic selection on improving feed conversion had not taken place in all these years of selection. Next to improving feed conversion, breeding companies are selecting laying hens that perform on agro-industrial by-products (AIBPs) or so-called low-opportunity-cost feeds (LCF), which are non-human-consumable residues that offer a sustainable alternative in animal nutrition, avoiding competition with food crops (Georganas et al., 2023). Among these AIBPs are Distillers Dried Grains with Solubles (DDGS), which are by-products of the ethanol industry (Świątkiewicz and Koreleski, 2008). Świątkiewicz et al. (2013) showed that DDGS are acceptable ingredients for poultry diets and that laying hens can be safely fed with diets containing up to 20% of DDGS without compromising egg laying performance. Van Hal et al. (2019) showed that when feeding free-range laying hens only with an LCF diet, the global warming potential was reduced by 48–58% and the land use by 34–47%. These results clearly demonstrate the unique capabilities of laying hens in transforming by-products or the so-called low-opportunity feeds into highly nutritious animal protein. It must be noted that little research has been carried out into the nutrition and metabolism of the long-life laying hen (van Eck et al., 2023; Kleyn and Ciacciarello, 2023), suggesting that there is further potential in improving the environmental impact of egg production via nutritional strategies.

5 Conclusion and future trends

The pursuit of extending the laying cycles of commercially kept laying hens has not only proven to be a biological reality but also holds significant implications for sustainability across multiple dimensions. Achieving extended laying cycles not only enables egg producers to maximize egg output per hen, reducing costs associated with flock replacements but also aligns with societal, environmental, and economic sustainability goals.

The multifaceted concept of sustainability, evolving over the years, emphasizes meeting present needs without compromising the ability of future generations to meet their own. Extended laying cycles contribute to sustainability by addressing animal welfare concerns, reducing environmental impact, and proving economically beneficial for egg producers. From a societal standpoint, the emphasis on animal health and well-being, coupled with genetic selection for robustness traits, showcases a commitment to responsible egg production. Environmental impacts vary across housing systems, with cage-free systems showing potential for reduced environmental footprints.LCA analyses underscore the importance of tailoring approaches to each unique flock. The potential reduction in environmental impact through fewer laying cycles, especially during the rearing period, demonstrates a positive stride towards sustainable egg production.

The economic impact of extended laying cycles is gaining attention, with studies indicating a positive net change in total profit for egg producers. Economic incentives, especially in volatile egg markets, have driven practical adoption of extended laying cycles. In light of global trends predicting an increase in egg production, breeders are urged to continue investing in sustainability, aligning with societal, environmental, and economic goals.

From the perspective of laying hen breeders, genetic improvement has been historically focused on economic and environmental traits, with extended laying cycles offering a continued avenue for benefit. The adoption of genomic selection has enhanced breeding programs, allowing for improved accuracy and shorter generation intervals. The industry's shift toward a more sustainable direction involves careful consideration of societal expectations, environmental concerns, and economic viability.

As breeding programs evolve, the challenge lies in balancing genetic progress with the need for behavioral traits associated with animal welfare. The introduction of cage-free systems, advancements in phenotypic recording tools, and a focus on reducing welfare issues like keel bone fractures underscore the commitment to improving both sustainability and animal welfare.

In conclusion, the extended laying cycle is not only a biological feat but a holistic approach that positions the laying hen itself at the focal point, while addressing societal, environmental, and economic dimensions of sustainability, with ongoing efforts required to ensure a well-balanced and forward-looking approach in laying hen breeding.

6 Conflict of interest statement

The author, Teun van de Braak, declares a conflict of interest as an employee of Hendrix Genetics. The company had no further role in the writing of the manuscript. However, readers are advised to consider this affiliation when interpreting the results.

7 Where to look for further information

For readers seeking deeper insights into extended laying cycles and sustainable egg production, the following resources offer valuable perspectives:

7.1 Scientific Journals and Articles

- *British Poultry Science* and *Frontiers in Animal Science* publish studies on genetics, welfare, and sustainability.

7.2 Industry Publications and Newsletters

- *Avinews, Poultryschool, Watt Global Media*, *The Poultry Site*, *Poultry World*, and *The Poultry Network* regularly feature articles on long-life layers, feeding strategies, and welfare-friendly systems. For example, Poultry World's article "Embrace the long-life layer and adapt accordingly" discusses nutritional adjustments and genetic advancements that support extended cycles.

7.3 Websites and Online Platforms

- Frontiers in Animal Science hosts open-access research on economic and welfare aspects.
- Everything you need to know about laying hens - Laying Hens website of the poultry breeding company with news and latest updates on poultry breeding.

7.4 Forums and Networks

- Engage with poultry science communities via the World's Poultry Science Association (WPSA) and LinkedIn groups focused on sustainable livestock production.

8 References

Albentosa, M. J., et al. (2003). Strain and age differences in behaviour, fear response and pecking tendency in laying hens. *British Poultry Science*, *44*(3), 333–344.

Alfonso-Carrillo, C., et al. (2021). Relationship between bone quality, egg production and eggshell quality in laying hens at the end of an extended production cycle (105 weeks). *Animals*, *11*(3), 623.

Anderson, K. E. & Havenstein, G. B. (2012). *History of the North Carolina layer tests detailed description of housing and husbandry changes made from 1958 through 2009*. North Carolina Cooperative Extension Service. https://poultry.ces.ncsu.edu/wp-content/uploads/2012/04/NCLPMT-1958-2009-Historical-Summary.pdf

Bain, M. M., et al. (2016). Increasing persistency in lay and stabilising egg quality in longer laying cycles. What are the challenges?. *British Poultry Science*, *57*(3), 330–338.

Barboza, P. S. & Hume, I. D. (2006). Physiology of intermittent feeding: integrating responses of vertebrates to nutritional deficit and excess. *Physiological and Biochemical Zoology*, *79*(2), 250–264. https://doi.org/10.1086/499984

Benavides-Reyes, C., et al. (2021). Research note: changes in eggshell quality and microstructure related to hen age during a production cycle. *Poultry Science*, *100*(9), 101287.

Blokhuis, H.J., et al. (2007). The LayWel project: welfare implications of changes in production systems for laying hens. *World's Poultry Science Journal*, *63*, 101–114.

Broom, D.M., et al. (2013). Sustainable, efficient livestock production with high biodiversity and good welfare for animals. *Proceedings of the Royal Society B: Biological Sciences*, *280*, 20132025. https://doi.org/10.1098/rspb.2013.2025

Dekalb 'top athletes' deliver 503.7 eggs for Niels Quicken - Dekalb. dekalb-poultry.com (Accessed 27-12-2023)

Dunn, I.C., et al. (2021). No evidence that selection for egg production persistency causes loss of bone quality in laying hens. *Genetics Selection Evolution*, *53*(1), 1–13.

Ellen, E. D., et al. (2008). Survival of laying hens: genetic parameters for direct and associative effects in three purebred layer lines. *Poultry Science*, *87*(2), 233–239.

FAO (2022). Agricultural production statistics 2000–2021. fao.org (Accessed 29-12-2023)

FAWC (2013). *An open letter to Great Britain Governments: keel bone fracture in laying hens*. FAWC.

Fernyhough, M., et al. (2020). The ethics of laying hen genetics. *Journal of Agricultural and Environmental Ethics*, *33*, 15–36. https://doi.org/10.1007/s10806-019-09810-2

Gautron, J., et al. (2021). What are the challenges facing the table egg industry in the next decades and what can be done to address them?. *Animal*, *15*, 100282.

Georganas, A., et al. (2023). Utilization of agro-industrial by-products for sustainable poultry production. *Sustainability*, *15*(4), 3679.

Hanlon, C., et al. (2022). Bone and eggshell quality throughout an extended laying cycle in three strains of layers spanning 50 years of selection. *Poultry Science*, *101*(3), 101672.

Hans-Wilhelm, W. (2022). *Zootecnica*, A projection of egg production until 2030, 8 March 2022, FAO data.

Hendrix Genetics (2021). Preparing for a good start, objectives for the brooding period - Laying hens. hendrix-genetics.com (Accessed 29-12-2023)

Hendrix Genetics (2024). Internal data.

Icken, W., et al. (2012). Genotype-environment interaction: breeding layers with different. In Alternative systems for poultry: Health, welfare and productivity (Vol. Symposium Series No. 30, pp. 316–339). CABI.

Jabalera Y., et al. (2022). Antimicrobial defenSes of table eggs: importance of antibacterial proteins in egg white as a function of hen age in an extended production cycle. *Food Microbiology*, *107*, Article 104068.
Julian, R. J. (2005). Production and growth related disorders and other metabolic diseases of poultry–a review. *The Veterinary Journal*, *169*(3), 350–369.
Keeling, L., et al. (2019). Animal welfare and the United Nations sustainable development goals. *Frontiers in Veterinary Science*, *6*, 336.
Kleyn, F. J. & Ciacciariello, M. (2023). An evaluation of the protein and energy requirements in long-life laying hens. *Journal of Applied Poultry Research*, *32*(1), 100319.
Kwakkel, R. P., et al. (2017). How to survive a 'challenging childhood': rearing the laying hen!. *Proceedings of the 21st European Symposium on Poultry Nutrition*, *21*, 70.
Lay Jr., D. C., et al. (2011). Hen welfare in different housing systems. *Poultry Science*, *90*(1), 278–294.
Lohmann (2021). LTZ_MG_management-systems_EN.pdf lohmann-breeders.com (Accessed 29-12-2023)
MEG (2024). DE: Ausblick Eiermarkt 2024 - MEG. marktinfo-eier-gefluegel.de
MEG, Marktbilanz Eier und Geflügel (2023). MEG Marktbilanz Eier und Geflügel 2023 - Verlag Eugen Ulmer: Bücher & Zeitschriften für Garten, Gartenbau & Landwirtschaft | Ulmer.de
Mench, J. A., et al. (2011). Sustainability of egg production in the United States–The policy and market context. *Poultry Science*, *90*(1), 229–240.
Mine, Y. & Guyonnet, V. (2024). *Handbook of egg science and technology*. CRC Press.
Mollenhorst, H., et al. (2006). On-farm quantification of sustainability indicators: an application to egg production systems. *British Poultry Science*, *47*(4), 405–417.
Molnár, A., et al. (2018). Effect of different split-feeding treatments on performance, egg quality, and bone quality of individually housed aged laying hens. *Poultry Science*, *97*(1), 88–101.
Mulder, N. D. (2022). Rabobank, eggflation: what happens after egg prices reach historic high ACCESSED 27-12-2023 eggflation: what happens after egg prices reach historic highs. rabobank.com
Nys, Y. (1986). Relationships between age, shell quality and individual rate and duration of shell formation in domestic hens. *British Poultry Science*, *27*, 253–259.
Nys, Y. (2017). Laying hen nutrition: optimising hen performance and health, bone and eggshell quality. *Achieving Sustainable Production of Eggs*, *2*, 47–74.
Oliveira, J. L., et al. (2018). Effects of litter floor access and inclusion of experienced hens in aviary housing on floor eggs, litter condition, air quality, and hen welfare. *Poultry Science*, *98*, 1664–1677.
Peeters, K. (2015). *Genetics of social interactions in laying hens: improving survival and productivity* (Doctoral dissertation). Wageningen University and Research.
Picard Druet, D., et al. (2020). Reliability of genomic evaluation for egg quality traits in layers. *BMC Genet*, *21*, 17. https://doi.org/10.1186/s12863-020-0820-2
Preisinger, R. & Flock, D. K. (2000). Genetic changes in layer breeding: historical trends and future prospects. *BSAP Occasional Publication*, *27*, 20–28. https://doi.org/10.1017/S1463981500040504
Rahmani, D., et al. (2019). Are consumers' egg preferences influenced by animal-welfare conditions and environmental impacts? *Sustainability*, *11*, 6218. https://doi.org/10.3390/su11226218

Roberts, J. R. (2004). Factors affecting egg internal quality and egg shell quality in laying hens. *The Journal of Poultry Science*, *41*(3), 161–177.

Rudolf Preisinger (2018). Innovative layer genetics to handle global challenges in egg production. *British Poultry Science*, *59*(1), 1–6. https://doi.org/10.1080/00071668.2018.1401828

Rufener, C., et al. (2019). Keel bone fractures are associated with individual mobility of laying hens in an aviary system. *Applied Animal Behaviour Science*, *217*, 48–56.

Sandilands, V. & Hocking, P. M. (2012). Is there a future for alternative production systems? In J. A. Mench (Ed.), *Alternative systems for poultry: health, welfare and productivity* (pp. 340–348). CABI.

Sitzenstock, F., et al. (2013). Efficiency of genomic selection in an established commercial layer breeding program. *Genetics Selection Evolution*, *45*, 29. https://doi.org/10.1186/1297-9686-45-29

Star, L., et al. (2008). A plea to implement robustness into a breeding goal: poultry as an example. *Journal of Agricultural and Environmental Ethics*, *21*, 109–125.

Świątkiewicz, S. & Koreleski, J. (2008). The use of distillers dried grains with solubles (DDGS) in poultry nutrition. *World's Poultry Science Journal*, *64*(2), 257–266.

Świątkiewicz, S., et al. (2013). Effects of selected feed additives on the performance of laying hens given a diet rich in maize dried distiller's grains with solubles (DDGS). *British Poultry Science*, *54*(4), 478–485.

Thøfner, I. C., et al. (2021). Keel bone fractures in Danish laying hens: prevalence and risk factors. *PLoS One*, *16*(8), e0256105.

Thompson, P. B., et al. (2011). Values and public acceptability dimensions of sustainable egg production. *Poultry Science*, *90*(9), 2097–2109.

Tixier-Boichard, M., et al. (2012). A Century of poultry genetics. *World's Poultry Science Journal*, *68*(2), 307–321.

Tona, K., et al. (2002) Effect of induced molting on albumen quality, hatchability, and chick body weight from broiler breeders. *Poultry Science*, *81*, 327–332.

Toscano, M. J., et al. (2020). Explanations for keel bone fractures in laying hens: are there explanations in addition to elevated egg production? *Poultry Science*, *99*(9), 4183–4194.

Traore, O. Z. & Doyon, M. (2023) Economic sustainability of extending lay cycle in the supply managed Canadian egg industry. *Frontiers in Animal Science*, *4*, 1201771.

Tůmová, E., et al. (2017). Age related changes in laying pattern and egg weight of different laying hen genotypes. *Animal Reproduction Science*, *183*, 21–26.

Turner, I., et al. (2022). Life cycle assessment of contemporary Canadian egg production systems during the transition from conventional cage to alternative housing systems: update and analysis of trends and conditions. *Resources, Conservation and Recycling*, *176*, 105907.

Underwood, G., et al. (2021). Advances in genetic selection and breeder practice improve commercial layer hen welfare. *Animal Production Science*, *61*(10), 856–866. https://doi.org/10.1071/AN20383

Van Asselt, E. D., et al. (2015). Assessing the sustainability of egg production systems in The Netherlands. *Poultry Science*, *94*(8), 1742–1750.

van Eck, L. M., et al. (2023). Lipid metabolism and body composition in long-term producing hens. *World's Poultry Science Journal*, *79*(2), 243–264.

Van Hal, O., et al. (2019). Accounting for feed-food competition in environmental impact assessment: towards a resource efficient food-system. *Journal of Cleaner Production, 240*, 118241.

Van Sambeek, F. (2010). *Progress in layer genetics "Longer production cycles from a genetic perspective*. Article Master. positiveaction.info

WCED (World Commission on Environment and Development) (1987). *Our common future*. Oxford University Press. sustainabledevelopment.un.org/content/documents/5987our-common-future.pdf

Widowski, T. M., et al. (2016). Laying hen welfare I. Social environment and space. *World's Poultry Science Journal, 72*(2), 333–342.

Widowski, T. M. & Rentsch, A. K. (2022). *Routledge handbook of animal welfare, Cahpter 5*. Farming Poultry. https://doi.org/10.4324/9781003182351-7

Wilkins, L. J., et al. (2011). Influence of housing system and design on bone strength and keel bone fractures in laying hens. *Veterinary Record, 169*(16), 414-414.

WUR, Bird flu at poultry farms, updates 2021/2022 - WUR Bird flu at poultry farms, updates 2021/2022 - WUR & Bird flu at poultry farms, updates 2022/2023 - WUR.

Xin, H., et al. (2011). Environmental impacts and sustainability of egg production systems. *Poultry Science, 90*(1), 263–277.

Zita, L., et al. (2009). Effects of genotype, age and their interaction on egg quality in brown-egg laying hens. *Acta Veterinaria Brno, 78*(1), 85–91.

Chapter 17

Strategies and solutions to reduce emissions from layer production systems in Europe

Cécile M. Levrault and Peter W. G. Groot Koerkamp, Wageningen University and Research, The Netherlands

1 Introduction

Over the years, demand for eggs has continued to increase steadily, simultaneously stimulating production. In 2020, there were more than 350 million laying hens in the European Union with production close to 6.7 million tonnes of eggs per year (European Commission, 2024). In so-called commercial, intensive production systems, animals are either reared in cages,

http://dx.doi.org/10.19103/AS.2024.0150.25

aviaries or free-range houses. By EU law (Council Directive 1999/74/EC), all laying hens should be provided access to perching and nesting space, litter to allow for behaviours such as pecking, dustbathing and scratching behaviours, and have unrestricted access to water and feed, irrespective of what housing system is used. As of 2012, conventional cages have been prohibited by the European Union, and animals can now only be kept in the so-called enriched cages which guarantee that each animal has at least 750 cm^2 of (cage) area. The alternative systems for their part (aviaries and free range systems) allow the birds to move freely with stocking densities that do not exceed nine hens per square meter (equating to at least 1111 cm^2/bird) (Council Directive 1999/74/EC).

In design terms, these production systems essentially differ in whether they are cage or non-cage systems, are single or multi-tier, and provide or do not provide access to the outdoors (Table 1). In 2021, enriched cages represented 39.2% of the European Union's laying hens production (151.8 million birds), aviaries 38.8% (150.3 M birds) and free range 22.0% (58.9 M birds standard free range, and 26.4 M organic). While cage-based systems still remain the main housing system in layers, a clear increase of the share of non-cage systems has been observed since the battery cages ban of 2012. Comparatively, cages (with or without enrichment) represented 65.6% of layer housing systems in the EU in 2010, aviaries 20.8%, and free range (including organic) 13.6% (European Commission, 2024). The trend towards aviaries and free range systems will be further driven by the wish to ban cage systems in general in the EU (European Commission, 2021a, 2022). While this trend has positive effects on the welfare of the animals, it poses challenges in terms of emissions and air quality in animal houses.

Table 1 Key characteristics of laying hens production systems currently in use in the EU.

	Enriched cages	Aviaries	Free range
Cage system	Yes	No (loose)	No (loose)
Tier	Multi	Multi	Single
Outdoor access	No	Optional	Yes
European egg marketing code	3	2 (1 if outdoor access; 0 if organic)	1 (0 if organic)
Manure and litter types produced	Manure on dropping plates or belts	Manure on belts; litter on the floor	Manure under the tier; litter on the floor
Manure handling	Belts under cages, regular removal from the house, option of air drying	Belts under tiers, regular removal from the house, option of air drying	Accumulated storage under tier, option of air drying

2 Environmental issues related to layer production

Each of the housing system used in layer production results in different excreta types and manure management (Table 1). In general terms, excreta from laying hens, also called faeces or fresh droppings, is essentially excreted as a (semi) solid with a moist content of about 75–80%. After excretion, excreta becomes slurry (liquid), manure (dry and stackable) or litter (mixed with bedding material on the floor and used for scratching and dust bathing by the hens). The practice of mixing faeces with water and further manage, store and apply as slurry has become rare. As manure or litter, it then dries out to varying degrees of dry matter, mainly depending on the water evaporation rate, and depending on the housing and ventilation systems, air conditions, bird density, bedding material, manure management and storage processes. The end product usually contains between 20% and 60% of moisture.

In 2021, the European Climate Law was outlined by the European Commission with the objective that the EU would reach carbon neutrality by 2050 (European Commission, 2021b). Earlier, a National Emission Ceilings Directive sought to provide clear objectives for the reduction of the emission of certain atmospheric pollutants (EU Directive 2016/2284). With respect to these targets, layer production too must reduce the emissions originating from its sector. In this chapter, we focus on the emissions of greenhouse gases, ammonia, particulate matter, odour and volatile organic compounds.

Different manure management practices lead to different emission levels of diverse pollutants, such as carbon dioxide (CO_2), ammonia (NH_3), methane (CH_4) and particulate matter (PM). Generally speaking, those that favour the quick drying of manure (e.g. on belts in cages and aviary systems, or in special drying tunnels) lead to significant reductions in ammonia losses (Rosa et al., 2020).

Understanding the processes behind these emissions and the complexity of their interactions is essential to mitigate the various emissions arising from this growing sector. The applicability of mitigation techniques, as well as their drawbacks, should be very well understood when striving to implement them to reduce emissions. Therefore, the goal of this chapter is to provide an overview of the options to mitigate emissions that are currently available for layer production, while clearly defining the shortcomings, knowledge gaps and barriers associated with each of them. This chapter is not meant as an in-depth review presenting and summarising all available literature. For readers wanting to know more, we list recent review papers at the end of this chapter.

3 Direct emissions originating from layer production

This section reviews:

- Ammonia emissions and other nitrogenous compounds
- Greenhouse gas emissions
- Particulate matter and dust emissions
- Volatile organic compounds emissions
- Odour emissions

These are discussed in more detail below.

3.1 Ammonia emissions and other nitrogenous compounds

Overall, poultry production contributes to approximately 13% of the NH_3 emissions from agriculture, itself emitting about 42 Tg N year^{-1} worldwide (Crippa et al., 2018; Olivier et al., 2020). In a study conducted in Spain, ammonia emissions from enriched cage systems were found to vary between 33 g and 53 g per hen per year (Alberdi et al., 2016). In The Netherlands, loose layers farms without outdoor access, ammonia (NH_3) emissions were estimated to equal 402 g per hen per year in single-tiered systems, and 129 g per hen per year in aviaries (Mosquera et al., 2009; Winkel et al., 2009). Emissions are therefore substantially increased by the presence of litter (aviaries and free range) and long term storage of manure (free range) in the housing system, and vary depending on amongst others the management of the manure on the belts.

Ammonia gas is produced as a by-product of the microbial decomposition of organic-nitrogenous substances in faeces, manure and litter (EPA, 2004b; Moum et al., 1969). Its effects on the environment are adverse, typically resulting in the acidification and eutrophication of ecosystems (Breemen et al., 1982). In poultry, it originates from the decomposition of the undigested proteins and uric acids that are contained in the faeces and urine, making them the two prime nitrogen components of birds excreta (Bussink and Oenema, 1998; Groot Koerkamp, 1994). The main processes responsible for the emission of NH_3 are the decomposition of uric acid by microbial action (uricase enzyme), and in case of undigested proteins by bacterial metabolisms, uricase and urease enzymes (Behera et al., 2013). During these degradation processes, other nitrogenous compounds such as nitrous oxides (NO_x) and nitrogen gas (N_2) may be produced as well (EPA, 2004b).

3.2 Greenhouse gas emissions

Greenhouse gas (GHG) emissions originating from poultry production essentially consist of carbon dioxide (CO_2), methane (CH_4) and nitrous oxide

(N_2O). With respective global warming potentials of 1, 28 and 265 (Myhre et al., 2013), these gases contribute to the enhanced greenhouse effect. Carbon dioxide produced by the hens with their respiration is not considered to contribute to additional global warming as it is part of the short carbon cycle. Typical emission levels for CH_4 and N_2O from enriched cages systems are 33 and 1 g per hen per year (Alberdi et al., 2016). In loose housing systems without access to the outdoors, CH_4 and N_2O emissions originate from the manure and litter and are higher in single-tiered (40 g CH_4 and 16 g N_2O per hen per year) than in aviary systems (27 g CH_4 and 11 g N_2O per hen per year) (Mosquera et al., 2009; Winkel et al., 2009). The different emission levels observed between housing systems are essentially due to their varying manure characteristics and management (MacLeod et al., 2013). Methane is formed and emitted during the decomposition of manure under anaerobic conditions, with production levels increasing with higher volatile solid contents. N_2O for its part is emitted as an intermediate product of the nitrification/denitrification process occurring in the excreta under low-oxygen conditions (Fabbri et al., 2007).

3.3 Particulate matter and dust emissions

Dust is produced in layer houses and contains faces, feed particles, feathers, skin fragments, microorganisms and diverse chemicals, such as pharmaceutical residues (EPRS, 2019). It also contains particulate matters (PM), a precursor of which is ammonia. These PM can have sizes ranging from 0.001 to 100 μm of aerodynamic diameter (Brook et al., 2003). PM that are larger than 10 μm are for the most part trapped in the humans nasal passages, while smaller particles will deposit further in the respiratory track (ICRP, 1994). Inhaling such particles can lead to respiratory diseases and inflammations, impacting not only the birds' health but also the one of the workers and surrounding population (Rimac et al., 2010; Viegas et al., 2013). Based on their aerodynamic diameter, PM that are smaller than 10 μm are classified into different fractions: PM_{10} (aerodynamic diameter ≤10 μm), $PM_{2.5}$ (≤2.5 μm) and $PM_{0.1}$ (≤0.1 μm) (EPA, 2004a).

In comparison to other livestock productions, poultry houses have been reported by Winkel et al. (2015) to have the highest mean PM concentrations in the Dutch farms monitored in their research. The Northern-European survey reported the mean inhalable PM concentrations for the laying hens held in aviaries to equate 8780 $\mu g/m^3$ (Takai et al., 1998), which is consistent with the findings of Winkel et al. (2015) who obtained concentrations of 8175 $\mu g/m^3$ in single-tiered houses. In a study conducted in two multi-layered poultry houses in the USA, emissions of 7.3 g per hen per year and 59.5 g per hen per year were found for $PM_{2.5}$ and PM_{10}, respectively (Lin et al., 2017). In conventional cages, emissions of PM have been estimated to be two to three times lower than in cage-free systems, with emissions oscillating from 1 to 5 for $PM_{2.5}$, and 7 to 17 g per hen per year for PM_{10} (Li et al., 2011). This results from the presence of litter on the floor and the occurrence of related bird activities (e.g. dustbathing) in the cage-free systems (Hayes et al., 2013).

3.4 Volatile organic compound emissions

Volatile organic compounds (VOC) have been defined as the organic chemical compounds that evaporate easily at ambient temperature (Anand and Mehendale, 2005). The volatility of each of these compounds is generally higher the lower its boiling point temperature is. The higher the volatility, the more likely a compound will be emitted into the air (World Health Organization, 1989). The respective effects of VOC on human health are adverse (World Health Organization, 1989), and have detrimental impacts on the environment (e.g. ozone formation in the atmosphere, acid deposition, aerosol formation) (Malley et al., 2015; Mazzuca et al., 2016; Williams and Koppmann, 2007). As a result, strict regulations on VOC have been set in place (e.g. European Union Directive 1999/13/EC in Europe). In poultry, VOC essentially originates from the waste products of the birds, primarily their manure and bedding material. As listed by Huanhong et al., the main VOC in free range poultry are methane, aromatics, alcohols, aldehydes, ketones, carboxylic acids, terpenes, hydrocarbons and sulphur, some of which additionally leading to odour nuisance (Huanhong et al., 2023).

3.5 Odour emissions

Odours originating from layers production can lead to odour nuisance for farm workers and surrounding population. This is notably due to the odours of hydrogen sulphide, dimethylamine, aldehydes, ketones and organic acids (Carey et al., 2004; Nowak et al., 2016). Most of these odourous compounds result from the natural aerobic and anaerobic microbial activities that are taking place in the manure and litter, while some odour is directly emitted by the birds themselves (Lacey et al., 2004; Mackie et al., 1998).

Odours can be monitored and characterised using methods that are either instrumental or sensorial (Zarra et al., 2013). Instrumental techniques (e.g. gas chromatography–mass spectrometry, proton transfer reaction–mass spectrometry) tend to be more objective, accurate and repeatable but do not provide insights on how the odour is perceived by humans (Akdeniz et al., 2012; Lebrero et al., 2011). Sensorial techniques for their part (e.g. dilution olfactometry, field-based odour panels) allow the characterisation of odours in terms of both concentration and perception (Lebrero et al., 2011). The assessed odour concentration can then be used together with the momentary air exchange in the barn and the total animal body mass to determine the odour emission factor (Jacobson et al., 2005; Mielcarek and Rzeźnik, 2015). A literature review from Mielcarek and Rzeźnik (2015) showed that, when converted to livestock units (one LU = 500 kg), poultry has the highest odour emission factors compared to swine and cattle. Broilers ranked the highest with an odour emission factor of 178 OU_E/s per LU, followed by laying hens with 141. As an indication, the other livestock production included in the study had values ranging from 20 to 89 OU_E/s per LU.

4 Emissions targets: the European Union

In 2016, the reduction of the emission of certain atmospheric pollutants has been defined by in the National Emission Ceilings Directive (EU Directive 2016/2284). This directive established clear reduction objectives for the member states. It was amended in 2024 with new targets to be reached by 2030. Before then, the ammonia emissions of the EU-28 (in comparison to the 2005 levels) must be curbed by 6% (European Commission, 2016).

In 2021, the European Climate Law was defined by the European Commission with the objective that the EU would reach carbon neutrality by 2050. As an intermediary goal, it was established that the EU's GHG emissions should be reduced by at least 55% before 2030 in comparison to the emission levels of 1990 (European Green Deal). For the agricultural sector, this represents a diminution of the GHG emissions of roughly 3000 Mt CO_2-eq by 2050 (European Commission, 2021b). At the meantime, the emissions of the EU-28 must be reduced by 59% for SO_2 and by 42% for $NO_{x,}$ before 2030 (European Commission, 2016).

Non-methane volatile organic compounds (NMVOC), for their part, must see their emissions decreased by 28% within the same time frame. The emissions of particulate matter (in particular $PM_{2.5}$) must be curbed by 22% (European Commission, 2016). Complementarily, the Industrial Emissions Directive (IED) establishes the required enforcement of clean air and water rules for industrial installations, including intensive livestock farms whose emissions affect air quality and lead to health problems (pigs, poultry). For layers specifically, this directive applies to the farms rearing more than 300 livestock units (about 21400 hens), and sets the most strict achievable emissions levels (European Parliament, 2010).

For these levels of reduction to be reached, the implementation of mitigation strategies anywhere possible is required. In layers production, mitigation strategies targeting ammonia and PM have priority.

5 Animal-related factors affecting ammonia and particulate matter (PM) emissions

5.1 Diet

Layers reared in commercial houses are fed with diets containing high crude protein (CP) content to meet their nutritional needs for growth and egg production (Emous et al., 2019; Summers, 1993). Unfortunately, high CP diets tend to increase the nitrogen content of the excreta, and thus favour NH_3 emissions (Emous et al., 2019; Hernandez et al., 2012; Groot Koerkamp, 1994). However, the emission of NH_3 does not solely depend on its CP content. In fact, the production of NH_3 depends on both the amount of uric acid (substrate), and on the enzyme and microbial activity, itself depending on the temperature, pH,= and water content of the excreta (Michaelis and Menten, 1913; Visek, 1968; Weaver and Meijerhof, 1991). Therefore, the true levers for NH_3 reduction are either of these limiting factors, that it is the amount of substrate available, or the enzyme activity. For instance, it is correct that lower CP diet can lead to lower NH_3 emissions, either if it leads to less uric acid in the excreta, or if it lessens enzyme activity. For example, studies in broilers showed that a decreased CP content in the diet led to a lower water intake (van Harn et al., 2019; Hernandez et al., 2012), itself caused by the reduced need for the bird's body to excrete protein surplus (Emous et al., 2019). This decreases both the uric acid and the water contents of the excreta, simultaneously curbing the enzyme activity (due to the lower water content) and the substrate available for the conversion. Optimal formulation of diets can therefore be an approach to reducing NH_3 emissions, but must be approached with respect to all the processes involved.

5.2 Genetics

Although research on this subject is still scarce, there is growing interest in the effect of breed and strain on NH_3 emissions from laying hens. The underlying hypotheses are that breed and strain could have an effect on manure quality and properties (i.e. pH, moisture content, nitrogen content) (Rahman et al., 2012), and lead to the expression (or not) of certain genes that could be linked to the presence or absence of nitrogen-degrading bacteria (Xiao et al., 2021). While these correlations have not been consistently validated yet, they do place genetics as an interesting approach to NH_3 and odour mitigation.

5.3 Bird age

Increasing bird age has appeared to be linked to higher PM and total suspended particles production (Bist et al., 2023b). However, the effect of age on NH_3

emissions seems to vary. For instance, Bist et al. (2023b) did not observe any increase in NH_3 production over 16 weeks for growing pullets, but attributed this result to the dry litter conditions. Conversely, Barreto Mendes et al. (2012) showed that daily NH_3 emissions increased exponentially with birds age as manure accumulation time increased. The effect of birds age on NH_3 emissions might therefore be mainly correlated with manure and litter accumulation in the house, change in quality and properties, and with the ambient climatic conditions (relative humidity, temperature). Bird activity is likely to play a role too. Young adults (aged around 17 to 27 weeks) tend initially to become more active and spend more time in the litter, where they defecate more. From roughly 20 weeks of age, the birds gradually start to lay eggs, become less active, and spend less time in the litter. Droppings are thus excreted less frequently into the litter, leading to a reduction of its water content, and to a reduction in ammonia emissions (Groot Koerkamp et al., 1998b).

5.4 Stocking density

A study conducted in respiration chambers by Barreto Mendes et al. (2012) showed that higher stocking densities (155–206 versus 413–620 cm^2 per bird) led to 22% higher NH_3 emissions per bird in pullets (<18 weeks of age). One of the hypothesis made was that this could be the result of the higher indoor relative humidity that was observed for the higher density levels (significant, $p = 0.03$). The authors suggested that at lower stocking density, a greater surface area per unit weight of manure was exposed to the ambient air, therefore facilitating moisture evaporation (more on this in "Physicochemical factors").

5.5 Bird activity

Studies have linked emissions to the diurnal changes of birds activities. When birds are most active, ammonia and PM emissions have been shown to increase (Li et al., 2008), with bird activities in the litter having the most profound effect on ammonia and PM emissions (Groot Koerkamp et al., 1998b). However, these are natural behavioural variations which should preferably not be interfered with.

6 Physicochemical factors affecting ammonia and particulate matter emissions

6.1 Climate, housing system and management

The physicochemical properties of manure and air play a key role in the emissions of NH_3 and PM originating from layer houses. The moisture content of manure affects the reproduction and activity of microorganisms as well as the functioning of enzymes that convert uric acid into urea. In general terms, higher moisture concentrations in the air lead to wetter manure, reduced evaporation, and thus to higher emissions of NH_3 (Dunlop et al., 2016b; Ni et al., 2017; Schefferle, 1965) with the exception of caked (compacted) litters. Although they can contain more water, studies carried out with broilers have shown that they lead to a reduction in ammonia emissions, probably because the caked layer acts as a seal that alters the production and volatilisation of ammonia (Dong et al., 2024; Miles et al., 2008, 2011b). Additionally, higher pH and temperature favour degradation and NH_3 volatilisation processes from the litter and manure. The water content of manure and litter is the major factor in the microbial activity and thus formation of NH_3 (Groot Koerkamp et al., 1995), which explains the large impact of air temperature, velocity and humidity on the ammonia emission from layer houses through their impact on the volatilisation rate of water from manure and litter. Consequentially, housing systems designs in general have shown to affect NH_3 emissions as they affect indoor air speed, relative humidity, pH and temperatures levels (Bist et al., 2023a). This also connects to the geolocation of the farm and its local climatic conditions. Seasonal variations (and their degree) of NH_3 emissions have repeatedly been observed. For instance, lower ventilation rates applied during winter to minimise heating expenses simultaneously lead to higher indoor humidity levels and higher NH_3 emissions (Nimmermark et al., 2009). Conversely, while summer temperatures should lead to the greater conversion rates of nitrogenous compounds into ammonia, it is usually not the case due to the higher air speeds and drier bedding and manure (Green et al., 2009; Liang et al., 2003; Liu et al., 2016). Similarly to NH_3, greater emissions of CO_2, CH_4 and N_2O are observed in the wintertime (Huang and Guo, 2019). The increase in relative humidity during winter possibly promotes the anaerobic conditions in the manure, leading to the generation of more CH_4. Simultaneously, the increase of microbial activity boosts the ammonification and nitrification processes, themselves leading to an augmentation of N_2O production (Meda et al., 2011).

7 Practical solutions for emissions mitigation: diet and genetics

7.1 Diet

Dietary manipulations can be used to reduce NH_3 emissions by reducing N losses in manure (in the form of uric acid) and/or by curbing enzyme activity. For this, several approaches are possible: nutrient utilisation and nitrogen efficiency can be optimised to reduce the amount of uric acid excreted by the bird (Bist et al., 2023a), or enzymatic activity can be moderated through changes in temperature, pH and water content of the litter and manure.

Decreasing the crude protein content of the diet can lead to significant reductions of nitrogen excretion in the manure and to lower water content, but must be used with care to ensure that the birds' nutritional requirements are met. For instance, a reduction of the dietary crude protein content from 17 to 13.5% resulted in a 30 to 35% reduction of nitrogen excreted in the manure (Blair et al., 1999). Additionally, feed products with high fibre-content, like the so-called distiller dried grains plus soluble (DDGS), can also be used to decrease NH_3 emissions. Indeed, dietary fibre enhances the growth of the bacterial populations of the large intestine. As a result, more N remains in the manure as bacterial protein instead of uric acid, which is a precursor to NH_3 (Kirchgessner et al., 1994; Roberts et al., 2007). In addition, the bacterial fermentation of fibres leads to the production of short-chain fatty acids that lower the pH of the manure (Roberts et al., 2007). The effect of fibre content showed in the studies conducted by Matterson et al. (1966) and Wu-Haan et al. (2010) in which the use of DDGS in the diet reduced ammonia emissions without any impact on the egg production. Wu-Haan et al. (2010) showed that feeding diets containing 20% DDGS to laying hens decreased daily NH_3 emissions by 24%, and H_2S emissions by 58%. A study published by Roberts et al. (2007) showed that including dietary fibre to laying hens' diet with either 10% corn DDGS, 7% wheat middling, or 5% soybean hulls lowered the total NH_3 emission from manure by 49%, 54% and 59%, respectively. However, these experiments were conducted at laboratory scale and over short time periods. Further experimentations in commercial settings are needed to confirm these findings.

7.2 Genetics

Genetics is another approach to reducing NH_3 emissions from laying hens, with certain breeds or strains having potentially lower innate emissions. For instance, a study conducted by Rahman et al. (2012) in a cage system with four white-egg strains and four brown-egg strains (24 chicks per strain) showed differences in NH_3 emissions between the white and the brown layers. White layers showed emissions per hen that were 19 to 37% higher NH_3 than the brown ones. Differences between strains within the white or brown hens were however not detected. The authors suggested that the lower NH_3 emissions found for the brown-egg hens could possibly be (partially) due to a greater nitrogen deposition in their eggs, to a lower dietary crude protein content in the fed diet (as a result of the different nutritional requirement between strains), and to a higher moisture content of the manure (Rahman et al., 2012). In another study, differences in NH_3 and H_2S emissions were found to be present between six breeds of laying hens (Hyline, Lohmann, Nongda, Jingfen, Xinghua and Zhusi). Overall, Lohman and Xinghua had the lowest NH_3 and H_2S emissions, and Hyline and Zhusi the highest (Lu et al., 2020). These results suggest that certain breeds could be selected to reduce both NH_3 emissions and odour pollution. Similarly, manure from hens that have certain bacteria with ammonia nitrogen-degrading abilities (e.g. *Enterococcus faecium*, *Bacillus coagulants*) in their digestive system have been shown to have lower ammonia emission rates (31–53% lower) (Xiao et al., 2021). This could be connected to the expression of certain genes (*gdh*A and *gln*A) that might be involved in the bacterial-mediated regulation of ammonia emissions by laying hens. Regulating the expression of certain genes could thus be an approach to reducing NH_3 emissions (Xiao et al., 2021) if further validated.

8 Practical solutions for emissions mitigation: manure and litter management

The implementation of manure drying systems that frequently remove and dry the manure and litter with the hens excrements allow to reach high levels of dry matter content in the litter, and to minimise the degradation of nitrogenous compounds into ammonia (Groot Koerkamp et al., 1998c). In general, the quicker the manure dries, the less NH_3 is produced and emitted (Fabbri et al., 2007). For instance, the frequent removal of manure has shown to exponentially reduce NH_3 emissions. For example, the use of belts yielded a decrease in NH_3 emissions of 68% when manure collected on a third of the total belt length was removed every day instead of every 4 days (Rosa et al., 2022). Additionally, opting for drier bedding materials such as wood shavings with low moisture content (<10%) is another approach to reducing the NH_3 emissions from layers (Bist et al., 2022). The humidity-regulating characteristics of these materials and their antibacterial properties have been shown to curb the culturing of NH_3-generating bacteria, and thus to have a mitigating effect on NH_3 emissions (Cabrera et al., 2018; D.M. Miles et al., 2011a). In addition, it has been suggested that a global, optimal management of bedding materials that aims at reducing the humidity of the litter is likely to decrease odour emissions. Indeed, if microbial activity is reduced, so should be the odour production (Dunlop et al., 2016a). Active drying of litter in an aviary system also reduced the moisture content by 5.8% and consequently the emission from the litter in an experimental house (Groot Koerkamp et al., 1998a), which was as low as the range of emissions in cages. In parallel, amendments can be used to acidify (lower) the pH of litter and manure so that the decomposition of uric acid into urea, and the conversion of ammonium into NH_3 can be (partially) prohibited. While this approach has yielded inconsistent results for the reduction of volatile fatty acids (Kim et al., 2011), it has shown to considerably reduce ammonia volatilisation (Ullman et al., 2004) and emissions (Wang et al., 2021; Zhang et al., 2020, 2011). For example, reducing of the pH of poultry litter from 7 to 3 by spraying acidic electrolysed water on it led to ammonia reductions of 60–80% in the lab-scale study of Chai et al. (2018). However, the potential side-effects on the birds have yet to be investigated.

Manure drying tunnels can be implemented on farms as an addition to the drying belts and used as end of the pipe PM filters. However, a study conducted in 16 Dutch laying hens farms showed that while they did significantly reduce the PM emissions of the farms, both odour and NH_3 emissions increased. The removal efficiency for PM_{10} increased linearly with the thickness of the manure layer, going from about 35% reduction at 4 cm to 84% at 17 cm (Winkel et al., 2017). Ammonia and odour concentrations increased from 5.5 to 13.9 ppm and 822 to 1178 OU_E m^3, respectively, after passing through the tunnel. Overall, the use of these tunnels reduced the emission rates of PM_{10} by 44%, but increased NH_3 by 20% and odour by 40%. The authors suggested that reducing the manure accumulation time in the house could be a way to reduce the emissions from drying tunnels (Winkel et al., 2017). The duration between excretion and tunnel drying appears to be crucial, as shorter accumulation time and faster drying allow to better prevent the degradation of nitrogenous compounds into NH_3 (Chepete et al., 2011).

The emissions observed during manure storage have also been demonstrated to be highly dependent on the water content of the manure. Comparisons between fresh and dried (70–80% DM) manure have shown clear differences in NH_3 emissions from the storage phase, with dry manure allowing for reductions of 73–75% percent in NH_3 emissions in comparison with fresh manure (Rosa et al., 2022). In this study, the highest percentage of reduction was observed at the lower temperatures inside the laboratory (17.5 ± 2.7°C versus 24.0 ± 5.7°C) (Rosa et al., 2022). Emissions from manure can therefore be curbed further by affecting the enzymatic activity during storage.

9 Practical solutions for emissions mitigation: ventilation, filtration and acid scrubbing

9.1 Ventilation systems

Ventilation systems are used to regulate the temperature of layer houses. However, the airflows induced in the building also affect the internal relative humidity of the air (Bist et al., 2023a). As previously stated, relative humidity levels in the barns have a direct effect on the water content of the manure, and thus on its emissions. Increasing air speed and turbulences in the barn can be used to reduce moisture levels in the bedding, but the windchill induced at bird height gives little room for action. Increases in ventilation rates or inflows (mechanical or natural ventilation) dilute the NH_3 and PM concentrations within the barn and improve indoor climate (Wicklen and Allison, 1989; Wood and

Heyst, 2016), but can increase ammonia volatilisation and lead to discharges in the environment, only shifting the issue. Yet, a study from Vranken and Claes (2003) suggests that optimizing the ventilation rate in accordance with the thermal neutral zone of the birds could yield reductions in NH_3 emissions of 8 to 13%. Ventilation could therefore be used as another lever to decrease the NH_3 emissions from layer houses by affecting the enzymatic activity in the litter, but the range of possible actions is limited and drawbacks (increased in volatilisation and discharges) can easily arise.

9.2 Filtration of exhaust air

Considering end of the pipe solutions, (bio)filters can be used to reduce NH_3 emissions, PM and odour from the indoor air of layer houses. These technologies can be used in mechanically ventilated barns, in which air is constrained to move through a confined volume (Wood and Heyst, 2016). Filtration is a physico-mechanical process, while biofiltration is based on the biological conversion of NH_3 into an inert form. Both approaches also aim to separate or trap PM from the air stream (Groot Koerkamp, 1994; Tan and Zhang, 2004). Filtration is usually done through dry processes (no water used), typically using a diffusion process for the removal of small particles ($\leq$0.5 μm), and using interception, impaction or gravitation for the larger ones (Wood and Heyst, 2016). In either cases, particles are separated based on their size and removed from the airstream (Tan and Zhang, 2004).

Biofilters are generally made of an organic media (e.g. wood chips, compost, soil, activated carbon) containing a colony of bacteria and through which the air is directed to filter the pollutants of interest (Ullman et al., 2004). This type of biofilter is often also referred to as bio-bed. Bio-trickling filters are filled with an inert packing material with a large specific area and a high porosity and continuously sprayed with recirculated water. Besides the presence of nitrifying bacteria, biofilters contain bacteria that use odourous compounds as a substrate (Melse and Ogink, 2005). The structure, porosity, humidity, temperature and sizes of biofilters and bio-trickling filters are some of the parameters that must thoroughly be considered and controlled so that they can operate effectively (Chen and Hoff, 2009; Ullman et al., 2004; Williams and Miller, 1992). Bio-trickling filters have been shown to have reduction efficiencies of about 70% for ammonia (Melse and Ogink, 2005).

9.3 Acid scrubbing of exhaust air

Acid scrubbers function similarly to biofilters, as they are filled with an inert packing material with a large specific area and a high porosity. As the air from the barn passes through it, it comes into contact with an acid solution with a typical pH of 3–5. The NH_3 present in the air is then transferred to the (acid) liquid solution in which it is dissolved and bound as ammonium (NH_4^+) (Melse and Ogink, 2005). Acid scrubbers differ from (bio)filters in that scrubbers have foul air dispersed through a synthetic media and contact an acid solution, while bio-beds and biofilters have colonies of bacteria in the media (natural and water, respectively).

In poultry, reduction efficiencies of 69–99% have been observed for NH_3 with acid scrubbers (Melse and Ogink, 2005), while lower reductions levels have been yielded for $PM_{2.5}$ and PM_{10}, ranging from 28% to 33%, and from 33% to 44%, respectively (Mosquera et al., 2011). Odour reduction from 26% to 64% can be attained with acid scrubbers (Melse and Ogink, 2005; Ogink et al., 2008).

In The Netherlands, also multistage scrubbers are already in use. These possess multiple layers, each targeting a pollutant, and combine the principles of biofilters and acid scrubbers. For instance, dust can be filtered through a water bed, while NH_3 is scrubbed with acid, and odour tempered by a woody mat (Zhao et al., 2011). In general, design, dimensioning, operation and regular maintenance of any filter type is crucial for proper functioning and achieving continues high reduction efficiencies (Melse and Ogink, 2005).

10 Case study: the ECO Air Care ventilation system

Developed by the Vencomatic Group (Eersel, The Netherlands), the ECO Air Care ventilation system is a new technology that can be used to treat the air of poultry houses. It consists of treatment units that combine the heating or cooling of the air entering the barn with acid scrubbing of the air exiting it (Fig. 1). Each unit includes a sedimentation tank, a clean water tank, an acid dosing and storage unit, and a silo for the processed water. The NH_3 captured by the system is then converted into liquid fertiliser which can then be used on the land. Complementarily, the ECO Air Care includes a live, continuous measurement system allowing to monitor the N emission of the barn.

In The Netherlands, this treatment system has been implemented in (among others) a broiler breeder farm in Hoogeloon where its effect on ammonia emissions was evaluated (van Wagenberg and Groot Koerkamp, 2024). In this study, four ECO Air Care units were placed next to the barn (105 x 24 m, 18500 broiler breeder places) and were used as the sole inlet and outlet ventilation system. The results of the study showed that the system allowed to reduce the ammonia emissions of the poultry house by approximately 90%, which is in line with other acid scrubbers. Regardless of the method used to estimate ammonia reduction (aerial ammonia concentration measurements, acid use, or process water measurements), this research showed that the ECO Air Care can be used to significantly reduce ammonia and PM emissions of poultry houses in

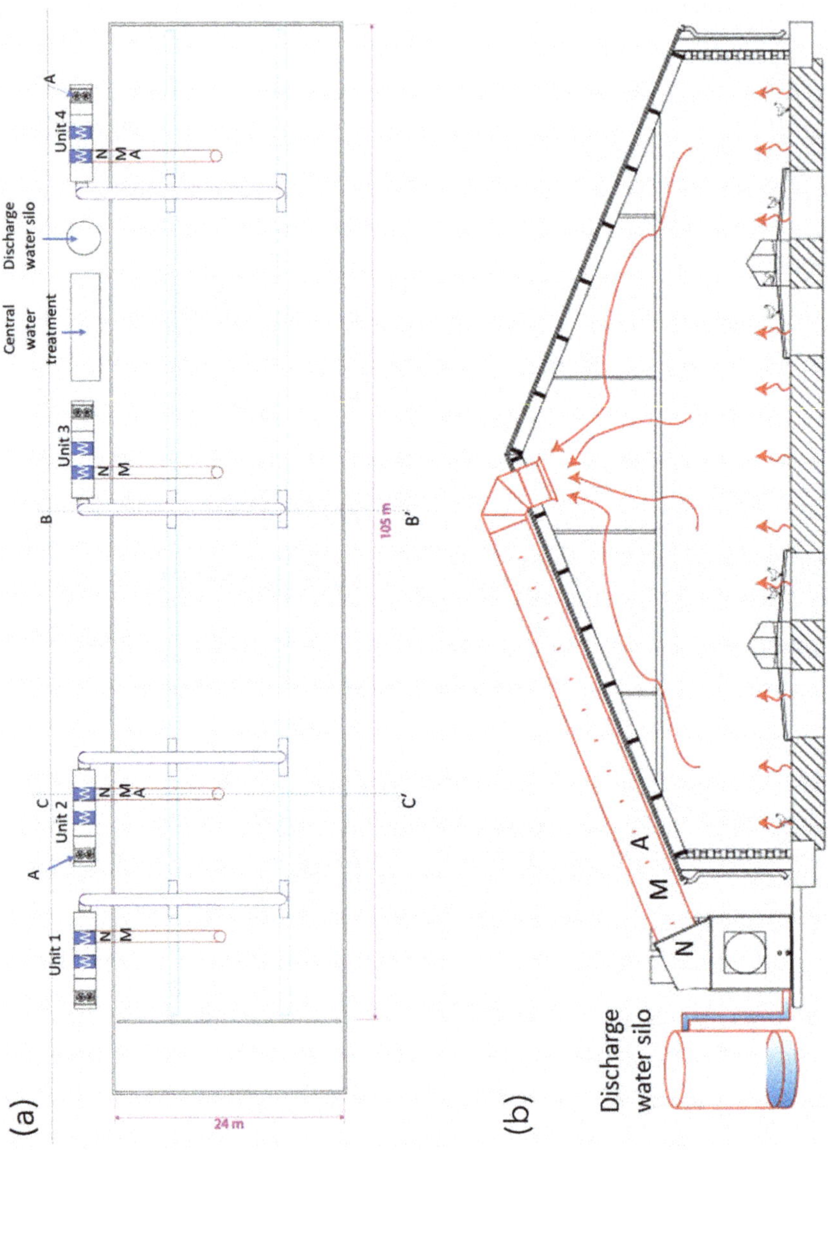

Figure 1 Top view (A) and cross section (B) of a commercial poultry house equipped with four ECO Air Care treatment units. The arrows indicate the outgoing airflow with gaseous water vapor and particulate matter; M = position of measuring fan, N = position of Nozzle for spraying process water, W = section of the unit where process water is sprayed, A= ammonia concentration sensor (© Vencomatic Group, 2023; van Wagenberg and Groot Koerkamp, 2024).

commercial conditions, which makes it an interesting end of the pipe mitigation strategy. The additional investment and operational costs can (partly) be compensated by the substantially improved indoor climate conditions through sufficient ventilation in winter and cooling in summer. More information can be found in the paper of van Wagenberg and Groot Koerkamp (2024).

11 Conclusion: the options available

An overview of the options that are available and an indication of their potential to reduce emissions from layers production is presented in Table 2.

In 2017, the European Bureau for Research on Industrial Transformation and Emissions reported the outcomes of their study on the best available techniques (BAT) in poultry production (Santonja et al., 2017). These include – amongst others – the BAT for the reduction of emissions within and from poultry houses. One of the main conclusions to be drawn from this report is that many of the BAT that could be used to reduce emissions from layers production come at high costs, and often require the extensive modification of existing farm buildings. This strongly limits the number of reduction strategies that can currently be implemented within the current economic possibilities of poultry farms but provides information on the techniques that could be included in future farm projects. However, other (lower costs) management strategies (change in diet, bedding, stocking densities) can be implemented more easily and remain options available for existing facilities. However, most of them are not yet well tested and implemented in practice, and their reduction potential is highly dependent on management practices. The line between the positive or negative outcomes a mitigation strategy can have appears to be very thin.

The report also stresses the need for an integrated approach when it comes to optimizing the environmental impact of whole-farms, which is essential for two reasons. Firstly, because mitigation strategies that are implemented at one specific stage of the production influence the emission potential of the next stage(s). For instance, reducing the NH_3 emissions of layer housing generally leads to higher nitrogen concentration in the manure, itself increasing the emissions from manure in a later stage during storage and field application (Santonja et al., 2017). Secondly, pollution swaps can occur between pollutants, which lead to a reduction in emissions of one component, and the undesired increase in another one. For example, manure drying tunnels enable a reduction of dust emissions, but lead to an increase in both ammonia emissions and odour nuisance (Winkel et al., 2017). This applies not only to pollutants but also to other key elements, such as energy use, resource consumption, animal welfare, farmers' income, etc. It is therefore essential to approach the reduction of emissions as a whole, rather than targeting isolated elements, in order to

Table 2 Summary table of the options that are available to reduce the emissions of layer production systems.

Mitigation method	Principle	Emissions that can be reduced	Reduction potential	Applicability
Diet	Decrease crude protein (CP) content of the ration (while still meeting the nutritional requirements)	N losses in general, and therefore NH_3	Decrease of CP content from 17% to 13.5%: 30–35% less nitrogen excreted	Generally applicable. Attention should be given to meeting the nutritional requirements of the birds. May increase feeding costs
	Addition of distiller dried grains plus soluble (DDGS)	NH_3, H_2S	20% DDGS in diet: NH_3 reduced by 24%, H_2S by 58%	
	Increase fibre content	NH_3	Reduction of NH_3 emissions up to 50%	
Genetics	Genetic selection of breeds or strains with specific manure quality and properties (to be further validated)	NH_3, odour	To be further investigated	Generally applicable. Effect on other variables (e.g. productivity) should be considered
Stocking density	Lower densities for a greater surface area per bird, and thus facilitate moisture evaporation	Indirect effect on NH_3	Stocking densities decreased from 155–206 to 413–620 cm^2 per bird: 22% NH_3 reduction	Generally applicable, but housing costs per bird likely increase
Building and aeration design	Barns designs that favour ventilation (lower humidity) and decrease temperatures	NH_3, PM Caution: can improve barn's climate, but lead to discharges in the environment	In accordance with the thermal neutral zone of the birds, NH_3 emissions can be reduced by 8–13%	Applicable to planned barns, but hardly to existing ones as a complete revision of the housing system may be required
Litter and manure management	Choice of (drier) bedding materials that favour aeration and drainage	NH_3	The drier the bedding material, the less NH_3 emissions	Generally applicable
	Use of amendments reducing the pH of the bedding	NH_3	A pH reduction from 7 to 3 reduced NH_3 by 60–80%	Generally applicable, possible side-effects on birds

	Manure drying systems that frequently remove and dry litter and excrements	NH_3, PM Caution: in some cases, PM emissions decreased, but odour and NH_3 increased	Frequent manure removal: NH_3 emissions decreased by 68%; Manure drying tunnels: PM_{10} emissions decreased by 44%; NH_3 and odour increased by 20% and 40%	Most systems are equipped with belts and generally also with additional drying
Filtration	<u>End of the pipe solution:</u> (Bio) filtration of the air of the barn	NH_3, odour	NH_3: 70% reduction; Odour: 51% reduction	Implementation is limited due to the high implementation costs. Only applicable to layer houses equipped with a central ventilation system
Acid scrubbing	<u>End of the pipe solution:</u> NH_3 present in the barn air transfers to the acid contained in the scrubber, in which it is dissolved	NH_3, $PM_{2.5}$, PM_{10}, odour	Reductions efficiencies of: NH_3: 69–99%;$PM_{2.5}$: 28%–33%;PM_{10}: 33%–44%;Odour: 26–64%	

avoid a situation where the reduction of a single element leads to an increase in other harmful aspects.

The take-home messages are as follows:

- Mitigation strategies are available to reduce the emissions from layers production.
- However, their readiness level, ease of implementation, costs, and reduction potential vary greatly.
- Reducing the environmental impact of layers farms should be approached in an integrated manner to avoid pollution swapping.

12 Future trends and research needs

As highlighted in this chapter, the effectiveness of certain mitigation strategies on reducing emissions from laying hens production has not been validated. The link between genetics and emissions, for instance, is not entirely certain. Similarly, the formulation of diets that reduce emissions while meeting the nutritional requirements of the birds and at reasonable costs for the farmer should be instigated further. The effects of management of bedding, manure and ventilation on emissions, for their part, have been and remain the subject of extensive research and need better integration and implementation in practice. Scrubbers and (bio)filters are proving effective, but these are end-of-pipe techniques that only improve outdoor air quality and not indoor, therefore not improving bird's and worker's health and welfare. Pollution swaps of each mitigation technique should be assessed in depth and in each specific context. They should be taken into account when adopting a given reduction strategy, on which little information is available in the literature to this date. Lastly, information on the costs (financial and resources) of each strategy should be clearly reported, so that more realistic plans can be made for the sector to reach the reduction target set by the European Union.

13 Where to look for further information

Complementary information on the topic of emissions reduction in layers production can be found in the following documents:

- "Ammonia emissions, impacts, and mitigation strategies for poultry production: A critical review" (Bist et al., 2023a);
- "Gaseous emissions from aviaries and nitrogen and phosphorus losses in the outdoor run of organic laying hens" (Dekker et al., 2012);
- "Reduction of nitrogen excretion and emissions from poultry: A review for conventional poultry" (Chalova et al., 2016);

- "Strategies to mitigate ammonia and nitrous oxide losses across the manure management chain for intensive laying hen farms" (Rosa et al., 2022);
- "Best available techniques (BAT) reference document for the intensive rearing of poultry or pigs" (Santonja et al., 2017).

14 References

Akdeniz, N., et al., 2012. Odor and odorous chemical emissions from animal buildings: part 4. Correlations between sensory and chemical measurements. *Trans. ASABE* 55, 2347–2356. https://doi.org/10.13031/2013.42496

Alberdi, O., et al., 2016. Ammonia and greenhouse gas emissions from an enriched cage laying hen facility. *Biosyst. Eng.* 144, 1–12. https://doi.org/10.1016/j.biosystemseng.2016.01.009

Anand, S.S. and Mehendale, H.M., 2005. *Encyclopedia of toxicology* (Second Edition). Artic. Titles: V 450–455. https://doi.org/10.1016/b0-12-369400-0/01015-2

Barreto-Mendes, L., et al., 2012. Ammonia emissions of pullets and laying hens as affected by stocking density and manure accumulation time. *Trans. ASABE* 55, 1067–1075. https://doi.org/10.13031/2013.41511

Behera, S.N., et al., 2013. Ammonia in the atmosphere: a review on emission sources, atmospheric chemistry and deposition on terrestrial bodies. *Environ. Sci. Pollut. Res.* 20, 8092–8131. https://doi.org/10.1007/s11356-013-2051-9

Bist, R.B., et al., 2022. *Air quality in cage-free houses during pullets production*. Houston, TX. https://doi.org/10.13031/aim.202200329

Bist, R.B., et al., 2023a. Ammonia emissions, impacts, and mitigation strategies for poultry production: a critical review. *J. Environ. Manag.* 328, 116919. https://doi.org/10.1016/j.jenvman.2022.116919

Bist, R.B., et al., 2023b. Temporal variations of air quality in cage-free experimental pullet houses. *Poultry* 2, 320–333. https://doi.org/10.3390/poultry2020024

Blair, R., et al., 1999. A quantitative assessment of reduced protein diets and supplements to improve nitrogen utilization. *J. Appl. Poult. Res.* 8, 25–47. https://doi.org/10.1093/japr/8.1.25

Breemen, N. et al., 1982. Soil acidification from atmospheric ammonium sulphate in forest canopy throughfall. *Nature* 299, 548–550. https://doi.org/10.1038/299548a0

Brook, R.D., et al., 2003. Air pollution: the "heart" of the problem. *Curr. Hypertens. Rep.* 5, 32–39. https://doi.org/10.1007/s11906-003-0008-y

Bussink, D.W. and Oenema, O., 1998. Ammonia volatilization from dairy farming systems in temperate areas: a review. *Nutr. Cycl. Agroecosystems* 51, 19–33. https://doi.org/10.1023/a:1009747109538

Cabrera, M.L., et al., 2018. Litter type and number of flocks affect sex hormones in broiler litter. *J. Environ. Qual.* 47, 156–161. https://doi.org/10.2134/jeq2017.08.0301

Carey, J.B., et al., 2004. A review of literature concerning odors, ammonia, and dust from broiler production facilities: 2. Flock and house management factors. *J. Appl. Poult. Res.* 13, 509–513. https://doi.org/10.1093/japr/13.3.509

Chai, L., et al., 2018. Mitigating airborne bacteria generations from cage-free layer litter by spraying acidic electrolysed water. *Biosyst. Eng.* 170, 61–71. https://doi.org/10.1016/j.biosystemseng.2018.03.013

Chalova, V.I., et al., 2016. Reduction of nitrogen excretion and emissions from poultry: a review for conventional poultry. *World's Poult. Sci. J.* 72, 509–520. https://doi.org/10.1017/s0043933916000477

Chen, L. and Hoff, S.J., 2009. Mitigating odors from agricultural facilities: a review of literature concerning biofilters. *Appl. Eng. Agric.* 25, 751–766. https://doi.org/10.13031/2013.28854

Chepete, J.H., et al., 2011. Ammonia emissions of laying-hen manure as affected by accumulation time. *J. Poult. Sci.* 48, 133. https://doi.org/10.2141/jpsa.010087

European Commission, 2016. Directive (EU) 2016/2284 of the European Parliament and of the Council of 14 December 2016 on the reduction of national emissions of certain atmospheric pollutants, amending Directive 2003/35/EC and repealing Directive 2001/81/EC, OJ L 344, vol 59, 17.12.2016pp. 1–31.

European Commission, 2021a. Inception impact assessment on the revision of the EU legislation on animal welfare. Legislative proposal of the European Parliament and of the Counci Ares(2021)4402058, pp 1–11.

European Commission, 2021b. Regulation (EU) 2021/1119 of the European Parliament and of the Council of 30 June 2021 establishing the framework for achieving climate neutrality and amending Regulations (EC) No 401/2009 and (EU) 2018/1999 ('European Climate Law'), OJ L 243, vol. 64, 09.07.2021, pp. 1–17.

European Commission, 2022. Commission staff working document executive summary of the fitness check EU Animal Welfare legislation {SWD(2022) 328 final}. Brussels, 04.10.2022, pp. 1–3.

European Commission, 2024. Eggs [WWW document]. https://agriculture.ec.europa.eu/farming/animal-products/eggs_en#:~:text=There%20are%20more%20than%20350%20million%20laying%20hens,standards%20and%20occasionally%20through%20some%20market%20support%20measures.

Crippa, M., et al., 2018. Gridded emissions of air pollutants for the period 1970–2012 within EDGAR v4.3.2. *Earth Syst. Sci. Data* 10, 1987–2013. https://doi.org/10.5194/essd-10-1987-2018

Dekker, S.E.M. et al., 2012.Gaseous emissions from aviaries and nitrogen and phosphorus losses in the outdoor run of organic laying hens. *IX International Livestock Environment Symposium (ILES IX) ILES12-0731*. https://doi.org/10.13031/2013.41642.

Dong, S., et al., 2024. *Effects of peat as a litter material on ammonia emissions from experimental broiler houses*. Wageningen Livestock Research, Public Report 1528. https://doi.org/10.18174/679212.

Dunlop, M.W., et al., 2016a. Odour emissions from poultry litter: a review litter properties, odour formation and odorant emissions from porous materials. *J. Environ. Manag.* 177, 306–319. https://doi.org/10.1016/j.jenvman.2016.04.009

Dunlop, M.W., et al., 2016b. Water activity of poultry litter: relationship to moisture content during a grow-out. *J. Environ. Manag.* 172, 201–206. https://doi.org/10.1016/j.jenvman.2016.02.036

Emous, R.A. et al., 2019. Effects of dietary crude protein levels on ammonia emission, litter and manure composition, N losses, and water intake in broiler breeders. *Poult. Sci.* 98, 6618–6625. https://doi.org/10.3382/ps/pez508

EPA, 2004a. *Air quality criteria for particulate matter: Volume II of II*. Environmental Protection Agency (EPA), Research Triangle Park, NC.

EPA, 2004b. National Emission Inventory – Ammonia emissions from animal husbandry operations, pp. 131. Environmental Protection Agency (EPA), Research Triangle Park, NC.

EPRS, 2019. The EU poultry meat and egg sector: main features, challenges and prospects. European Parliamentary Research Service (EPRS), Policy Commons in depth-analysis (No. 10.2861/33350), PE 644.195, November 2019. https://doi.org/10.2861/33350.

European Parliament, 2010. Directive 2010/75/EU of the European Parliament and of the Council of 24 November 2010 on industrial emissions (integrated pollution prevention and control), OJ L 344, vol. 53, 17.12.2010, pp. 17-119 doi:10.3000/17252555.L_2010.334.eng.

Fabbri, C., et al., 2007. Ammonia, methane, nitrous oxide and particulate matter emissions from two different buildings for laying hens. *Biosyst. Eng.* 97, 441–455. https://doi.org/10.1016/j.biosystemseng.2007.03.036

Green, A.R., et al., 2009. Air quality and bird health status in three types of commercial egg layer houses. *J. Appl. Poult. Res.* 18, 605–621. https://doi.org/10.3382/japr.2007-00086

Groot Koerkamp, P.W.G.G., 1994. Review on emissions of ammonia from housing systems for laying hens in relation to sources, processes, building design and manure handling. *J. Agric. Eng. Res.* 59, 73–87. https://doi.org/10.1006/jaer.1994.1065

Groot Koerkamp, P.W.G., et al., 1995. The effect of manure and litter handling and indoor climatic conditions on ammonia emissions from a battery cage and an aviary housing system for laying hens. *Netherlands J. Agric. Sc.* 43, 351–373.

Groot Koerkamp, P.W.G., et al., 1998a. PhD thesis Ammonia emission from aviary housing systems for laying hens: inventory, characteristics and solutions – Chapter 6: litter composition and ammonia emission in aviary houses. Part I: performance of a litter drying system, pp. 88–99. ISBN 90-5485-885-0.

Groot Koerkamp, P.W.G., et al., 1998b. Ammonia emission from aviary housing systems for laying hens: inventory, characteristics and solutions – Chapter 8: litter composition and ammonia emission in aviary houses. Part III: Flow of water to the litter through fresh droppings, pp. 116–128. ISBN 90-5485-885-0

Groot Koerkamp, P.W.G., et al., 1998c. Litter composition and ammonia emission in aviary houses for laying hens. Part 1: performance of a Litter Drying System. *J. Agric. Eng. Res.* 70, 375–382. https://doi.org/10.1006/jaer.1998.0290

Harn, J., et al., 2019. Effect of low protein diets supplemented with free amino acids on growth performance, slaughter yield, litter quality, and footpad lesions of male broilers. *Poult. Sci.* 98, 4868–4877. https://doi.org/10.3382/ps/pez229

Hayes, M.D., et al., 2013. Ammonia, greenhouse gas, and particulate matter emissions of aviary layer houses in the Midwestern U.S. *Trans. ASABE* 56, 1921–1932. https://doi.org/10.13031/trans.56.10053

Hernandez, F., et al., 2012. Effect of dietary protein level on retention of nutrients, growth performance, litter composition and NH_3 emission using a multi-phase feeding programme in broilers. *Span. J. Agric. Res.* 11, 736–746. https://doi.org/10.5424/sjar/2013113-3597

Huang, D. and Guo, H., 2019. Diurnal and seasonal variations of greenhouse gas emissions from a commercial broiler barn and cage-layer barn in the Canadian

Prairies. *Environ. Pollut.* 248, 726–735. https://doi.org/10.1016/j.envpol.2019.02.065

Huanhong, K., et al., 2023. Volatile organic compound emissions in free-range chicken production: impacts on environment, welfare and sustainability. *AIMS Agric. Food* 8, 1071–1091. https://doi.org/10.3934/agrfood.2023058

ICRP, 1994. *Human respiratory tract model for radiological protection*. Publication 66. ICRP, Pergamon.

Jacobson, L.D., et al., 2005. Development of the offset model for determination of odor-annoyance-free setback distances from animal production sites: part I. Review and experiment. *Trans. ASAE* 48, 2259–2268. https://doi.org/10.13031/2013.20089

Kim, S.C., et al., 2011. Reducing volatile fatty acids in poultry litter with chemical additives: a laboratory experiment. *J. Poult. Sci.* 48, 139. https://doi.org/10.2141/jpsa.010071

Kirchgessner, M., et al., 1994. Evidence for a high efficiency of bacterial protein synthesis in the digestive tract of adult sows fed supplements of fibrous feedstuffs. *Anim. Feed Sci. Technol.* 46, 293–306. https://doi.org/10.1016/0377-8401(94)90147-3

Lacey, R.E., et al., 2004. A review of literature concerning odors, ammonia, and dust from broiler production facilities: 1. Odor concentrations and emissions. *J. Appl. Poult. Res.* 13, 500–508. https://doi.org/10.1093/japr/13.3.500

Lebrero, R., et al., 2011. Odor assessment and management in wastewater treatment plants: a review. *Crit. Rev. Environ. Sci. Technol.* 41, 915–950. https://doi.org/10.1080/10643380903300000

Li, H., et al., 2008. Effects of bird activity, ventilation rate and humidity on PM10 concentration and emission rate of a Turkey barn. *Livest. Environ.* VIII, 31 August–4 September, Iguassu Falls, Braz. https://doi.org/10.13031/2013.25487

Li, S., et al., 2011. Particulate matter concentrations and emissions of a high-rise layer house in Iowa. *Trans. ASABE* 54, 1093–1101. https://doi.org/10.13031/2013.37101

Liang, Y., et al., 2003. *Ammonia emissions from U.S. poultry houses: Part II. Layer houses*. Presented at the Proceedings 3rd international conference on air pollution from agricultural operations, St, Joseph, MO, pp. 147–158.

Lin, X., et al., 2017. Emissions of ammonia, carbon dioxide and particulate matter from cage-free layer houses in California. *Atmos. Environ.* 152, 246–255. https://doi.org/10.1016/j.atmosenv.2016.12.018

Liu, S., et al., 2016. Modeling of dynamic ammonia concentrations in two commercial layer hen houses. *J. Environ. Inform.* https://doi.org/10.3808/jei.201700360

Lu, D., et al., 2020. Effects of different laying hen species on odour emissions. *Animals* 10, 2172. https://doi.org/10.3390/ani10112172

Mackie, R.I., et al., 1998. Biochemical identification and biological origin of key odor components in livestock waste. *J. Anim. Sci.* 76, 1331. https://doi.org/10.2527/1998.7651331x

MacLeod, M., et al., 2013. *Greenhouse gas emissions from pig and chicken supply chains: a global life cycle assessment*. Food and Agriculture Organization of the United Nations (FAO), Rome.

Malley, C.S., et al., 2015. The impact of speciated VOCs on regional ozone increment derived from measurements at the UK EMEP supersites between 1999 and 2012. *Atmos. Chem. Phys.* 15, 8361–8380. https://doi.org/10.5194/acp-15-8361-2015

Matterson, L.D., et al., 1966. Corn distillers dried grains with solubles in rations for high-producing hens 1 2. *Poult. Sci.* 45, 147–151. https://doi.org/10.3382/ps.0450147

Mazzuca, G.M., et al., 2016. Ozone production and its sensitivity to NOx and VOCs: results from the DISCOVER-AQ field experiment, Houston 2013. *Atmos. Chem. Phys.* 16, 14463–14474. https://doi.org/10.5194/acp-16-14463-2016

Meda, B., et al., 2011. Influence of rearing conditions and manure management practices on ammonia and greenhouse gas emissions from poultry houses. *World's Poult. Sci. J.* 67, 441–456. https://doi.org/10.1017/s0043933911000493

Melse, R.W. and Ogink, N.W.M., 2005. Air scrubbing techniques for ammonia and odor reduction at livestock operations: review of on-farm research in the Netherlands. *Trans. ASAE* 48, 2303–2313. https://doi.org/10.13031/2013.20094

Michaelis, L. and Menten, M.L., 1913. The kinetics of invertase action. *Biochem. Zeitung* 49, 333–369.

Mielcarek, P. and Rzeźnik, W., 2015. Odor emission factors from livestock production. *Pol. J. Environ. Stud.* 24, 27–35. https://doi.org/10.15244/pjoes/29944

Miles, D.M., et al., 2008. Winter broiler litter gases and nitrogen compounds: temporal and spatial trends. *Atmos. Environ.* 42, 3351–3363. https://doi.org/10.1016/j.atmosenv.2006.11.056

Miles, D.M., et al., 2011a. Litter ammonia generation: moisture content and organic versus inorganic bedding materials 1. *Poult. Sci.* 90, 1162–1169. https://doi.org/10.3382/ps.2010-01113

Miles, D. M., et al., 2011b. Spatial contrasts of seasonal and intraflock broiler litter trace gas emissions, physical and chemical properties. *J. Environ. Qual.* 40, 176–187. https://doi.org/10.2134/jeq2010.0055

Mosquera, J., et al., 2009. *Fine dust emission from animal houses; Laying hens with floor housing* [Fijnstofemissie uit stallen: leghennen in scharrelhuisvesting]. Wageningen UR Livestock Research, Lelystad.

Mosquera, J., et al., 2011. *Dust emission from animal houses: air scrubbing techniques* (in Dutch with English summary). Report 295. Wageningen UR Livestock Research, Lelystad.

Moum, S.G., et al., 1969. A simple method of determining concentrations of ammonia in animal quarters. *Poult. Sci.* 48, 347–348. https://doi.org/10.3382/ps.0480347

Myhre, G., et al., 2013. *Climate change 2013: The Physical Science Basis. Contribution of Working Group I to the fifth assessment report of the Intergovernmental Panel on Climate Change*. Cambridge University Press, Cambridge and New York.

Ni, J.-Q., et al., 2017. Emission factors and characteristics of ammonia, hydrogen sulfide, carbon dioxide, and particulate matter at two high-rise layer hen houses. *Atmos. Environ.* 154, 260–273. https://doi.org/10.1016/j.atmosenv.2017.01.050

Nimmermark, S., et al., 2009. Ammonia, dust and bacteria in welfare-oriented systems for laying hens. *Ann. Agric. Environ. Med.* 16, 103–113.

Nowak, A., et al., 2016. Cytotoxicity of odorous compounds from poultry manure. *Int. J. Environ. Res. Public Heal.* 13, 1046. https://doi.org/10.3390/ijerph13111046

Ogink, N.W.M., et al., 2008. Multi-pollutant and one-stage scrubbers for removal of ammonia, odor, and particulate matter from animal house exhaust air. Livest. Environ. VIII, 31 August–4 September, Iguassu Falls, Braz. https://doi.org/10.13031/2013.25508

Olivier, J.G.J., et al., 2020. Emission database for global atmospheric research (Edgar) 2020 v5.0. *Environ. Monit. Assess.* 31, 93–106. https://doi.org/10.1007/bf00547184

Rahman, S., et al., 2012. Effects of laying-hen strain on manure properties and ammonia emission. *Trans ASABE* 55, 1059–1065.

Rimac, D., et al., 2010. Exposure to poultry dust and health effects in poultry workers: impact of mould and mite allergens. *Int. Arch. Occup. Environ. Heal.* 83, 9–19. https://doi.org/10.1007/s00420-009-0487-5

Roberts, S.A., et al., 2007. Effects of dietary fiber and reduced crude protein on ammonia emission from laying-hen manure. *Poult. Sci.* 86, 1625–1632. https://doi.org/10.1093/ps/86.8.1625

Rosa, E., et al., 2020. Ammonia emission from a manure-belt laying hen facility equipped with an external manure drying tunnel. *J. Clean. Prod.* 251, 119591. https://doi.org/10.1016/j.jclepro.2019.119591

Rosa, E., et al., 2022. Strategies to mitigate ammonia and nitrous oxide losses across the manure management chain for intensive laying hen farms. *Sci. Total Environ.* 803, 150017. https://doi.org/10.1016/j.scitotenv.2021.150017

Santonja, G.G., et al., 2017. *Best Available Techniques (BAT) reference document for the intensive rearing of poultry or pigs*; EUR 28674 EN. Publications Office of the European Union, Luxembourg. https://doi.org/10.2760/020485

Schefferle, H.E., 1965. The decomposition of uric acid in built up poultry litter. *J. Appl. Microbiol.* 28, 412–420. https://doi.org/10.1111/j.1365-2672.1965.tb02171.x

Summers, J.D., 1993. Reducing nitrogen excretion of the laying hen by feeding lower crude protein diets. *Poult. Sci.* 72, 1473–1478. https://doi.org/10.3382/ps.0721473

Takai, H., et al., 1998. Concentrations and emissions of airborne dust in livestock buildings in Northern Europe. *J. Agric. Eng. Res.* 70, 59–77. https://doi.org/10.1006/jaer.1997.0280

Tan, Z. and Zhang, Y., 2004. A review of effects and control methods of particulate matter in animal indoor environments. *J. Air Waste Manag. Assoc.* 54, 845–854. https://doi.org/10.1080/10473289.2004.10470950

Ullman, J.L., et al., 2004. A review of literature concerning odors, ammonia, and dust from broiler production facilities: 4. Remedial management practices. *J. Appl. Poult. Res.* 13, 521–531. https://doi.org/10.1093/japr/13.3.521

Union, T.C. of the E., Council Directive 1999/74/EC.

VencomaticGroup, 2023. ECO air care [WWW document]. https://www.vencomaticgroup.com/ecoaircare (accessed 5.8. 2024).

Viegas, S., et al., 2013. Occupational exposure to poultry dust and effects on the respiratory system in workers. *J. Toxicol. Environ. Heal., Part A* 76, 230–239. https://doi.org/10.1080/15287394.2013.757199

Visek, W.J., 1968. Some aspects of ammonia toxicity in animal cells. *J. Dairy Sci.* 51, 286–295. https://doi.org/10.3168/jds.s0022-0302(68)86976-0

Vranken, E. and Claes, S., 2003. Reduction of ammonia emission from livestock buildings by the optimization of ventilation control. In: *Air Pollution from Agricultural Operations - III, North Carolina USA*, pp. 167–173. https://doi.org/10.13031/2013.15508

Wagenberg, A.V. van and Groot Koerkamp, P.W.G.G., 2024. Validation methods for the ammonia removal of an air scrubber on a poultry house using the acid use and the process water nitrogen balance. *T*Journal of the ASABE. 67(3): 761–774.. https://doi.org/10.13031/ja.15865

Wang, Z., et al., 2021. Acidic aerobic digestion of anaerobically-digested sludge enabled by a novel ammonia-oxidizing bacterium. *Water Res.* 194, 116962. https://doi.org/10.1016/j.watres.2021.116962

Weaver, W.D. and Meijerhof, R., 1991. The effect of different levels of relative humidity and air movement on litter conditions, ammonia levels, growth, and carcass quality for broiler chickens. *Poult. Sci.* 70, 746–755. https://doi.org/10.3382/ps.0700746

Wicklen, G.L.V. and Allison, J.M., 1989. Aerosol and ammonia concentrations in broiler houses using mechanical and natural ventilation. *J. Agric. Eng. Res.* 42, 97–109. https://doi.org/10.1016/0021-8634(89)90043-7

Williams, J. and Koppmann, R., 2007. *Volatile Organic Compounds in the Atmosphere: An Overview*. Blackwell Publishing Ltd, Oxford, pp. 1–32. https://doi.org/10.1002/9780470988657.ch1

Williams, T.O. and Miller, F.C., 1992. Biofilters and facility operations: part 2. Design and operational factors related to the use of biofilters at composting facilities include moisture, loading rates, oxygen content, pH, and pressure drop. *Biocycle* 33, 72–77.

Winkel, A., et al., 2009. *Fine dust emission from animal houses: laying hens in aviary systems* [Fijnstofemissie uit stallen: leghennen in volièrehuisvesting]. Wageningen UR Livestock Research, Lelystad.

Winkel, A., et al., 2015. Emissions of particulate matter from animal houses in the Netherlands. *Atmos. Environ.* 111, 202–212. https://doi.org/10.1016/j.atmosenv.2015.03.047

Winkel, A., et al., 2017. Evaluation of manure drying tunnels to serve as dust filters in the exhaust of laying hen houses: emissions of particulate matter, ammonia, and odour. *Biosyst. Eng.* 162, 81–98. https://doi.org/10.1016/j.biosystemseng.2017.07.006

Wood, D.J. and Heyst, B.J.V., 2016. A review of ammonia and particulate matter control strategies for poultry housing. *Trans. ASABE* 59, 329–344. https://doi.org/10.13031/trans.59.10836

World Health Organization, 1989. Indoor air quality: organic pollutants. *Environ. Technol. Lett.* 10, 855–858. https://doi.org/10.1080/09593338909384805

Wu-Haan, W., et al., 2010. The use of distillers dried grains plus solubles as a feed ingredient on air emissions and performance from laying hens. *Poult. Sci.* 89, 1355–1359. https://doi.org/10.3382/ps.2009-00471

Xiao, S., et al., 2021. Ammonia reduction by the gdhA and glnA genes from bacteria in laying hens. *Ecotoxicol. Environ. Saf.* 222, 112486. https://doi.org/10.1016/j.ecoenv.2021.112486

Zarra, T., et al., 2013. *Odour Impact Assessment Handbook*, pp. 31–83. https://doi.org/10.1002/9781118481264.ch3

Zhang, C., et al., 2020. The impact of absorbents on ammonia recovery in a capacitive membrane stripping system. *Chem. Eng. J.* 382, 122851. https://doi.org/10.1016/j.cej.2019.122851

Zhang, H., et al., 2011. Effect of alum-amended litter and stocking density on ammonia release and footpad and hock dermatitis of broilers. *Agric. Sci. China* 10, 777–785. https://doi.org/10.1016/s1671-2927(11)60062-x

Zhao, Y., et al., 2011. Effectiveness of multi-stage scrubbers in reducing emissions of air pollutants from pig houses. *Trans. ASABE* 54, 285–293. https://doi.org/10.13031/2013.36256

Index

www.ingramcontent.com/pod-product-compliance
Ingram Content Group UK Ltd.
Pitfield, Milton Keynes, MK11 3LW, UK
UKHW021824100926
8556IPUK00003B/143

9 781801 468756